# Frontiers in Mathematics

Advisory Editorial Board

John Clark
Christian Lomp
Narayanaswami Vanaja
Robert Wisbauer

# Lifting
# Modules

## Supplements and Projectivity
## in Module Theory

rlag
· Berlin

Authors:

John Clark
Department of Mathematics and Statistics
University of Otago
PO Box 56
Dunedin
New Zealand
e-mail: jclark@maths.otago.ac.nz

Narayanaswami Vanaja
Department of Mathematics
University of Mumbai
Vidyanagari Marg
Mumbai 400098
India
e-mail: vanaja.narayanaswami@gmail.com

Christian Lomp
Departamento de Matemática Pura
Faculdade de Ciências
Universidade do Porto
Rua Campo Alegre 687
4169-007 Porto
Portugal
e-mail: clomp@fc.up.pt

Robert Wisbauer
Institute of Mathematics
Heinrich Heine University Düsseldorf
Universitätsstr. 1
40225 Düsseldorf
Germany
e-mail: wisbauer@math.uni-duesseldorf.de

2000 Mathematical Subject Classification 16D10, 16D40, 16D50, 16D70, 16D90, 16L30, 16L60, 16P20, 16P60, 16S50, 16S90

A CIP catalogue record for this book is available from the
Library of Congress, Washington D.C., USA

Bibliographic information published by Die Deutsche Bibliothek
Die Deutsche Bibliothek lists this publication in the Deutsche Nationalbibliografie;
detailed bibliographic data is available in the Internet at <http://dnb.ddb.de>.

© 2006 Birkhäuser Verlag, P.O. Box 133, CH-4010 Basel, Switzerland
Part of Springer Science+Business Media
Cover design: Birgit Blohmann, Zürich, Switzerland
Printed on acid-free paper produced from chlorine-free pulp. TCF ∞

ISBN 978-3-7643-7572-0
DOI 10.1007/978-3-7643-7573-6
9 8 7 6 5 4 3 2 1

ISBN 978-3-7643-757

# Contents

# Preface

*Extending modules* are generalizations of injective modules and, dually, *lifting modules* generalise projective supplemented modules. However, while every module has an injective hull it does not necessarily have a projective cover. This creates a certain asymmetry in the duality between extending modules and lifting modules; it stems from the fact that in any module $M$ there exist, by Zorn's Lemma, (intersection) complements for any submodule $N$ but, by contrast, supplements for $N$ in $M$ need not exist. (A module is *extending*, or *CS*, if every complement submodule is a direct summand, and it is *lifting* if it is amply supplemented and every supplement is a direct summand.) The terms *extending* and *lifting* were coined by Harada and Oshiro (e.g., [156, 272]).

The monograph *Continuous and Discrete Modules* by Mohamed and Müller [241] considers both extending and lifting modules while the subsequent *Extending Modules* [85] presents a deep insight into the structure of the former class.

The purpose of our monograph is to provide a similarly up-to-date account of lifting modules. However, this involves more than a routine dualisation of the results on extending modules in [85] since, as alluded to in our opening paragraph, it becomes necessary to be mindful of the existence of supplements and their properties. An investigation of supplements leads naturally to the topics of *semilocal modules* and *hollow (dual Goldie) dimension*, as studied in [56, 226, 228].

A number of results concerning lifting modules have appeared in the literature in recent years. Major contributions to their theory came, for example, from research groups in Japan, Turkey, and India. Moreover, progress in other branches of module theory has also had a flow-on effect, providing further enrichment to lifting module theory. Many of the sections in our text end with comments which highlight the development of the concepts and results presented and give further details of the contributors and their work.

One aspect of these investigations is the (co)torsion-theoretic point of view as elaborated in, for example, [329]. This provides a way of studying the corresponding notions in the category of comodules over coalgebras (or corings) which behaves similarly to the category $\sigma[M]$ of modules subgenerated by some given module $M$.

Roughly speaking, the various classes of modules associated with extending modules are obtained by weakening the injectivity properties correspondingly. Consequently, our dual concept of lifting modules is pursued through relaxed projectivity conditions. However, since the existence of supplements in a module $M$ is not an immediate consequence of (weak) projectivity conditions, we must frequently rely also on either finiteness conditions or structural conditions on the lattice of submodules of $M$ (such as the AB5* condition).

The authors gratefully acknowledge the contributions of many colleagues and friends to the present monograph. A considerable part of the material consists of reports on their results and some effort was made to coordinate this information and profit from the induced synergy.

They also want to express their thanks for the financial support behind the project. In particular, Christian Lomp was partially supported by the Centro de Matemática da Universidade do Porto, financed by FCT (Portugal) through the programs POCTI and POSI, with national and European Community structural funds. The cooperation between the partners in Germany and India was supported by the German Research Foundation (DFG) and the German Academic Exchange Service (DAAD).

February 2006,

John Clark, Christian Lomp, Narayanaswami Vanaja, Robert Wisbauer

# Introduction

In the first chapter, we present basic notions and techniques from module theory which are employed in the rest of the text. While proofs are often provided, the reader is referred to standard texts for the background details for the more common concepts. Section 1 considers the fundamentals of injectivity, setting the scene for the dual concepts to be introduced later. Similarly, the reader is reminded here of independent families of submodules, prior to the dual notion of coindependence. Small submodules, the radical, hollow modules, local modules, and max modules are considered in Section 2. The next section sees the introduction of coclosed submodules and coclosure (both key ingredients of Section 20 onwards) while Section 4 looks at an assortment of projectivity conditions and how these can influence the endomorphism ring of a module. These weaker forms of projectivity play important rôles throughout the rest of the monograph. The chapter ends with a section on the hollow dimension of a module, this being the dual of the more familiar Goldie or uniform dimension.

Chapter 2 considers torsion-theoretic aspects, the first two sections looking at preradicals, colocalisation, and torsion theories associated with small submodules and projectivity. Section 8 introduces small modules. Here, given two modules $M$ and $N$ over the ring $R$, we say that $N$ is $M$-*small* if $N$ is small in some $L$ in $\sigma[M]$. The preradical and torsion theories (co)generated by the class of $M$-small modules are examined in detail. Corational modules and in particular minimal corational submodules make their appearance in Section 9, leading to the definition of a copolyform module — these are duals of notions used in the more familiar construction and theory of rings and modules of quotients. The last section of this chapter looks briefly at the theory of proper classes of short exact sequences in $\sigma[M]$ and supplement submodules relative to a radical for $\sigma[M]$, the hope being that in the future these general notions will provide further insight to the theory of lifting modules.

Much of module theory is concerned with the presence (or absence) of decomposability properties and Chapter 3 presents a treatment of those properties which are particularly relevant to the theory of lifting modules. This begins with a section on the exchange property, including characterizations of exchange rings, exchangeable decompositions and the (more general) internal exchange property. This is followed by Section 12 which initially looks at LE-modules, that is, modules with a local endomorphism ring, and modules which decompose as direct sums of LE-modules. After an overview of the background to Azumaya's generalization of the Krull–Schmidt Theorem, this section then looks at Harada's work on the interplay between the exchange property and local semi-T-nilpotency. The main result presented here is an important characterization of the exchange property for a module with an indecomposable decomposition due to Zimmermann-Huisgen and Zimmermann (following earlier results by Harada).

Section 13 introduces the concept of local summands, showing how this is intimately related to local semi-T-nilpotency, in particular through a theorem of Dung which also generalises earlier work by Harada. The next section defines partially invertible homomorphisms and the total, concepts pioneered by Kasch. The coincidence of the Jacobson radical and the total for the endomorphism ring of a module with an LE-decomposition provides yet another link with local semi-T-nilpotency and the exchange property. Section 15 considers rings of stable range 1, unit-regular rings and the cancellation, substitution, and internal cancellation properties — these concepts are inter-related for exchange rings by a result of Yu. Chapter 3 ends with a section on decomposition uniqueness, in particular results of a Krull–Schmidt nature due to Facchini and coauthors, and biuniform modules (modules which are both uniform and hollow).

A *complement* of a submodule $N$ of a module $M$ can be defined to be any submodule $L$ of $M$ maximal with respect to the property that $L \cap N = 0$. Dually, a *supplement* $K$ of $N$ in $M$ can be defined to be any submodule $K$ of $M$ minimal with respect to the property that $K + N = M$. Chapter 4 looks at supplements in modules, a keystone concept for the rest of the monograph. As emphasised in our Preface, in contrast to complements of submodules in a module, supplements of submodules need not exist (consider the $\mathbb{Z}$-module $\mathbb{Z}$, for example). The first section of the chapter introduces and investigates the class of semilocal modules and, more specifically, semilocal rings. Weak supplements are also introduced here and, among other things, weakly supplemented modules (in which every submodule has a weak supplement) are shown to be semilocal (and so, in spite of their more relaxed definition, weak supplements also do not always exist). Section 18 looks at the connection between finite hollow dimension and the existence of weak supplements; in particular, it is shown that the rings over which every finitely generated left (or right) module has finite hollow dimension are those for which every finitely generated module is weakly supplemented and these are precisely the semilocal rings. Modules having semilocal endomorphism rings are investigated in Section 19 and shown to have the $n$-th root uniqueness property.

Section 20 formally introduces supplements and the classes of supplemented, finitely supplemented, cofinitely supplemented, and amply supplemented modules; the relationships between these classes and hollow submodules and hollow dimension also receive attention. As with their essential closure counterpart, the coclosure (if it exists) of a submodule in a module need not be unique. Modules for which every submodule has a unique coclosure are called *UCC* modules and investigated in the last section of Chapter 4. In particular, as a dual to the relationship between polyform modules and modules with unique complements, an amply supplemented copolyform module $M$ is UCC, and, if $M \oplus M$ is a UCC weakly supplemented module, then $M$ is copolyform and $M \oplus M$ is amply supplemented.

The opening section of the final Chapter 5 introduces the modules of our title, lifting modules. As already mentioned, these are the duals of extending modules. More specifically, a module $M$ is *lifting* if, given any submodule $N$ of $M$, $N$ lies over a direct summand of $M$, that is, there is a direct summand $X$ of $M$ such that $X \subseteq N$ and $N/X$ is small in $M/X$. The building blocks of this class, in other words the indecomposable lifting modules, are precisely the hollow modules. Indeed, any lifting module of finite uniform or finite hollow dimension is a finite direct sum of hollow modules and several other conditions are established that guarantee an indecomposable decomposition for a lifting module. A weaker version of lifting, namely hollow-lifting, is also considered, as well as the effect of chain conditions on the lifting property.

The direct sum of two lifting modules need not be lifting, so Section 23 looks at when the direct sum of finitely many lifting modules is also lifting. The results here involve projectivity conditions defined in Section 4 and exchangeable decompositions from Section 11. Section 24 looks at the lifting property for infinite direct sums, in particular for modules with LE-decompositions (with material from Chapter 3 used extensively) and for direct sums of hollow modules. A module $M$ is said to be $\Sigma$-lifting if the direct sum of any family of copies of $M$ is lifting. If $M$ is a $\Sigma$-lifting LE-module then it is local and self-projective. Section 25 investigates $\Sigma$-lifting modules $M$ and their indecomposable summands, in particular when $M$ is also copolyform or self-injective.

The fairly long Section 26 looks at four subclasses of lifting modules, namely the weakly discrete, quasi-discrete, discrete, and strongly discrete modules (and we have the strict hierarchy: lifting $\Rightarrow$ weakly discrete $\Rightarrow$ quasi-discrete $\Rightarrow$ discrete $\Rightarrow$ strongly discrete). Roughly speaking, each of these may be defined by requiring the module to be supplemented, its supplement submodules to be further constrained, and certain relative projectivity conditions to be satisfied. Each class is characterized and examined in depth, in particular with regard to decomposition, factor modules, and stability under direct sums. The section ends with a look at two particular types of strongly discrete modules, namely the self-projective modules $M$ which are perfect and, more generally, semiperfect in $\sigma[M]$. A module $M$ is called $\Sigma$-*lifting* ($\Sigma$-*extending*) if the direct sum of any family of copies of $M$ is lifting (extending, respectively).

The study of extending modules and lifting modules has its roots in the module theory of quasi-Frobenius rings. It is well-known that these rings are characterized as those over which every injective module is projective. Section 27 looks at the more general situation of modules $M$ for which every injective module in $\sigma[M]$ is lifting. These are known as *Harada* modules and if $_R R$ is Harada then $R$ is called a *left Harada ring*. A module $M$ is called *co-Harada* if every projective module in $\sigma[M]$ is $\Sigma$-extending, and results due to Harada and Oshiro showing the interplay between Harada, co-Harada, and quasi-Frobenius rings are generalised here to a module setting. Finally, Section 28 carries this duality theme further by investigating modules $M$ for which extending, or indeed all, modules in $\sigma[M]$ are lifting. We give extensive characterizations of both conditions. Specialising to the module $_RR$, we see that every extending $R$-module is lifting precisely when every self-projective $R$-module is extending and this characterizes artinian serial rings $R$. Similarly, every left $R$-module is extending if and only if all left $R$-modules are lifting, and this happens precisely when every left $R$-module is a direct sum of uniserial modules of length at most 2.

In the Appendix details are provided of two graph-theoretical techniques, namely Hall's Marriage Theorem and König's Graph Theorem. These are used in Chapter 3.

We have included a collection of exercises for most sections, in the hope that these will encourage the reader to further pursue the section material.

# Notation

| | |
|---|---|
| $\mathbb{Z}\ (\mathbb{Q})$ | the integers (rationals) |
| $\mathrm{card}(I)$ | the cardinality of the set $I$ |
| ACC (DCC) | the ascending (descending) chain condition |
| id, $\mathrm{id}_M$, $1_M$ | identity map (of the set $M$) |
| $\mathrm{Ke}\,f$ | the kernel of a linear map $f$ |
| $\mathrm{Coke}\,f$ | the cokernel of a linear map $f$ |
| $\mathrm{Im}\,f$ | the image of a map $f$ |
| $g \circ f,\ f \diamond g$ | composition of the maps $M \xrightarrow{f} N \xrightarrow{g} L$ |
| $M_K$ | short for $\bigoplus_{k \in K} M_k$ where the $M_k$ are modules |
| $\mathrm{Gen}(M)$ | the class of $M$-generated modules, 1.1 |
| $\mathrm{Tr}(M, N)$ | the trace of the module $M$ in the module $N$, 1.1 |
| $\mathrm{Cog}(M)$ | the class of $M$-cogenerated modules, 1.2 |
| $\mathrm{Re}(N, M)$ | the reject of the module $M$ in the module $N$, 1.2 |
| $\sigma[M]$ | the full subcategory of $R$-Mod subgenerated by $M$, 1.3 |
| $K \trianglelefteq M$ | $K$ is a large (essential) submodule of $M$, 1.5 |
| $\mathrm{Soc}(M)$ | the socle of the module $M$, 1.20 |
| $K \ll M$ | $K$ is a small submodule of the module $M$, 2.2 |
| $\nabla(M, N)$ | $\{ f \in \mathrm{Hom}(M, N) \mid \mathrm{Im}\,f \ll N \}$, 2.4 |
| $\nabla(M)$ | $\{ f \in \mathrm{End}(M) \mid \mathrm{Im}\,f \ll N \}$, 2.4 |
| $\mathrm{Rad}\,(M)$ | the radical of the module $M$, 2.8 |
| $\mathrm{Jac}\,(R),\ J(R)$ | the Jacobson radical of the ring $R$, 2.11 |
| $K \overset{cs}{\underset{M}{\hookrightarrow}} L$ | $K \subset L$ is cosmall in $M$, 3.2 |
| $N \overset{cc}{\hookrightarrow} M$ | $N$ is a coclosed submodule of $M$, 3.7 |
| $K \overset{scc}{\hookrightarrow} M$ | $K$ is a strongly coclosed submodule of $M$, 3.12 |
| $\mathrm{u.dim}(M)$ | the uniform dimension of the module $M$, 5.1 |
| $\mathrm{h.dim}(M)$ | the hollow dimension of the module $M$, 5.2 |
| $\mathcal{C}_M^{\trianglelefteq}$ | the class of singular modules in $\sigma[M]$, 7.8 |
| $\mathrm{Z}_M(N)$ | the sum of the $M$-singular submodules of $N$, 7.8 |
| $\mathrm{Re}_M^{\trianglelefteq}(N)$ | the reject of the $M$-singular modules in $N$, 7.9 |
| $\mathbb{S}[M],\ \mathbb{S}$ | the class of small modules in $\sigma[M]$, 8.2 |
| $\mathrm{Tr}_{\mathbb{S}}(N)$ | $\mathrm{Tr}(\mathbb{S}, N)$, the sum of the $M$-small submodules of $N$, 8.5 |
| $\mathrm{Re}_{\mathbb{S}}(N)$ | $\mathrm{Re}\,(N, \mathbb{S})$, reject of the $M$-small modules in $N$, 8.9 |
| $K \overset{cr}{\underset{M}{\hookrightarrow}} L$ | $K$ is a corational submodule of $L$ in $M$, 9.7 |

# Chapter 1

# Basic notions

## 1  Preliminaries

For basic definitions, theorems and notation, we refer the reader to texts in Module Theory, mainly to [363] and [85] and occasionally to [13], [100], and [189]. In this section, we recall some of the notions which will be of particular interest to us.

Throughout $R$ will denote an associative ring with unit 1 and $M$ will usually stand for a nonzero unital left $R$-module; if necessary we will write $_R M$ or $M_R$ to indicate if we mean a left or right module over $R$. $R$-Mod and Mod-$R$ denote the category of all unital left and right modules, respectively.

For $R$-modules $M, N$, we denote the group of $R$-module homomorphisms from $M$ to $N$ by $\mathrm{Hom}_R(M, N)$, or simply $\mathrm{Hom}(M, N)$, and the endomorphism ring of $M$ by $\mathrm{End}_R(M)$, or $\mathrm{End}(M)$. The *kernel* of any $f \in \mathrm{Hom}(M, N)$ is denoted by Ke $f$, and its *image* by Im $f$.

It is of some advantage to write homomorphisms of left modules on the right side of the (elements of the) module and homomorphisms of right modules on the left and we will usually do so. In this case we denote by $f \diamond g$ the composition of two left $R$-module morphisms $f : M \to N$, $g : N \to L$, and so $((m)f)g = (m)f \diamond g$, for any $m \in M$. If these are morphisms of right $R$-modules we write the morphisms on the left and their composition is (as usual) $g \circ f$, and $g(f(m)) = g \circ f(m)$, for any $m \in M$. When no confusion is likely we may take the liberty to delete the symbols $\circ$ or $\diamond$, respectively.

With this notation, a left $R$-module $M$ is a right module over $S = \mathrm{End}_R(M)$ and indeed an $(R, S)$-bimodule. $(R, S)$-submodules of $M$ are referred to as *fully invariant* or *characteristic* submodules.

**1.1. Trace.** Let $M$ be an $R$-module. Then an $R$-module $N$ is called *$M$-generated* if it is a homomorphic image of a direct sum of copies of $M$. The class of all $M$-generated $R$-modules is denoted by $\mathrm{Gen}(M)$.

The *trace of $M$ in $N$* is the sum of all $M$-generated submodules of $N$; this is a fully invariant submodule of $N$ which we denote by $\mathrm{Tr}(M, N)$. In particular, $\mathrm{Tr}(M, R)$, the trace of $M$ in $R$, is a two-sided ideal.

Given a class $\mathbb{K}$ of $R$-modules, the *trace of $\mathbb{K}$ in $N$* is the sum of all $K$-generated submodules of $N$, for all $K \in \mathbb{K}$, and we denote it by $\mathrm{Tr}(\mathbb{K}, N)$.

Given a left ideal $I$ of $R$, we have $IM \in \mathrm{Gen}(I)$, hence $IM \subset \mathrm{Tr}(I, M)$. If $I$ is idempotent, then $IM = \mathrm{Tr}(I, M)$.

**1.2. Reject.** Dually, an $R$-module $N$ is called *M-cogenerated* if it is embeddable in a direct product of copies of $M$. The class of all $M$-cogenerated modules is denoted by $\mathrm{Cog}(M)$.

The *reject of $M$ in $N$* is the intersection of all submodules $U \subset N$ for which $N/U$ is $M$-cogenerated; this is a fully invariant submodule of $N$ which we denote by $\mathrm{Re}(N, M)$. For any class $\mathbb{K}$ of $R$-modules, the *reject of $\mathbb{K}$ in $N$* is the intersection of all submodules $U \subset N$ such that $N/U$ is cogenerated by modules in the class $\mathbb{K}$, and we denote this by $\mathrm{Re}(N, \mathbb{K})$. It follows that (see [363, 14.4])

$$\mathrm{Re}(N, \mathbb{K}) = \bigcap \{\mathrm{Ke}\, f \mid f \in \mathrm{Hom}(N, K), K \in \mathbb{K}\}.$$

**1.3. The category $\sigma[M]$.** An $R$-module $N$ is said to be *subgenerated by $M$* if $N$ is isomorphic to a submodule of an $M$-generated module. We denote by $\sigma[M]$ the full subcategory of $R$-Mod whose objects are all $R$-modules subgenerated by $M$ (see [363, §15]). A module $N$ is called a *subgenerator* in $\sigma[M]$ if $\sigma[M] = \sigma[N]$.

$\sigma[M]$ is closed under direct sums (coproducts), kernels and cokernels. The product of any family of modules $\{N_\lambda\}_\Lambda$ in the category $\sigma[M]$ is denoted by $\prod_\Lambda^M N_\lambda$. It is obtained as the trace of $\sigma[M]$ in the product in $R$-Mod, that is,

$$\prod_\Lambda^M N_\lambda = \mathrm{Tr}(\sigma[M], \prod_\Lambda N_\lambda).$$

**1.4. Finitely presented and coherent modules in $\sigma[M]$.** Let $N \in \sigma[M]$. Then $N$ is said to be *finitely presented in $\sigma[M]$* if it is finitely generated and, in every exact sequence $0 \to K \to L \to N \to 0$ in $\sigma[M]$ where $L$ is finitely generated, $K$ is also finitely generated (see [363, 25.1]).

$N$ is called *locally coherent in $\sigma[M]$* if all its finitely generated submodules are finitely presented in $\sigma[M]$ (see [363, 26.1]).

$N$ is called *locally noetherian* if all its finitely generated submodules are noetherian (see [363, § 27]).

While $N$ is finitely generated in $\sigma[M]$ if and only if it is finitely generated in $R$-Mod, modules which are finitely presented in $\sigma[M]$ need not be finitely presented in $R$-Mod. For example, let $M$ be a simple module which is not finitely $R$-presented; then $M$ is finitely presented in $\sigma[M]$ but not in $R$-Mod.

In general there need not exist any nonzero finitely presented modules in $\sigma[M]$ (see [284, Example 1.7]). On the other hand, every finitely generated module is finitely presented in $\sigma[M]$ if and only if $M$ is locally noetherian (see [363, 27.3]).

**1.5. Large (essential) submodules.** A submodule $K$ of a nonzero module $M$ is said to be *large* or *essential* (we write $K \trianglelefteq M$) if $K \cap L \neq 0$ for every nonzero submodule $L \subset M$. If all nonzero submodules of $M$ are large in $M$, then $M$ is called *uniform.* A monomorphism $f : N \to M$ is called *large* or *essential* if Im $f \trianglelefteq M$. This can be characterised by the property that any $g : M \to L$ for which $f \diamond g$ is a monomorphism has to be a monomorphism.

**1.6. Characterisation of uniform modules.** *For $M$ the following are equivalent:*

(a) *$M$ is uniform;*

(b) *every nonzero submodule of $M$ is indecomposable;*

(c) *for any module morphisms $K \xrightarrow{f} M \xrightarrow{g} N$, where $f \neq 0$, $f \diamond g$ injective implies that $f$ and $g$ are injective.*

*Proof.* (a)$\Leftrightarrow$(b). For any nonzero submodules $U, V \subset M$ with $U \cap V = 0$, the submodule $U + V = U \oplus V$ is decomposable.

(a)$\Rightarrow$(c). Let $f \diamond g$ be injective. Then clearly $f$ is injective. Assume $g$ is not injective. Then Im $f \cap$ Ke $g \neq 0$ and this implies Ke $f \diamond g \neq 0$, a contradiction.

(c)$\Rightarrow$(a). Let $K, V \subset M$ be submodules with $K \neq 0$ and $K \cap V = 0$. Then the composition of the canonical maps $K \to M \to M/V$ is injective and now (c) implies that $M \to M/V$ is injective, that is, $V = 0$. $\qquad\square$

**1.7. Singular modules.** A module $N \in \sigma[M]$ is called *$M$-singular* or *singular in $\sigma[M]$* provided $N \simeq L/K$ for some $L \in \sigma[M]$ and $K \trianglelefteq L$. The class of all $M$-singular modules is closed under submodules, factor modules and direct sums and we denote it by $\mathcal{C}_M^{\trianglelefteq}$. Notice that this notion is dependent on the choice of $M$ (i.e., the category $\sigma[M]$) and obviously $\mathcal{C}_M^{\trianglelefteq} \subseteq \mathcal{C}_R^{\trianglelefteq}$.

**1.8. $M$-injectivity.** A module $N$ is called *$M$-injective* if, for every submodule $K \subset M$, any morphism $f : K \to N$ can be extended to a morphism $M \to N$. In case $M$ is $M$-injective, then $M$ is also called *self-injective* or *quasi-injective*.

Any $M$-injective module is also $L$-injective for any $L \in \sigma[M]$. For $N \in \sigma[M]$, the *injective hull* $\widehat{N}$ in $\sigma[M]$ is an essential extension $N \trianglelefteq \widehat{N}$ where $\widehat{N}$ is $M$-injective. It is related to the injective hull $E(N)$ of $N$ in $R$-Mod by the trace functor. More specifically, $\widehat{N} = \mathrm{Tr}(\sigma[M], E(N)) = \mathrm{Tr}(M, E(N))$ since $\mathrm{Tr}(\sigma[M], Q) = \mathrm{Tr}(M, Q)$ for any $R$-injective module $Q$ (see, e.g., [363, §16, 17]).

**1.9. Complement submodules.** Given a submodule $K \subset M$, a submodule $L \subset M$ is called a *complement of $K$ in $M$* if it is maximal in the set of all submodules $L' \subset M$ with $K \cap L' = 0$. By Zorn's Lemma, every submodule has a complement in $M$. A submodule $L \subset M$ is called a *complement submodule* provided it is the complement of some submodule of $M$. If $L$ is a complement of $K$ in $M$, then there is a complement $\overline{K}$ of $L$ in $M$ that contains $K$. By construction, $K \trianglelefteq \overline{K}$ and $\overline{K}$ has no proper essential extension in $M$; thus $\overline{K}$ is called an *essential closure* of $K$ in $M$. In general essential closures need not be uniquely determined.

A submodule $K \subset M$ is called *(essentially) closed* if $K = \overline{K}$.

The following characterises closed submodules (see [363] and [85]).

**1.10. Characterisation of closed submodules.** *For a submodule $K \subset M$, the following are equivalent:*

(a) *$K$ is a closed submodule;*

(b) *$K$ is a complement submodule;*

(c) *$K = \widehat{K} \cap M$, where $\widehat{K}$ is a maximal essential extension of $K$ in $\widehat{M}$.*

Notice that in (c), $\widehat{K}$ is a direct summand in the $M$-injective hull $\widehat{M}$.

**1.11. Factors by closed submodules.** *Let the submodule $K \subset M$ be a complement of $L \subset M$. Then:*

(1) *$(K + L)/K \trianglelefteq M/K$  and  $K + L \trianglelefteq M$.*

(2) *If $U \trianglelefteq M$ and $K \subset U$, then $U/K \trianglelefteq M/K$.*

*Proof.* (1) is shown in [363, 17.6].

(2) First notice that $L \cap U \trianglelefteq L$ and hence $K$ is also a complement of $L \cap U$. By (1), this implies that $(K + L \cap U)/K \trianglelefteq M/K$. Since $(K + L \cap U)/K \subset U/K$ we also get $U/K \trianglelefteq M/K$. $\qquad\square$

**1.12. Weakly $M$-injective modules.** The module $N$ is called *weakly $M$-injective* if every diagram in $R$-Mod

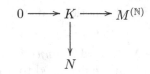

with exact row and $K$ finitely generated, can be extended commutatively by a morphism $M^{(\mathbb{N})} \to N$, that is, the functor $\mathrm{Hom}(-, N)$ is exact with respect to the given row. Any weakly $M$-injective module $N \in \sigma[M]$ is $M$-generated and also $\widehat{M}$-generated (see [363, 16.11]). Notice that, putting $M = R$, weakly $R$-injective modules are just *FP-injective* modules (see [363, 35.8]).

**1.13. $\pi$-injective and direct injective modules.** A module $M$ is said to be *$\pi$-injective* if, for any submodules $K, L \subset M$ with $K \cap L = 0$, the canonical monomorphism $M \to M/K \oplus M/L$ splits. Furthermore, $M$ is called *direct injective* if for any direct summand $K \subset M$, any monomorphism $K \to M$ splits. Notice that any self-injective module $M$ has both these properties (e.g., [85], [363]).

Regarding injective modules in $\sigma[M]$ and the locally noetherian property of $M$ we have the following proved in [363, 27.3, 27.5].

**1.14. Decomposition of injective modules.** *For $M$ the following are equivalent:*

(a) *$M$ is locally noetherian;*

(b) *every weakly $M$-injective module is $M$-injective;*

(c) *every injective module in* $\sigma[M]$ *is a direct sum of indecomposables;*

(d) *any direct sum of injective modules in* $\sigma[M]$ *is injective in* $\sigma[M]$;

(e) *every (countable) direct sum of M-injective hulls of simple modules in* $\sigma[M]$ *is M-injective.*

**1.15. Semisimple modules.** A module is *simple* if it is nonzero and does not properly contain any nonzero submodule. The module $M$ is called *semisimple* if it is a direct sum of simple submodules. This is equivalent to the property that every submodule of $M$ is a direct summand, or that every $R$-module is $M$-injective. Thus a semisimple $R$-module $M$ is always self-injective but need not be $R$-injective.

$M$ is said to be *homogeneous semisimple* if it is generated by a single simple module; that is, it is a direct sum of simple modules which are isomorphic to each other. This is the case if and only if $\sigma[M]$ has a simple (sub)generator.

Note that the module $_RR$ is simple if and only if $R$ is a division ring, $_RR$ is semisimple if and only if $R$ is a (finite) direct sum of matrix rings over a division ring, and $_RR$ is homogeneous semisimple if and only if $R$ is a matrix ring over a division ring. Consequently all three notions are left-right symmetric. If $_RR$ is semisimple, then $R$ is called *left* (or *right*) *semisimple* or *artinian semisimple*.

**1.16. Extending modules.** A module $M$ is called an *extending* or *CS module* if every submodule of $M$ is essential in a direct summand of $M$. The notation CS refers to the fact that these modules are characterised by the property that every complement submodule is a direct summand in $M$.

It is well-known that any (self-)injective module is an extending module and hence semisimple modules are extending. Any uniform module is extending because every nonzero submodule is essential and, conversely, any indecomposable extending module must be uniform. Direct summands of extending modules are extending but direct sums of uniform modules need not be extending. For example the $\mathbb{Z}$-module $\mathbb{Z}/8\mathbb{Z} \oplus \mathbb{Z}/2\mathbb{Z}$ is not an extending module. For a detailed presentation of extending modules we refer to [85].

**1.17. Almost $M$-injective modules.** Let $M$ and $N$ be two $R$-modules. Then $N$ is called *almost $M$-injective* if, for any $g \in \mathrm{Hom}(K, N)$ and monomorphism $f \in \mathrm{Hom}(K, M)$, either there exists $h \in \mathrm{Hom}(M, N)$ yielding a commutative diagram

$$
\begin{array}{ccc}
0 \longrightarrow K & \overset{f}{\longrightarrow} & M \\
{\scriptstyle g}\downarrow & {\scriptstyle h} \nearrow & \\
N & &
\end{array} \ ,
$$

or there exists a direct summand $A$ of $M$ with projection $p : M \to A$ and some map $\overline{h} : N \to A$ such that $g\overline{h} = fp \neq 0$, that is we get the commutative diagram

$$
\begin{array}{ccc}
0 \longrightarrow K & \xrightarrow{\ f\ } & M \\
\quad\downarrow{\scriptstyle g} & & \vdots\,{\scriptstyle p} \\
N & \xrightarrow[\ \overline{h}\ ]{} & A\,.
\end{array}
$$

**1.18. Almost $M$-injective uniform modules.** Let $M$ and $N$ be uniform modules. Then $N$ is almost $M$-injective if and only if either any map $g : K \to N$, where $K \subseteq M$, can be extended to $M$ or the map $g$ is injective and $g^{-1} : (K)g \to M$ can be extended to $N$. Hence if $g$ is not injective and $M$ is almost $N$-injective, then there always exists an extension of $g$ to $M$.

The almost injectivity property can help to make direct sums of extending modules extending. It provides a necessary and sufficient condition for a finite direct sum of uniform modules with local endomorphism ring to be extending.

**1.19. Finite direct sums of uniform modules extending.** *Let $M = \bigoplus_{i=1}^{n} M_i$ where each $M_i$ is a uniform module with local endomorphism ring. Then the following are equivalent:*

(a) *$M$ is extending;*

(b) *$M_i$ is almost $M_j$-injective, for $i \neq j$.*

*Proof.* By [85, Lemma 8.5], $M$ is extending if and only if $M$ is uniform extending. By [183, Lemma 11], $M$ is uniform extending if and only if $M_i \oplus M_j$ is extending for any $i \neq j$. Following [30, Lemma 8], this in turn is equivalent to $M_i$ and $M_j$ being mutually almost relative injective.                                         $\square$

**1.20. The socle of a module.** The *socle* of $M$, denoted by $\mathrm{Soc}(M)$ or $\mathrm{Soc}\,M$, is defined as the sum of all its simple submodules and can be shown to coincide with the intersection of all the essential submodules of $M$. It is a fully invariant submodule of $M$ and $\mathrm{Soc}\,K = K \cap \mathrm{Soc}\,M$ for any submodule $K \subset M$. Note that $M$ is semisimple precisely when $M = \mathrm{Soc}\,M$.

**1.21. Cosemisimple modules.** The module $M$ is called *cosemisimple* if all simple modules (in $\sigma[M]$) are $M$-injective. Such modules can be characterised by the property that there exists a semisimple cogenerator in $\sigma[M]$. We will encounter more properties of these modules later on (see also [363, 23.1]).

Clearly semisimple modules are cosemisimple but the converse need not be true. For example, commutative rings $R$ are cosemisimple (as $R$-modules) if and only if they are von Neumann regular (see [363, 37.11]).

**1.22. Definitions.** A module $P \in \sigma[M]$ is called *cohereditary* in $\sigma[M]$ if every factor module of $P$ is $M$-injective. The module $M$ is said to be *$\sigma$-cohereditary* if every injective module in $\sigma[M]$ is cohereditary.

Notice that the ring $R$ is left hereditary if and only if $_R R$ is $\sigma$-cohereditary.

**1.23. Properties of cohereditary modules.**

(1) *If $P_1, ..., P_k$ are cohereditary in $\sigma[M]$, then so is $\bigoplus_{i=1}^{k} P_i$.*

(2) *If $M$ is locally noetherian, then any direct sum of cohereditary modules in $\sigma[M]$ is cohereditary in $\sigma[M]$, and the following are equivalent:*

    (a) *$\widehat{M}$ is cohereditary in $\sigma[M]$;*

    (b) *$M$ is $\sigma$-cohereditary;*

    (c) *every indecomposable injective module is cohereditary in $\sigma[M]$.*

*Proof.* (1) Let $P_1, P_2$ be cohereditary in $\sigma[M]$ and $P = P_1 \oplus P_2$. For any submodule $K \subset P$, $P_1/(K \cap P_1)$ is $M$-injective and we have

$$P/K \simeq P_1/(K \cap P_1) \oplus \overline{P}_2,$$

where $\overline{P}_2$ is some factor module of $P_2$ and hence is also $M$-injective. Hence $P/K$ is $M$-injective and so $P$ is cohereditary.

The assertion for arbitrary finite direct sums follows by induction.

(2) Let $\{P_\lambda\}_\Lambda$ be a family of cohereditary modules in $\sigma[M]$ and $K \subset \bigoplus_\Lambda P_\lambda$. Now $\bigoplus_\Lambda P_\lambda$ is the direct limit of its finite partial sums, and hence $\bigoplus_\Lambda P_\lambda/K$ is the direct limit of $M$-injective modules (by (1)) and so is $M$-injective (by [363, 27.3]). The equivalences follow easily from [363, 27.5]. $\square$

The following technical observation will be helpful.

**1.24. Lemma.** *Let $K$, $L$ and $N$ be submodules of $M$. Assume $K + L = M$ and $(K \cap L) + N = M$. Then*

$$K + (L \cap N) = L + (K \cap N) = M.$$

*Proof.* First observe that

$$
\begin{aligned}
K + (L \cap N) &= K + (L \cap K) + (L \cap N) \\
&= K + (L \cap ((L \cap K) + N)) \\
&= K + (L \cap M) = K + L = M.
\end{aligned}
$$

Applying the same arguments to $L + (K \cap N)$ yields $L + (K \cap N) = M$. $\square$

**1.25. Independent submodules.** A family $\{N_\lambda\}_\Lambda$ of nonzero submodules of $M$ is called *independent* if, for any $\lambda \in \Lambda$ and nonempty (finite) subset $F \subseteq \Lambda \setminus \{\lambda\}$,

$$N_\lambda \cap \sum_{i \in F} N_i = 0.$$

This generalises the notion of linearly independent elements in modules. Clearly $\{N_\lambda\}_\Lambda$ is an independent family of submodules of $M$ if and only if $\sum_\Lambda N_\lambda = \bigoplus_\Lambda N_\lambda$ (internal direct sum).

Dualising the notion of independent submodules yields

**1.26. Coindependent submodules.** A family $\{N_\lambda\}_\Lambda$ of proper submodules of $M$ is called *coindependent* if, for any $\lambda \in \Lambda$ and any finite subset $F \subseteq \Lambda \setminus \{\lambda\}$,

$$N_\lambda + \bigcap_{i \in F} N_i = M,$$

and it is called *completely coindependent* if, for every $\lambda \in \Lambda$,

$$N_\lambda + \bigcap_{\mu \neq \lambda} N_\mu = M.$$

Clearly $\{N_\lambda\}_\Lambda$ is coindependent if and only if each of its finite subfamilies is coindependent.

Completely coindependent families of submodules are coindependent, but the converse is not true in general. For example, the family of submodules of $\mathbb{Z}$ given by $\{\mathbb{Z}p : p \text{ is a prime}\}$ is coindependent but not completely coindependent.

Obviously, for finite families of submodules coindependent and completely coindependent are the same properties.

Any family $\{N\}$ of one submodule is trivially coindependent.

If $|\Lambda| \geq 2$ any coindependent family $\{N_\lambda\}_\Lambda$ consists of *comaximal submodules* of $M$, that is, $N_\lambda + N_\mu = M$ for any $\mu \neq \lambda$. However, a family of mutually comaximal submodules need not be coindependent. For example, for any field $K$, $\{K(1,0), K(0,1), K(1,1)\}$ is a family of mutually comaximal submodules of $K^2$, but $K(1,0) \cap K(0,1) = (0,0)$ yields $K(1,1) + (K(1,0) \cap K(0,1)) \neq K^2$.

**1.27. Remarks.** Coindependent families are defined in [327]; they are called *d-independent* in [235], and *independent* in [377]. In [269, p. 361] the name *co-independent* is used for $\{N_\lambda\}_\Lambda$ if it is completely coindependent and $M/\bigcap_\Lambda N_\lambda \simeq \bigoplus_\Lambda M/N_\lambda$.

**1.28. Properties of coindependent families.** *Let $\{N_\lambda\}_\Lambda$ be a coindependent family of submodules of $M$.*

(1) *If $\{N'_\lambda\}_\Lambda$ is a family of proper submodules of $M$, such that for every $\lambda \in \Lambda$, $N_\lambda \subseteq N'_\lambda$, then $\{N'_\lambda\}_\Lambda$ is coindependent.*

(2) *If $L \subset M$ is a submodule such that $\bigcap_F N_\lambda + L = M$ for every finite subset $F$ of $\Lambda$, then $\{N_\lambda\}_\Lambda \cup \{L\}$ is coindependent.*

(3) *If $M$ is finitely generated and $\Lambda = \{1, 2, \ldots, m\}$, then there exist finitely generated submodules $L_i \subseteq N_i$, for each $i \in \{1, \ldots, m\}$, such that the family $\{L_1, \ldots, L_m\}$ is coindependent.*

*Proof.* (1) By assumption, for any $\lambda \in \Lambda$ and finite subset $F \subset \Lambda \setminus \{\lambda\}$,

$$M = N_\lambda + \bigcap_{i \in F} N_i \subseteq N'_\lambda + \bigcap_{i \in F} N'_i \subseteq M,$$

thus proving the assertion.

(2) Let $F$ be a finite subset of $\Lambda$ and $\mu \in \Lambda \setminus F$. Let $N = \bigcap_F N_i$. Then, by hypothesis, $N + N_\mu = M$ and $(N \cap N_\mu) + L = M$. Hence, by Lemma 1.24, we get $M = (N \cap L) + N_\mu$ as $N + L = M$. From this it follows easily that $\{N_\lambda\}_\Lambda \cup \{L\}$ is a coindependent family of submodules of $M$.

(3) Since $M$ is finitely generated, for each $1 \leq i \leq m$ there exist finitely generated submodules $X_i \subseteq N_i$ and $Y_i \subseteq \bigcap_{j \neq i} N_j$ such that $X_i + Y_i = M$. Let $L_i := X_i + \sum_{j \neq i} Y_j \subseteq N_i$. By construction, $M = X_i + Y_i \subseteq L_i + \bigcap_{j \neq i} L_j$, and the assertion follows. $\qquad\square$

**1.29. Chinese Remainder Theorem.** *Let $\{K_\lambda\}_\Lambda$ be a family of nonzero modules of $M$ and $\{f_\lambda : M \to M/K_\lambda\}_\Lambda$ the family of induced projections. Then there is a morphism $f : M \to \prod_\Lambda M/K_\lambda$ such that*

(1) *Ke $f = \bigcap_\Lambda K_\lambda$;*

(2) *if $f$ is an epimorphism, then $\{K_\lambda\}_\Lambda$ is a completely coindependent family;*

(3) *if $\Lambda$ is finite and $\{K_\lambda\}_\Lambda$ is a coindependent family, then $f$ is an epimorphism and*

$$M / {\textstyle\bigcap_\Lambda} K_\lambda \simeq {\textstyle\prod_\Lambda} M/K_\lambda.$$

*Proof.* (1) The morphism $f : M \to \prod_\Lambda M/K_\lambda$ exists by the universal property of the product and obviously Ke $f = \bigcap_\Lambda K_\lambda$.

(2) Let $\lambda \in \Lambda$. We prove that

$$M = K_\lambda + \bigcap_{\mu \in \Lambda \setminus \{\lambda\}} K_\mu.$$

Let $m \in M$. If $m \notin K_\lambda$, then $(m)f_\lambda \neq 0$. Hence $(\delta_{\mu\lambda} (m)f_\lambda)_{\mu \in \Lambda}$ is an element of $\prod_\Lambda M/K_\lambda$, where $\delta_{\mu\lambda}$ denotes the Kronecker symbol, that is,

$$\delta_{\mu\lambda} = \begin{cases} 1_R & \text{if } \mu = \lambda \\ 0_R & \text{if } \mu \neq \lambda \end{cases} , \text{ for any } \mu, \lambda \in \Lambda.$$

Since $f$ is epimorph, there exists $m_\lambda \in M$ such that $(m_\lambda)f = (\delta_{\mu\lambda} (m)f_\lambda)_{\mu \in \Lambda}$. Thus, for all $\mu \in \Lambda$,

$$(m_\lambda)f_\mu = \delta_{\mu\lambda} (m)f_\lambda.$$

Hence for all $\mu \neq \lambda$, $m_\lambda \in K_\mu$ and so $m_\lambda \in \bigcap_{\mu \in \Lambda \setminus \{\lambda\}} K_\mu$. For $\mu = \lambda$, $(m_\lambda)f_\lambda = (m)f_\lambda$ yields $m - m_\lambda \in K_\lambda$. This produces the desired outcome:

$$m = m - m_\lambda + m_\lambda \in K_\lambda + {\textstyle\bigcap_{\mu \in \Lambda \setminus \{\lambda\}}} K_\mu.$$

(3) We prove this by induction on $n = |\Lambda|$. For $n = 1$ our claim is trivial. Let $n > 1$ and suppose that our claim holds for all coindependent families $\{L_1, \ldots, L_{n-1}\}$ of submodules of $M$. Let $\{K_1, \ldots, K_n\}$ be a coindependent family of cardinality $n$. Then $\{K_1, \ldots, K_{n-1}\}$ is also coindependent.

Set $K := \bigcap_{i=1}^{n-1} K_i$. By induction, we have $M/K \simeq \bigoplus_{i=1}^{n-1} M/K_i$. Furthermore, $K + K_n = M$, so

$$
\begin{aligned}
M/\bigcap_{i=1}^{n} K_i \;&=\; M/(K \cap K_n) \;=\; K/(K \cap K_n) \oplus K_n/(K \cap K_n) \\
&\simeq\; M/K_n \oplus M/K \;\simeq\; \bigoplus_{i=1}^{n} M/K_i,
\end{aligned}
$$

as required.                                                                                           $\square$

**1.30. Corollary.** *Let $K, N \subset M$ be submodules with $K + N = M$. Let*

$$p_N : M \to M/N, \; p_{N \cap K} : M \to M/(N \cap K) \text{ and } \eta : M/(N \cap K) \to M/N$$

*denote the natural maps. Then $M/(N \cap K) \simeq M/N \oplus M/K$, $p_{N \cap K} \diamond \eta = p_N$, and for the map*

$$\phi : M/N \to M/(N \cap K), \quad m + N = n + k + N \mapsto k + (K \cap N),$$

*where $m = n + k \in M$, $n \in N$ and $k \in K$, we have $\phi \diamond \eta = \mathrm{id}_{M/N}$.*

**References.** Brodskiĭ and Wisbauer [45]; Lomp [226]; Miyashita [235]; Oshiro [269]; Reiter [295]; Renault [297]; Takeuchi [327]; Wisbauer [363]; Zelinsky [377].

# 2 Small submodules and the radical

A fundamental part of our theory is based on the notions of *small submodules* and *small epimorphisms*, which dualise large submodules and large monomorphisms.

**2.1. Definitions.** A submodule $K \subseteq M$ is called *small* or *superfluous in* $M$, written $K \ll M$, if, for every submodule $L \subset M$, the equality $K + L = M$ implies $L = M$. An epimorphism $f : M \to N$ is said to be *small* or *superfluous* if $\operatorname{Ke} f \ll M$; such an $f$ is also called a *small cover*.

Thus $K \ll M$ if and only if the canonical projection $M \to M/K$ is a small epimorphism. From [363, Section 19] we recall:

**2.2. Small submodules and epimorphisms.** *Let $K$, $L$, $N$ and $M$ be $R$-modules.*

(1) *An epimorphism $f : M \to N$ in $R$-Mod is small if and only if every (mono)morphism $h : L \to M$ for which $h \diamond f$ is an epimorphism is itself an epimorphism.*

(2) *If $f : M \to N$ and $g : N \to L$ are two epimorphisms, then $f \diamond g$ is small if and only if both $f$ and $g$ are small.*

(3) *If $K \subset L \subset M$, then $L \ll M$ if and only if $K \ll M$ and $L/K \ll M/K$.*

(4) *If $K_1, \ldots, K_n$ are small submodules of $M$, then $K_1 + \cdots + K_n$ is also small in $M$.*

(5) *For any $K \ll M$ and $f : M \to N$, $Kf \ll N$. The converse is true if $f$ is a small epimorphism.*

(6) *If $K \subset L \subset M$ and $L$ is a direct summand in $M$, then $K \ll M$ if and only if $K \ll L$.*

(7) *If $K \ll M$, then $M$ is finitely generated if and only if $M/K$ is finitely generated.*

To provide easy examples of these notions observe that the $\mathbb{Z}$-module $\mathbb{Z}$ has no nonzero small submodules and $\mathbb{Z}$ is a small submodule in the rationals $\mathbb{Q}$.

**2.3. Small submodules in factor modules.** *Let $K, L$ and $N$ be submodules of $M$.*

(1) *If $M = K + L$, $L \subseteq N$ and $N/L \ll M/L$, then*

$$(K \cap N)/(K \cap L) \ll M/(K \cap L).$$

(2) *If $M = K \oplus L$, $M = N + L$ and $(N + K)/K \ll M/K$, then*

$$(N + K)/N \ll M/N.$$

*Proof.* (1) Consider a submodule $X$ such that $K \cap L \subseteq X \subseteq M$ and

$$M/(K \cap L) = (K \cap N)/(K \cap L) + X/(K \cap L).$$

Then $M = (K \cap N) + X$. By 1.24, $M = N + (K \cap X)$. Since $N/L \ll M/L$, this implies $M = L + (K \cap X)$. Again by 1.24, $M = X + (K \cap L)$ and hence $M = X$. Thus $(K \cap N)/(K \cap L) \ll M/(K \cap L)$.

(2) The obvious maps

$$M/K \simeq L \to L/(L \cap N) \simeq M/N$$

send the small submodule $(N + K)/K$ of $M/K$ onto $(N + K)/N$ which is hence small in $M/N$.                                                                              $\square$

**2.4. The small ideal of modules.** The *small ideal* (or *cosingular ideal*) of two modules $M$ and $N$ is defined as

$$\nabla(M, N) = \{f \in \mathrm{Hom}(M, N) \mid \mathrm{Im}\, f \ll N\}.$$

If $M = N$ we set $\nabla(M) = \nabla(M, M)$.

**2.5. $\nabla$ and direct sums.**

(1) *For direct sums of modules $M = \bigoplus_I M_i$ and $N = \bigoplus_J N_j$,*

$$\nabla(M, N) = 0 \text{ if and only if } \nabla(M_i, N_j) = 0 \text{ for all } i \in I,\, j \in J.$$

(2) *For modules $X$ and $M = M_1 \oplus \cdots \oplus M_n$, the following are equivalent:*

   (a) $\nabla(X, M/K) = 0$ *for all* $K \ll M$;

   (b) $\nabla(X, M_i/U_i) = 0$ *for all* $i \le n$ *and* $U_i \ll M_i$.

*Proof.* (1) Clearly $\nabla(M, N) = 0$ implies $\nabla(M_i, N_j) = 0$ for all $i \in I, j \in J$, since for any $f \in \nabla(M_i, N_j)$, $\pi_i f \varepsilon_j \in \nabla(M)$ where $\pi_i : M \to M_i$ and $\varepsilon_j : N_j \to N$ denote the canonical projection and inclusion, respectively. On the other hand, if $0 \ne f \in \nabla(M, N)$, then there exist $i \in I$ and $j \in I$ such that $0 \ne \varepsilon_i f \pi_j \in \nabla(M_i, N_j)$.

(2) (a)$\Rightarrow$(b) is obvious.

(b)$\Rightarrow$(a). We prove this by induction on $n$. For $n = 1$ the assertion is clear.

Let $n \ge 2$ and assume the statement holds for all $m \le n$. Consider any $f \in \nabla(X, M/K)$ for $K \ll M$. Set $M' = \bigoplus_{i=1}^{n-1} M_i$.

Let $g : M/K \to M'/K' \oplus M_n/K_n$ be the natural map where $K'$ and $K_n$ are the images of $K$ under the projection maps onto $M'$ and $M_n$ respectively. By the induction hypothesis and (1) we get $f \diamond g = 0$. Hence

$$\mathrm{Im}\, f \subseteq (K' + K_n)/K = (K' + K)/K \ll (M' + K)/K.$$

By the induction hypothesis $f = 0$.                                              $\square$

**2.6. Lemma.** *Let $P \to M$ be an epimorphism and assume $P$ to be cogenerated by $M$. Then $\nabla(P, M) = 0$ implies $\nabla(P) = 0$.*

*Proof.* Assume $\nabla(P, M) = 0$. If $f \in \nabla(P)$, then for any $g \in \text{Hom}(P, M)$, we have $f \diamond g \in \nabla(P, M) = 0$, that is, $\text{Im}\, f \subseteq \text{Ke}\, g$. Hence $\text{Im}\, f \subseteq \text{Re}(P, M) = 0$ and thus $\nabla(P) = 0$. $\qquad\qquad\square$

**2.7. The radical of a module.** The *radical of an R-module M*, denoted by $\text{Rad}\, M$ or $\text{Rad}(M)$, is defined as the intersection of all maximal submodules of $M$, taking $\text{Rad}(M) = M$ when $M$ has no maximal submodules. Notice that this notion is dual to the socle of $M$, the sum of the minimal submodules of $M$ (see 1.20).

It can be shown (see [363, 21.5]) that $\text{Rad}(M)$ is the sum of all small submodules of $M$. Nevertheless $\text{Rad}(M)$ need not itself be small in $M$. To illustrate this, note that every finitely generated submodule of the $\mathbb{Z}$-module $\mathbb{Q}$ is small but their sum is $\mathbb{Q}$ (hence not small in $\mathbb{Q}$).

Note that $\text{Rad}(M) = 0$ if and only if the intersection of all maximal submodules of $M$ is zero or, equivalently, $M$ is a subdirect product of simple modules.

**2.8. Properties of the radical.**

(1) *For any morphism $f : M \to N$, $(\text{Rad}\, M)f \subset \text{Rad}\, N$; moreover, if $\text{Ke}\, f \subset \text{Rad}\, M$, then $(\text{Rad}\, M)f = \text{Rad}(Mf)$.*

(2) $\text{Rad}(M/\text{Rad}\, M) = 0$ *and $\text{Rad}(M)$ is the smallest submodule $U \subset M$ such that $M/U$ is cogenerated by simple modules.*

(3) *If $M$ is finitely cogenerated and $\text{Rad}\, M = 0$, then $M$ is semisimple and finitely generated.*

(4) $\text{Rad}\, M$ *is an $\text{End}(M)$-submodule of $M$, indeed a fully invariant submodule of $M$.*

(5) *If $M = \bigoplus_\Lambda M_\lambda$, then*
$$\text{Rad}\, M = \bigoplus_\Lambda \text{Rad}\, M_\lambda \ \text{ and } \ M/\text{Rad}\, M \simeq \bigoplus_\Lambda M_\lambda/\text{Rad}\, M_\lambda.$$

(6) *$M$ is finitely generated if and only if $\text{Rad}\, M \ll M$ and $M/\text{Rad}\, M$ is finitely generated.*

(7) *If every proper submodule of $M$ is contained in a maximal submodule (for example, if $M$ is finitely generated), then $\text{Rad}\, M \ll M$.*

(8) *If $M/\text{Rad}\, M$ is semisimple and $\text{Rad}\, M \ll M$, then every proper submodule of $M$ is contained in a maximal submodule.*

(9) $\text{Soc}\,(\text{Rad}\, M) \ll M$.

*Proof.* The properties (1)–(8) are well known (e.g., [363]).

(9) Put $U = \text{Soc}\,(\text{Rad}\, M)$ and assume $M = U + K$ for some submodule $K \subseteq M$. Setting $V = U \cap K$, we get $U = V \oplus V'$ for some $V' \subseteq U$ and $M = U + K = (V \oplus V') + K = V' \oplus K$. Now any simple submodule of $V'$ is a direct summand of $(V'$ and$)$ $M$ and is small in $M$ and hence is zero. Therefore $V' = 0$ and so $K = M$, showing $U \ll M$. $\qquad\qquad\square$

**2.9. Essential radical.** *Assume* $\operatorname{Rad} M \trianglelefteq M$.

(1) *Let* $K \subseteq L \subseteq M$ *be submodules and assume* $K$ *to be a direct summand of* $M$. *Then* $\operatorname{Rad} K = \operatorname{Rad} L$ *if and only if* $K = L$.

(2) *If* $\operatorname{Rad} M$ *has ACC (DCC) on direct summands, then* $M$ *has ACC (DCC) on direct summands.*

*Proof.* (1) Let $M = K \oplus K'$. Then $L = K \oplus (L \cap K')$ and $\operatorname{Rad} L = \operatorname{Rad} K \oplus \operatorname{Rad}(L \cap K')$. If $\operatorname{Rad} K = \operatorname{Rad} L$, then

$$0 = \operatorname{Rad}(L \cap K') = (\operatorname{Rad} M) \cap L \cap K'.$$

Thus $L \cap K' = 0$, since $\operatorname{Rad} M \trianglelefteq M$, and so $K = L$.

(2) This is a consequence of (1).                                          $\square$

**2.10. Chain conditions on small submodules.** *Let* $M$ *be an* $R$-*module.*

(1) $M$ *has ACC on small submodules if and only if* $\operatorname{Rad} M$ *is noetherian.*

(2) $M$ *has DCC on small submodules if and only if* $\operatorname{Rad} M$ *is artinian.*

*Proof.* (1) Let $M$ satisfy ACC on small submodules. Since every finitely generated submodule of $\operatorname{Rad} M$ is small in $M$, this implies ACC on finitely generated submodules of $\operatorname{Rad} M$. Hence $\operatorname{Rad} M$ is noetherian. The converse implication is obvious.

(2) Let $M$ satisfy DCC on small submodules. Since $\operatorname{Soc}(\operatorname{Rad} M) \ll M$ (see 2.8 (9)), this implies that $\operatorname{Soc}(\operatorname{Rad} M)$ is finitely generated. Moreover all finitely generated submodules of $\operatorname{Rad} M$ are artinian and so it has essential socle showing that $\operatorname{Rad} M$ is finitely cogenerated.

Suppose that $\operatorname{Rad} M$ is not artinian. Then the set $\Omega$ of all submodules $L \subset \operatorname{Rad} M$ such that $(\operatorname{Rad} M)/L$ is not finitely cogenerated is nonempty. Consider any chain of submodules $\{L_\lambda\}_\Lambda$ and let $L = \bigcap_\Lambda L_\lambda$. If $L \notin \Omega$, then $(\operatorname{Rad} M)/L$ is finitely cogenerated and hence $L = L_\lambda$, for some $\lambda \in \Lambda$. Thus $L \in \Omega$ and, by Zorn's Lemma, there exists a minimal element $P$ in $\Omega$.

We prove that this $P$ is small in $M$. For this assume $M = P + Q$ for some $Q \subset M$. If $P \cap Q \neq P$, then $(\operatorname{Rad} M)/(P \cap Q)$ is finitely cogenerated (by the choice of $P$). Write $\operatorname{Soc}[(\operatorname{Rad} M)/P)] = S/P$, where $P \subset S \subset \operatorname{Rad} M$. Then $S = P + (S \cap Q)$ and

$$S/P \simeq (S \cap Q)/(P \cap Q) \subset \operatorname{Soc}[(\operatorname{Rad} M)/(P \cap Q)],$$

and hence $S/P$ is finitely generated, a contradiction. Thus $P \cap Q = P$ and $Q = M$ showing that $P \ll M$.

Now it follows from property (3) in 2.2 that $M/P$ has DCC on small submodules and hence $M/P$ is finitely cogenerated as shown above. This contradicts the construction of $P$ and the assumption that $\Omega$ is non-empty. Thus every factor module of $\operatorname{Rad} M$ is finitely cogenerated and so $\operatorname{Rad} M$ is artinian.      $\square$

**2.11. The radical of a ring.** The radical of the left $R$-module $R$ is called the *Jacobson radical* of $R$ and we denote it by $\mathrm{Jac}(R)$ or $J(R)$. This is a two-sided ideal of $R$, is equal to the radical of the right $R$-module $R$, and consists of all elements of $R$ which annihilate all simple left (or right) $R$-modules.

By 2.8, $\mathrm{Jac}(R)$ is the sum of small left ideals and such ideals can be characterised by *quasi-regularity* (see [363, 21.11]).

**2.12. Hollow and local modules.** A nonzero $R$-module $M$ is called

$\qquad$ *hollow* $\quad$ if every proper submodule is small in $M$,

$\quad$ *semi-hollow* $\quad$ if every proper finitely generated submodule is small in $M$,

$\qquad$ *local* $\quad$ if it has a largest proper submodule (namely $\mathrm{Rad}(M)$).

Notice that $M$ is hollow if and only if it has no nontrivial coindependent family of submodules.

**2.13. Characterisation of hollow modules.** *For $M$ the following are equivalent:*

(a) *$M$ is hollow;*

(b) *every factor module of $M$ is indecomposable;*

(c) *for any nonzero modules $K$ and $N$ and any morphisms $K \xrightarrow{f} M \xrightarrow{g} N$, if $f \diamond g$ is surjective, then both $f$ and $g$ are surjective.*

*Proof.* (a)$\Rightarrow$(b). If $M$ is hollow, then every factor module of $M$ is hollow and hence indecomposable.

(b)$\Rightarrow$(a). Let $U, V \subset M$ be proper submodules with $U + V = M$. Then $M/(U \cap V) \simeq M/U \oplus M/V$. This contradicts (b).

(a)$\Rightarrow$(c). Let $f \diamond g$ be surjective. Then clearly $g$ is surjective and $\mathrm{Im}\, f + \mathrm{Ke}\, g = M$. Now $\mathrm{Ke}\, g \ll M$ implies $\mathrm{Im}\, f = M$.

(c)$\Rightarrow$(a). Let $K, V \subset M$ be proper submodules with $K + V = M$. Then the composition of the canonical maps $K \to M \to M/V$ is surjective and hence (c) implies that the inclusion $K \to M$ is surjective, that is, $K = M$. $\qquad\square$

Clearly a local module is hollow and its unique maximal submodule has to be the radical. Examples of local $\mathbb{Z}$-modules are simple modules and the modules $\mathbb{Z}_{p^k}$, where $p$ is prime and $k \in \mathbb{N}$. The $\mathbb{Z}$-modules $\mathbb{Z}_{p^\infty}$, for any prime $p$, are hollow but not local.

Any finitely generated module is hollow if and only if it is local. In particular the ring $R$ as a left (right) module is hollow if and only if it is local. Such rings are called *local rings* and have the following characterisations (e.g. [363, 19.8]).

**2.14. Local rings.** *For a ring $R$ the following are equivalent:*

(a) *$R$ is a local ring;*

(b) *$\mathrm{Jac}(R)$ is a maximal left (right) ideal in $R$;*

(c) *the sum of two non-units is a non-unit in $R$;*

(d) *R has a maximal submodule which is small in R.*

## 2.15. Properties of hollow modules.

(1) *If M is (semi-)hollow, then every factor module is (semi-)hollow.*

(2) *If $K \ll M$ and $M/K$ is (semi-)hollow, then M is (semi-)hollow.*

(3) *M is semi-hollow if and only if M is local or $\mathrm{Rad}(M) = M$.*

(4) *The following statements are equivalent:*

   (a) *M is local;*

   (b) *M is (semi-)hollow and cyclic (or finitely generated);*

   (c) *M is (semi-)hollow and $\mathrm{Rad}(M) \neq M$.*

*Proof.* These assertions are easily verified (e.g., [363, 41.4]).                    □

Semi-hollow modules need not be hollow, nor indeed indecomposable. To see this, observe that every finitely generated submodule of the $\mathbb{Z}$-modules $\mathbb{Q}$ and $\mathbb{Q}/\mathbb{Z}$ is small and hence these modules are semi-hollow but neither of them is hollow and $\mathbb{Q}/\mathbb{Z}$ is not indecomposable.

The $\mathbb{Z}$-modules $\mathbb{Q} \oplus \mathbb{Z}/p\mathbb{Z}$ and $\mathbb{Q}/\mathbb{Z} \oplus \mathbb{Z}/p\mathbb{Z}$, $p$ any prime, show that a module may have a unique maximal submodule without being local. They also show that the direct sum of two (semi-)hollow modules need not be semi-hollow.

Over a left cosemisimple ring $R$ a module $M$ is (semi-)hollow if and only if $M$ is simple.

## 2.16. Existence of hollow factor modules. *Assume that the R-module M does not contain an infinite coindependent family of submodules. Then every nonzero factor module of M has a hollow factor module.*

*Proof.* Let $N$ be a proper submodule of $M$ and assume that $M/N$ has no hollow factor module. To obtain the desired contradiction, we construct by induction a sequence $L_1, L_2, \ldots$ of proper submodules such that $\{L_1, L_2, \ldots\}$ is an infinite coindependent family of submodules of $M$ and for any $k \in \mathbb{N}$, the intersection $(L_1/N) \cap \cdots \cap (L_k/N)$ is not small in $M/N$.

First, since $M/N$ is not hollow, there is a proper submodule $L_1$ of $M$ such that $N \subset L_1$ and $L_1/N$ is not small in $M/N$. Now let us assume that, for some $k > 1$, we have constructed proper submodules $L_1, \ldots, L_{k-1}$ of $M$, each containing $N$ and such that $L_1/N \cap \cdots \cap L_{k-1}/N$ is not small in $M/N$. Then there exists a proper submodule $K$ properly containing $N$ and such that $(L_1 \cap \cdots \cap L_{k-1}) + K = M$. By assumption, $M/K = (M/N)/(K/N)$ is not hollow. Hence there exist proper submodules $K_1, K_2$ of $M$ both containing $K$ such that $K_1 + K_2 = M$. Set $L_k = K_1$. Clearly $\{L_1, \ldots, L_k\}$ is coindependent (see 1.28) and $L_1/N \cap \cdots \cap L_k/N$ is not small in $M/N$ since $(L_1 \cap \cdots \cap L_k) + K_2 = M$ by Lemma 1.24. Continuing in this way produces the contradictory infinite coindependent family of submodules.    □

A module $M$ is called *uniserial* if its lattice of submodules is linearly ordered by inclusion. If $M$ can be written as a direct sum of uniserial modules, then it is called *serial*. The ring $R$ is said to be *left uniserial* or *left serial* provided $R$ has the corresponding property as left $R$-module. (See [363, Section 55].)

**2.17. Characterisations of uniserial modules.** *For $M$ the following are equivalent:*

(a) *$M$ is uniserial;*

(b) *every factor module of $M$ is uniform;*

(c) *every factor module of $M$ has zero or simple socle;*

(d) *every submodule of $M$ is hollow;*

(e) *every finitely generated submodule of $M$ is local;*

(f) *every submodule of $M$ has at most one maximal submodule.*

*Proof.* This follows from [363, 55.1]. $\qquad\square$

**2.18. Noetherian uniserial modules.** *Let $M$ be a noetherian uniserial module. Then every submodule of $M$ is fully invariant in $M$ and*

(1) *if $M$ is self-injective, then every submodule $X$ of $M$ is self-injective and $\mathrm{End}(X)$ is a local ring;*

(2) *if $M$ is self-projective, then for every submodule $X$ of $M$, $M/X$ is self-projective and $\mathrm{End}(M/X)$ is a local ring.*

*Proof.* Let $0 \neq f : M \to M$ be an endomorphism. Suppose there exists $X_1 \subseteq M$ such that $(X_1)f = X_2 \supsetneq X_1$. Since $\mathrm{Ke}\, f \subset X_1$ it is easy to see that $(X_2)f = X_3 \supsetneq X_2$. Thus we get a properly ascending chain $X_1 \subsetneq X_2 \subsetneq X_3 \subsetneq \cdots$ of submodules of $M$, a contradiction since $M$ is noetherian. Thus every submodule of $M$ is fully invariant in $M$.

Assertion (1) follows as fully invariant submodules of self-injective modules are self-injective (see [363, 17.11]) and the endomorphism rings of indecomposable self-injective modules are local rings (see [363, 22.1]). The proof of assertion (2) is similar. $\qquad\square$

**2.19. Bass and max modules.** A module $M$ is called a

$\quad$ *max module* $\quad$ if every nonzero submodule of $M$ has a maximal submodule,

$\quad$ *Bass module* $\quad$ if every nonzero module in $\sigma[M]$ has a maximal submodule.

For convenience we list some elementary observations.

(1) *$M$ is a max module if and only if $\mathrm{Rad}\,N \neq N$, for any nonzero $N \subset M$.*

(2) *$M$ is a Bass module if and only if $\mathrm{Rad}\,N \neq N$ (or $\mathrm{Rad}\,N \ll N$), for any nonzero $N \in \sigma[M]$.*

(3) *Every submodule of a max module is again a max module.*

(4) *If $M$ is a Bass module, then any module $N \in \sigma[M]$ is a Bass module.*

(5) *If some M-generated module contains a maximal submodule, then M contains a maximal submodule.*

The module $_RR$ is a max module if every nonzero left ideal contains a maximal submodule. $_RR$ is a left Bass module if every nonzero $R$-module has a maximal submodule; such rings are also called *left Bass rings*.

Over a left Bass ring $R$, semi-hollow, hollow and local are equivalent properties for any $R$-module and it is well known that left perfect rings are left Bass rings (e.g., [363, 43.5 and 43.9], see also Exercise 2.22 (2)).

Notice that the meaning of the terminology used here varies widely in the literature.

Obvious examples of Bass modules are semisimple and cosemisimple modules $M$ since in these cases $\mathrm{Rad}\,N = 0$ for any $N \in \sigma[M]$.

**2.20. Direct sums and products of max modules.** *Let $\{M_\lambda\}_\Lambda$ be a family of left R-modules. Then the following are equivalent:*

(a) *$M_\lambda$ is a max module, for each $\lambda \in \Lambda$;*

(b) *$\bigoplus_\Lambda M_\lambda$ is a max module;*

(c) *$\prod_\Lambda M_\lambda$ is a max module.*

*Proof.* (c)$\Rightarrow$(b)$\Rightarrow$(a) are trivial implications (by (3) above).

(a)$\Rightarrow$(c). Let $0 \neq N \subset \prod_\Lambda M_\lambda$ and for each $\mu \in \Lambda$ let $\pi_\mu : \prod_\Lambda M_\lambda \to M_\mu$ denote the canonical projection. Then $0 \neq (N)\pi_\mu \subseteq M_\mu$ for some $\mu \in \Lambda$ and hence $(N)\pi_\mu$ has some maximal submodule, say $X$. Now $(X)\pi_\mu^{-1}$ is a maximal submodule of $N$. $\hspace{2cm}\square$

**2.21. Characterisations of Bass modules.** *For M the following are equivalent:*

(a) *M is Bass module;*

(b) *for every $0 \neq N \in \sigma[M]$, $\mathrm{Rad}\,N \ll N$;*

(c) *every self-injective module in $\sigma[M]$ has a maximal submodule;*

(d) *the M-injective hull of any simple module in $\sigma[M]$ is a max module;*

(e) *there is a cogenerator in $\sigma[M]$ that is a max module.*

*Proof.* (a)$\Leftrightarrow$(b) follows immediately from the properties given in 2.19.

(b)$\Rightarrow$(c) is trivial.

(c)$\Rightarrow$(d). For any simple module $S \in \sigma[M]$, let $\widehat{S}$ denote its $M$-injective hull. For any $N \subset \widehat{S}$, $N\mathrm{End}(\widehat{S}) \subset \widehat{S}$ is a fully invariant submodule and hence is self-injective (see [363, 17.11]). By (c), $N\mathrm{End}(\widehat{S})$ has a maximal submodule and hence $N$ has a maximal submodule (by 2.19 (5)).

(d)$\Rightarrow$(e). The product of the injective hulls of simples in $\sigma[M]$ is a cogenerator in $\sigma[M]$ (see [363, 17.12]) and is a max module by 2.20.

(e)$\Rightarrow$(a). Let $Q$ be any cogenerator in $\sigma[M]$ that is a max module. Then any nonzero $N \in \sigma[M]$ can be embedded in a product of copies of $Q$ which is also a max module by 2.20. Thus $N$ is a max module.                                    $\square$

**2.22. Exercises.**

(1) Let $R$ be a ring such that every maximal left ideal is a two-sided ideal. Prove that for any left $R$-module $M$ ([118, Lemma 3]),

$$\mathrm{Rad}\, M = \bigcap \{IM \mid I \subset R \text{ is a maximal left ideal}\}.$$

(2) A subset $I$ of a ring $R$ is called *left T-nilpotent* in case for every sequence $a_1, a_2, \ldots$ in $I$ there is an $n \in \mathbb{N}$ such that $a_1 a_2 \ldots a_n = 0$.
   Prove that for $R$, the following are equivalent:

   (a) $R$ is a left Bass ring;

   (b) for every nonzero left $R$-module $N$, $\mathrm{Rad}\, N \ll N$;

   (c) every self-injective left $R$-module has a maximal submodule;

   (d) $R/\mathrm{Jac}\, R$ is a left Bass ring and $\mathrm{Jac}\, R$ is left T-nilpotent.

(3) Prove that for a left noetherian ring $R$, the following are equivalent ([298]):

   (a) $R$ is a left Bass ring;

   (b) $R$ is left artinian.

(4) Prove that for a commutative ring $R$, the following are equivalent:

   (a) $R$ is a Bass ring;

   (b) $R/\mathrm{Jac}\, R$ is von Neumann regular and $\mathrm{Jac}\, R$ is T-nilpotent.

**References.** Al-Khazzi and Smith [8]; Faith [102]; Golan [121]; Generalov [118]; Harada [151]; Hirano [170]; Inoue [175]; Leonard [220]; Pareigis [282]; Rayar [294]; Renault [298]; Takeuchi [327]; Talebi-Vanaja [330]; Tuganbaev [336]; Wisbauer [363].

# 3   Cosmall inclusions and coclosed submodules

Dualising essential extensions and submodules we are led to the following notion, called a *coessential extension* in  Takeuchi [327] and a *superfluous contraction* in Chamard [60].

**3.1. Definition.** Given submodules $K \subset L \subset M$, the inclusion $K \subset L$ is called *cosmall in* $M$ if $L/K \ll M/K$; we denote this by $K \overset{cs}{\underset{M}{\hookrightarrow}} L$.

Note that this is equivalent to saying that $M/K$ is a small cover of $M/L$ and that, trivially, $L \ll M$ if and only if $0 \overset{cs}{\underset{M}{\hookrightarrow}} L$.

**3.2. Cosmall inclusions.** *Let* $U \subset K \subset L \subset M$ *and* $G \subset H \subset M$ *be any submodules.*

(1) *The following are equivalent:*

    (a) $K \subset L$ *is cosmall in* $M$;

    (b) *for any submodule* $X \subset M$, $L + X = M$ *implies* $K + X = M$.

(2) $U \overset{cs}{\underset{M}{\hookrightarrow}} L$ *if and only if* $U \overset{cs}{\underset{M}{\hookrightarrow}} K$ *and* $K \overset{cs}{\underset{M}{\hookrightarrow}} L$.

(3) *If* $K \overset{cs}{\underset{M}{\hookrightarrow}} L$ *and* $K + X = M$ *for a submodule* $X \subset M$, *then* $K \cap X \overset{cs}{\underset{M}{\hookrightarrow}} L \cap X$.

(4) *If* $G \overset{cs}{\underset{M}{\hookrightarrow}} H$, $K \overset{cs}{\underset{M}{\hookrightarrow}} L$ *and* $G + K = M$, *then* $K \cap G \overset{cs}{\underset{M}{\hookrightarrow}} L \cap H$.

(5) *If* $K \subset L \subset M$ *and* $L = K + S$, *where* $S \ll M$, *then* $K \overset{cs}{\underset{M}{\hookrightarrow}} L$.

(6) *If* $M = K + L$, $K \subseteq C \subset M$ *and* $C \cap L \ll M$, *then* $K \overset{cs}{\underset{M}{\hookrightarrow}} C$.

(7) *Let* $f : M \to N$ *be an epimorphism.*

    (i) *If* $K \subset L \subset M$ *and* $K \overset{cs}{\underset{M}{\hookrightarrow}} L$, *then* $(K)f \overset{cs}{\underset{N}{\hookrightarrow}} (L)f$.

    (ii) *If* $C \subseteq D \subseteq N$, *then* $C \overset{cs}{\underset{N}{\hookrightarrow}} D$ *if and only if* $(C)f^{-1} \overset{cs}{\underset{M}{\hookrightarrow}} (D)f^{-1}$.

*Proof.* (1) (a)$\Rightarrow$(b). Let $L + X = M$. Then

$$M/K = L/K + (X + K)/K = (X + K)/K$$

and hence $M = X + K$.

    (b)$\Rightarrow$(a). Let $K \subset X \subset M$ be such that $L/K + X/K = M/K$. Then $L + X = M = K + X$ and hence $X/K = M/K$. This shows $L/K \ll M/K$.

    (2) This follows easily from (1).

    (3) If there is a submodule $U \subset M$ containing $K \cap X$ such that

$$M/(K \cap X) = (L \cap X)/(K \cap X) + U/(K \cap X),$$

then $(L \cap X) + U = M$. Applying Lemma 1.24 twice, we get

$$M = L + (X \cap U) = K + (X \cap U) = (K \cap X) + U = U.$$

Thus $K \cap X \subset L \cap X$ is cosmall in $M$.

(4) Applying (1) and (3) twice, we get $M = L + H = K + H = K + G$ and $K \cap H \subset L \cap H$ and $K \cap G \subset K \cap H$ are cosmall in $M$. So, by (2), $K \cap G \subset L \cap H$ is cosmall in $M$.

(5) Let $M = L + X$. Then $M = K + S + X = K + X$, since $S \ll M$, so, by (1), $K \overset{cs}{\underset{M}{\hookrightarrow}} L$.

(6) This follows from (5) as $C = K + (C \cap L)$.

(7) This follows easily from the definition of cosmall inclusion. $\square$

**3.3. Cosmall and coindependent submodules.** *Let $\{K_\lambda\}_\Lambda$ be a coindependent family of submodules and $\{L_\lambda\}_\Lambda$ a family of submodules of $M$ such that $L_\lambda \overset{cs}{\underset{M}{\hookrightarrow}} K_\lambda$ for every $\lambda \in \Lambda$. Then, for every finite subset $F$ of $\Lambda$,*

(i) $\{K_\lambda\}_{\Lambda \setminus F} \cup \{L_i\}_{i \in F}$ *is coindependent,*

(ii) $\{L_\lambda\}_\Lambda$ *is a coindependent family of submodules, and*

(iii) $\bigcap_F L_i \overset{cs}{\underset{M}{\hookrightarrow}} \bigcap_F K_i$.

*Proof.* (i) By 1.28 (2) and 3.2 (1), for any $i \in \Lambda$, $\{K_\lambda\}_{\Lambda \setminus \{i\}} \cup \{L_i\}$ is coindependent and assertion (i) follows by induction on the cardinality of $F$.

(ii) $\{L_\lambda\}_\Lambda$ is coindependent, since for every finite subset $F$ and $\lambda \in \Lambda \setminus F$, (i) implies that $\{L_i\}_F \cup \{L_\lambda\}$ is coindependent, and thus $M = L_\lambda + \bigcap_F L_i$.

(iii) By induction and 3.2 (4), it is easy to see that (iii) holds. $\square$

**3.4. Small submodules and coindependent families.** *Assume that there exists a coindependent family $\{N_1, \ldots, N_n\}$ in $M$ such that $M/N_i$ is hollow for all $i \le n$ and $N_1 \cap \cdots \cap N_n$ is small in $M$. Then a submodule $K$ is small in $M$ if and only if $N_i + K \ne M$ for every $i \in \{1, \ldots, n\}$.*

*Proof.* The necessity is clear. Conversely assume that $N_i + K \ne M$ for all $i \in \{1, \ldots, n\}$. Then $(K + N_i)/N_i$ is small in $M/N_i$ since $M/N_i$ is hollow, that is, $N_i \overset{cs}{\underset{M}{\hookrightarrow}} (K + N_i)$ for all $i \le n$. By 3.3, it follows that $\bigcap_{i=1}^n N_i \overset{cs}{\underset{M}{\hookrightarrow}} \bigcap_{i=1}^n (K + N_i)$. Since $\bigcap_{i=1}^n N_i \ll M$, 3.2 (2) implies that $K \subseteq \bigcap_{i=1}^n (K + N_i) \ll M$. $\square$

As an application we obtain

**3.5. Lemma.** *Let $\{N_1, \ldots, N_n\}$ be a family of coindependent submodules of $M$ such that $\bigcap_{i=1}^n N_i \ll M$ and each $M/N_i$ is hollow. Then, for any coindependent family of submodules $\{L_1, \ldots, L_k\}$ of $M$, we have $k \le n$.*

*Proof.* Assume that $\{L_1, \ldots, L_k\}$ is coindependent for some $k \le n$. We show by induction that, rearranging $N_1, \ldots, N_n$ if necessary,

(*) for any $0 \le j \le n$, the set $\{N_1, \ldots, N_j, L_{j+1}, \ldots, L_k\}$ is coindependent.

For $j = 0$ $(*)$ is clear. Now assume that $j > 0$ and that $(*)$ is true for $j - 1$. Set $K := N_1 \cap \cdots \cap N_{j-1} \cap L_{j+1} \cap \cdots \cap L_k$. Then $K + L_j = M$, so $K$ is not small in $M$. By 3.4, $K + N_s = M$ holds for some $1 \leq s \leq n$. Clearly $s \geq j$ otherwise $N_s = M$. By rearranging $\{N_j, \ldots, N_n\}$ if necessary, we may take $s = j$. Then the set $\{N_1, \ldots, N_j, L_{j+1}, \ldots, L_n\}$ is coindependent. Thus $(*)$ holds for $j$, completing our induction argument.

For $j = n$ we have that $\{N_1, \ldots, N_n, L_{n+1}, \ldots, L_k\}$ is coindependent. Since $\bigcap_{i=1}^{n} N_i \ll M$ this implies $k = n$.                                                                   $\square$

**3.6. Definition.** A submodule $L \subset M$ is called *coclosed in $M$*, we write $L \overset{cc}{\hookrightarrow} M$, if $L$ has no proper submodule $K$ for which $K \subset L$ is cosmall in $M$, that is, $K \overset{cs}{\underset{M}{\hookrightarrow}} L$ implies $K = L$. Thus $L$ is coclosed in $M$ if and only if, for any proper submodule $K \subset L$, there is a submodule $N$ of $M$ such that $L + N = M$ but $K + N \neq M$.

Obviously, any direct summand $L$ of $M$ is coclosed in $M$.

**3.7. Properties of coclosed submodules.** *Let $K \subset L \subset M$ be submodules.*

(1) *If $L$ is coclosed in $M$, then $L/K$ is coclosed in $M/K$.*

(2) *If $K \ll L$ and $L/K$ is coclosed in $M/K$, then $L$ is coclosed in $M$.*

(3) *If $L \subset M$ is coclosed, then $K \ll M$ implies $K \ll L$;*
    *hence $\mathrm{Rad}(L) = L \cap \mathrm{Rad}(M)$.*

(4) *If $L$ is hollow, then either $L \subset M$ is coclosed in $M$ or $L \ll M$.*

(5) *If $f : M \to N$ is a small epimorphism and $L$ is coclosed in $M$, then $(L)f$ is coclosed in $N$.*

(6) *If $K$ is coclosed in $M$, then $K$ is coclosed in $L$ and the converse is true if $L$ is coclosed in $M$.*

*Proof.* (1) Suppose there exists a proper submodule $N \subset L$ such that $N/K \subset L/K$ is cosmall in $M/K$. Then $(L/K)/(N/K) \ll (M/K)/(N/K)$ and so $L/N \ll M/N$, showing that $N \overset{cs}{\underset{M}{\hookrightarrow}} L$. This contradicts the assumption that $L$ is coclosed in $M$.

(2) If $N \overset{cs}{\underset{M}{\hookrightarrow}} L$, then $(N + K)/K \subseteq L/K$ is cosmall in $M/K$ by 3.2 7(i). Hence $N + K = L$ and since $K \ll L$, $N = L$. Thus $L$ is coclosed in $M$.

(3) Consider $K \subset L$ with $K \ll M$. Assume $K + K' = L$ for some $K' \subset L$. Choose $K' \subset L' \subset M$ such that $M/K' = L/K' + L'/K'$. Then $M = L + L' = K + K' + L' = L'$ and this shows that $L/K' \ll M/K'$. Since $L$ is coclosed in $M$, we conclude $L = K'$ and so $K \ll L$.

(4) Assume $L$ is not coclosed in $M$. Then there exists $K \subset L$ such that $L/K \ll M/K$. Then, since $L$ is hollow, $L \ll M$ by 2.2 (3).

(5) Suppose $X \overset{cs}{\underset{N}{\hookrightarrow}} (L)f$. Put $K = (X)f^{-1}$. Then $K \overset{cs}{\underset{M}{\hookrightarrow}} (L + \mathrm{Ke}\,f)$ by 3.2 (7). Since $\mathrm{Ke}\,f \subseteq K$, $K = \mathrm{Ke}\,f + (L \cap K)$. As $\mathrm{Ke}\,f \ll M$, $(L \cap K) \overset{cs}{\underset{M}{\hookrightarrow}} \mathrm{Ke}\,f + (L \cap K)$

by 3.2 (5). Hence $(L \cap K) \overset{cs}{\underset{M}{\to}} (L + \mathrm{Ke}\, f)$ by 3.2 (2). Therefore $(L \cap K) \overset{cs}{\underset{M}{\to}} L$. As $L$ is coclosed in $M$, $L \cap K = L$ and this implies $X = (L)f$.

(6) Let there exist $X \subseteq K$ such that $K/X \ll L/X$. Then $K/X \ll M/X$ also. But $K$ is coclosed in $M$ implies $K = X$, proving $K$ is coclosed in $L$.

Now suppose $K$ is coclosed in $L$ and $L$ is coclosed in $M$. Let $X \subset K$ with $K/X \ll M/X$. By (1), $L/X$ is coclosed in $M/X$ and by (3) applied to $M/X$, $K/X \ll L/X$. As $K$ is coclosed in $L$, $X = K$. $\qquad\square$

It is well known that a module $M$ is semisimple if and only if each of its submodules is (essentially) closed. Dually we observe:

**3.8. Coclosed characterisation of cosemisimple modules.** *Every submodule of a module $M$ is coclosed in $M$ if and only if $M$ is cosemisimple.*

*Proof.* If $M$ is cosemisimple, then the radical of any factor module of $M$ is zero (see, e.g., [363, 23.1]) and so every nonzero factor of $M$ has no proper small submodule. Hence there are no cosmall extensions $L \subset N$ in $M$, that is, every submodule is coclosed.

For the converse, first note that if $M/N$ is a factor module of $M$ and $K/N \ll M/N$, then $N \subseteq K$ is cosmall in $M$. Thus if every submodule of $M$ is coclosed, $\mathrm{Rad}\, M/N = 0$ for all $N \subseteq M$. $\qquad\square$

Note that if $N$ is coclosed in $M$, then for any submodule $L \subseteq N$ we have $L \ll M$ if and only if $L \ll N$. Since $N/K$ is coclosed in $M/K$ we deduce:

**3.9. Lemma.** *Let $N \subset M$ be a coclosed submodule. Then for any $K \subset L \subset N$,*

$$L/K \ll M/K \quad \text{if and only if} \quad L/K \ll N/K.$$

*In particular* $\mathrm{Rad}\, N/K = (\mathrm{Rad}\, M/K) \cap N/K$ *and, if* $\mathrm{Rad}\, M/K \ll M/K$, *then* $\mathrm{Rad}\, N/K \ll N/K$.

The following notion was introduced in Keskin [199] as *s-closure*.

**3.10. Coclosure of submodules.** Let $L \subset M$ be a submodule. A submodule $K \subset L$ is called a *coclosure of $L$ in $M$* if $K \subset L$ is cosmall in $M$ and $K \subset M$ is coclosed; that is, $L/K \ll M/K$, and for $U \subset K$, $K/U \ll M/U$ implies $U = K$. Thus coclosures are characterised by a minimality property with respect to cosmall inclusion.

Notice that in general coclosures need not exist. For example, the $\mathbb{Z}$-submodule $2\mathbb{Z} \subset \mathbb{Z}$ has no coclosure in $\mathbb{Z}$. One of our concerns will be to find conditions which guarantee the existence of coclosures. If coclosures exist they need not be unique as shown by an example in the exercises 3.14.

Let $K, N \subset M$ be submodules where $K$ is coclosed in $M$. If $N \subset K$, then $K/N$ is coclosed in $M/N$ (see 3.7). In general $(K + N)/N$ need not be coclosed in $M/N$. We consider the case when this also holds.

**3.11. Strongly coclosed submodules.** A submodule $K \subset M$ is called *strongly coclosed* in $M$ if, for every $L \subset M$ with $K \nsubseteq L$, we have $L \overset{cs}{\underset{M}{\nrightarrow}} (K + L)$. We denote this property by $K \overset{scc}{\hookrightarrow} M$.

Strongly coclosed submodules are coclosed but not every coclosed submodule is strongly coclosed. In fact a direct summand of a module is always coclosed but need not be strongly coclosed (see 17.17). We have the following

**3.12. Characterisations of strongly coclosed submodules.** *For a submodule $K$ of $M$, the following are equivalent:*

(a) *$K$ is strongly coclosed in $M$;*

(b) *$(K)f \not\ll \operatorname{Im} f$, for any $X \in R$-Mod and $f \in \operatorname{Hom}(M, X)$ with $K \nsubseteq \operatorname{Ke} f$;*

(c) *$(K)f$ is coclosed in $\operatorname{Im} f$, for any $X \in R$-Mod and $f \in \operatorname{Hom}(M, X)$;*

(d) *$(K + N)/N$ is coclosed in $M/N$, for any submodule $N \subset M$.*

*Proof.* (a)$\Rightarrow$(b). Let $K$ be strongly coclosed in $M$ and $f : M \to X$ such that $(K)f \neq 0$. For $L = \operatorname{Ke} f$, $f$ induces an isomorphism $\overline{f} : M/L \to \operatorname{Im} f$. Denoting by $p : M \to M/L$ the projection map, we have

$$(K)f = (K)p\overline{f} = ((K + L)/L)\overline{f} \not\ll \operatorname{Im} f.$$

(b)$\Rightarrow$(c). Suppose $f : M \to X$ is a homomorphism such that $(K)f \neq 0$. If $L \subseteq (K)f$ is cosmall in $\operatorname{Im} f$, then $(K)fp_L \ll \operatorname{Im} fp_l$, where $p_L : \operatorname{Im} f \to \operatorname{Im} f/L$ denotes the projection. This contradicts our hypothesis. Hence $(K)f$ is coclosed in $\operatorname{Im} f$.

(c)$\Rightarrow$(a). Let $L \subseteq M$ with $K \nsubseteq L$ and let $p_L : M \to M/L$ denote the projection. Then $(K)p_L = (K + L)/L$ is coclosed in $M/L$, hence non-small in $M/L$.

(a)$\Rightarrow$(d). Let $p_N : M \to M/N$ be the projection and $K \nsubseteq N$. Then for any $f : M/N \to M/N$, $((K + N)/N)f = (K)p_N f$ is coclosed in $\operatorname{Im} p_N f = \operatorname{Im} f$. Thus $(K + N)/N \overset{cc}{\hookrightarrow} M/N$.

(d)$\Rightarrow$(a) follows as a coclosed submodule is small if and only if it is the zero submodule.                                                                                      $\square$

**3.13. Sum of strongly coclosed submodules.** *Let $\{K_i\}_I$ be a family of strongly coclosed submodules of $M$. Then $\sum_I K_i \overset{scc}{\hookrightarrow} M$.*

*Proof.* Let $L \subset M$ be such that $\sum_{i \in I} K_i \nsubseteq L$. Then there exists an $i \in I$ such that $K_i \nsubseteq L$. Now $K_i \overset{scc}{\hookrightarrow} M$ implies $L \overset{cs}{\underset{M}{\nrightarrow}} (L + K_i)$. Since $L \subseteq L + K_i \subseteq L + \sum_{i \in I} K_i$,

3.2 gives $L \overset{cs}{\underset{M}{\nrightarrow}} (L + \sum_{i \in I} K_i)$.                                                                     $\square$

### 3.14. Exercises.

(1) For a field $K$ consider the ring $R = \begin{pmatrix} K & 0 \\ K & K \end{pmatrix}$.

For any $x \in K$ define the right $R$-module $D_x = (e_{11} + xe_{21})R$, where the $e_{ij}$ denote the matrix units. Prove ([114]):

   (i) $D_x$ is a direct summand of $R$.

   (ii) $\operatorname{Soc}(R_R) = e_{11}R \oplus e_{21}R = D_x \oplus e_{21}R$.

   (iii) For each $x \in K$, $D_x$ is a coclosure of $\operatorname{Soc}(R_R)$.

(2) Let $N \subset M$ be a coclosed submodule and assume every factor module of $M$ has a maximal submodule. Prove that every factor module of $N$ has a maximal submodule.

**References.** Chamard [60]; Ganesan and Vanaja [114]; Golan [121]; Keskin [199]; Lomp [226]; Reiter [295]; Takeuchi [327]; Talebi and Vanaja [330]; Wisbauer [363]; Zöschinger [394].

# 4   Projectivity conditions

It is well-known that projectivity conditions on a module $M$ often allow us to
relate properties of $M$ to properties of its endomorphism ring. In this section we
review various forms of projectivity.

**4.1. $M$-projective modules.** An $R$-module $P$ is called *$M$-projective* if every diagram in $R$-Mod with exact row

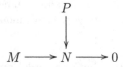

can be extended commutatively by some morphism $P \to M$, or — equivalently —
the functor $\mathrm{Hom}(P, -)$ is exact with respect to all exact sequences of $R$-modules
of the form

$$0 \longrightarrow K \longrightarrow M \longrightarrow N \longrightarrow 0 \ .$$

**4.2. Properties of $M$-projective modules.** *Let $P$ be $M$-projective.*

(1) *If $0 \to M' \to M \to M'' \to 0$ is exact, then $P$ is $M'$- and $M''$-projective.*

(2) *$P$ is $K$-projective for submodules $K$ of finitely $M$-generated modules.*

(3) *If $P$ is finitely generated, then $P$ is $K$-projective for any $K \in \sigma[M]$.*

*Proof.* See [363, §18].                                                         $\square$

**4.3. Direct sums and projectivity.**

(1) *For every family $\{P_i\}_I$ of $R$-modules, $\bigoplus_I P_i$ is $M$-projective if and only if
$P_i$ is $M$-projective, for every $i \in I$.*

(2) *For every finite family $\{M_i\}_{i=1}^n$ of $R$-modules, $P$ is $\bigoplus_{i=1}^n M_i$-projective if
and only if $P$ is $M_i$-projective for $i = 1, 2, \cdots, n$.*

*Proof.* See [363, §18].                                                         $\square$

   Note that 4.3(2) is not true for infinite families. For example, if $R = \mathbb{Z}$, then
$\mathbb{Q}$ is trivially $\mathbb{Z}$-projective but is not $\mathbb{Z}^{(\mathbb{N})}$-projective.

**4.4. $M$-projectivity of factor modules.** *Let $p : P \to L$ be an epimorphism where $P$
is $M$-projective.*

(1) *If $\mathrm{Ke}\, p \subseteq \mathrm{Re}(P, M)$, then $L$ is $M$-projective.*

(2) *If $\mathrm{Ke}\, p \ll P$ and $L$ is $M$-projective, then $\mathrm{Ke}\, p \subseteq \mathrm{Re}(P, M)$.*

*Proof.* (1) For any epimorphism $g : M \to N$ and $f : L \to N$, the diagram

$$0 \longrightarrow \operatorname{Ke} p \longrightarrow P \overset{p}{\longrightarrow} L \longrightarrow 0$$
$$\Big\downarrow f$$
$$M \overset{g}{\longrightarrow} N \longrightarrow 0,$$

can be extended commutatively by some $h : P \to M$. By assumption, $\operatorname{Ke} p \subseteq \operatorname{Ke} h$ and hence $h$ factors via $L$, that is, there is some $h' : L \to M$ with $h = ph'$. This yields $pf = hg = ph'g$ and, since $p$ is an epimorphism, we conclude $f = h'g$. Thus $L$ is $M$-projective.

(2) For any $h : P \to M$ and $K = (\operatorname{Ke} p)h$, construct the diagram

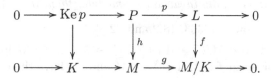

By $M$-projectivity of $L$, there exists $\alpha : L \to M$ with $\alpha g = f$. Now consider the map $\tilde{h} := h - p\alpha : P \to M$. By construction, $\tilde{h}g = 0$ and hence $P = \operatorname{Ke}(\tilde{h}g) = K\tilde{h}^{-1}$. Since $(\operatorname{Ke} p)\tilde{h} = (\operatorname{Ke} p)h$, we conclude $\operatorname{Ke} p + \operatorname{Ke}\tilde{h} = P$ and hence $\operatorname{Ke}\tilde{h} = P$. From this we obtain $0 = (\operatorname{Ke} p)\tilde{h} = (\operatorname{Ke} p)h$ which implies $\operatorname{Ke} p \subseteq \operatorname{Re}(P, M)$ as required. $\qquad\square$

From this we obtain an interesting corollary.

**4.5. Small epimorphisms.** *Let* $p : \tilde{P} \to P$, $q : \tilde{M} \to M$ *be small epimorphisms where* $\tilde{P}$, $P$ *are $M$-projective, and* $\tilde{M}$, $M$ *are $P$-projective. If there exists an isomorphism* $g : \tilde{P} \to \tilde{M}$, *then* $P \simeq M$.

*Proof.* Consider the diagram

$$0 \longrightarrow \operatorname{Ke} p \longrightarrow \tilde{P} \overset{p}{\longrightarrow} P \longrightarrow 0$$
$$\Big\downarrow g$$
$$0 \longrightarrow \operatorname{Ke} q \longrightarrow \tilde{M} \overset{q}{\longrightarrow} M \longrightarrow 0.$$

Since $P$ is $M$-projective, by 4.4, $\operatorname{Ke} p \subseteq \operatorname{Re}(\tilde{P}, M) \subseteq \operatorname{Ke}(gq)$ and hence there exist morphisms $\operatorname{Ke} p \to \operatorname{Ke} q$ and $P \to M$ leading to a commutative diagram.

Replacing $g$ by $g^{-1}$ in the diagram we obtain, by similar arguments, morphisms $\operatorname{Ke} q \to \operatorname{Ke} p$ and $M \to P$ which are inverse to the ones constructed before, respectively. $\qquad\square$

**4.6. Projectives in** $\sigma[M]$**.** A module $P$ in $\sigma[M]$ is called *projective in* $\sigma[M]$ if it is $N$-projective for every $N \in \sigma[M]$.

**Characterisations.** *The following are equivalent:*

(a) *$P$ is projective in $\sigma[M]$;*

(b) *$P$ is $M^{(\Lambda)}$-projective for every index set $\Lambda$;*

(c) *every epimorphism $N \to P$ in $\sigma[M]$ splits.*

*Proof.* See [363, 18.3].                                                                            □

**Properties.** *Let $M$ be projective in $\sigma[M]$.*

(1) *$M$ is a direct summand of a direct sum of cyclic modules and has a maximal submodule.*

(2) *If $0 \neq U \subset M$ is a fully invariant submodule, then $M \notin \sigma[M/U]$.*

*Proof.* (1) This is shown in [363, 18.2 and 22.3].

(2) Assume $M \in \sigma[M/U]$ and let $i : M \to \widehat{M}$ be the $M$-injective hull. Then there is an epimorphism $f : (M/U)^{(\Lambda)} \to \widehat{M}$ and we have the diagram

$$
\begin{array}{ccc}
M^{(\Lambda)} & & M \\
{\scriptstyle p_U^{(\Lambda)}}\Big\downarrow & & \Big\downarrow{\scriptstyle i} \\
(M/U)^{(\Lambda)} & \xrightarrow{\ f\ } \widehat{M} \longrightarrow & 0,
\end{array}
$$

where $p_U : M \to M/U$ denotes the canonical projection. By projectivity of $M$, there exists $h : M \to M^{(\Lambda)}$ extending the diagram commutatively.

However, $(U)h \diamond \pi_\lambda \subseteq U$ for each $\lambda \in \Lambda$ and projection $\pi_\lambda : M^{(\Lambda)} \to M$; hence $(U)h \diamond p_U^{(\Lambda)} = 0$ and so $U \subset \mathrm{Ke}\, i$, a contradiction.                      □

As seen in 4.1, if $P$ is finitely generated and $M$-projective, then it is projective in $\sigma[M]$. In general $M$-projectivity need not imply projectivity in $\sigma[M]$. This does not even hold for $M = R$ in general. For $M = \mathbb{Z}$ this is known as *Whitehead's problem* (see [303, p. 238]).

**4.7. Projective covers in** $\sigma[M]$**.** Let $N, P \in \sigma[M]$. A small epimorphism $p : P \to N$ where $P$ is projective in $\sigma[M]$ is called *a projective cover of $N$ in $\sigma[M]$* or a $\sigma[M]$-*projective cover of $N$*.

Projective covers are unique up to isomorphisms (see, e.g., 4.5). Moreover, if $p : P \to N$ is a projective cover in $\sigma[M]$ and $Q \to N$ is an epimorphism where $Q$ is a projective module in $\sigma[M]$, then there is a decomposition $Q = Q_1 \oplus Q_2$ with $Q_1 \simeq P$, $Q_2 \subseteq \mathrm{Ke}\, f$, and $f|_{Q_1} : Q_1 \to N$ is a projective cover in $\sigma[M]$.

In general projective covers need not exist in $\sigma[M]$. Their existence will be investigated in the section on semiperfect modules.

**4.8. Self-projective modules.** An $M$-projective module $M$ is called *self-projective* (or *quasi-projective*). As easily seen, finite direct sums of self-projective modules are again self-projective. Notice that $M$ is projective in $\sigma[M]$ if and only if *any* direct sum of copies of $M$ is self-projective and hence such modules are also called $\Sigma$-*self-projective*.

**4.9. Self-projective factor modules.** *Let $M$ be self-projective and $U \subset M$ a submodule.*

(1) *If $U$ is fully invariant, then $M/U$ is self-projective.*

(2) *If $U \ll M$ and $M/U$ is self-projective, then $U$ is fully invariant.*

*Proof.* (1) is shown in [363, 18.2 (4)].

(2) Putting $P = M$, $L = M/U$ and $p : M \to M/U$ in 4.4 (2), we obtain $U = \mathrm{Re}(M, M/U)$. This means in particular that $U$ is fully invariant. $\qquad\square$

**4.10. Properties of self-projective modules.** *Let $M$ be self-projective.*

(1) *Let $U, K \subset M$ be submodules with $U$ fully invariant. If there exists an epimorphism $f : M/U \to M/K$, then $U \subset K$.*

(2) *If $M$ is finitely generated, then $M$ is projective in $\sigma[M]$.*

(3) *For any finitely generated left ideal $I \subset \mathrm{End}(M)$, $I = \mathrm{Hom}(M, MI)$.*

(4) *If $M = N_1 + \cdots + N_k$ where the $N_i$ are submodules, then there exist $f_1, \ldots, f_k$ in $\mathrm{End}(M)$ such that*
$$(M)f_i \subseteq N_i, \text{ for each } i \le k, \text{ and } f_1 + \cdots + f_k = \mathrm{id}_M.$$

(5) *Let $M_1$ be a direct summand and $N$ a submodule of $M$ such that $M = M_1 + N$. Then there exists $M_2 \subseteq N$ with $M = M_1 \oplus M_2$.*

*Proof.* (1) With the canonical projections $p_U, p_K$, we have a commutative diagram

$$
\begin{array}{ccc}
M & \xleftarrow{\ \ g\ \ } & M \\
{\scriptstyle p_U}\downarrow & & \downarrow{\scriptstyle p_K} \\
M/U & \xrightarrow{\ f\ } M/K & \longrightarrow 0,
\end{array}
$$

where $g : M \to M$ exists by self-projectivity of $M$. Since $(U)g \subset U$, we obtain $U \subset \mathrm{Ke}\, p_K = K$.

(2) and (3) are shown in [363, 18.2 and 18.4].

(4) We have a diagram with the canonical epimorphisms

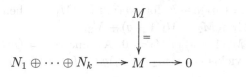

which can be extended commutatively by some morphism $M \to N_1 \oplus \cdots \oplus N_k$. From this the assertion follows.

(5) Choose an idempotent $e \in \mathrm{End}(M)$ with $M_1 = Me$ and let $p : M \to M/N$ be the canonical projection. By self-projectivity of $M$ there exists $\alpha \in \mathrm{End}(M)$ with $p = \alpha \diamond e \diamond p|_{M_1}$. Then $d = e + (1 - e)\alpha e$ is an idempotent in $\mathrm{End}(M)$, $Md = Me = M_1$ and

$$M_2 = \mathrm{Ke}\, d = M(1 - e)(1 - \alpha e) \subseteq N$$

is a direct summand of $M$ satisfying our claim.                                 □

### 4.11. Self-projectives with noetherian self-injective hull.

(1) *Let $M$ be a self-projective left $R$-module whose self-injective hull $\widehat{M}$ is noetherian. Then $\mathrm{End}(M)$ is left artinian.*

(2) *If the injective hull $E(R)$ of $_R R$ is noetherian, then $R$ is left artinian.*

(3) *If $R$ is commutative, then every self-injective noetherian module is artinian.*

*Proof.* (1) and (2) are shown in [85, 5.16].

(3) Let $R$ be a commutative ring and $M$ a noetherian self-injective $R$-module. Without loss of generality we may $M$ assume to be faithful. Then $R$-Mod $= \sigma[M]$, that is $R \subset M^k$ for some $k \in \mathbb{N}$, and $M$ is $R$-injective. This implies that the injective hull $E(R)$ is a direct summand of $M^k$ and hence is finitely generated. By (2), $R$ is artinian and hence $M$ is also artinian.                                 □

A close analysis of the proof of [363, 41.14] yields:

### 4.12. Relative projectives. *For $M = M_1 \oplus M_2$ the following are equivalent.*

(a) *$M_1$ is $M_2$-projective;*

(b) *for every submodule $N \subset M$ such that $M = N + M_2$, there exists a submodule $N'$ of $N$ such that $M = N' \oplus M_2$.*

*Proof.* (a)$\Rightarrow$(b). Let $M = N + M_2$. With canonical mappings, we obtain the diagram

$$
\begin{array}{ccccc}
 & & M_1 & & \\
 & & \downarrow & & \\
M_2 & \longrightarrow & M & \longrightarrow & M/N,
\end{array}
$$

which can be commutatively extended by some morphism $g : M_1 \to M_2$. This means $m_1 + N = (m_1)g + N$ for every $m_1 \in M_1$ and hence $M_1(1 - g) \subset N$. Now we have $M = M_1 + M_2 \subset M_1(1 - g) + M_2$.

We show $M_1(1 - g) \cap M_2 = 0$. Assume $m_2 = (m_1)(1 - g)$, for $m_2 \in M_2$, $m_1 \in M_1$. This yields $m_1 = m_2 - (m_1)g \in M_2 \cap M_1 = 0$.

(b)⇒(a). Let $p : M_2 \to W$ be an epimorphism and $f : M_1 \to W$ be any morphism. We form

$$N = \{m_2 - m_1 \in M \mid m_2 \in M_2, \, m_1 \in M_1 \text{ and } (m_2)p = (m_1)f\}.$$

Since $p$ is epic, $M = N + M_2$. Therefore, by (b), $M = N' \oplus M_2$ with $N' \subset N$. Let $e : M \to M_2$ be the projection along $N'$. This yields a morphism $M_1 \to M \to M_2$.

Since $M_1(1 - e) \subset N' \subset N$ we have, for every $m_1 \in M_1$, $m_1 - (m_1)e \in N$, and hence $(m_1)f = (m_1e)p$, that is, $f = ep$. Thus $M_1$ is $M_2$-projective. $\qquad\square$

An interesting generalisation of self-projectives are the

**4.13. $\pi$-projective modules.** The $R$-module $M$ is called *$\pi$-projective* if, for any two submodules $U, V \subset M$ with $U + V = M$, the following equivalent conditions are satisfied:

(a) *There exists $f \in \mathrm{End}(M)$ with $\mathrm{Im}\,(f) \subset U$ and $\mathrm{Im}\,(1 - f) \subset V$;*

(b) *the canonical epimorphism $U \oplus V \to M$, $(u, v) \mapsto u + v$, splits;*

(c) $\mathrm{End}(M) = \mathrm{Hom}(M, U) + \mathrm{Hom}(M, V)$.

Clearly any factor $M/N$ of a $\pi$-projective module $M$ by a fully invariant submodule $N$ is $\pi$-projective. Also every direct summand of a $\pi$-projective module is again $\pi$-projective; however, the direct sum of two $\pi$-projective modules need not be $\pi$-projective.

**4.14. Properties of $\pi$-projective modules.** *Let $M$ be $\pi$-projective and let $U, V \subset M$ be submodules with $M = U + V$.*

(1) *If $U$ is a direct summand in $M$, then there exists $V' \subset V$ with $M = U \oplus V'$.*

(2) *If $U \cap V = 0$, then $V$ is $U$-projective (and $U$ is $V$-projective).*

(3) *If $U \cap V = 0$ and $V \simeq U$, then $M$ is self-projective.*

(4) *If $U$ and $V$ are direct summands of $M$, then $U \cap V$ is also a direct summand of $M$.*

*Proof.* See [363, 41.14]. $\qquad\square$

The following observation is an easy consequence of 4.13.

**4.15. $\pi$-projectives and coindependent families.** *Let $M$ be $\pi$-projective and $\{K_\lambda\}_\Lambda$ be a coindependent family of proper submodules of $M$. Then $\{\mathrm{Hom}(M, K_\lambda)\}_\Lambda$ is a coindependent family of left ideals of $\mathrm{End}(M)$.*

**4.16. $\pi$-projectives and cosmall inclusions.** *Let $M$ be a $\pi$-projective module. Then, for any submodule $N \subset M$,*

$$N\mathrm{Hom}(M, N) \overset{cs}{\underset{M}{\hookrightarrow}} N.$$

*Proof.* Assume $N+L = M$. Then $\mathrm{Hom}(M, N)+\mathrm{Hom}(M, L) = \mathrm{End}(M)$ and hence $N\mathrm{Hom}(M, N) + N\mathrm{Hom}(M, L) = N\mathrm{End}(M)$. Thus

$$N\mathrm{Hom}(M, N) + L = N\mathrm{End}(M) + L = M$$

and hence $N\mathrm{Hom}(M, N) \overset{cs}{\underset{M}{\hookrightarrow}} N$ by 3.2 (1).                                  □

As seen above, if $N$ is a submodule of a $\pi$-projective module $M$, then the product $N\mathrm{Hom}(M, N) \subseteq N$ is cosmall in $M$. Now, if $I$ is a left ideal $I$ of the ring $R$, then $I\mathrm{Hom}(R, I) = I^2$. This leads us to call a submodule $N \subset M$ *idempotent* if $N = N\mathrm{Hom}(M, N)$. Hence

**4.17. Corollary.** *A coclosed submodule of a $\pi$-projective module is idempotent.*

It follows that any coclosed left ideal of a ring is idempotent. In particular the only coclosed ideals of a principal ideal domain are the trivial ideals.

We now consider further generalisations of self-projectivity. The first of these was introduced by Brodskiĭ in [43].

**4.18. Intrinsically projective modules.** A module $M$ is called *intrinsically projective* if every diagram with exact row

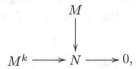

where $k \in \mathbb{N}$ and $N \subset M$, can be extended commutatively by some $M \to M^k$.

**4.19. Characterisation of intrinsically projective modules.** *Let $M$ be any module and set $S = \mathrm{End}(M)$. The the following statements are equivalent:*

 (a) *$M$ is intrinsically projective;*

 (b) *for every finitely generated left ideal $I \subset S$, $I = \mathrm{Hom}(M, MI)$.*

*If $_RM$ is finitely generated, then* (a), (b) *are equivalent to:*

 (c) *for every left ideal $I \subset S$, $I = \mathrm{Hom}(M, MI)$;*

 (d) *the map $I \mapsto MI$ from left ideals in $S$ to submodules of $M$ is injective.*

*Proof.* See [364, 5.7].                                                          □

**4.20. Semi-projective modules.** The module $M$ is called *semi-projective* if the condition in 4.18 holds simply for $k = 1$.

As easily seen, $M$ is semi-projective if and only if, for every cyclic left ideal $I \subset \mathrm{End}(M)$, $I = \mathrm{Hom}(M, MI)$, or — equivalently — for any $f \in \mathrm{End}(M)$, $\mathrm{End}(M)f = \mathrm{Hom}(M, Mf)$. Notice that this implies in particular that any epimorphism in $\mathrm{End}(M)$ is a retraction.

**4.21. Direct projective modules.** The module $M$ is called *direct projective* if, for every direct summand $X$ of $M$, every epimorphism $M \to X$ splits. This is obviously equivalent to the condition that, if $N$ is a submodule of $M$ for which $M/N$ is isomorphic to a direct summand of $M$, then $N$ is a direct summand of $M$.

**4.22. Properties of direct projective modules.** *Let $M$ be direct projective and $S =$ End$(M)$.*

(1) *If $U$ and $V$ are direct summands of $M$, then every epimorphism $U \to V$ splits.*

(2) *Every direct summand of $M$ is direct projective.*

(3) *If $U$ and $V$ are direct summands with $U + V = M$, then $U \cap V$ is a direct summand in $U$ and $M = U \oplus V'$, for some $V' \subset V$.*

(4) $\nabla(M) = \{f \in S \mid \operatorname{Im} f \ll M\} \subseteq \operatorname{Jac} S.$

*Proof.* These are given in [363, 41.18, 41.19]. $\qquad\square$

**4.23. Epi-projective modules.** A module $L$ is said to be *$M$-epi-projective* if, for any epimorphisms $p : M \to N$ and $f : L \to N$, there exists a morphism $h : L \to M$ such that $f = h \diamond p$, that is, the diagram with exact row and column

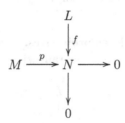

can be extended commutatively by some $h : L \to M$.

We call $M$ *epi-projective* if it is $M$-epi-projective.

**4.24. Properties of epi-projective modules.**

(1) *If $L \oplus M$ is $M$-epi-projective, then $L$ is $M$-projective.*

(2) *If $M \oplus L$ is $M \oplus L$-epi-projective, then $M$ and $L$ are epi-projective and are relative projective to each other.*

(3) *$M$ is self-projective if and only if $M \oplus M$ is $M \oplus M$-epi-projective.*

(4) *If $M$ is epi-projective, then $M$ is direct projective.*

(5) *If $M$ is hollow and epi-projective, then $M$ is self-projective.*

*Proof.* (1) Let $p : M \to N$ be any epimorphism and $f : L \to N$ be any morphism. Then $f + p : L \oplus M \to N$ is an epimorphism and — by assumption — there exists a morphism $h : L \oplus M \to M$ such that $f + p = h \diamond p$. Then for any $l \in L$, $(l)f = (l)(f + p) = (l)h \diamond p$. This shows that $L$ is $M$-projective.

(2), (3) are obvious from (1).

(4) Let $X \subset M$ be a direct summand with inclusion $i : X \to M$ and projection $p : M \to X$. For any epimorphism $f : M \to X$ there exists $h \in \text{End}(M)$ with $h \diamond f = p$. Then $i \diamond h \diamond f = i \diamond p = \text{id}_X$ showing that $f$ splits.

(5) (D. Keskin) Let $M$ be hollow and epi-projective, $K \subsetneq M$ a submodule, and $\pi : M \to M/K$ the natural projection. Consider any homomorphism $f : M \to M/K$. If $f$ is surjective, then $f$ can be lifted to $M$ as $M$ is epi-projective. If $f$ is not surjective, then (since $M/K$ is hollow) $\pi - f : M \to M/K$ is an onto map and hence there exists $g : M \to M$ such that $g\pi = \pi - f$. This implies $f = (1 - g)\pi$ showing that $M$ is self-projective.                                                                    $\square$

**4.25. Endomorphism rings of epi-projective modules.** *Let $M$ be epi-projective and $S = \text{End}(M)$. Then*

(1) $\text{Jac}\,(S) = \nabla(M) = \{f \in S \,|\, \text{Im}\, f \ll P\}$ *and*

(2) *if $M$ is hollow, then $S$ is a local ring.*

*Proof.* (1) This follows from the proof of [363, 22.2].

(2) Clearly every surjective $f \in S$ is an isomorphism. Hence the equality $\text{Im}\, f + \text{Im}\, (1 - f) = M$ implies that $f$ or $1 - f$ is an isomorphism, that is, $S$ is a local ring.                                                                    $\square$

**4.26. Image summand and coclosed projectives.** A module $L$ is called *im-summand (im-coclosed) $M$-projective*, if every diagram with exact row

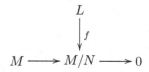

can be extended commutatively by some morphism $L \to M$, provided $\text{Im}\, f$ is a direct summand (coclosed) in $M/N$.

Clearly im-coclosed $M$-projective modules are im-summand $M$-projective and the latter are $M$-epi-projective.

**4.27. Hopfian modules.** The module $M$ is said to be *(semi-)Hopfian* if every surjective endomorphism of $M$ is an isomorphism (a retraction).

Obviously direct-projective modules are semi-Hopfian and every indecomposable semi-Hopfian module is Hopfian. Furthermore, over a commutative ring $R$, any finitely generated $R$-module is Hopfian (see, e.g., [130], [349]).

We now show that, under certain projectivity conditions, hollowness of modules is reflected by their endomorphism rings. Note that hollow modules are trivially $\pi$-projective but need not be direct projective.

**4.28. Hollow modules and projectivity.** *Let $M$ be a module with endomorphism ring $S = \mathrm{End}(M)$. Recall that $\nabla(M) = \{f \in S \mid \mathrm{Im}\, f \ll M\}$.*

(1) *If $M$ is semi-Hopfian, then $\nabla(M) \subseteq \mathrm{Jac}\, S$. In particular, if $M$ is also hollow, then $S$ is local.*

(2) *If $M$ is $\pi$-projective and $N$ is a submodule of $M$ such that $\mathrm{Hom}(M, N) \ll S$ (as an $S$-submodule), then $N \ll M$. In particular, if in this case $S$ is local, then $M$ is hollow.*

(3) *If $M$ is $\pi$-projective and semi-projective, then $\nabla(M) = \mathrm{Jac}\, S$.*

(4) *If $M$ is self-projective, then the following statements are equivalent:*

    (a) *$M$ is hollow;*

    (b) *$\mathrm{End}(M)$ is a local ring.*

(5) *If $M$ is projective in $\sigma[M]$, the following statements are equivalent:*

    (a) *$M$ is local;*

    (b) *$M$ is (semi-)hollow;*

    (c) *$\mathrm{End}(M)$ is a local ring.*

*Proof.* (1) Let $M$ be semi-Hopfian and $f \in \nabla(M)$, so that $\mathrm{Im}\, f \ll M$. Suppose there exists $g \in S$ with $Sf + Sg = S$. Then $M = \mathrm{Im}\, f + \mathrm{Im}\, g = \mathrm{Im}\, g$ since $\mathrm{Im}\, f$ is small. Thus $g$ is an epimorphism and by hypothesis, $Sg = S$. This implies that $Sf \ll S$ and so $f \in \mathrm{Jac}\, S$.

If $M$ is also hollow, then $M$ is Hopfian. Now, for any $f \in \mathrm{End}(M)$, either $f$ or $\mathrm{id}_M - f$ belongs to $\mathrm{Jac}\, S$ and so one of them is an epimorphism and hence an isomorphism. Thus $\mathrm{End}(M)$ is local.

(2) Let $N$ be a submodule of $M$ such that $\mathrm{Hom}(M, N) \ll S$. If $N + L = M$ for some submodule $L$, then $S = \mathrm{Hom}(M, N) + \mathrm{Hom}(M, L) = \mathrm{Hom}(M, L)$. Hence $L = M$ and so $N \ll M$.

(3) Since semi-projective modules are semi-Hopfian, we have $\nabla(M) \subseteq \mathrm{Jac}\, S$ by (1). Let $f \in \mathrm{Jac}\, S$. Since $M$ is semi-projective, $\mathrm{Hom}(M, \mathrm{Im}\, f) = Sf$ is small in $S$. By (2), $\mathrm{Im}\, f \ll M$, that is, $f \in \nabla(M)$.

(4) If $M$ is self-projective, then $M$ is $\pi$-projective and epimorphisms $M \to M$ are retractions. Now the assertion follows from (1) and (2).

(5) If $M$ is projective in $\sigma[M]$, then $\mathrm{Rad}(M) \neq M$ and (4) applies. $\qquad\square$

The next observation was pointed out in [201, Lemma 2.1].

**4.29. ∩-direct projectives.** *For a module $M$ the following are equivalent.*

(a) *For any direct summands $K, L$ of $M$ with $M = K + L$, $K \cap L$ is a direct summand of $M$;*

(b) *for any direct summands $K, L$ of $M$ with $M = K + L$, any diagram with*
    *projection $p$,*

$$M$$
$$\downarrow f$$
$$M \xrightarrow{\ \ p\ \ } M/(K \cap L) \longrightarrow 0,$$

*can be extended commutatively by some $h : M \to M$ such that $h \diamond p = f$.*

If this holds $M$ is called $\cap$-*direct projective.*

*Proof.* (a)$\Rightarrow$(b) is obvious.

(b)$\Rightarrow$(a). Let $K, L$ be direct summands of $M$ with $M = K + L$ and write
$M = K \oplus K' = L \oplus L'$. Consider the morphism

$$f : M \to M/(K \cap L), \quad k + l \mapsto k + K \cap L, \ \text{ for } k \in K, l \in L.$$

Then there is a morphism $h : M \to M$ such that $h \diamond p = f$. Thus we have
$(k + l)f = k + K \cap L = (k + l)h + K \cap L$ and hence

$$(M)h \subseteq K, \ (M)(\mathrm{id}_M - h) \subseteq L \ \text{ and } \ (L)h \subseteq K \cap L.$$

Now $M = (M)h + (M)(\mathrm{id}_M - h) = (M)h + L$ and $(M)h = (L)h + (L')h$, thus
$M = (L)h + (L')h + L = (L')h + L$.

Let $y \in (L')h \cap L = 0$, that is, $y = (u)h$ for some $u \in L'$. Write $u = k + l$
for some $k \in K$, $l \in L$. Then $(u)h - (k)h = (l)h \in K \cap L$ and $(u)h + K \cap L = (k)h + K \cap L = k + K \cap L$, that is, $y - k = (u)h - k \in K \cap L$. This implies

$$u = k + l = (y + l) - (y - k) \in L' \cap L = 0,$$

hence $(L')h \cap L = 0$ and $M = (L')h \oplus L$. Since $(L')h \subseteq K$ we derive from this
that $K = (L')h \oplus (K \cap L)$ and

$$M = K \oplus K' = (L')h \oplus (K \cap L) \oplus K',$$

showing that $K \cap L$ is a direct summand of $M$.                                    $\square$

**4.30. Direct projectives are $\cap$-direct projective.** *If $M$ is a direct projective module,*
*then it is $\cap$-direct projective.*

*Proof.* Let $K$ and $L$ be direct summands of $M$ with $K + L = M$ and $M = K \oplus K' = L \oplus L'$. Then $K' \simeq (K + L)/K \simeq L/K \cap L$ and hence the epimorphism
$M \to L \to L/(K \cap L)$ splits by direct projectivity. Since this epimorphism has
kernel $(K \cap L) \oplus L'$, it follows that $K \cap L$ is a direct summand in $M$.        $\square$

Replacing direct summands by coclosed submodules (see 3.7) in 4.29 we
define

**4.31. ∩-coclosed projectives.** The module $M$ is called ∩-*coclosed projective* if, for any coclosed submodules $K$ and $L$ of $M$ with $M = K + L$, any diagram with projection $p$,

$$
\begin{array}{c}
M \\
\downarrow f \\
M \xrightarrow{\ p\ } M/K \cap L \longrightarrow 0,
\end{array}
$$

can be extended commutatively by some $h : M \to M$ such that $h \diamond p = f$.

Obviously ∩-coclosed projective modules are ∩-direct projective but the converse implication only holds under additional assumptions, for example, if every coclosed submodule of $M$ is a direct summand.

**4.32. Small and Rad-projective modules.** A module $L$ is called *small $M$-projective* if any diagram

$$
\begin{array}{c}
L \\
\downarrow f \\
M \longrightarrow M/K \longrightarrow 0
\end{array}
$$

can be extended commutatively by some morphism $L \to M$, provided $K \ll M$.

$L$ is said to be *Rad-$M$-projective* provided the above diagram can be extended commutatively for all submodules $K \subseteq \mathrm{Rad}(M)$.

$M$ is called *small self-projective* if it is small $M$-projective and is called *Rad-self-projective* if it is Rad-$M$-projective.

A module $L$ is called *im-summand (im-coclosed) small $M$-projective*, if the above diagram can be extended commutatively by some morphism $L \to M$, provided $\mathrm{Im}\, f$ is a direct summand (coclosed) in $M/N$ and $K \ll M$.

Obviously Rad-$M$-projective modules are small $M$-projective.

We note that for weakly supplemented modules $M$, small $M$-projective modules are $M$ projective (see 17.14).

There are some more variations of projectivity which are of particular relevance to decomposition properties of modules. Notice that, although the name is very similar, the next notion is quite different from that defined in 4.32.

**4.33. im-small projectives.** A module $L$ is called *image small $M$-projective*, for short *im-small $M$-projective*, if every diagram with exact row

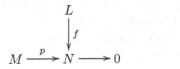

can be extended commutatively by some morphism $L \to M$, provided Im $f \ll N$. The modules $L$ and $M$ are said to be *relatively im-small projective* if $L$ is im-small $M$-projective and vice versa.

Combining this with the conditions in 4.32, we call the module $L$ *small im-small M-projective* provided the extension of the above diagram is possible if Im $f \ll N$ and Ke $p \ll M$.

Notice that over a left hereditary ring $R$, every injective left $R$-module $Q$ is trivially im-small $M$-projective for any $R$-module $M$, since all factor modules of $Q$ are injective and hence any small image is zero.

### 4.34. Properties of im-small projective modules.

(1) *Suppose $L$ is im-small $M$-projective, $L'$ is a direct summand of $L$ and $M'$ is a direct summand of $M$. Then $L'$ is im-small $M'$-projective.*

(2) *If $L_i$ is im-small $M$-projective, for all $i \in I$, then $\oplus_{i \in I} L_i$ is im-small $M$-projective.*

*Proof.* This is easily seen by standard arguments.                          $\square$

### 4.35. im-small projectives in direct sums. *Let $M = M_1 \oplus M_2$. The following statements are equivalent.*

(a) *$M_1$ is im-small $M_2$-projective;*

(b) *for every submodule $N$ of $M$ such that $(N + M_1)/N \ll M/N$, there exists a submodule $N'$ of $N$ such that $M = N' \bigoplus M_2$.*

*Proof.* (a)$\Rightarrow$(b). Let $N \subset M$ such that $(N + M_1)/N \ll M/N$. Then $M = N + M_2$. Consider the morphisms

$$g : M_1 \to M/N, \; m_1 \mapsto m_1 + N, \quad f : M_2 \to M/N, \; m_2 \mapsto m_2 + N.$$

Since Im $g = (N + M_1)/N \ll M/N$ and $f$ is an epimorphism, there is a morphism $p : M_1 \to M_2$ such that $p \diamond f = g$. Then $N' = \{\, a - (a)p \mid a \in M_1 \,\}$ is a submodule of $N$ with $M = N' \oplus M_2$.

(b)$\Rightarrow$(a). Let $A \subset M_2$ and $f : M_1 \to M_2/A$ be a morphism with Im $f \ll M_2/A$. Let $p : M_2 \to M_2/A$ denote the canonical projection and put

$$N = \{\, a + b \in M_1 \oplus M_2 \mid (a)f = -(b)p, \; a \in M_1, \; b \in M_2 \,\}.$$

Clearly $A \subseteq N$ and $M = N + M_2$. Let Im $f = X/A$, where $X \subseteq M_2$, and consider the morphism

$$h : M_2/A \to M/N, \quad m_2 + A \mapsto m_2 + N.$$

Then $(X/A)h = (X + N)/N$, and so $(X + N)/N \ll M/N$. Since $(N + M_1)/N \subseteq (X + N)/N$, we know that $(N + M_1)/N \ll M/N$. Hence, by hypothesis, there is a submodule $N'$ of $N$ such that $M = N' \oplus M_2$. With the canonical projection $\alpha : N' \oplus M_2 \to M_2$, the morphism $f$ can be lifted to $\alpha|_{M_1} : M_1 \to M_2$. Thus $M_1$ is im-small $M_2$–projective.                          $\square$

**4.36. im-small projectivity and hollow modules.** *Let $L, M$ and $H$ be modules.*

(1) *If $L$ is im-small $M$-projective and im-small $H$-projective where $H$ is hollow, then $L$ is im-small $M \oplus H$-projective.*

(2) *Let $H_1, \ldots, H_k$ be hollow modules such that $L$ is im-small $H_i$-projective for each $1 \le i \le k$. Then $L$ is im-small $\bigoplus_{i=1}^{k} H_i$-projective.*

*Proof.* (1) For any submodule $K \subset M \oplus H$, let $\eta : M \oplus H \to (M \oplus H)/K$ be the natural map and $f : L \to (M \oplus H)/K$ be a map such that $(L)f \ll (M \oplus H)/K$. Let $\pi_M : M \oplus H \to M$ and $\pi_H : M \oplus H \to H$ be the projections along $H$ and $M$ respectively.

Suppose first that $(K)\pi_H = H$. Then $M \oplus H = M + K$ and $(M)\eta = (M \oplus H)\eta$. Then, since $L$ is im-small $M$-projective, there exists $h : L \to M$ such that $h \diamond \eta = f$, as required.

Next assume that $(K)\pi_H \neq H$. Now $K \subseteq (K)\pi_M + (K)\pi_H$. Let

$$\varphi : (M \oplus H)/K \to (M \oplus H)/[(K)\pi_M + (K)\pi_H]$$

be the natural map. We have

$$(M \oplus H)/[(K)\pi_M + (K)\pi_H] \simeq M/(K)\pi_M \oplus H/(K)\pi_H.$$

With the projection maps

$$p_1 : (M \oplus H)/[(K)\pi_M + (K)\pi_H] \to M/(K)\pi_M \quad \text{and}$$

$$p_2 : (M \oplus H)/[(K)\pi_M + (K)\pi_H] \to H/(K)\pi_H$$

we obtain

$$(L)f\varphi p_1 \ll M/(K)\pi_M \quad \text{and} \quad (L)f\varphi p_2 \ll H/(K)\pi_H.$$

Since $L$ is im-small $M$-projective and im-small $H$-projective, there are homomorphisms $g_1 : L \to M$ and $g_2 : L \to H$ such that

$$g_1 \diamond \eta \diamond \varphi \diamond p_1 = f \diamond \varphi \diamond p_1 \quad \text{and} \quad g_2 \diamond \eta \diamond \varphi \diamond p_2 = f \diamond \varphi \diamond p_2.$$

It can be easily checked that

$$(f - (g_1 + g_2) \diamond \eta) \diamond \varphi = 0$$

and hence

$$\operatorname{Im}(f - (g_1 + g_2) \diamond \eta) \subseteq [(K)\pi_M + (K)\pi_H]/K.$$

We have $[(K)\pi_M + (K)\pi_H]/K = (K + (K)\pi_H)/K$. By assumption $(K)\pi_H \neq H$. Now $H$ being hollow implies that $(K + (K)\pi_H)/K \ll (H + K)/K$. Put $t = f - (g_1 + g_2) \diamond \eta$. Then $t : L \to (H + K)/K$ is a map with $\operatorname{Im} t \ll (H + K)/K$. Since $L$ is im-small $H$-projective we get a map $g_3 : L \to H$ such that $g_3 \diamond \eta = t$, that is, $g_3 \diamond \eta = f - (g_1 + g_2) \diamond \eta$ and hence $f = (g_1 + g_2 + g_3) \diamond \eta$. Thus $L$ is im-small $H \oplus M$-projective.

(2) follows from (1) by finite induction. $\square$

The preceding proof also applies to small im-small projectives. In this case we need not require $H$ to be hollow.

**4.37. Small im-small projectivity.** *Let $L, M$ and $H$ be modules. If $L$ is small im-small $M$-projective and small im-small $H$-projective, then $L$ is small im-small $M \oplus H$-projective.*

**4.38. Almost projective modules.** Let $M$ and $N$ be two $R$-modules. Then $N$ is called *almost $M$-projective* if every diagram with exact row

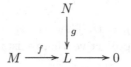

can be either extended commutatively by a morphism $h : N \to M$ or there exists a nonzero direct summand $M_1$ of $M$ and $h : M_1 \to N$ such that $hg = e_1 f \neq 0$ where $e_1 : M_1 \to M$ is the canonical inclusion map.

Almost projectives were introduced by Harada and his collaborators. The following observation is proved in Harada-Mabuchi [161].

**4.39. Indecomposables and almost projectives.** *Suppose $M$ is an indecomposable but not a hollow module. If $N$ is almost $M$-projective, then $N$ is $M$-projective.*

*Proof.* Let $K$ be any submodule of $M$, $g : M \to M/K$ be the canonical map and $f : N \to M/K$ be any map. First assume that $K$ is not small in $M$. Then $M = K + K_1$ for some proper submodule $K_1$ of $M$. Let $g' : M \to M/(K_1 \cap K)$ be the canonical map. There exists an isomorphism $\phi : M/(K_1 \cap K) \to M/K \oplus M/K_1$. Let $\pi : M/K \oplus M/K_1 \to M/K$ be the projection map. Then $g = g'\phi\pi$. By hypothesis, the map $f : N \to M/K \subseteq M/K \oplus M/K_1$ can be lifted to a map $h : N \to M$ such that $hg'\phi = f$. Then $f = hg'\phi\pi = hg$.

Next suppose that $K \ll M$. Since $M$ is not hollow, we have $M = K_1 + K_2$, where $K_1$ and $K_2$ are proper submodules of $M$. Since $K \ll M$ we can assume $K \subseteq K_1$ and $K \subseteq K_2$. Let $\eta : M/K \to M/(K_1 \cap K_2)$ be the natural map. Then $M/(K_1 \cap K_2) \simeq M/K_1 \oplus M/K_2$. Let $\pi_1 : M/(K_1 \cap K_2) \to M/K_1$ be the projection and $i_1 : M/K_1 \to M/(K_1 \cap K_2)$ be the inclusion map. Then $g\eta : M \to M/(K_1 \cap K_2)$ is an epimorphism and $f\eta\pi_1 i_1 : N \to M/(K_1 \cap K_2)$ is not an epimorphism. By hypothesis, there exists a map $h' : N \to M$ such that $h'g\eta = f\eta\pi_1 i_1$. Thus $(N)(f - h'g) \subseteq \mathrm{Ke}\,(\eta - \eta\pi_1 i_1) = K_1/K$ and we have a diagram with exact row

$$
\begin{array}{c}
N \\
\big\downarrow{\scriptstyle f-h'g} \\
M \xrightarrow{\ g\ } M/K \longrightarrow 0.
\end{array}
$$

Since $(N)(f - h'g) \subseteq K_1/K \neq M/K$, there exists a map $h'' : N \to M$ such that $h''g = f - h'g$ and so $f = (h'' + h')g$. Therefore $N$ is $M$-projective. $\square$

**4.40. Almost projective hollow modules.** *Let $M$ and $N$ be hollow modules such that $N$ is almost $M$-projective and there exists no epimorphism from $M$ onto $N$. Then $N$ is $M$-projective.*

*Proof.* Let $f : M \to X$ be an epimorphism and $g : N \to X$ be any morphism. First suppose $g$ is not onto. Then there is no map $h : M \to N$ such that $hg = f$ since $hg$ cannot be an epimorphism. Hence there exists $h : N \to M$ such that $hf = g$.

Next suppose $g$ is onto and there does not exist $h : N \to M$ such that $hf = g$. Then there exists $h : M \to N$ such that $hg = f$. Now $\operatorname{Im} g = \operatorname{Im} f = X$. Since the image under $g$ of any small submodule of $N$ cannot be $X$ and $N$ is hollow, $h$ must be surjective. This contradicts our assumption and hence $N$ is $M$-projective. $\square$

**4.41. Generalised projectivity.** A module $K$ is said to be *generalised $M$-projective* if, for any epimorphism $g : M \to X$ and morphism $f : K \to X$, there exist decompositions $K = K_1 \oplus K_2$, $M = M_1 \oplus M_2$, a morphism $h_1 : K_1 \to M_1$ and an epimorphism $h_2 : M_2 \to K_2$, such that $h_1 \diamond g = f|_{K_1}$ and $h_2 \diamond f = g|_{M_2}$, that is, every diagram with exact row

$$
\begin{array}{c}
K \\
\downarrow f \\
M \xrightarrow{\;g\;} X \longrightarrow 0
\end{array}
$$

can be embedded in a diagram

$$
\begin{array}{ccccccc}
K_1 & \oplus & K_2 & = & & K & \\
\vert & & \wedge & & & & \\
h_1 \vert & & h_2 \vert & & & \downarrow f & \\
\vee & & \vert & & & & \\
M_1 & \oplus & M_2 & = M & \xrightarrow{\;g\;} & X & \longrightarrow 0
\end{array}
$$

such that $h_2$ is an epimorphism, $h_1 \diamond g = f|_{K_1}$ and $h_2 \diamond f = g|_{M_2}$.

The following gives a characterisation of relative generalised projective modules which will be used repeatedly in the sequel.

**4.42. Characterisation of generalised projectivity.** *Let $M = K \oplus L$. Then $K$ is generalised $L$-projective if and only if, whenever $M = N + L$ for some $N \subset M$, we have*

$$M = N' \oplus K' \oplus L' = N' + L \text{ with } N' \subseteq N, K' \subseteq K \text{ and } L' \subseteq L.$$

*Proof.* Assume $K$ is generalised $L$-projective and let $M = N + L$. For the natural epimorphism $\eta : M \to M/N$, define $g = \eta|_L$ and $f = \eta|_K$. By hypothesis there are decompositions $K = K_1 \oplus K_2$, $L = L_1 \oplus L_2$, and maps $h_1 : K_1 \to L_1$ and $h_2 : L_2 \to K_2$ such that $h_2$ is an epimorphism, $h_1 g = f|_{K_1}$ and $h_2 f = g|_{L_2}$. Define submodules $C, D$ of $M$ by

$$C = \{k_1 - (k_1)h_1 \mid k_1 \in K_1\} \quad \text{and} \quad D = \{l_2 - (l_2)h_2 \mid l_2 \in L_2\}.$$

Clearly $M = C \oplus D \oplus L_1 \oplus K_2$. As $h_2$ is onto, $K_2 \subseteq D + L_2$.

Putting $N' = C \oplus D$ we have

$$M = (K_1 \oplus L_1) \oplus (K_2 \oplus L_2) = (C \oplus L_1) \oplus (D \oplus L_2) = N' \oplus L.$$

It remains to show that $N' \subseteq N$. For $k_1 \in K_1$ and $l_2 \in L_2$ we compute

$$(k_1 - (k_1)h_1)\eta \;=\; (k_1)\eta - (k_1)h_1\eta = (k_1)f - (k_1)h_1 g = (k_1)f - (k_1)f = 0,$$
$$(l_2 - (l_2)h_2)\eta \;=\; (l_2)\eta - (l_2)h_2\eta = (l_2)g - (l_2)h_2 f = (l_2)g - (l_2)g = 0.$$

This shows $k_1 - (k_1)h_1 \in \mathrm{Ke}\,\eta$ and $l_2 - (l_2)h_2 \in \mathrm{Ke}\,\eta$ and therefore we conclude $N' = C + D \subseteq \mathrm{Ke}\,\eta = N$.

Conversely, assume the conditions hold. Given an epimorphism $g : L \to X$ and a morphism $f : K \to X$, define $\eta : M \to X$ by $(k + l)\eta = (k)f + (l)g$. Let $N = \mathrm{Ke}\,\eta$. For arbitrary $k \in K$, $(k)f = (l)g$ for some $l \in L$. Hence $k - l \in N$ and hence $K \subseteq N + L$. Therefore $M = K \oplus L = N + L$ and

$$M = N' \oplus K' \oplus L' = N' \oplus L \quad \text{with} \quad N' \subseteq N, K' \subseteq K \text{ and } L' \subseteq L.$$

Define $K'' = K \cap (N' \oplus L')$ and $L'' = L \cap (N' \oplus K')$. Then $K = K' \oplus K''$ and $L = L' \oplus L''$. Let $p_1 : M \to L'$ denote the projection along $N' \oplus K'$ and $p_2 : M \to K'$ the projection along $N' \oplus L'$. Define $h_1 = p_1|_{K''}$ and $h_2 = p_2|_{L''}$.

First we show that $h_2$ is onto. Given $k' \in K'$, write $k' = n' + l$, with $n' \in N'$ and $l \in L$. Then $l = k' - n' \in L \cap (K' \oplus N') = L''$ and $(l)h_2 = k'$.

Now, for $k'' \in K''$, we have $k'' = n' + l'$ for some $n' \in N'$, $l' \in L'$. Then $(k'')f = (k'')\eta = (l')\eta = (l')g = (k'')h_1 g$. Hence $h_1 g = f|_{K''}$.

Furthermore, for $l'' \in L''$, we have $l'' = n' + k'$ with $n' \in N'$ and $k' \in K'$. Then $(l'')g = (k' + n')\eta = (k')\eta = (k')f = (l'')h_2 f$. Hence $h_2 f = g|_{L''}$. $\qquad\square$

**4.43. Generalised projectivity and direct summands.** *Let $K, L$ be modules with $K$ generalised $L$-projective. Then, for any direct summand $L^*$ of $L$, the module $K$ is generalised $L^*$-projective.*

*Proof.* Put $M = K \oplus L$ and $L = L^* \oplus L^{**}$. Define $N = K \oplus L^*$, and let $X$ be a submodule of $N$ such that $N = X + L^*$. Then $M = N + L^{**} = X + L$. Since $K$ is generalised $L$-projective,

$$M = X' \oplus K' \oplus L' = X' + L \quad \text{with} \quad X' \subseteq X, K' \subseteq K, L' \subseteq L.$$

Hence $N = X' \oplus K' \oplus (N \cap L') = X' + (N \cap L)$. Now

$$N \cap L = (K \oplus L^*) \cap L = L^* \oplus (K \cap L) = L^*$$

and the result follows. $\qquad\qquad\qquad\qquad\qquad\qquad\qquad\qquad\qquad\qquad\square$

Notice that we have not shown that a direct summand of a generalised $L$-projective module is again generalised $L$-projective. This can only be shown in special situations (see 20.33).

The following can be easily deduced from the definitions.

**4.44. Almost and generalised relative projectives.** *Let $M, N$ be $R$-modules.*

(1) *If $M$ is indecomposable, then $M$ is almost $N$-projective if and only if $M$ is generalised $N$-projective.*

(2) *If $\{M_i\}_I$ is a family of almost $N$-projective modules, then $\bigoplus_I M_i$ is almost $N$-projective.*

Summarising we note that the various projectivity conditions of a module $M$ considered in this section are related in the following way:

$$\begin{array}{ll} \text{projective in } \sigma[M] & \Rightarrow \text{self-projective} \\ & \Rightarrow \text{intrinsically projective} \\ & \Rightarrow \text{semi-projective} \\ & \Rightarrow \text{direct projective} \\ & \Rightarrow \cap\text{-direct projective} \\ \text{self-projective} & \Rightarrow \text{epi-projective} \\ & \Rightarrow \text{direct projective} \\ & \Rightarrow \text{semi-Hopfian.} \end{array}$$

**4.45. Exercises.**

(1) Let $P$ be projective in $\sigma[M]$ and $K \subset P$ a submodule. Prove that $P/K$ is $M$-singular if and only if $K \trianglelefteq P$.

(2) Let $M$ be projective in $\sigma[M]$. Prove that $M$ is $\sigma$-cohereditary if and only if every submodule of $M$ is projective in $\sigma[M]$ (i.e., $M$ is hereditary in $\sigma[M]$).

(3) Let $f : M \to M$ be an epimorphism with $\mathrm{Ke}\, f \subset \mathrm{Rad}\, M$. Prove that if $\mathrm{Rad}\, M$ is a Hopfian module, then $f$ is an automorphism. (See [166].)

(4) Prove that a module $M$ is $\cap$-direct projective if and only if for any idempotents $e$ and $f$ of $S = \mathrm{End}(M)$, there exists an idempotent $g$ of $S$ such that $Mg = Mh$ and $M(1 - g) \subseteq Mf$. (See [327].)

(5) Recall that an $R$-module $M$ is *epi-projective* if it is $M$-epi-projective. Prove that the following are equivalent for the ring $R$.

    (a) The direct sum of epi-projective left $R$-modules is epi-projective;

    (b) every self-pseudo-projective left $R$-module is projective;

(c) $R \oplus M$ is epi-projective for every simple left $R$-module $M$;

(d) every (finitely generated) left $R$-module is self-projective.

(e) $R$ is left (and right) semisimple.

(6) Let $M$ and $N$ be $R$-modules such that $N$ is $M$-epi-projective.

    (i) Show that every epimorphism $f : M \to N$ splits.

    (ii) Let $A$ be a submodule of $M$. Show that $N$ is $M/A$-epi-projective.

    (iii) Let $K$ be a direct summand of $N$. Show that $K$ is $M$-epi-projective.

(Compare [80] for the dual notions and properties.)

(7) Show that an $R$-module $N$ is projective if and only if $N$ is $M$-epi-projective for all $R$-modules $M$.

(8) Let $M$ be an $R$-module with the following submodule lattice where $M/N_1 \not\simeq M/N_2$:

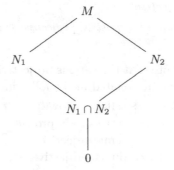

    (i) Prove that $M$ is not self-projective.

    (ii) Prove that $M$ is epi-projective if and only if $\mathrm{End}(M/N_1) \simeq \mathrm{End}(M/N_2) \simeq \mathbb{Z}_2$.

(This is dual to [179, Lemma 2].)

(9) Prove that for a ring $R$ the following are equivalent (f.g. = finitely generated).

    (a) $R$ is left semihereditary (every f.g. left ideal is projective);

    (b) every f.g. submodule of a projective left $R$-module is self-projective;

    (c) every f.g. submodule of a (f.g.) projective left $R$-module is direct projective;

    (d) every f.g. (principal) left ideal of the ring $S = \mathrm{End}(R^n)$ is direct projective for all $n \geq 1$. (See [369].)

(10) Prove that for a ring $R$ the following are equivalent.

    (a) $R$ is left hereditary (every left ideal is projective);

    (b) every submodule of a projective left $R$-module is self-projective;

    (c) every submodule of a projective left $R$-module is direct projective;

    (d) every principal left ideal of the ring $S = \mathrm{End}(F)$ is direct projective, for any free $R$-module $F$. (See [369].)

(11) Let $R$ be a ring for which (finitely generated) submodules of self-projective left $R$-modules are self-projective. Prove that every factor ring of $R$ is left (semi-) hereditary ([122]).

(12) If $R$ is a ring over which submodules of $\Sigma$-self-projective modules are direct projective, then show that every factor ring of $R$ is hereditary. (See [369].)

(13) Prove that the following are equivalent for a ring $R$. (See [369].)

    (a) $R$ is left (and right) semisimple;

    (b) every (finitely generated) $R$-module is direct projective;

    (c) every 2-generated $R$-module is direct projective;

    (d) the direct sum of two direct projective $R$-modules is direct projective;

    (e) the direct sum of two self-projective $R$-modules is direct projective;

    (f) for all $n \geq 1$, every cyclic $\mathrm{End}(R^n)$-module is direct projective;

    (g) there exists some $n \geq 1$ such that every cyclic $\mathrm{End}(R^n)$-module is direct projective.

(14) Let $R$ be a left duo ring and $I$ be an ideal of $R$. Prove that $I$ is coclosed as a left ideal in $R$ if and only if $I$ is idempotent and every factor module of $_R I$ has a maximal submodule.

(15) Show that the coclosed ideals of a commutative noetherian ring are precisely the idempotent ideals.

(16) An epimorphism $p : P \to M$ is called a *quasi-projective cover* of the module $M$, if $P$ is self-projective (= quasi-projective), $\mathrm{Ke}\, p \ll P$, and for each submodule $0 \neq U \subset \mathrm{Ke}\, p$, $P/U$ is not self-projective. Prove ([121, 104]):

    (i) If $g : L \to M$ is a projective cover of $M$ in $R$-Mod (or in $\sigma[M]$), then there exists a submodule $U \subset \mathrm{Ke}\, g$ such that the induced map $g' : L/U \to M$ is a quasi-projective cover. (Notice that if $L$ is not in $\sigma[M]$, then $L/U$ need not be in $\sigma[M]$.)

    (ii) If $P \to M$ is a quasi-projective cover, then $\mathrm{An}_R(M) = \mathrm{An}_R(P)$.

(17) Prove that a self-projective module $M$ is Hopfian if and only if $M$ is not isomorphic to a proper direct summand of $M$ ([166]).

(18) Let $M$ be a self-projective module and $N \subset M$ be a fully invariant submodule with $N \ll M$. Prove that $M$ is Hopfian if and only if $M/N$ is Hopfian.

(19) For any field $K$ consider the ring $R = \begin{pmatrix} K & 0 \\ K[X] & K[X] \end{pmatrix}$.

Prove that every finitely generated left $R$-module is Hopfian ([130]).

(20) A ring $R$ is called *Hopfian* if every surjective ring homomorphism $R \to R$ is an isomorphism. Prove that for any vector space $V$ over a division ring $D$, the ring $\mathrm{End}_D(V)$ is Hopfian ([348]).

(21) A module $P$ is called *strongly $M$-projective* if it is $M^\Lambda$-projective for any index set $\Lambda$. Prove that this is equivalent to $P$ being projective as an $R/\mathrm{An}(M)$-module.

(22) Let $H_1, \ldots, H_n$ be hollow modules. Prove that $H_i$ and $\oplus_{j \neq i} H_j$ are relatively generalised projective if and only if they are relatively almost projective.

(23) For $R$-modules $M, N$ put $M \rightsquigarrow N$ if $\nabla(M, N') = 0$ for all factor modules $N'$ of $N$.

    (i) Show that if $M \rightsquigarrow N$, then $M$ is im-small $N$-projective.

(ii)  Let $M_i$, $i = 1, \ldots, n$, and $M$ be $R$-modules such that $M_i \rightsquigarrow M_j$ for $i \neq j$, and $M \rightsquigarrow M_i$ for all $i$. Show (by induction on $n$) that $M \rightsquigarrow \bigoplus_{i=1}^{n} M_i$.

(iii)  Let $A$ be a direct summand of $M$ that is strongly coclosed. Prove $A \rightsquigarrow M/A$.

**References.** Faticoni [104]; Golan [121, 122]; Goodearl [130]; Harada and Mabuchi [161]; Keskin [201]; Kuratomi [214]; Lomp [226]; Mohamed and Müller [245]; Talebi and Vanaja [330]; Tiwary and Pandeya [332]; Varadarajan [348]; Vasconcelos [349]; Wisbauer [363].

# 5 Hollow dimension of modules

In this section we consider a *dimension* for modules which is dual to the uniform (or Goldie) dimension. As the next result shows, the latter can be characterised in various ways (see, e.g., [85]). As before, $M$ will denote any left $R$-module.

**5.1. Uniform dimension.** *For $M$ the following are equivalent:*

(a) *$M$ does not contain an infinite independent set of submodules;*

(b) *for any ascending chain $N_1 \subset N_2 \subset \cdots$ of submodules of $M$ there exists $j \in \mathbb{N}$ such that for all $k \geq j$, $N_j \trianglelefteq N_k$.*

(c) *$\sup\{k \in \mathbb{N} \mid M$ contains $k$ independent submodules$\} = n$ for some $n \in \mathbb{N}$;*

(d) *$M$ contains independent uniform submodules $N_1, \ldots, N_n$ with $\bigoplus_{i=1}^n N_i \trianglelefteq M$;*

(e) *for some $n \in \mathbb{N}$, there exists an essential monomorphism from a direct sum of $n$ uniform modules to $M$.*

Notice that in this setting the natural numbers $n$ in (c), (d) and (e) are all the same and if the conditions are satisfied, $M$ is said to have *uniform* (or *Goldie*) *dimension $n$* and this is denoted by u.dim$(M) = n$. If $M = 0$ we set u.dim$(M) = 0$ and if $M$ does not have finite uniform dimension we set u.dim$(M) = \infty$.

Replacing uniform modules by hollow modules and dualising the corresponding conditions, we obtain the following characterisations of hollow dimension.

**5.2. Hollow dimension.** *For $M$ the following are equivalent:*

(a) *$M$ does not contain an infinite coindependent family of submodules;*

(b) *for any descending chain $N_1 \supset N_2 \supset \cdots$ of submodules of $M$ there exists $j$ such that for all $k \geq j$, $N_k \subseteq N_j$ is cosmall in $M$;*

(c) *$\sup\{k \in \mathbb{N} \mid M$ has $k$ coindependent submodules$\} = n$, for some $n \in \mathbb{N}$;*

(d) *$M$ has coindependent submodules $N_1, \ldots, N_n$ such that $\bigcap_{i=i}^n N_i \ll M$ and $M/N_i$ is hollow for every $1 \leq i \leq n$;*

(e) *for some $n \in \mathbb{N}$, there exists a small epimorphism from $M$ to a direct sum of $n$ hollow modules;*

(f) *for some $n \in \mathbb{N}$, there exists an epimorphism from $M$ to a direct sum of $n$ nonzero modules but no epimorphism from $M$ to a direct sum of more than $n$ nonzero modules.*

*Proof.* (a)$\Rightarrow$(d). Note that a coindependent family of submodules of $M$ can be actually seen as a subset of the lattice of submodules of $M$, since pairwise elements of such a family are proper comaximal submodules of $M$ and cannot be equal. Hence the set $\mathcal{M}$ of all coindependent families of submodules of $M$ is partially ordered by set-theoretic inclusion inherited from the set of submodules of $M$. Also the subset

$$\mathcal{M}_h := \{X \in \mathcal{M} \mid \text{for all } N \in X, M/N \text{ is hollow}\} \subseteq \mathcal{M}$$

is partially ordered by set inclusion and contains the union of each of its chains. Thus, by Zorn's Lemma, there exists a maximal element $X \in \mathcal{M}_h$. By (a) $X$ is finite, say $X = \{N_1, \ldots, N_n\}$. To establish (d), it suffices to show that $N_1 \cap \cdots \cap N_n$ is small in $M$. Assume that $(N_1 \cap \cdots \cap N_n) + K = M$ for some proper submodule $K$ of $M$. By Lemma 2.16, there exists a proper submodule $L$ containing $K$ such that $M/L$ is hollow. Obviously the set $\{N_1, \ldots, N_n, L\}$ is also coindependent, contradicting the maximality of $X$. Hence $N_1 \cap \cdots \cap N_n$ is small in $M$, as required.

(d)$\Rightarrow$(c). Let $\{L_1, \ldots, L_k\}$ be a coindependent family of $M$. By Lemma 3.5, $k \leq n$. Thus every coindependent family of $M$ has at most $n$ elements.

(c)$\Rightarrow$(b). If (b) is not satisfied, then there is a chain $M \neq N_1 \supset N_2 \supset \cdots$ of submodules of $M$ such that for any $j \geq 1$ there exists an integer $k(j) > j$ such that $N_j/N_{k(j)}$ is not small in $M/N_{k(j)}$. Let $\{j_m\}_{m \in \mathbb{N}}$ be the sequence of indices defined as follows: $j_1 := 1$ and $j_m := k(j_{m-1})$ for all $m > 1$. By the foregoing, for each $m$ there exists a proper submodule $K_{j_m}$ such that $N_{j_{m+1}} \subset K_{j_m}$ with $N_{j_m} + K_{j_m} = M$. Thus

$$(N_1 \cap K_{j_1} \cap \cdots \cap K_{j_{m-1}}) + K_{j_m} \supseteq N_{j_m} + K_{j_m} = M$$

for all $m > 1$. Hence, by 1.28, we get that $\{N_1, K_{j_1}, K_{j_2}, \ldots, K_{j_m}, \ldots\}$ is coindependent. This contradicts (c).

(b)$\Rightarrow$(a). If (a) is not satisfied, then $M$ contains an infinite coindependent family of submodules $\{N_1, N_2, \ldots\}$. Then $N_1 \supset N_1 \cap N_2 \supset N_1 \cap N_2 \cap N_3 \supset \cdots$ and, for any $k \in \mathbb{N}$, $(N_1 \cap \cdots \cap N_k) + N_{k+1} = M$. From this it follows that $(N_1 \cap \cdots \cap N_k)/(N_1 \cap \cdots \cap N_l)$ is not small in $M/(N_1 \cap \cdots \cap N_l)$ for all $l > k$, contradicting (b).

(d)$\Rightarrow$(e). Let $N_1, \ldots, N_n$ be given as in (d). Then the projection yields

$$M \longrightarrow M/\bigcap_{i=1}^n N_i \simeq \bigoplus_{i=1}^n M/N_i \,,$$

where each factor module $M/N_i$ is hollow.

(e)$\Rightarrow$(f). Let $M \to \bigoplus_{i=1}^n M_i$ be a small epimorphism with each $M_i$ hollow. Then there exists a coindependent family $\{N_1, \ldots, N_n\}$ such that $M_i \simeq M/N_i$ and $\bigcap_{i=1}^n N_i \ll M$. Let $M \to \bigoplus_{i=1}^k M_i'$ be an epimorphism of nonzero submodules and let $\{L_1, \ldots, L_k\}$ be a coindependent family of submodules such that $M_i' \simeq M/L_i$. By Lemma 3.5, $k \leq n$.

(f)$\Rightarrow$(d) is clear by the Chinese Remainder Theorem 1.29.

(c)$\Rightarrow$(a) is trivial.                                                                $\square$

Similar to 5.1, the natural numbers $n$ in (c)–(f) of 5.2 are all equal and if $M$ satisfies these conditions we say that $M$ has *hollow dimension* $n$, denoting this by h.dim$(M) = n$.

If $M = 0$ we put $\mathrm{h.dim}(M) = 0$ and if $M$ does not have finite hollow dimension we set $\mathrm{h.dim}(M) = \infty$.

Obviously every artinian module has finite hollow dimension. Clearly a module $M$ is hollow if and only if $\mathrm{h.dim}(M) = 1$.

**5.3. Coclosed modules and hollow dimension.** *A module $M$ with finite hollow dimension satisfies ACC and DCC on coclosed submodules.*

*Proof.* Let $N_1 \supset N_2 \supset \cdots$ be a descending chain of coclosed submodules of $M$. Then, since $M$ has finite hollow dimension, by 5.2 (b) there is an integer $i$ such that $N_k \subset N_i$ is cosmall in $M$ for every $k \geq i$. Since $N_i$ is coclosed, $N_k = N_i$ for all $k \geq i$.

Now assume that $0 =: P_0 \subset P_1 \subset P_2 \subset \cdots$ is a strictly ascending chain of coclosed submodules of $M$. Then, since $P_{k-1} \subset P_k$ is not cosmall in $M$ for all $k > 0$, it follows by 5.2 that $M$ has infinite hollow dimension. $\square$

In general the converse of 5.3 need not hold. To see this, recall that the only coclosed ideals of a principal ideal domain $R$ are 0 and $R$. Hence any principal ideal domain $R$ with infinitely many maximal ideals is an example of a ring satisfying ACC and DCC on coclosed ideals, but does not have hollow dimension since it is not semilocal (as shown later in 18.7). Of course, $R = \mathbb{Z}$ is such an example.

Let $N \subset M$ be a submodule and $p : M \to M/N$ the canonical projection. Let $g : M/N \to \bigoplus_{i=1}^{k} N_i$ be an epimorphism with $N_i \neq 0$, yielding the diagram

$$0 \longrightarrow N \longrightarrow M \overset{p}{\longrightarrow} M/N \longrightarrow 0$$
$$\downarrow g$$
$$\bigoplus_{i=1}^{k} N_i.$$

Thus there is an epimorphism from $M$ to a direct sum of $k$ nonzero modules and hence $\mathrm{h.dim}(M) \geq \mathrm{h.dim}(M/N)$. This shows that the hollow dimension of a factor module $M/N$ never exceeds the hollow dimension of $M$. In particular, if $\mathrm{h.dim}(M/N) = \infty$, then $\mathrm{h.dim}(M) = \infty$.

Now assume that $\mathrm{h.dim}(M/N) = k$ and $N \ll M$. Then, by 5.2 (e), we may choose all the $N_i$ to be hollow. Moreover, $\mathrm{Ke}\, pg \subset N$ and so $pg$ is a small epimorphism. Hence, again by 5.2 (e), $\mathrm{h.dim}(M) = k = \mathrm{h.dim}(M/N)$. Thus $\mathrm{h.dim}(M) = \mathrm{h.dim}(M/N)$ whenever $N \ll M$.

**5.4. Properties of hollow dimension.** *Let $N$ be a submodule of $M$.*

(1) *If $M = M_1 \oplus \cdots \oplus M_k$, then $\mathrm{h.dim}(M) = \mathrm{h.dim}(M_1) + \cdots + \mathrm{h.dim}(M_k)$.*

(2) *If $N \ll M$, then $\mathrm{h.dim}(M) = \mathrm{h.dim}(M/N)$; conversely, if $\mathrm{h.dim}(M)$ is finite and equal to $\mathrm{h.dim}(M/N)$, then $N \ll M$.*

(3) *If* h.dim$(M) < \infty$, *then any epimorphism* $f : M \to M$ *is small.*
    *If moreover* $M$ *is semi-Hopfian, then* $f$ *is an isomorphism.*

*Proof.* (1) Clearly h.dim$(M) \geq$ h.dim$(M_i)$. Thus if h.dim$(M_i) = \infty$ for some $i$, then h.dim$(M) = \infty$. Now assume h.dim$(M_i) = n_i < \infty$ for each $i \leq k$. Then, by 5.2(e), there exist small epimorphisms $f_i : M_i \to \bigoplus_{j=1}^{n_i} H_{ij}$ with $H_{ij}$ hollow for all $1 \leq j \leq n_i$. These yield a small epimorphism $f = (f_1, \ldots, f_k)$,

$$M \xrightarrow{f} \bigoplus_{i=1}^{k} \left( \bigoplus_{j=1}^{n_i} H_{ij} \right) \longrightarrow 0.$$

Thus h.dim$(M) =$ h.dim$(M_1) + \cdots +$ h.dim$(M_k)$.

(2) The first part was pointed out above.

Now let h.dim$(M) =$ h.dim$(M/N) = n < \infty$ and $f : M \to \bigoplus_{i=1}^{n} M/K_i$ be a small epimorphism with each $M/K_i$ hollow. Assume that $N \not\ll M$. Then $N + K_i = M$ for some $i$, by 3.4, and so $M/(N \cap K_i) \simeq M/N \oplus M/K_i$. Thus, in view of (1),

$$\text{h.dim}(M) \geq \text{h.dim}(M/N) + \text{h.dim}(M/K_i) > \text{h.dim}(M/N) = \text{h.dim}(M),$$

a contradiction. Hence $N \ll M$.

(3) Since $M$ has finite hollow dimension and h.dim$(M) =$ h.dim$(\text{Im } f) =$ h.dim$(M/\text{Ke } f)$, we get from (2) that Ke$f \ll M$. If $M$ is semi-Hopfian, then Ke$f$ is a direct summand and hence 0.                                               □

Let h.dim$(M) = n$ be finite and $\{K_1, \ldots, K_n\}$ be a maximal coindependent family of submodules of $M$. If $N \ll M$, then $L_i := N + K_i$ is a proper submodule of $M$ for each $i \leq n$. Hence, by 1.28 (2), $\{L_1, \ldots, L_n\}$ is also a maximal coindependent family in $M$ and $N \subseteq L_1 \cap \cdots \cap L_n$. Thus every small submodule $N$ of a module $M$ with finite hollow dimension is contained in an intersection $L_1 \cap \cdots \cap L_n \ll M$ where $\{L_1, \ldots, L_n\}$ is a coindependent family of submodules in $M$ (see [295]).

By 2.16, any nonzero module with finite hollow dimension has a hollow factor module. Thus any nonzero artinian module has a hollow factor module.

Recall that a module $M$ is called *conoetherian* if every finitely cogenerated module in $\sigma[M]$ is artinian (see [363, 31.6]). Since every nonzero module has a nonzero cocyclic factor module we conclude that any nonzero module in $\sigma[M]$ has a hollow factor module if $M$ is conoetherian. Consequently, the same holds for all nonzero modules over a conoetherian ring.

**5.5. Hollow factor modules.** *Let $M$ be such that each of its nonzero factor modules has a hollow factor module. Then $M$ contains a coindependent family $\{K_\lambda\}_\Lambda$ of submodules such that $M/K_\lambda$ is hollow for every $\lambda \in \Lambda$ and $\bigcap_\Lambda K_\lambda$ is small in $M$.*

*Proof.* Let $\mathcal{M}_h$ be as in the proof of 5.2 and choose $\{K_\lambda\}_\Lambda$ to be a maximal coindependent family of proper submodules, such that $M/K_\lambda$ is hollow for every

$\lambda \in \Lambda$. Let $K = \bigcap_\Lambda K_\lambda$. If $K$ is not small in $M$, then there is a proper submodule $L$ of $M$ such that $K + L = M$. By hypothesis, $M/L$ has a hollow factor module $M/N$. Then $L \subseteq N$ and so $K + N = M$. By 1.28 (2), $\{K_\lambda\}_\Lambda \cup \{N\}$ is a coindependent family and hence it is an element of $\mathcal{M}$, a contradiction to the maximality of $\{K_\lambda\}_\Lambda$. Hence $K$ must be small, as required. $\qquad\square$

From 5.5 it follows that, for any conoetherian module $M$, every nonzero module $N \in \sigma[M]$ contains a maximal coindependent family $\{K_\lambda\}_\Lambda$ of submodules such that $N/K_\lambda$ is hollow for every $\lambda \in \Lambda$ and $\bigcap_\Lambda K_\lambda$ is small in $N$.

**5.6. Duality and hollow dimension.** We now illustrate the duality existing between uniform and hollow dimensions by considering *annihilator conditions* on $M$ and $\mathrm{Hom}(M, Q)$, where $Q$ is an *injective cogenerator* in $\sigma[M]$.

Let $T = \mathrm{End}(_R Q)$ and for $N \in \sigma[M]$ consider $N^* = \mathrm{Hom}(N, Q)$ as a right $T$-module. For any $R$-submodule $K \subseteq N$ and $T$-submodule $X \subseteq N^*$, define

$$
\begin{aligned}
\mathrm{An}(K) &= \{f \in N^* \mid (K)f = 0\} \subseteq N^*, \\
\mathrm{Ke}\,(X) &= \bigcap \{\mathrm{Ke}\, g \mid g \in X\} \subseteq N.
\end{aligned}
$$

Then obviously $\mathrm{An}(K_1 + K_2) = \mathrm{An}(K_1) \cap \mathrm{An}(K_2)$ for all $K_1, K_2 \subseteq N$, and, since $_R Q$ is an injective cogenerator in $\sigma[M]$ we have (by [363, 28.1])

(AC1)  $\mathrm{Ke}\,(\mathrm{An}(K)) = K$ for all $K \subseteq N$;

(AC2)  $\mathrm{An}(\mathrm{Ke}\,(X)) = X$ for every finitely generated $T$-submodule $X \subseteq N^*$;

(AC3)  $\mathrm{An}(K_1 \cap K_2) - \mathrm{An}(K_1) + \mathrm{An}(K_2)$ for all $K_1, K_2 \subseteq N$.

Using these identities, we now prove the

**5.7. Dual dimension formula.** *Let $M$ be an $R$-module, $_R Q$ an injective cogenerator in $\sigma[M]$, and $T = \mathrm{End}(_R Q)$. For any module $N \in \sigma[M]$,*

$$
\mathrm{h.dim}(_R N) = \mathrm{u.dim}(\mathrm{Hom}(N, Q)_T).
$$

*Proof.* Assume that $H_1, \ldots, H_k$ are nonzero factor modules of $N$ allowing an exact sequence

$$
N \longrightarrow \bigoplus_{i=1}^k H_i \longrightarrow 0.
$$

The functor $\mathrm{Hom}(-, Q)$ yields the exact sequence

$$
0 \longrightarrow \bigoplus_{i=1}^k \mathrm{Hom}(H_i, Q) \longrightarrow N^*
$$

where, as above, $N^* = \mathrm{Hom}(N, Q)$. Moreover, each $\mathrm{Hom}(H_i, Q)$ is a nonzero submodule of $N^*$ since $H_i$ is nonzero and $Q$ is a cogenerator in $\sigma[M]$. Hence $N^*$ contains a direct sum of $k$ submodules and so $\mathrm{h.dim}(_R N) \leq \mathrm{u.dim}(N_T^*)$.

Now assume that $N^*$ contains a $T$-submodule $X$ which is a direct sum of $k$ nonzero (cyclic) submodules, that is, $X = f_1 T \oplus \cdots \oplus f_k T$ with $0 \neq f_i \in N^*$. Obviously $\mathrm{Ke}\,(f_i T) = \mathrm{Ke}\, f_i$ is a proper submodule of $N$ for every $1 \leq i \leq k$. We show that $\{\mathrm{Ke}\, f_1, \ldots, \mathrm{Ke}\, f_k\}$ form a coindependent family of submodules of $N$. Applying (AC1)–(AC3) we get, for all $1 \leq i \leq k$,

$$
\begin{aligned}
0 &= f_i T \cap \textstyle\sum_{j \neq i} f_j T \\
&= \mathrm{An}(\mathrm{Ke}\,(f_i T)) \cap \mathrm{An}(\mathrm{Ke}\,(\textstyle\sum_{j \neq i} f_j T)) \quad \text{by (AC2)} \\
&= \mathrm{An}(\mathrm{Ke}\, f_i + \mathrm{Ke}\,(\textstyle\sum_{j \neq i} f_j T)) \\
&= \mathrm{An}(\mathrm{Ke}\, f_i + \mathrm{Ke}\,(\textstyle\sum_{j \neq i} \mathrm{An}(\mathrm{Ke}\, f_j))) \quad \text{by (AC2)} \\
&= \mathrm{An}(\mathrm{Ke}\, f_i + \mathrm{Ke}\,(\mathrm{An}(\textstyle\bigcap_{j \neq i} \mathrm{Ke}\, f_j))) \quad \text{by (AC3)} \\
&= \mathrm{An}(\mathrm{Ke}\, f_i + \textstyle\bigcap_{j \neq i} \mathrm{Ke}\, f_j) \quad \text{by (AC1)}.
\end{aligned}
$$

Applying (AC1) yields

$$
N = \mathrm{Ke}\,(0) = \mathrm{Ke}\,(\mathrm{An}(\mathrm{Ke}\, f_i + \bigcap_{j \neq i} \mathrm{Ke}\, f_j)) = \mathrm{Ke}\, f_i + \bigcap_{j \neq i} \mathrm{Ke}\, f_j.
$$

Hence $\{\mathrm{Ke}\, f_1, \ldots, \mathrm{Ke}\, f_k\}$ is coindependent and so $\mathrm{u.dim}(N_T^*) \leq \mathrm{h.dim}(_R N)$.   □

**5.8. Morita invariance of hollow dimension.** Assume $M$ to be a finitely generated projective generator in $\sigma[M]$ and put $S = \mathrm{End}(M)$. Then the functors

$$
\mathrm{Hom}_R(M, -) : \sigma[M] \to S\text{–Mod}, \qquad M \otimes_S - : S\text{–Mod} \to \sigma[M]
$$

define an equivalence of categories (e.g., [363, 46.2]). From this it follows easily that under this condition, for every $N \in \sigma[M]$ and left $S$-module $X$,

$$
\mathrm{h.dim}(N) = \mathrm{h.dim}(_S \mathrm{Hom}_R(M, N)) \quad \text{and} \quad \mathrm{h.dim}(X) = \mathrm{h.dim}(M \otimes_S X).
$$

**5.9. Exercise.**

Let $M$ be an $R$-module admitting an ascending chain of submodules

$$
0 =: N_0 \subset N_1 \subset N_2 \subset N_3 \subset \cdots,
$$

such that, for no $k \geq 1$, $N_{k-1} \subset N_k$ is cosmall in $M$. Prove that $M$ contains an infinite coindependent family of submodules ([299, Theorem 5]). (Hint: 5.2 (c)).

**5.10. Comments.** An early attempt to define the dual of Goldie dimension was made in Fleury [106] in 1974. He considered the *spanning dimension* of a module and introduced *hollow* modules as the duals of the uniform modules appearing in Goldie's work. Fleury's spanning dimension dualises chain condition 5.1(b) by asking for any descending chain $N_1 \supset N_2 \supset \cdots$ of submodules of $M$ the existence of some $k \in \mathbb{N}$ such that $N_k \ll M$. Modules with finite spanning dimension are closely related to artinian modules and projective modules with finite spanning dimension are either local or artinian ([293, Proposition 3.5]).

Takeuchi [327] introduced coindependent families of submodules as duals to independent families of submodules in 1975. With this notion he dualised 5.1(a) and defined *cofinite-dimensionality* (as a dual to Goldie dimension) by property 5.2(a). His definition was based on the paper Miyashita [235] from 1966, where a dimension notion for modular lattices was introduced.

In [344] Varadarajan approached the dualisation of the Goldie dimension (in 1979) in a more categorical way by dualising 5.1(e). He defined the *corank* of a module by property 5.2(f). Comparing his definition with the one given by Fleury he showed that "finite spanning dimension" implies "finite corank". Reiter [295] introduced a dual of the Goldie dimension (in 1981) as *codimension* of a module by property 5.2(b) and he dualised property 5.1(b) in the same way as Fleury.

In 1984 Grzeszczuk and Puczyłowski [135] compared all the previous approaches and proved 5.2. They defined the Goldie dimension of a modular lattice and from this they derived the dual Goldie dimension of a modular lattice as the Goldie dimension of the dual lattice. They then applied these notions to the lattice of submodules of a module and found that modules with finite spanning dimension also satisfy their definition of dual Goldie dimension.

Theorem 5.3 is taken from Takeuchi's [327]. Properties 5.4 were shown by several authors. The duality between Goldie and dual Goldie dimension provided by an injective cogenerator was first observed in Page [281].

**References.** Fleury [106]; Grzeszczuk and Puczyłowski [135]; Hanna and Shamsuddin [146, 147]; Miyashita [235]; Page [281]; Park and Rim [283]; Rangaswamy [293]; Reiter [295]; Rim and Takemori [299]; Sarath and Varadarajan [307]; Takeuchi [327]; Varadarajan [344, 345].

# Chapter 2

# Preradicals and torsion theories

## 6 Preradicals and colocalisation

In this section we recall some fundamental facts from preradical theory and give an outline on colocalisation. Throughout the section let $M$ be a left $R$-module. The following properties of module classes will be of interest.

**6.1. Definitions.** A class $\mathbb{A}$ of modules in $\sigma[M]$
is said to be    if it is closed under
     *hereditary*    submodules,
  *cohereditary*    factor modules,
     *pretorsion*    direct sums and quotients,
       *torsion*    direct sums, quotients and extensions,
  *torsion free*    direct products, submodules, and extensions.

**6.2. Definitions.** A $\tau$ *for* $\sigma[M]$ is a subfunctor of the identity functor, that is,

(i) $\tau(N) \subseteq N$ for each $N \in \sigma[M]$, and

(ii) for each morphism $f : N \to N'$, $\tau(f) = f\mid_{\tau(N)}\colon \tau(N) \to \tau(N')$.

Any preradical $\tau$ for $\sigma[M]$ induces an endofunctor

$$\tau^{-1} : \sigma[M] \to \sigma[M], \quad N \mapsto N/\tau(N).$$

Notice that for all $N$, $\tau(N) \subseteq N$ is a fully invariant submodule of $N$. Moreover, for any submodule $K \subset L$, and modules $\{N_\lambda\}_\Lambda$ in $\sigma[M]$,

(1) if $\tau(K) = K$, then $K \subseteq \tau(L)$;

(2) if $\tau(L/K) = 0$, then $\tau(L) \subseteq K$;

(3) $\tau(\bigoplus_\Lambda N_\lambda) = \bigoplus_\Lambda \tau(N_\lambda)$;

(4) $\tau(\prod_\Lambda N_\lambda) \subseteq \prod_\Lambda \tau(N_\lambda)$.

**6.3. Associated module classes.** To any preradical $\tau$ for $\sigma[M]$, we associate the module classes

$$\mathbb{T}_\tau = \{N \in \sigma[M] \mid \tau(N) = N\}, \quad \mathbb{F}_\tau = \{N \in \sigma[M] \mid \tau(N) = 0\}.$$

$\mathbb{T}_\tau$ is called the *preradical* (or *(pre)torsion class*) of $\tau$; it is a pretorsion class as defined in 6.1. $\mathbb{F}_\tau$ is referred to as the *preradical free* or *(pre)torsion free class* of $\tau$; it is closed under direct products and submodules.

It follows from the definitions that $\mathbb{T}_\tau \cap \mathbb{F}_\tau = 0$, and $\mathrm{Hom}(K, L) = 0$ for all $K \in \mathbb{T}_\tau$ and $L \in \mathbb{F}_\tau$.

**6.4. Definitions.** A preradical $\tau$ for $\sigma[M]$ is said to be
 *idempotent* if $\tau(\tau(N)) = \tau(N)$, for each $N \in \sigma[M]$,
 *a radical* if $\tau(N/\tau(N)) = 0$, for each $N \in \sigma[M]$.

For any $P \in \sigma[M]$, the trace $\mathrm{Tr}(P, -) : \sigma[M] \to \sigma[M]$ (see 1.1) defines a preradical for $\sigma[M]$ which is always idempotent, and it is a radical provided $P$ is projective in $\sigma[M]$.

**6.5. Definitions.** Given any class $\mathbb{A}$ of modules in $\sigma[M]$, we define a *preradical* $\tau^{\mathbb{A}}$ by setting for each $N \in \sigma[M]$,

$$\tau^{\mathbb{A}}(N) = \mathrm{Tr}(\mathbb{A}, N) = \sum\{\mathrm{Im}\, f \mid f : A \to N, \, A \in \mathbb{A}\},$$

and a *preradical* $\tau_{\mathbb{A}}$ is defined by setting for each $N \in \sigma[M]$,

$$\tau_{\mathbb{A}}(N) = \mathrm{Re}(N, \mathbb{A})) = \bigcap\{\mathrm{Ke}\, f \mid f : N \to A, \, A \in \mathbb{A}\}.$$

Well-known examples are obtained by taking $\mathbb{A}$ to be the class of all simple modules; then $\tau^{\mathbb{A}}(N) = \mathrm{Soc}\,(N)$ and $\tau_{\mathbb{A}}(N) = \mathrm{Rad}\,(N)$ for any $N \in \sigma[M]$.

**6.6. Proposition.** *Let $\mathbb{A}$ be a class of modules in $\sigma[M]$.*

(1) *If $\mathbb{A}$ is a pretorsion class, then $\tau^{\mathbb{A}}$ is an idempotent preradical.*

(2) *If $\mathbb{A}$ is a torsion class, then $\tau^{\mathbb{A}}$ is a radical.*

(3) *If $\mathbb{A}$ is a hereditary pretorsion class, then for all $K \subseteq N$,*

$$\tau^{\mathbb{A}}(K) = K \cap \tau^{\mathbb{A}}(N).$$

*Proof.* See [41, Chapters I, II] or [323, Chapter VI].      $\square$

**6.7. Preradical generated by singular modules.** Let $\mathcal{C}_M^{\trianglelefteq}$ denote the class of singular modules in $\sigma[M]$. It is closed under submodules, factor modules and direct sums, that is, it is a hereditary pretorsion class. The *M-singular submodule* of any $N$ in $\sigma[M]$ is defined as

$$\mathrm{Z}_M(N) = \mathrm{Tr}(\mathcal{C}_M^{\trianglelefteq}, N) = \sum\{U \subset N \mid U \text{ is } M\text{-singular}\}.$$

If $\mathrm{Z}_M(N) = 0$ then $N$ is called *non-M-singular*. Furthermore, $\mathrm{Z}_M(N) = N$ holds if and only if $N$ is $M$-singular. The induced functor $\mathrm{Z}_M$ is a hereditary preradical but need not be a radical (see [85], [364]).

If $M = R$ we simply write $\mathrm{Z}(N)$ instead of $\mathrm{Z}_R(N)$; clearly $\mathrm{Z}_M(N) \subseteq \mathrm{Z}(N)$.

**6.8. Proposition.** *Let $\tau$ be a preradical for $\sigma[M]$ with associated classes $\mathbb{T}_\tau$ and $\mathbb{F}_\tau$ (see 6.3).*

(1) *The following are equivalent:*

 (a) *$\tau$ is idempotent;*

 (b) *for any $N \in \sigma[M]$, $\tau(N) = \mathrm{Tr}(\mathbb{T}_\tau, N)$;*

 (c) *$\mathbb{F}_\tau$ is closed under extensions.*

(2) *The following are equivalent:*

 (a) *$\tau$ is a radical;*

 (b) *for any $N \in \sigma[M]$, $\tau(N) = \bigcap\{K \subseteq N \mid N/K \in \mathbb{F}_\tau\} = \mathrm{Re}(N, \mathbb{F}_\tau)$;*

 (c) *$\mathbb{T}_\tau$ is closed under extensions.*

*Proof.* See [41, Chapter I, II] or [323, Chapter VI]. □

Any preradical $\tau$ is a functor that preserves monomorphisms, and the functor $\tau^{-1}$ preserves epimorphisms. In general these functors need not be exact.

**6.9. Proposition.** *Let $\tau$ be a preradical for $\sigma[M]$.*

(1) *The following are equivalent:*

 (a) *$\tau$ is left exact;*

 (b) *for any $K \subseteq N \in \sigma[M]$, $\tau(K) = K \cap \tau(N)$;*

 (c) *$\tau$ is idempotent and $\mathbb{T}_\tau$ is closed under submodules.*

 In this case $\tau$ is said to be *hereditary.*

(2) *The following are equivalent:*

 (a) *$\tau$ preserves epimorphisms;*

 (b) *$\tau^{-1}$ is right exact;*

 (c) *for any $K \subseteq N \in \sigma[M]$, $\tau(N/K) = (\tau(N) + K)/K$;*

 (d) *$\tau$ is a radical and $\mathbb{F}_\tau$ is closed under quotients.*

 In this case $\tau$ is said to be *cohereditary.*

*Proof.* We only prove (2), the proof of (1) being similar.

(a)⇔(b) is obvious.

(c)⇒(d). Condition (c) implies $\tau(N/\tau(N)) = (\tau(N) + \tau(N))/\tau(N) = 0$, that is, $\tau$ is a radical. Let $L \in \mathbb{F}_\tau$. Then $\tau(L) = 0$ and for any $K \subset L$, $\tau(L/K) = (\tau(L) + K)/K = 0$, thus $L/K \in \mathbb{F}_\tau$.

(d)⇒(c). Clearly $(\tau(N) + K)/K \subseteq \tau(N/K)$. By (d), $\tau(N/(\tau(N) + K)) = 0$ and hence $\tau(N/K) \subseteq (\tau(N) + K)/K$.

(b)⇔(d). Consider the commutative diagram in $\sigma[M]$ with exact middle row

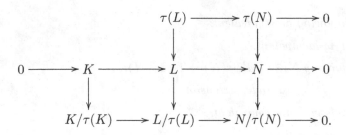

It is not difficult to see (e.g., [234, 2.9]) that the first row is exact if and only if the bottom row is exact (in $L/\tau(L)$, it is always exact in $N/\tau(N)$).                    □

**6.10. Proposition.** *Let $\tau$ be a cohereditary preradical for $\sigma[M]$ with associated module classes $\mathbb{T}_\tau$ and $\mathbb{F}_\tau$.*

(1) *If $K \subseteq N$ such that $K \in \mathbb{F}_\tau$ and $N \in \mathbb{T}_\tau$, then $K \ll N$.*

(2) *If $f : N \to L$ where $N \in \mathbb{T}_\tau$ and $\mathrm{Ke}\, f \in \mathbb{F}_\tau$, then $\mathrm{Ke}\, f \ll N$.*

(3) *The class $\mathbb{T}_\tau$ is closed under small epimorphisms in $\sigma[M]$.*

*Proof.* (1) Let $K + H = N$ for some $H \subset N$. Since $\mathbb{T}_\tau$ is closed under quotients and $\tau$ is a cohereditary preradical, we have

$$N/H = \tau(N/H) = \tau((K+H)/H) = \tau(K/(H \cap K)) = 0$$

since $K \in \mathbb{F}_\tau$ and hence $\tau(K) = 0$. Thus $N = H$ and $K \ll N$.

(2) follows from (1).

(3) Let $K \ll L$ for $L \in \sigma[M]$ and assume $\tau(L/K) = L/K$. Then

$$L/K = \tau(L/K) = (\tau(L) + K)/K.$$

This implies $L = \tau(L) + K = \tau(L)$, that is, $L \in \mathbb{T}_\tau$.                    □

There is a generalisation of projectivity in $\sigma[M]$ which is of interest in the context of cohereditary preradicals.

**6.11. Definition.** An $R$-module $P$ is called *pseudo-projective in $\sigma[M]$* if any diagram in $\sigma[M]$ with exact row and nonzero $f$

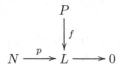

can be non-trivially extended by some $s \in \mathrm{End}(P)$ and $g : P \to N$ to a commutative diagram

$$\begin{array}{ccc} P & \xrightarrow{\;s\;} & P \\ {\scriptstyle g}\downarrow & & \downarrow{\scriptstyle f} \\ N & \xrightarrow{\;p\;} & L \longrightarrow 0, \end{array}$$

that is, $g \diamond p = s \diamond f \neq 0$.

An epimorphism $p : P \to N$ is a *pseudo-projective cover* of $N$ in $\sigma[M]$ if $P$ is pseudo-projective in $\sigma[M]$ and $\mathrm{Ke}\, p \ll P$.

Interesting features of these modules are given by properties of the related trace functor.

**6.12. Pseudo-projective modules in $\sigma[M]$.** *For $P \in \sigma[M]$ the following assertions are equivalent:*

(a) *$P$ is pseudo-projective in $\sigma[M]$;*

(b) *the trace functor $\mathrm{Tr}(P, -) : \sigma[M] \to \sigma[M]$ preserves epimorphisms;*

(c) *for any $N \in \sigma[M]$ and $K \subset N$, $\mathrm{Tr}(P, N/K) = (\mathrm{Tr}(P, N) + K)/K$;*

(d) *$\mathrm{Tr}(P, N/\mathrm{Tr}(P, N)) = 0$ for all $N \in \sigma[M]$, and the class $\{X \in \sigma[M] \mid \mathrm{Tr}(P, X) = 0\}$ is closed under factor modules;*

(e) *the functor $\sigma[M] \to \sigma[M]$ induced by $N \mapsto N/\mathrm{Tr}(P, N)$ is right exact.*

*Proof.* The equivalence of (b), (c), (d) and (e) follows from 6.9.

(a)$\Rightarrow$(d). Assume that for some $N \in \sigma[M]$ there is a nonzero homomorphism $f : P \to N/\mathrm{Tr}(P, N)$. Consider the diagram with canonical projection $p$,

$$\begin{array}{c} P \\ \downarrow{\scriptstyle f} \\ N \xrightarrow{\;p\;} N/\mathrm{Tr}(P, N) \longrightarrow 0. \end{array}$$

By (a), there are $s : P \to P$ and $g : P \to N$ such that $s \diamond f = g \diamond p \neq 0$. However, since $\mathrm{Im}\, g \subset \mathrm{Tr}(P, N)$, we always have $g \diamond p = 0$, a contradiction.

A similar argument shows that the given class is closed under factor modules.

(b)$\Rightarrow$(a). Consider any diagram in $\sigma[M]$ with exact row and nonzero $f$

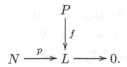

Forming the pullback and applying the trace functor yields the commutative diagram with exact rows

$$
\begin{array}{ccc}
\mathrm{Tr}(P,Q) \xrightarrow{\ s'\ } & P & \longrightarrow 0 \\
\ \ \downarrow{\scriptstyle g'} & \ \ \downarrow{\scriptstyle f} & \\
\mathrm{Tr}(P,N) \xrightarrow{\ p\ } & \mathrm{Tr}(P,L) & \longrightarrow 0.
\end{array}
$$

Since $\mathrm{Tr}(P,Q)$ is $P$-generated, there exists some $t : P \to \mathrm{Tr}(P,Q)$ such that $t \diamond s' \diamond f = t \diamond g' \diamond p \neq 0$. Thus $s = t \diamond s' : P \to P$ and $g = t \diamond g' : P \to N$ are the morphisms required to make $P$ pseudo-projective. $\qquad\square$

**6.13. Cohereditary preradicals and pseudo-projectives.** *For a preradical $\tau$ for $\sigma[M]$, the following are equivalent:*

(a) *$\tau$ is cohereditary;*

(b) *there is a pseudo-projective module $P \in \sigma[M]$ such that $\mathbb{T}_\tau = \mathrm{Gen}(P)$.*

*Proof.* (b)$\Rightarrow$(a) follows from 6.12.

(a)$\Rightarrow$(b). Let $\tau$ be cohereditary and $G$ a generator in $\sigma[M]$. Then for any $N \in \mathbb{T}_\tau$ there is an epimorphism $G^{(\Lambda)} \to N$, for some index set $\Lambda$. This induces an epimorphism $\tau(G)^{(\Lambda)} \to N$ and hence $P = \tau(G)$ is a generator for $\mathbb{T}_\tau$, that is, $\mathbb{T}_\tau = \mathrm{Gen}(P)$. Then $\tau = \mathrm{Tr}(P,-)$ preserves epimorphisms and hence $P$ is pseudo-projective in $\sigma[M]$ by 6.12. $\qquad\square$

Since the module ${}_R R$ is a generator in $R$-Mod, any cohereditary preradical $\tau$ for $R$-Mod is generated by the ideal $\tau(R)$.

**6.14. Cohereditary preradicals in $R$-Mod.** *Let $\tau$ be a preradical for $R$-Mod. Then the following are equivalent:*

(a) *$\tau$ preserves epimorphisms;*

(b) *$\tau^{-1}$ is a right exact functor;*

(c) *for any $N \in R$-Mod, $\tau(N) = \tau(R)N$;*

(d) *$\tau$ is a radical and $\mathbb{F}_\tau$ is cohereditary.*

In particular, pseudo-projective modules $P$ in $R$-Mod can be characterised by their trace ideals $\mathrm{Tr}(P,R)$ and in this case 6.12 reads as follows.

**6.15. Pseudo-projective modules in $R$-Mod.** *For an $R$-module $P$ with trace ideal $T = \mathrm{Tr}(P,R)$, the following are equivalent:*

(a) *$P$ is pseudo-projective in $R$-Mod;*

(b) *the trace functor $\mathrm{Tr}(P,-) : R$-Mod $\to R$-Mod preserves epimorphisms;*

(c) *for any (free) $R$-module $N$ and $K \subset N$, $\mathrm{Tr}(P,N/K) = (\mathrm{Tr}(P,N)+K)/K$;*

(d) *for every $R$-module $L$, $\mathrm{Tr}(P,L) = TL$;*

(e) $P = TP$.

*If these conditions hold, then $T = T^2$ and $\mathrm{Gen}(P) = \mathrm{Gen}(T)$.*

We now consider injectivity and projectivity relative to a preradical.

**6.16. Relative projectivity.** Let $\tau$ be an idempotent preradical for $\sigma[M]$ with associated classes $\mathbb{T}_\tau$, $\mathbb{F}_\tau$ and $N \in \sigma[M]$.

A module $P \in \sigma[M]$ is called $\mathbb{F}_\tau$-*N*-*projective* if every diagram with exact row in $\sigma[M]$

$$
\begin{array}{ccccccccc}
 & & & & & & P & & \\
 & & & & & & \downarrow & & \\
0 & \longrightarrow & K & \longrightarrow & N & \longrightarrow & L & \longrightarrow & 0
\end{array}
$$

can be extended commutatively by some morphism $P \to N$, provided $K \in \mathbb{F}_\tau$.

$P$ is called $\mathbb{F}_\tau$-*projective* (or $\tau$-*projective*) in $\sigma[M]$ if it is $\mathbb{F}_\tau$-*N*-projective for all $N \in \sigma[M]$.

Notice that the terminology $\tau$-*projective* (used widely in the literature) may cause confusion with similar notions considered here like *small projective*, etc. Hence we prefer to use the name $\mathbb{F}_\tau$-*projective* for this property.

**6.17. Small projectives are $\mathbb{F}_\tau$-projective.** *Let $\tau$ be a cohereditary preradical for the category $\sigma[M]$.*

(1) *Let $N \in \mathbb{T}_\tau$. If $P \in \sigma[M]$ is small $N$-projective, then $P$ is $\mathbb{F}_\tau$-$N$-projective.*

(2) *If $P \in \mathbb{T}_\tau$ and is small projective in $\sigma[M]$, then $P$ is $\mathbb{F}_\tau$-projective in $\sigma[M]$.*

*Proof.* (1) This follows by 6.10 where it is shown that under the given conditions $\tau$-free submodules of $N$ are small in $N$.

(2) Let $0 \to K \to N \to L \to 0$ be an exact sequence in $\sigma[M]$ where $K \in \mathbb{F}_\tau$. For any $f \in \mathrm{Hom}(P, L)$, $\mathrm{Im}\, f \in \mathbb{T}$ and we have the diagram with exact rows

$$
\begin{array}{ccccccccc}
 & & & & & & P & & \\
 & & & & & & \downarrow{\scriptstyle f} & & \\
0 & \longrightarrow & K' & \longrightarrow & \tau(N) & \longrightarrow & \tau(L) & \longrightarrow & 0
\end{array}
$$

where $K' \subset K \in \mathbb{F}_\tau$. By (1) this can be extended commutatively as desired. $\quad\square$

As we will see shortly, $\mathbb{F}_\tau$-projectivity plays an important rôle in colocalisation and hence we provide basic facts for this notion.

**6.18. Properties of $\mathbb{F}_\tau$-projective modules.** *We use the notation as in 6.16. Let $P \in \sigma[M]$ be $\mathbb{F}_\tau$-projective in $\sigma[M]$ and $S \subseteq \mathrm{End}(P)$.*

(1) *For any submodule $U \subset P$ with $U \in \mathbb{T}_\tau$, the factor module $P/U$ is also $\mathbb{F}_\tau$-projective in $\sigma[M]$.*

(2) *If $P \in \mathbb{T}_\tau$, then for any left $S$-module $X$, the left $R$-module $P \otimes_S X$ is in $\mathbb{T}_\tau$ and is $\mathbb{F}_\tau$-projective in $\sigma[M]$.*

*Proof.* (1) Consider the diagram with exact row

$$
\begin{array}{ccc}
 & & P \xrightarrow{\ p\ } P/U \\
 & & \downarrow{\scriptstyle f} \\
0 \longrightarrow K \longrightarrow N \xrightarrow{\ g\ } L \longrightarrow 0
\end{array}
$$

where $K \in \mathbb{F}_\tau$, $U \in \mathbb{T}_\tau$ and $p$ is the projection map. By assumption, there is a morphism $h : P \to N$ such that $p \diamond f = h \diamond g$. By construction, $0 = (Up)f = (Uh)g$ and $Uh \subseteq \tau(K) = 0$. Thus $h$ factorises over $P/U$ showing that $P/U$ is $\mathbb{F}_\tau$-projective.

(2) Since $P \otimes_S X$ is $P$-generated it belongs to $\mathbb{T}_\tau$.

Let $0 \to K \to N \to L \to 0$ be an exact sequence in $\sigma[M]$ with $\tau(K) = 0$. Then $\mathrm{Hom}(P, N) \simeq \mathrm{Hom}(P, L)$ and the Hom-tensor relations yield the commutative diagram

$$
\begin{array}{ccc}
0 \longrightarrow \mathrm{Hom}_R(P \otimes_S X, N) \xrightarrow{\ \simeq\ } \mathrm{Hom}_R(P \otimes_S X, L) \\
\downarrow{\scriptstyle \simeq} \hspace{3cm} \downarrow{\scriptstyle \simeq} \\
0 \longrightarrow \mathrm{Hom}_S(X, \mathrm{Hom}(P, N)) \xrightarrow{\ \simeq\ } \mathrm{Hom}_S(X, \mathrm{Hom}(P, L)) \longrightarrow 0.
\end{array}
$$

From this we see that $P \otimes_S X$ is $\mathbb{F}_\tau$-projective. $\qquad\qquad\square$

**6.19. Uniqueness property.** *Let $\tau$ be a preradical for $\sigma[M]$ and $P \in \sigma[M]$. Then the following are equivalent:*

(a) *$P$ belongs to $\mathbb{T}_\tau$ and is $\mathbb{F}_\tau$-projective;*

(b) *for any $K \subset N \in \sigma[M]$ where $K \in \mathbb{F}_\tau$, any diagram with exact row*

$$
\begin{array}{ccc}
 & & P \\
 & & \downarrow \\
0 \longrightarrow K \longrightarrow N \longrightarrow N/K \longrightarrow 0
\end{array}
$$

*can be uniquely extended commutatively by some morphism $g : P \to N$, that is*

$$
\mathrm{Hom}(P, N) \simeq \mathrm{Hom}(P, N/K).
$$

*Proof.* (a)$\Rightarrow$(b). The existence of maps extending the diagram is clear by the $\mathbb{F}_\tau$-projectivity of $P$. Assume both $g$ and $h$ are such extensions. Then the image of $g - h : P \to N$ is in $\mathbb{T}_\tau$ and also a submodule of $K \in \mathbb{F}_\tau$. Hence $\mathrm{Im}\,(g - h) = 0$ and thus $g = h$.

(b)⇒(a). Clearly $P$ is $\mathbb{F}_\tau$-projective. By the uniqueness condition, the diagram

$$
\begin{array}{c}
P \\
\downarrow \\
P/\tau(P) \longrightarrow 0 \longrightarrow 0
\end{array}
$$

can be extended uniquely; this must be by the zero map, and so $P = \tau(P)$. □

**6.20. Colocalisation.** Let $\tau$ be an idempotent preradical for $\sigma[M]$ with associated classes $\mathbb{T}_\tau$ and $\mathbb{F}_\tau$. A morphism $g : P \to N$ is called a $\tau$-colocalisation if

$$P \in \mathbb{T}_\tau, \ P \text{ is } \mathbb{F}_\tau\text{-projective in } \sigma[M], \text{ and } \mathrm{Ke}\, g, \mathrm{Coke}\, g \in \mathbb{F}_\tau.$$

Notice that the conditions imply $\tau(N) = \mathrm{Im}\, g$ (see 6.21). Furthermore, it follows from 6.10 that $\mathrm{Ke}\, g \ll P$ if $\tau$ is cohereditary.

Before investigating the existence of $\tau$-colocalisations for modules $N \in \sigma[M]$, we prove a crucial property.

**6.21. Property of colocalisation.** *Let $\tau$ be an idempotent preradical for $\sigma[M]$ and $g_1 : P_1 \to N_1$, $g_2 : P_2 \to N_2$ be two $\tau$-colocalisations in $\sigma[M]$. Then for any morphism $h : N_1 \to N_2$, the diagram*

$$
\begin{array}{ccc}
P_1 & \xrightarrow{g_1} & N_1 \\
& & \downarrow{\scriptstyle h} \\
P_2 & \xrightarrow{g_2} & N_2
\end{array}
$$

*can be extended uniquely to a commutative diagram by some $\tilde{h} : P_1 \to P_2$.*

*Proof.* By the assumptions, $N/(P_1)g_1 \in \mathbb{F}_\tau$, thus $\tau(N_1) \subseteq (P_1)g_1$. Moreover, $(P_1)g_1 \in \mathbb{T}_\tau$ and hence $\tau(N_1) = (P_1)g_1$. Similar relations hold for $N_2$. Thus we obtain the diagram

$$
\begin{array}{ccccccc}
0 & \longrightarrow & \mathrm{Ke}\, g_1 & \longrightarrow & P_1 & \xrightarrow{g_1} \tau(N_1) & \subset & N_1 \\
& & & & & \downarrow{\scriptstyle \tau(h)} & & \downarrow{\scriptstyle h} \\
0 & \longrightarrow & \mathrm{Ke}\, g_2 & \longrightarrow & P_2 & \xrightarrow{g_2} \tau(N_2) & \subset & N_2 \, .
\end{array}
$$

By $\mathbb{F}_\tau$-projectivity of $P_1$, there exists some $\tilde{h} : P_1 \to P_2$ yielding a commutative diagram. By 6.19, the map $\tilde{h}$ is uniquely determined. □

**6.22. Uniqueness of colocalisation.** *Let $\tau$ be an idempotent preradical for $\sigma[M]$ and $g_1 : P_1 \to N$, $g_2 : P_2 \to N$ be two $\tau$-colocalisations of $N$ in $\sigma[M]$. Then there is an isomorphism $h : P_1 \to P_2$ such that $g_1 = h \diamond g_2$.*

*Proof.* As shown in the proof of 6.21, $(P_1)g_1 = \tau(N) = (P_2)g_2$. It follows from 6.19 that the diagram

$$
\begin{array}{ccc}
P_1 & \xrightarrow{\ g_1\ } & N \\
& & \Big\downarrow{\scriptstyle =} \\
P_2 & \xrightarrow{\ g_2\ } & N
\end{array}
$$

can be extended to commutative diagrams by a unique $h : P_1 \to P_2$ and a unique $k : P_2 \to P_1$. The uniqueness property also implies that $h \diamond k$ and $k \diamond h$ yield the identities on $P_1$ and $P_2$, respectively. Hence $h$ and $k$ are isomorphisms.     $\square$

**6.23. Colocalisation functor.** *Let $\tau$ be an idempotent preradical for $\sigma[M]$ and assume that every $N \in \sigma[M]$ has a $\tau$-colocalisation $g_N : \mathcal{C}_\tau(N) \to N$. Then the assignment*

$$
\mathcal{C}_\tau : \sigma[M] \to \sigma[M], \quad N \mapsto \mathcal{C}_\tau(N),
$$

*induces an endofunctor.*

*Proof.* This follows from the fact that, by 6.21, any morphism $N \to L$ in $\sigma[M]$ can be uniquely extended to a morphism $\mathcal{C}_\tau(N) \to \mathcal{C}_\tau(L)$.     $\square$

**6.24. Lemma.** *Let $\tau$ be an idempotent preradical for $\sigma[M]$ with associated module classes $\mathbb{T}_\tau$, $\mathbb{F}_\tau$, and let $N \in \sigma[M]$.*

(1) *If $N \in \mathbb{F}_\tau$ and every factor module of $N$ has a $\tau$-colocalisation, then every factor module of $N$ belongs to $\mathbb{F}_\tau$.*

(2) *If $\tau(N)$ is the image of an $\mathbb{F}_\tau$-projective module and $\tau$ is cohereditary, then $N$ has a $\tau$-colocalisation.*

*Proof.* (1) Let $K \subset N \in \mathbb{F}_\tau$ and choose $K \subseteq L \subset N$ such that $L/K = \tau(N/K) \in \mathbb{T}_\tau$. Then $N/L \in \mathbb{F}_\tau$.

Let $g : \mathcal{C}_\tau(N/K) \to N/K$ be a $\tau$-colocalisation. Then $\mathrm{Im}\, g = \tau(N/K) = L/K$ and, by the projectivity property of $\mathcal{C}_\tau(N/K)$, the diagram

$$
\begin{array}{ccccccccc}
& & & & \mathcal{C}_\tau(N/K) & & & & \\
& & & & \Big\downarrow{\scriptstyle g} & & & & \\
0 & \longrightarrow & K & \longrightarrow & L & \longrightarrow & L/K & \longrightarrow & 0
\end{array}
$$

can be extended commutatively by some morphism $h : \mathcal{C}_\tau(N/K) \to L$. However, $\mathcal{C}_\tau(N/K) \in \mathbb{T}_\tau$ and $L \in \mathbb{F}_\tau$ imply $h = 0$, thus $g = 0$ and $K = L$. By construction this means $N/K \in \mathbb{F}_\tau$.

(2) Let $\tau$ be cohereditary. By assumption, there is an exact sequence in $\sigma[M]$ of the form

$$
0 \longrightarrow K \longrightarrow P \xrightarrow{\ \tilde{f}\ } \tau(N) \longrightarrow 0,
$$

where $P$ is $\mathbb{F}_\tau$-projective. This induces the exact sequence

$$0 \longrightarrow K/\tau(K) \longrightarrow P/\tau(K) \xrightarrow{f} \tau(N) \longrightarrow 0,$$

where $P/\tau(K)$ is $\mathbb{F}_\tau$-projective by 6.18. Since $\tau$ preserves epimorphisms, we obtain the commutative diagram with exact rows

$$
\begin{array}{ccccccccc}
0 & \longrightarrow & \operatorname{Ke}\tau(f) & \longrightarrow & \tau(P/\tau(K)) & \xrightarrow{\tau(f)} & \tau(N) & \longrightarrow & 0 \\
 & & \downarrow{\scriptstyle i} & & \downarrow{\scriptstyle i} & & \downarrow{\scriptstyle =} & & \\
0 & \longrightarrow & K/\tau(K) & \longrightarrow & P/\tau(K) & \xrightarrow{f} & \tau(N) & \longrightarrow & 0,
\end{array}
$$

where both $i$ denote inclusions. Notice that $\operatorname{Ke}\tau(f) \in \mathbb{F}_\tau$ and hence, by 6.10, $\operatorname{Ke}\tau(f) \ll \tau(P/\tau(K))$. Moreover, by $\mathbb{F}_\tau$-projectivity of $P/\tau(K)$, there exists some morphism $h : P/\tau(K) \to \tau(P/\tau(K))$ with $h \diamond \tau(f) = f$. This implies $i \diamond h \diamond \tau(f) = i \diamond f = \tau(f)$, that is, $(i \diamond h - \operatorname{id}) \diamond \tau(f) = 0$ and $\operatorname{Im}(i \diamond h - \operatorname{id}) \subseteq \operatorname{Ke}\tau(f)$. Thus $\operatorname{Im}(i \diamond h - \operatorname{id}) \in \mathbb{F}_\tau$ and so, since obviously $\operatorname{Im}(i \diamond h - \operatorname{id}) \in \mathbb{T}_\tau$, we get $i \diamond h = \operatorname{id}$. Thus $\tau(P/\tau(K))$ is a direct summand of $P/\tau(K)$ and so is $\mathbb{F}_\tau$-projective. $\qquad\square$

As an immediate consequence of the above lemma we observe:

**6.25. Existence of colocalisations.** *Suppose that there is a projective generator in* $\sigma[M]$. *Then the following are equivalent for an idempotent preradical* $\tau$ *for* $\sigma[M]$.

(a) *Every module in* $\sigma[M]$ *has a* $\tau$-*colocalisation, that is, there is a* $\tau$-*colocalisation functor* $\sigma[M] \to \sigma[M]$;

(b) $\tau$ *is cohereditary, that is,* $\mathbb{F}_\tau$ *is a cohereditary module class.*

In particular, every cohereditary radical $\tau$ in $R$-Mod allows the construction of a $\tau$-colocalisation functor for $R$-Mod. As observed in 6.14, in this case all information about $\tau$ is encoded in the ideal $I = \tau(R)$ and we consider this case in more detail. In preparation for this we recall and review the following useful notion.

**6.26. Morita context.** Let $_RP_S$ and $_SQ_R$ be bimodules over the rings $R$ and $S$ with bilinear maps

$$(-,-) : P \otimes_S Q \to R, \quad [-,-] : Q \otimes_R P \to S,$$

satisfying the conditions (for $p_1, p_2 \in P$ and $q_1, q_2 \in Q$)

$$q_1(p_2, q_2) = [q_1, p_2]q_2, \quad (p_1, q_1)p_2 = p_1[q_1, p_2].$$

The quadruple $(P, Q, (-,-), [-,-])$ is called a *general Morita context*.

With these data $P \otimes_S Q$ is a ring without unit with multiplication induced by

$$(p_1 \otimes q_1)(p_2 \otimes q_2) = p_1 \otimes q_1(p_2, q_2) (= p_1[q_1, p_2] \otimes q_2)$$

and $(-,-) : P \otimes_S Q \to R$ is a ring morphism.

**6.27. Standard Morita context.** Given an $R$-module $P$ with $S = \mathrm{End}(P)$, then $P^* = \mathrm{Hom}(P, R)$ is an $(S, R)$-bimodule and the maps

$$(-, -) : P \otimes_S P^* \to R, \quad p \otimes g \mapsto (p)g,$$
$$[-, -] : P^* \otimes_R P \to S, \quad g \otimes p \mapsto [x \mapsto (x)g \cdot p],$$

have the properties (for $p_1, p_2 \in P$, $g_1, g_2 \in P^*$)

$$g_1(p_2, g_2) = [g_1, p_2]g_2, \quad (p_1, g_1)p_2 = p_1[g_1, p_2].$$

Thus $(P, P^*, (-, -), [-, -])$ is a Morita context, called the *standard Morita context* induced by $P$, and $P \otimes_S P^*$ is a ring with multiplication given by (for $p_1, p_2 \in P$ and $g_1, g_2 \in P^*$)

$$(p_1 \otimes g_1)(p_2 \otimes g_2) = p_1 \otimes g_1(p_2, g_2) = p_1 \otimes g_1(p_2)g_2 (= p_1[g_1, p_2] \otimes g_2 = (p_1)g_1 p_2 \otimes g_2).$$

The following technical observation will help to describe the colocalisation functor.

**6.28. Lemma.** *Let $P \in R$-Mod, $S = \mathrm{End}(P)$, and $I = \mathrm{Tr}(P, R)$. Consider the evaluation map*

$$\psi_N : P \otimes_S \mathrm{Hom}(P, N) \to N, \quad p \otimes f \mapsto (p)f.$$

*Then $I(\mathrm{Ke}\,\psi_N) = 0$ and $I(\mathrm{Coke}\,\psi_N) = 0$*

*Proof.* Obviously $I = (P)P^*$ is generated by elements of the form $(q)g$ where $q \in P$ and $g \in P^*$.

Consider any element $a$ in the kernel of $\psi_N$, that is, there are $p_i \in P$ and $f_i \in \mathrm{Hom}(P, N)$, with $a = \sum p_i \otimes f_i$, such that $\sum (p_i)f_i = 0$. Then

$$\begin{aligned}
(q, g)(\sum p_i \otimes_S f_i) &= \sum (q, g)p_i \otimes_S f_i = \sum q[g, p_i] \otimes_S f_i \\
&= \sum q \otimes_S [g, p_i]f_i = \sum q \otimes_S g(p_i, f_i) \\
&= q \otimes_S g(\sum (p_i)f_i) = 0,
\end{aligned}$$

hence $I(\mathrm{Ke}\,\psi_N) = 0$. Furthermore,

$$IN = P\mathrm{Hom}(P, R)N \subseteq P\mathrm{Hom}(P, N) = \mathrm{Tr}(P, N) = \mathrm{Im}\,\psi_N,$$

and hence $I(\mathrm{Coke}\,\psi_N) = I(N/\mathrm{Tr}(P, N)) = 0$.                                    $\square$

**6.29. $\tau$-colocalisation in $R$-Mod.** *Let $\tau$ be a cohereditary radical for $R$-Mod and $I = \tau(R)$. In this case $I$ is an idempotent ideal and*

$$\mathbb{T}_\tau = \{N \in R\text{-Mod} \,|\, IN = N\}, \quad \mathbb{F}_\tau = \{N \in R\text{-Mod} \,|\, IN = 0\}.$$

*Let $P \in \mathbb{T}_\tau$ be $\mathbb{F}_\tau$-projective with $\mathrm{Tr}(P, R) = I$ and put $S = \mathrm{End}(P)$.*

(1) *For any $N \in R$-Mod the evaluation map*

$$\psi_N : P \otimes_S \mathrm{Hom}(P, N) \to N, \quad p \otimes f \mapsto (p)f,$$

*is a $\tau$-colocalisation.*

(2) $\psi_R : P \otimes_S \mathrm{Hom}(P, R) \to R$ *is a $\tau$-colocalisation of $R$ and is an $(R, R)$-bimodule morphism.*

(3) $\Lambda = P \otimes_S P^*$ *has a ring structure (without unit) such that $\psi_R$ is a ring morphism, and $\Lambda = \Lambda^2$.*

(4) *If $R$ is commutative, then $\Lambda$ is also commutative.*

(5) *If $\mu : \Lambda \to R$ and $\mu' : \Lambda' \to R$ are two $\tau$-colocalisations of $R$ which are ring morphisms and $(R, R)$-bilinear maps, then there exists a bijection $h : \Lambda \to \Lambda'$ that is a ring morphism and is $(R, R)$-bilinear.*

*Proof.* (1) Since $P = IP$, we have $I(P \otimes_S \mathrm{Hom}(P, N)) = P \otimes_S \mathrm{Hom}(P, N)$.

By 6.18, $P \otimes_S \mathrm{Hom}(P, N)$ is $\mathbb{F}_\tau$-projective and it follows from 6.28 that $\mathrm{Ke}\,\psi_N$ and $\mathrm{Coke}\,\psi_N$ belong to $\mathbb{T}_\tau$.

(2) The first assertion follows from (1) and so $\psi_R$ is left $R$-linear. An easy check shows that $\psi_R$ is also right $R$-linear.

(3) The Morita context induced by $P$ and $P^*$ yields a product on $P \otimes_S P^*$ (see 6.27) and it is easy to verify that, for this product, $\psi_R$ is a ring morphism.

We claim that for any $\lambda, \lambda' \in \Lambda$,

$$(*) \qquad \lambda'\lambda = (\lambda')\mu\,\lambda - \lambda'(\lambda)\mu.$$

To see this consider for $\lambda \in \Lambda$ the map

$$\phi_\lambda : \Lambda \to \Lambda, \quad \lambda' \mapsto \lambda'\lambda - (\lambda')\mu\,\lambda.$$

This is a left $R$-linear map with $\mathrm{Im}\,\phi_\lambda \subseteq \mathrm{Ke}\,\mu \in \mathbb{F}_\tau$ but also $\mathrm{Im}\,\phi_\lambda \in \mathbb{T}_\tau$. Thus $\phi_\lambda = 0$ and $\lambda'\lambda = (\lambda')\mu\,\lambda$. The other equality follows by a similar argument.

Now recall that $(\Lambda)\mu = I$ and $\Lambda \in \mathbb{T}_\tau$. Hence $\Lambda = I\Lambda = (\Lambda)\mu\Lambda$ and

$$\Lambda^2 = (\Lambda)\mu\,\Lambda = I\Lambda = \Lambda.$$

(4) Assume $R$ to be commutative. For $a \in R$ consider the map

$$g_a : \Lambda \to \Lambda, \quad \lambda \mapsto a\lambda - \lambda a.$$

Since $R$ is commutative, $g_a$ is left $R$-linear and obviously $\mathrm{Im}\,g_a \subseteq \mathrm{Ke}\,\mu$. Thus $\mathrm{Im}\,g_a \in \mathbb{T}_\tau \cap \mathbb{F}_\tau = 0$, that is, $a\lambda = \lambda a$ for all $a \in R$. By $(*)$ in the proof of (3) above we get

$$\lambda'\lambda = (\lambda')\mu\,\lambda = \lambda(\lambda')\mu = \lambda\lambda'.$$

(5) The existence of an $R$-module isomorphism $h : \Lambda \to \Lambda'$ with $\mu = h \diamond \mu'$ follows from the uniqueness of the colocalisation. For $\lambda \in \Lambda$ consider the map

$$h_\lambda : \Lambda \to \Lambda', \quad x \mapsto (x\lambda)h - (x)h\,(\lambda)h,$$

that is obviously left $R$-linear. Since $\mu$ and $\mu'$ are ring morphisms we have

$$\begin{aligned}
(x)h_\lambda \mu' &= [(x\lambda)h - (x)h\,(\lambda)h]\mu' \\
&= (x\lambda)\mu - (x)\mu\,(\lambda)\mu \; = \; 0,
\end{aligned}$$

that is, $\operatorname{Im} h_\lambda \subseteq \operatorname{Ke} \mu' \in \mathbb{F}_\tau$. Now $\Lambda \in \mathbb{T}_\tau$ implies $\operatorname{Im} h_\lambda \in \mathbb{T}_\tau$ and hence $\operatorname{Im} h_\lambda = 0$ and $h_\lambda = 0$. This shows that $h$ is a ring morphism.

A similar argument proves that $h$ is also right $R$-linear.    □

The colocalisation in $R$-Mod can also be described in the following way.

**6.30. $\tau$-colocalisation in $R$-Mod by $\Lambda$.** *Let $\tau$ be a cohereditary radical for $R$-Mod, $I = \tau(R)$, and $\mu : \Lambda \to R$ be a colocalisation of $R$. Then for any $N \in R$-Mod, the $R$-linear map*

$$\varphi : \Lambda \otimes_R N \to N, \quad \sum \lambda_i \otimes m_i \mapsto \sum (\lambda_i)\mu\, n_i,$$

*is a $\tau$-colocalisation of $N$.*

*Proof.* As shown in 6.29, $\Lambda$ has a ring and $(R, R)$-bimodule structure and $\mu$ is a morphism for those.

Next observe that $I\Lambda = \Lambda$ implies $\Lambda \otimes_R N = I(\Lambda \otimes_R N) \in \mathbb{T}_\tau$. Since $_R\Lambda$ is $\mathbb{F}_\tau$-projective, $\Lambda \otimes_R N$ is also $\mathbb{F}_\tau$-projective by 6.18.

Since $\tau$ is cohereditary we have $\tau(N) = \tau(R)N = IN$ and hence

$$\operatorname{Im} \varphi = (\Lambda)\mu\, N = IN = \tau(N),$$

and therefore $\operatorname{Coke} \varphi = N/\tau(N) \in \mathbb{F}_\tau$.

It remains to show that $\operatorname{Ke} \varphi \in \mathbb{F}_\tau$, that is, $I\operatorname{Ke} \varphi = 0$. For this, take any element $\sum \lambda_i \otimes n_i \in \operatorname{Ke} \varphi$, in other words $\sum (\lambda_i)\mu\, n_i = 0$, and take any $x = (\lambda')\mu \in I = \operatorname{Im} \mu$. Then, referring to the identity $(*)$ in the proof of 6.29(3), we obtain

$$\begin{aligned}
x(\sum \lambda_i \otimes n_i) &= \sum (\lambda')\mu\, \lambda_i \otimes n_i = \sum \lambda'(\lambda_i)\mu \otimes n_i \\
&= \sum \lambda' \otimes (\lambda_i)\mu\, n_i = \lambda' \otimes \sum (\lambda_i)\mu\, n_i = 0.
\end{aligned}$$

This completes the proof of our assertion.    □

**6.31. Exercises.**

(1) Let $K$ be a fully invariant submodule of $L \in \sigma[M]$. For $N \in \sigma[M]$ put (see, e.g., [290])

$\alpha_K^L(N) = \sum \{f(K) \mid f : L \to N\} \subseteq \operatorname{Tr}(K, N)$, and

$\omega_K^L(N) = \bigcap \{g^{-1}(K) \mid g : N \to L\} \supseteq \operatorname{Re}(N, L/K)$.

Prove that $\alpha_K^L$ and $\omega_K^L$ induce preradicals for $\sigma[M]$ and that

(i) $\alpha_K^L(L) = K$;   $\alpha_L^L(N) = \mathrm{Tr}(L, N)$.

(ii) $\omega_K^L(L) = K$;   $\omega_0^L(N) = \mathrm{Re}(N, L)$.

(iii) If $N$ is injective in $\sigma[M]$, then $\alpha_K^L(N) = \mathrm{Tr}(K, N)$;

(iv) If $N$ is projective in $\sigma[M]$, then $\omega_K^L(N) = \mathrm{Re}(N, L/K)$.

(2) Let $I \subset R$ be an ideal. For $N \in R\text{-Mod}$ put

$$\tau_I(N) = \{n \in N \mid In = 0\}, \quad \text{and} \quad \tau^I(N) = IN.$$

(i) Prove that $\tau_I$ induces a hereditary preradical for $R$-Mod and

    ($\alpha$) $\tau_I(R) = \{a \in R \mid Ia = 0\}$;

    ($\beta$) $\tau_I$ is a radical if and only if $I^2 = I$.

(ii) Show that $\tau^I$ induces a cohereditary preradical for $R$-Mod and $\tau^I(R) = I$. $\tau^I$ is idempotent if and only if $I^2 = I$.

(3) Let $\tau$ be a cohereditary preradical for $\sigma[M]$. Prove that for any exact sequence

$$K \xrightarrow{f} N \xrightarrow{g} L \longrightarrow 0$$

in $\sigma[M]$, $\tau(g)$ is an epimorphism and $\tau(\mathrm{Ke}\,(\tau(g))) = \mathrm{Im}\,(\tau(f))$.

(4) Prove that $P \in \sigma[M]$ is pseudo-projective if and only if for any $N \in \sigma[M]$ and $\mathrm{Tr}(P, N) \subset L \subset N$ with $N/L$ cocyclic, $\mathrm{Tr}(P, N/L) = 0$.

(5) Prove that for any module $M$, the following are equivalent (compare [230, Theorem 5.1]).

(a) Every preradical for $\sigma[M]$ is cohereditary;

(b) Soc is a cohereditary preradical;

(c) $M$ is semisimple.

**References.** Bican, Kepka, and Nemec [41]; Jambor, Malcolmson, and Shapiro [181]; Kashu [196]; Lomp [226, 228]; Mbuntum and Varadarajan [230]; McMaster [232]; Ohtake [265]; Raggi, Ríos Montes, and Wisbauer [290]; Stenström [323].

# 7   Torsion theories

For any nonempty class $\mathbb{A}$ of modules in $\sigma[M]$ define the classes

$$
\begin{aligned}
\mathbb{A}^\circ &= \{B \in \sigma[M] \mid \mathrm{Hom}(B,A) = 0 \text{ for all } A \in \mathbb{A}\}, \\
&= \{B \in \sigma[M] \mid \mathrm{Re}(B,\mathbb{A}) = B\}, \text{ and} \\
\mathbb{A}^\bullet &= \{B \in \sigma[M] \mid \mathrm{Hom}(A,B) = 0 \text{ for all } A \in \mathbb{A}\} \\
&= \{B \in \sigma[M] \mid \mathrm{Tr}(\mathbb{A},B) = 0\}.
\end{aligned}
$$

**7.1. Torsion theories.** An ordered pair $(\mathbb{T}, \mathbb{F})$ of classes of modules from $\sigma[M]$ is called a *torsion theory* in $\sigma[M]$ if $\mathbb{T} = \mathbb{F}^\circ$ and $\mathbb{F} = \mathbb{T}^\bullet$. In this case, $\mathbb{T}$ is called the *torsion class* and its elements are the *torsion modules*, while $\mathbb{F}$ is the *torsion free class* and its elements are the *torsion free modules* of the given torsion theory.

The class $\mathbb{T}$ is closed under direct sums, factor modules and extensions in $\sigma[M]$, while $\mathbb{F}$ is closed under products, submodules and extensions.

The torsion theory $(\mathbb{T}, \mathbb{F})$ is called *hereditary* if the class $\mathbb{T}$ is closed under submodules; this is equivalent to $\mathbb{F}$ being closed under essential extensions (injective envelopes).

$(\mathbb{T}, \mathbb{F})$ is said to be *cohereditary* provided $\mathbb{F}$ is closed under factor modules; this implies that $\mathbb{T}$ is closed under small epimorphisms.

**7.2. Torsion theories induced by $\mathbb{A}$.** Let $\mathbb{A}$ be a class of modules in $\sigma[M]$.

 (i) The pair $(\mathbb{A}^{\bullet\circ}, \mathbb{A}^\bullet)$ is a torsion theory, called the *torsion theory generated by* $\mathbb{A}$, and the torsion class can be characterised by

$$\mathbb{A}^{\bullet\circ} = \{X \in \sigma[M] \mid \text{ for all } V \subsetneq X,\ \mathrm{Tr}(\mathbb{A}, X/V) \neq 0\}.$$

 (ii) The pair $(\mathbb{A}^\circ, \mathbb{A}^{\circ\bullet})$ is also a torsion theory, called the *torsion theory cogenerated by* $\mathbb{A}$, and the torsion free class is

$$\mathbb{A}^{\circ\bullet} = \{Y \in \sigma[M] \mid \text{ for all } 0 \neq U \subseteq Y,\ \mathrm{Re}\,(U, \mathbb{A}) \neq U\}.$$

By the definitions, $\mathbb{A} \subseteq \mathrm{Gen}(\mathbb{A}) \subseteq \mathbb{A}^{\bullet\circ}$ and $\mathbb{A} \subseteq \mathrm{Cog}(\mathbb{A}) \subseteq \mathbb{A}^{\circ\bullet}$.

Starting from the classes $\mathrm{Cog}(\mathbb{A})$ or $\mathrm{Gen}(\mathbb{A})$, the corresponding torsion classes can be obtained by a standard construction using transfinite induction (e.g., [323], VI, Proposition 1.5]).

**7.3. Characterisation of classes.** *Let $\mathbb{A}$ be a class of modules in $\sigma[M]$.*

 (1) *The following are equivalent:*

   (a) $\mathbb{A}$ *is closed under factor modules, direct sums and extensions;*

   (b) $\mathbb{A} = \mathbb{A}^{\bullet\circ}$.

(2) *The following are equivalent:*

(a) $\mathbb{A}$ *is closed under submodules, products, and extensions;*

(b) $\mathbb{A} = \mathbb{A}^{\circ \bullet}$.

*Proof.* This is, for example, shown in [323, Section IV.2]. □

**7.4. Hereditary classes.** For any class $\mathbb{A}$ of modules in $\sigma[M]$, denote

$$\mathbb{A}^{\triangleleft} := \{X \in \sigma[M] \mid \mathrm{Hom}(U, A) = 0 \text{ for all } U \subset X,\ A \in \mathbb{A}\} \subset \mathbb{A}^{\circ}.$$

This defines a hereditary pretorsion class of modules.

Notice that any hereditary pretorsion class in $\sigma[M]$ is of the form $\{E\}^{\circ}$ for some injective module $E \in \sigma[M]$ (e.g., [364, 9.5]). In fact a somewhat weaker injectivity condition suffices, given by the following dual of pseudo-projectivity as defined in 6.11.

**7.5. Pseudo-injective modules.** A module $Q$ is called *pseudo-injective in* $\sigma[M]$ if for any monomorphism $f : Y \to X$ in $\sigma[M]$ and nonzero $g : Y \to Q$ there exist $h \in \mathrm{End}(Q)$ and $k \in \mathrm{Hom}(X, Q)$ such that $f \diamond k = g \diamond h \neq 0$.

A module $Q$ is pseudo-injective in $\sigma[M]$ if and only if $\{Q\}^{\circ}$ is a hereditary torsion class in $\sigma[M]$. In this case $\{Q\}^{\circ} = \{\widehat{Q}\}^{\circ}$ where $\widehat{Q}$ denotes the $M$-injective hull of $Q$.

**7.6. Cohereditary classes.** For any class $\mathbb{A}$ of modules in $\sigma[M]$, denote

$$\mathbb{A}^{\blacktriangleright} := \{X \in \sigma[M] \mid \mathrm{Hom}(A, X/Y) = 0 \text{ for all } Y \subset X,\ A \in \mathbb{A}\} \subset \mathbb{A}^{\bullet}.$$

This defines a cohereditary class of modules. It is easy to verify that $\mathbb{A}^{\blacktriangleright}$ is also closed under extensions and submodules but it need not be closed under products (see 9.3).

**7.7. Relations for $\{M\}^{\bullet}$ and $\{M\}^{\blacktriangleright}$.** Let $f : P \to M$ be an epimorphism.

(1) $\{P\}^{\bullet} \subseteq \{M\}^{\bullet}$ *and* $\{P\}^{\blacktriangleright} \subseteq \{M\}^{\blacktriangleright}$.

(2) *If* $\mathrm{Ke}\, f \ll P$, *then* $\{P\}^{\blacktriangleright} = \{M\}^{\blacktriangleright}$.

(3) *If* $P$ *is pseudo-projective in* $\sigma[M]$, *then* $\{P\}^{\bullet} = \{P\}^{\blacktriangleright}$.

(4) *If* $P \to M$ *is a pseudo-projective cover of* $M$ *in* $\sigma[M]$, *then* $\{P\}^{\bullet} = \{M\}^{\blacktriangleright}$.

(5) *If* $M$ *is self-projective, then for any submodule* $K \subset M$, $K \in \{M\}^{\bullet}$ *if and only if* $K \in \{M\}^{\blacktriangleright}$.

*Proof.* (1) This assertion is obvious.

(2) Suppose $X \in \{M\}^{\blacktriangleright}$ and $K = \mathrm{Ke}\, f \ll P$. Any morphism $g : P \to X$ induces the commutative diagram

$$\begin{array}{ccc} P & \xrightarrow{\ f\ } & M \\ {\scriptstyle g}\downarrow & & \downarrow{\scriptstyle \bar{g}} \\ X & \longrightarrow & X/(K)g. \end{array}$$

Now $X \in \{M\}^{\blacktriangleright}$ implies $\bar{g} = 0$, that is, $\mathrm{Im}\, g = (K)g$. From this it follows that $P = K + \mathrm{Ke}\,(g)$ and hence $\mathrm{Ke}\,(g) = P$ since $K \ll P$. Thus $g = 0$ and $X \in \{P\}^{\bullet}$. Since a similar argument applies to any factor module of $X$, we conclude $X \in \{P\}^{\blacktriangleright}$ and $\{M\}^{\blacktriangleright} = \{P\}^{\blacktriangleright}$.

(3) This follows from 6.12.

(4) is a consequence of (2) and (3).

(5) If $M$ is self-projective, then for any submodules $L \subset K \subseteq M$, $M$ is also $K$-projective and so any homomorphism $f : M \to K/L$ lifts to a homomorphism $\bar{f} : M \to K$ such that $\bar{f}\pi = f$ where $\pi$ denotes the projection of $K$ onto $K/L$. Hence $\mathrm{Hom}(M, K) = 0$ if and only if $\mathrm{Hom}(M, K/L) = 0$ for all $L \subset K$.     $\square$

As observed above, for any projective module $P \in \sigma[M]$, $\{P\}^{\bullet}$ is a cohereditary torsion free class. In general, however, the cohereditary torsion free classes need not be of this form unless projective covers are available. This follows from 7.7.

**7.8. Torsion theory generated by singular modules.** Let $\mathcal{C}_M^{\trianglelefteq}$ denote the class of singular modules in $\sigma[M]$ and $\mathrm{Z}_M$ denote the related preradical for $\sigma[M]$ (see 6.7).

The torsion theory generated by $\mathcal{C}_M^{\trianglelefteq}$ is called the *Goldie torsion theory* in $\sigma[M]$. Its torsion free class consists of the non-$M$-singular modules. Its torsion class $(\mathcal{C}_M^{\trianglelefteq})^{\bullet\circ}$ is just the extension of $\mathcal{C}_M^{\trianglelefteq}$ by itself, that is, a module $N \in \sigma[M]$ is in $(\mathcal{C}_M^{\trianglelefteq})^{\bullet\circ}$ if and only if there is an exact sequence $0 \to K \to N \to L \to 0$ where $K$ and $L$ are $M$-singular.

If $M$ itself is non-$M$-singular (polyform), then the torsion class $(\mathcal{C}_M^{\trianglelefteq})^{\bullet\circ}$ is simply $\mathcal{C}_M^{\trianglelefteq}$. For more details the reader is referred to [85, Section 2] and [364, Section 10, 11].

**7.9. Torsion theory cogenerated by singular modules.** To study the torsion theory cogenerated by $\mathcal{C}_M^{\trianglelefteq}$, we consider the corresponding radical for $N \in \sigma[M]$,

$$\mathrm{Re}_M^{\trianglelefteq}(N) = \bigcap \{U \subset N \mid N/U \in \mathcal{C}_M^{\trianglelefteq}\}.$$

Then the torsion class $(\mathcal{C}_M^{\trianglelefteq})^{\circ}$ consists of modules $N$ with $N = \mathrm{Re}_M^{\trianglelefteq}(N)$, that is, modules which have no proper essential submodules and are not $M$-singular;

these are precisely the $M$-projective semisimple modules in $\sigma[M]$. The torsion free modules are those which do not contain any simple projective module.

**Torsion class.** *The torsion class* $(\mathcal{C}_M^{\trianglelefteq})^\circ$ *is generated by the simple $M$-projective modules in $\sigma[M]$ (i.e., consists of all $M$-projective semisimple modules).*

Thus $M$ itself is a torsion module if and only if it is semisimple (and then $(\mathcal{C}_M^{\trianglelefteq})^{\circ\bullet} = 0$).

**7.10. $M$-rational modules.** The torsion class of the torsion theory cogenerated by the $M$-injective hull $\widehat{M}$ of $M$ is

$$
\begin{aligned}
\mathbb{T}_M &= \{X \in \sigma[M] \mid \mathrm{Hom}(X, \widehat{M}) = 0\} \\
&= \{X \in \sigma[M] \mid \mathrm{Hom}(Y, M) = 0 \text{ for any submodule } Y \subseteq X\}.
\end{aligned}
$$

This is a hereditary torsion class and the modules in $\mathbb{T}_M$ are called *M-rational modules*. A submodule $K \subset M$ is said to be *rational in $M$* if $M/K \in \mathbb{T}_M$. Rational submodules in $M$ are always essential and $M$ is called *polyform* if every essential submodule is rational. $M$ is polyform if and only if $M$ does not contain an $M$-singular submodule and for such modules the class of singular modules in $\sigma[M]$ coincides with $M$-rational modules (e.g., [364, 10.2]).

The torsion theory induced by $\mathbb{T}_M$ is also called the *Lambek torsion theory* in $\sigma[M]$ (since for $M = R$ this theory has been investigated by J. Lambek). Notice that in $\sigma[M]$ this theory depends on the choice of a subgenerator and hence need not be uniquely determined (see [364, Section 10]).

**7.11. Exercise.**

Let $R$ be a left semi-artinian ring. Prove that $\mathcal{C}_R^{\trianglelefteq}$ corresponds to the hereditary torsion class generated by the class of small simple modules ([291]).

**References.** Bican, Kepka and Nemec [41] Kashu [196]; Lomp [226, 228]; Mishina and Skornjakov [234]; Ramamurthi [291]; Stenström [323].

# 8   Torsion theories related to small modules

The singular modules in $R$-Mod can be characterised as the factor modules of
projectives by essential submodules (see, e.g., [85, 4.6]). Dualisation leads to small
submodules of injectives. These are called *small modules* in the corresponding
category and we consider them also because of their relevance for the structure
theory of lifting modules.

**8.1. Small modules in $\sigma[M]$.** An $R$-module $N \in \sigma[M]$ is called *$M$-small* (or *small
in $\sigma[M]$*) if $N \ll L$ for some $L \in \sigma[M]$. We let $\mathbb{S}[M]$, or simply $\mathbb{S}$, denote the class
of all small modules in $\sigma[M]$.

The module $N \in \sigma[M]$ is said to be *non-$M$-small* if $N$ is not small in $\sigma[M]$,
that is, $N \notin \mathbb{S}[M]$.

**8.2. Properties of small modules.**

(1) *For $N \in \sigma[M]$ the following are equivalent:*

    (a) *$N$ is $M$-small;*

    (b) *$N$ is small in its $M$-injective hull $\widehat{N}$;*

    (c) *for any $M$-injective module $Q$ and any morphism $f : N \to Q$ in $\sigma[M]$,
we have $\operatorname{Im} f \ll Q$.*

(2) *Any $M$-small module is $R$-small.*

(3) *Any simple module in $\sigma[M]$ is either $M$-small or $M$-injective.*

(4) *$\mathbb{S}$ is closed under submodules, homomorphic images, and finite direct sums.*

(5) *$\mathbb{S}$ is closed under infinite (direct) sums if and only if every injective module
in $\sigma[M]$ has a small radical.*

(6) *$\mathbb{S} = 0$ if and only if $M$ is cosemisimple.*

(7) *If $M$ is $\sigma$-cohereditary (see 1.22), then*

    (i) *$\mathbb{S}$ is closed under extensions in $\sigma[M]$;*

    (ii) *$N \in \sigma[M]$ is $M$-small if and only if $N$ has no $M$-injective factor module.*

*Proof.* (1) (a)$\Rightarrow$(c). Let $N \ll L$ with $L \in \sigma[M]$ and $f : N \to Q$ be any morphism
in $\sigma[M]$ where $Q$ is $M$-injective. Then there exists some $g : L \to Q$ extending
$f$ to $L$. Assume $(N)g + U = (N)f + U = Q$, for some submodule $U \subset Q$. Then
$N + (U)g^{-1} = L$ and so $(U)g^{-1} = L$, implying $U = Q$. This shows that $(N)f \ll Q$.
The remaining implications are trivial.

(2) Since the $M$-injective hull $\widehat{N}$ of $N$ is embeddable in the $R$-injective hull
$E(N)$, $N \ll \widehat{N}$ implies $N \ll E(N)$. Notice that a simple module $M$ is never
$M$-small but may be $R$-small.

(3) Let $E$ be a simple module in $\sigma[M]$. If $E$ is not small in its $M$-injective
hull $\widehat{E}$, then $E = \widehat{E}$ and so $E$ is $M$-injective.

(4) Obviously submodules and finite direct sums of $M$-small modules are again $M$-small. A simple argument using (1) (c) shows that homomorphic images of $M$-small modules are small in $\sigma[M]$.

(5) For any injective $Q \in \sigma[M]$, the radical of $Q$ is a homomorphic image of a direct sum of cyclic submodules of $\mathrm{Rad}(Q)$ which are small submodules of $Q$. Thus if the class of $M$-small modules is closed under infinite direct sums, clearly $\mathrm{Rad}(Q) \ll Q$.

Now assume that all injectives in $\sigma[M]$ have small radicals. Any direct sum $\bigoplus_\Lambda N_\lambda$ of $M$-small modules is contained in the radical of the $M$-injective hull of $\bigoplus_\Lambda N_\lambda$ which is small by assumption. This means that $\bigoplus_\Lambda N_\lambda$ is $M$-small.

(6) $\mathbb{S} = 0$ means in particular that the simple modules are not $M$-small and hence are $M$-injective by (3); thus $M$ is cosemisimple.

Conversely, assume $M$ to be cosemisimple and let $N \in \mathbb{S}$. For any $n \in N$ let $K \subset Rn$ be a maximal submodule. Then $Rn/K$ is $M$-small and $M$-injective and thus $n = 0$, implying $N = 0$.

(7) (i) Let $0 \to K \to L \to L/K \to 0$ be an exact sequence such that $L \in \sigma[M]$ and $K, L/K \in \mathbb{S}$. Denote by $\widehat{L}$ the $M$-injective hull of $L$. Then $K \ll \widehat{L}$ and $L/K \ll \widehat{L}/K$, since the latter is $M$-injective. By 2.2 (3), this implies $L \ll \widehat{L}$, showing that $\mathbb{S}$ is closed under extensions.

(ii) A nonzero factor module $N/K$ of an $M$-small module $N$ is again $M$-small by (4) and so not $M$-injective by 1 (c).

Assume $N \notin \mathbb{S}$. Then $\widehat{N} = N + K$ for some proper submodule $K \subset \widehat{N}$. Then $N \to \widehat{N} \to \widehat{N}/K$ is surjective where $\widehat{N}/K$ is $M$-injective, a contradiction. □

**8.3. Small covers of an injective module.** *Let $M$ be an $R$-module and let $f : L \to N$ be a small cover of an injective module $N$ in $\sigma[M]$. Then any nonsmall submodule of $L$ is a non-$M$-small module. In particular, any nonzero direct summand of $L$ is a non-$M$-small module.*

*Proof.* Suppose $A \subseteq L$ is a nonsmall submodule of $L$. Assume that $A$ is an $M$-small module. Then $(A)f$ is an $M$-small module by 8.2 (4) and, since $N$ is injective, $(A)f \ll N$ by 8.2 (1) (c). Since $f$ is a small epimorphism, $(A)f \ll N$ implies that $A \ll L$, a contradiction. Hence $A$ is a non-$M$-small module. □

The class of small modules defines preradicals and torsion theories by the process described in 7.2. These are the subject of the rest of this section.

**8.4. Preradical generated by $M$-small modules.** Recall that the class $\mathbb{S}$ of small modules in $\sigma[M]$ is closed under submodules, factor modules and finite direct sums (see 8.2). The preradical generated by $\mathbb{S}$ is defined as the trace of $\mathbb{S}$ in any $N \in \sigma[M]$ and we denote it by

$$\begin{aligned}
\mathrm{Tr}_\mathbb{S}(N) &= \sum \{U \subset N \mid U \text{ is } M\text{-small}\} \\
&= \sum \{U \subset N \mid U \ll \widehat{N}\} = N \cap \mathrm{Rad}(\widehat{N}),
\end{aligned}$$

where $\widehat{N}$ denotes the $M$-injective hull of $N$. Notice that the class $\mathbb{S}$ depends on the module $M$ and if necessary we use the symbol $\mathrm{Tr}_{\mathbb{S}[M]}$.

The fact that small submodules of $N$ are also small in $\widehat{N}$ implies $\mathrm{Rad}(N) \subseteq \mathrm{Tr}_{\mathbb{S}}(N)$ and, by the same token, $\mathrm{Tr}_{\mathbb{S}[M]}(N) \subseteq \mathrm{Tr}_{\mathbb{S}[R]}(N)$. From the definitions we easily conclude:

**Radical modules.** *For $N \in \sigma[M]$ the following are equivalent:*

(a) $N = \mathrm{Tr}_{\mathbb{S}}(N)$;

(b) $N \subseteq \mathrm{Rad}(\widehat{N})$;

(c) $Rn \ll \widehat{N}$ *for all* $n \in N$;

(d) $N \in \mathrm{Gen}(\mathbb{S})$.

Notice that $\mathrm{Tr}_{\mathbb{S}}(N)$ need not belong to $\mathbb{S}$ unless $\mathbb{S}$ is closed under infinite direct sums, that is, injectives in $\sigma[M]$ have small radicals (cf. 8.2).

**8.5. Torsion theory generated by $M$-small modules.** For the torsion theory generated by $\mathbb{S}$ we have the following.

**Torsion and torsion free modules.**

(1) *The class* $\mathbb{S}^{\bullet} = \{N \in \sigma[M] \,|\, \mathrm{Tr}_{\mathbb{S}}(N) = 0\}$ *is cogenerated by the simple $M$-injective modules in* $\sigma[M]$.

(2) $\mathbb{S}^{\bullet\circ}$ $= \{N \in \sigma[M] \,|\, N$ *has no simple $M$-injective factor module$\}$*
$= \{N \in \sigma[M] \,|\,$ *for any proper* $U \subset N$, $\mathrm{Tr}(\mathbb{S}, N/U) \neq 0\}$.

(3) *If $M$ is $\sigma$-cohereditary (see 1.22), then* $\mathrm{Gen}(\mathbb{S}) = \mathbb{S}^{\bullet\circ}$.

*Proof.* (1) Since $N \trianglelefteq \widehat{N}$ it follows that $\mathrm{Tr}_{\mathbb{S}}(N) = 0$ if and only if $\mathrm{Rad}(\widehat{N}) = 0$. Thus, in this case, $\widehat{N}$ (and $N$) is cogenerated by simple modules, that is, there is an embedding $\gamma : \widehat{N} \to \prod_{\Lambda} E_{\lambda}$, where the $E_{\lambda}$ are simple modules in $\sigma[M]$. By the injectivity of $\widehat{N}$, there is a map $\delta : \prod_{\Lambda} E_{\lambda} \to \widehat{N}$ for which $\delta \circ \gamma = \mathrm{id}_{\widehat{N}}$. Then, for every $M$-small simple module $E_{\lambda}$, $\delta(E_{\lambda}) = 0$, and this implies that $\widehat{N}$ is embedded in a product of $M$-injective simple modules.

Now assume that $N$ is cogenerated by simple $M$-injective modules. Then, for any $n \in N$, there is a nonzero morphism $Rn \to E$ for some simple $M$-injective $E$. This means that $Rn$ cannot be $M$-small, that is, $\mathrm{Tr}_{\mathbb{S}}(N) = 0$.

(2) This follows from (1) and 7.2.

(3) Let $N \in \mathbb{S}^{\bullet\circ}$, that is, $\mathrm{Tr}(\mathbb{S}, N/U) \neq 0$ for each proper $U \subset N$. Assume $N \neq \mathrm{Tr}_{\mathbb{S}}(N) = N \cap \mathrm{Rad}(\widehat{N})$. Then $N/\mathrm{Tr}_{\mathbb{S}}(N) \subset \widehat{N}/\mathrm{Rad}(\widehat{N})$ and — by assumption — the left-hand side contains a nonzero $M$-small submodule which is small in the $M$-injective module $\widehat{N}/\mathrm{Rad}(\widehat{N})$, a contradiction.                                  $\square$

Notice that $\mathrm{Gen}(\mathbb{S})$ is always a subclass of $\mathbb{S}^{\bullet\circ}$ which in general is proper since it need not be closed under extensions.

From the properties mentioned above it follows immediately that if $M = \mathrm{Tr}_{\mathbb{S}}(M)$, then $N = \mathrm{Tr}_{\mathbb{S}}(N)$ for all $N \in \sigma[M]$. As a special case we have $\mathbb{Z} \ll \mathbb{Q}$ as a $\mathbb{Z}$-submodule and hence $\mathrm{Tr}_{\mathbb{S}[\mathbb{Z}]}(\mathbb{Z}) = \mathbb{Z}$ and $\mathrm{Tr}_{\mathbb{S}[\mathbb{Z}]}(N) = N$ for any $\mathbb{Z}$-module $N$.

From the general theory (see 6.2) we obtain

**8.6. Properties of $\mathrm{Tr}_{\mathbb{S}}$.** *Let $K, L$ and $\{N_\lambda\}_\Lambda$ be modules in $\sigma[M]$.*

(1) $\mathrm{Tr}_{\mathbb{S}}(K)$ *is a fully invariant submodule of $K$.*

(2) *For any morphism $f : K \to L$, $(\mathrm{Tr}_{\mathbb{S}}(K))f \subseteq \mathrm{Tr}_{\mathbb{S}}(L)$.*

(3) $\mathrm{Tr}_{\mathbb{S}}(\bigoplus_\Lambda N_\lambda) = \bigoplus_\Lambda \mathrm{Tr}_{\mathbb{S}}(N_\lambda)$.

(4) $\mathrm{Tr}_{\mathbb{S}}(\prod_\Lambda N_\lambda) \subseteq \prod_\Lambda \mathrm{Tr}_{\mathbb{S}}(N_\lambda)$.

**8.7. Example.** For a field $K$ consider the matrix ring $R = \begin{pmatrix} K & 0 \\ K & K \end{pmatrix}$. Then $\mathrm{Tr}_{\mathbb{S}[R]}(R_R) = \mathrm{Soc}(R_R)$ and $\mathrm{Tr}_{\mathbb{S}[R]}(_R R) = \mathrm{Soc}(_R R)$. Since $\mathrm{Soc}(R_R) \neq \mathrm{Soc}(_R R)$ we conclude $\mathrm{Tr}_{\mathbb{S}[R]}(R_R) \neq \mathrm{Tr}_{\mathbb{S}[R]}(_R R)$, showing that (in contrast to Rad) the preradical $\mathrm{Tr}_{\mathbb{S}[R]}$ is not left-right symmetric.

**8.8. Remarks.** The duality between the notions "singular" and "small" has led to different terminology in the literature; some authors use the term *cosmall* instead of *singular* (e.g., [282]), and others call *small* modules *cosingular* (e.g., [291]). The notion $\mathrm{Tr}_{\mathbb{S}}(N)$ (see 8.5) can be thought of as dual to the singular submodule of $N$ and hence some authors denote it by $Z^*(N)$ (e.g., [278]).

Now we turn to the torsion theory cogenerated by small modules.

**8.9. Torsion theory cogenerated by $M$-small modules.** Given $N \in \sigma[M]$, the torsion submodule of $N$ with respect to the torsion theory cogenerated by $\mathbb{S}$ is given by the reject of $\mathbb{S}$ in $N$, that is, by

$$\mathrm{Re}_{\mathbb{S}}(N) := \mathrm{Re}\,(N, \mathbb{S}) \quad = \bigcap \{U \subseteq N \mid N/U \text{ is } M\text{-small }\}$$
$$= \bigcap \{U \subseteq N \mid N/U \text{ is } M\text{-small and cocyclic }\}.$$

Recall that the class $\mathbb{S}$ depends on $M$ and to remember this we write $\mathbb{S}[M]$ and $\mathrm{Re}_{\mathbb{S}[M]}$ if appropriate. The last equality is derived from the fact that every factor module of $M$ is cogenerated by cocyclic factor modules (i.e., modules with simple essential socle) of $M$.

By definition, $\mathrm{Re}_{\mathbb{S}}(N)$ is the smallest submodule $K \subset N$ for which $N/K$ is cogenerated by $M$-small modules. The modules with $\mathrm{Re}_{\mathbb{S}}(N) = 0$ are precisely the $\mathbb{S}$-cogenerated modules. We call them *$M$-small-cogenerated* modules.

The general properties of radicals yield

**8.10. Properties of $\mathrm{Re}_{\mathbb{S}}$.** *Let $K$, $L$ and $\{N_\lambda\}_{\lambda \in \Lambda}$ be modules in $\sigma[M]$.*

(1) *If $K \subseteq L$, then $\mathrm{Re}_{\mathbb{S}}(K) \subseteq \mathrm{Re}_{\mathbb{S}}(L)$ and $(\mathrm{Re}_{\mathbb{S}}(L) + K)/K \subseteq \mathrm{Re}_{\mathbb{S}}(L/K)$.*

(2) *If $f : K \to L$ is a morphism, then $(\mathrm{Re}_{\mathbb{S}}(K))f \subseteq \mathrm{Re}_{\mathbb{S}}(L)$.*

(3) $\mathrm{Res}\,(L/\mathrm{Res}(L)) = 0$.

(4) $\mathrm{Res}\,(\bigoplus_\Lambda N_\lambda) = \bigoplus_\Lambda \mathrm{Res}(N_\lambda)$.

(5) $\mathrm{Res}\,(\prod_\Lambda N_\lambda) \subseteq \prod_\Lambda \mathrm{Res}(N_\lambda)$.

(6) *If $N = K + L$, where $L$ is an $M$-small module, then $\mathrm{Res}(K) = \mathrm{Res}(N)$.*

The torsion modules, that is, the modules in $\mathbb{S}^\circ$, are those $N \in \sigma[M]$ with $N = \mathrm{Res}(N)$. Condition (d) below shows that they can be characterised by a generalised injectivity condition.

**8.11. Fully non-$M$-small modules.** *For $N \in \sigma[M]$ the following are equivalent:*

(a) $N = \mathrm{Res}(N)$;

(b) *for any nonzero $f : N \to K$ in $\sigma[M]$, $\mathrm{Im}\,f$ is coclosed in $K$;*

(c) *for any nonzero $f : N \to K$ in $\sigma[M]$, $\mathrm{Im}\,f$ is not small in $K$;*

(d) *for any nonzero monomorphism $i : A \to B$ and morphism $f : A \to L$ where $L$ is a factor module of $N$, there exists an epimorphism $p : L \to \tilde{L}$ and some $g : B \to \tilde{L}$ such that $i \diamond g = f \diamond p \neq 0$;*

(e) *every nonzero homomorphic image of $N$ is (strongly) coclosed in any extension module in $\sigma[M]$;*

(f) $\nabla(N, X) = 0$ *for all $X \in \sigma[M]$.*

Modules $N$ satisfying these conditions are called *fully non-$M$-small*.

*Proof.* (a)$\Rightarrow$(b). Assume for some submodule $U \subset \mathrm{Im}\,f$, $\mathrm{Im}\,f/U \ll K/U$. Then the map $N \xrightarrow{f} K \to K/U$ has an $M$-small image and so has to be zero. This gives $U = \mathrm{Im}\,f$ and so $\mathrm{Im}\,f$ is coclosed in $K$.

(b)$\Rightarrow$(c)$\Rightarrow$(a) are obvious.

(a)$\Rightarrow$(d). The data given yield — by a pushout construction — the commutative diagram

$$
\begin{array}{ccc}
0 \longrightarrow A & \xrightarrow{\;\;i\;\;} & B \\
\quad\;\; f\downarrow & & \downarrow g' \\
0 \longrightarrow L & \xdashrightarrow{\;\;p'\;\;} & Q.
\end{array}
$$

By (a), $(L)p'$ is not small in $Q$ and so there exists a proper submodule $V \subset Q$ with $(L)p' + V = Q$. Letting $q : Q \to Q/V$ denote the projection, the composition $p = p' \diamond q : L \to Q/V$ is an epimorphism and $f \diamond p = i \diamond g' \diamond q \neq 0$.

(d)$\Rightarrow$(a). Let $N$ satisfy the conditions in (d) and let $L$ be a nonzero factor module of $N$. Then we construct the commutative diagram

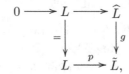

$$
\begin{array}{ccc}
0 \longrightarrow L & \longrightarrow & \widehat{L} \\
\;\;=\downarrow & & \downarrow g \\
L & \xrightarrow{\;\;p\;\;} & \tilde{L},
\end{array}
$$

where $g \neq 0$ and $\widehat{L} = L + \operatorname{Ke} g$. This shows that $L$ is not $M$-small.

(b)$\Leftrightarrow$(e) and (c)$\Leftrightarrow$(f) are obvious. $\qquad\square$

The above conditions suggest that the obvious candidates for fully non-$M$-small modules are the cohereditary modules (1.22). Further examples are provided by cosemisimple modules.

**8.12. Properties of fully non-$M$-small modules.** *Let $N \in \mathbb{S}^{\circ}$.*

(1) *Every $M$-small submodule $K \subset N$ is small in $N$, so $\operatorname{Rad}(N) = \operatorname{Tr}_{\mathbb{S}}(N)$.*

(2) *If $N \subset L$ for $L \in \sigma[M]$, then $N$ is coclosed in $L$.*

(3) *A submodule $K \subset N$ is coclosed if and only if $K \in \mathbb{S}^{\circ}$.*

(4) *If $N \simeq P/K$ with $K \ll P$, then $P \in \mathbb{S}^{\circ}$.*

(5) *$\mathbb{S}^{\circ}$ is closed under factor modules, direct sums, extensions, small covers and coclosed submodules.*

*Proof.* (1) Consider an $M$-small submodule $K \subset N$. Let $L \subset N$ with $K + L = N$. The factor module $K/(K \cap L) \simeq N/L$ is $M$-small and hence zero. Therefore $K \cap L = K$ and so $L = N$, proving $K \ll N$.

(2) Assume for some submodule $U \subset N$, $N/U \ll L/U$. Then $N/U$ is $M$-small and hence zero. Hence $N = U$.

(3) Sufficiency is clear by (2). For necessity, let $K \subset N$ be a coclosed submodule and assume $K/U$ is $M$-small for some $U \subset K$. Then, by (1), $K/U \ll N/K$ and thus $U = K$, proving that $K \in \mathbb{S}^{\circ}$.

(4) Let $N \simeq P/K$ with $K \ll P$. Assume there exists $L \subset P$ such that $P/L$ is $M$-small. Then $P/(K + L)$ is an $M$-small factor module of $N$ and hence zero, that is, $P = K + L$. Since $K \ll P$ this means $L = P$.

(5) The first three properties of $\mathbb{S}^{\circ}$ follow from general torsion theory, the others by (3) and (4). $\qquad\square$

**8.13. $M$-cosmall inclusions.** For $L \in \sigma[M]$, an inclusion $K \subset L$ is called *$M$-cosmall* if $L/K$ is $M$-small, that is, $L/K \in \mathbb{S}$.

**Properties.**

(1) *Any cosmall inclusion $K \subset L$ in $N$ is $M$-cosmall.*

(2) *If $N \in \mathbb{S}^{\circ}$, then for every submodule $L \subset N$, any $M$-cosmall inclusion $K \subset L$ is cosmall in $N$.*

*Proof.* (1) is obvious and (2) follows from the properties shown in 8.12. $\qquad\square$

**8.14. $M$-small modules and pseudo-projectives.** *Let $M$ be pseudo-projective in the category $\sigma[M]$.*

(1) *If $N \in \sigma[M]$ and $\operatorname{Hom}(M, N) = 0$, then $N$ is $M$-small.*

(2) *If $M \in \mathbb{S}^{\circ}$, then $\mathbb{S} = \{N \in \sigma[M] \mid \operatorname{Hom}(M, N) = 0\}$.*

*Proof.* (1) Suppose that $N$ is not $M$-small. Let $K \subset \widehat{N}$ be a proper submodule with $N + K = \widehat{N}$ and consider the composition $p : N \to \widehat{N} \to \widehat{N}/K$. There exists a nonzero morphism $f : M \to \widehat{N}/K$. By pseudo-projectivity, there are morphisms $g : M \to M$ and $h : M \to N$ such that $g \diamond f = h \diamond p \neq 0$. Thus $\mathrm{Hom}(M, N) \neq 0$.

(2) This is clear by (1) since $M$ has no nonzero $M$-small factor modules.  $\square$

Notice that for all $M$-small modules $N$, $\mathrm{Re}_{\mathbb{S}}(N) = 0$ and all modules with this property (i.e., those in $\mathrm{Cog}(\mathbb{S})$) belong to the torsion free class $\mathbb{S}^{\circ\bullet}$.

**8.15. Examples.** (1) Let $R = M = \mathbb{Z}$. Then $\mathrm{Tr}_{\mathbb{S}[\mathbb{Z}]}(\mathbb{Q}) = \mathbb{Q}$, although $\mathbb{Q}$ is a factor module of $\mathbb{Z}^{(\mathbb{N})}$ and $\mathrm{Tr}_{\mathbb{S}[\mathbb{Z}]}(\mathbb{Z}^{(\mathbb{N})}) = 0$. Clearly $\mathbb{Z}^{(\mathbb{N})}$ is not $\mathbb{Z}$-small. This shows that the class of modules $N$ with $\mathrm{Tr}_{\mathbb{S}}(N) = 0$ need not be closed under factor modules and that the (direct) sum of $\mathbb{Z}$-small modules need not be $\mathbb{Z}$-small.

(2) Let $R = \mathbb{Z}$ and $M = \mathbb{Z}/4\mathbb{Z}$. Then $\mathbb{Z}/2\mathbb{Z}$ is $M$-small and hence we have $\mathrm{Re}_{\mathbb{S}}(\mathbb{Z}/2\mathbb{Z}) = 0$ but $\mathrm{Re}_{\mathbb{S}}(M) \neq 0$. This shows that the class of modules $N$ with $\mathrm{Tr}_{\mathbb{S}}(N) = 0$ need not be closed under extensions.

Clearly $\mathrm{Re}_{\mathbb{S}}$ defines a radical in $\sigma[M]$. The largest idempotent radical smaller than $\mathrm{Re}_{\mathbb{S}}$ is obtained by transfinite induction (e.g., [323, Chapter VI]).

**8.16. $\mathrm{Re}_{\mathbb{S}}{}^\alpha(N)$.** For an $R$-module $N \in \sigma[M]$, we set

$$\mathrm{Re}_{\mathbb{S}}{}^0(N) = N, \quad \mathrm{Re}_{\mathbb{S}}{}^1(N) = \mathrm{Re}_{\mathbb{S}}(N),$$

and, for any ordinal $\alpha$ which is not a limit ordinal, we define

$$\mathrm{Re}_{\mathbb{S}}{}^\alpha(N) = \mathrm{Re}_{\mathbb{S}}(\mathrm{Re}_{\mathbb{S}}{}^{\alpha-1}(N)),$$

and for a limit ordinal $\alpha$,

$$\mathrm{Re}_{\mathbb{S}}{}^\alpha(N) = \bigcap\nolimits_{\beta < \alpha} \mathrm{Re}_{\mathbb{S}}{}^\beta(N).$$

This yields the descending sequence

$$N = \mathrm{Re}_{\mathbb{S}}{}^0(N) \supseteq \mathrm{Re}_{\mathbb{S}}{}^1(N) \supseteq \mathrm{Re}_{\mathbb{S}}{}^2(N) \supseteq \cdots$$

of submodules of $N$. For any fully non-$M$-small submodule of $L \subset N$, we have $L \subseteq \mathrm{Re}_{\mathbb{S}}{}^\alpha(N)$ for every ordinal $\alpha$. Hence the idempotent radical determined by the class $\mathbb{S}$ is

$$\varrho_{\mathbb{S}}(N) = \bigcap\nolimits_\alpha \mathrm{Re}_{\mathbb{S}}{}^\alpha(N).$$

**8.17. When is $\varrho_{\mathbb{S}}$ cohereditary?** *For $M$, the following are equivalent:*

(a) *$\varrho_{\mathbb{S}}$ is a cohereditary radical for $\sigma[M]$;*

(b) *for any $N \in \sigma[M]$, $\mathrm{Re}_{\mathbb{S}}(N)/\varrho_{\mathbb{S}}(N) \ll N/\varrho_{\mathbb{S}}(N)$;*

(c) *for any $N \in \mathbb{F}_{\varrho_{\mathbb{S}}}$, $\mathrm{Re}_{\mathbb{S}}(N) \ll N$.*

*If these conditions hold, then $\varrho_{\mathbb{S}} = \mathrm{Res}^2$.*

*Proof.* (a)$\Rightarrow$(b). Let $\varrho_{\mathbb{S}}(N) \subseteq X \subset N$ such that $\mathrm{Res}_{\mathbb{S}}(N) + X = N$. Then $N/X \simeq \mathrm{Res}_{\mathbb{S}}(N)/(X \cap \mathrm{Res}_{\mathbb{S}}(N))$, where $\varrho_{\mathbb{S}}(N) \subseteq X \cap \mathrm{Res}_{\mathbb{S}}(N)$. Hence $N/X$ is a factor module of $\mathrm{Res}_{\mathbb{S}}(N)/\varrho_{\mathbb{S}}(N) \in \mathbb{F}_{\varrho_{\mathbb{S}}}$ and thus $N/X \in \mathbb{F}_{\varrho_{\mathbb{S}}}$ (by cohereditariness). Consequently $\mathrm{Res}(N/X) \neq N/X$ and there is a proper submodule $Y \subset N$ such that $X \subset Y$ and $N/Y \in \mathbb{S}$. Now $\mathrm{Res}_{\mathbb{S}}(N) + Y = N$ but also $\mathrm{Res}_{\mathbb{S}}(N) \subset Y$, a contradiction.

Clearly in this case $\varrho_{\mathbb{S}} = \mathrm{Res}^2$.

(b)$\Rightarrow$(c) is obvious.

(c)$\Rightarrow$(a). Let $N \in \mathbb{F}_{\varrho_{\mathbb{S}}}$ and $f : N \to N'$ be any epimorphism such that $N' \notin \mathbb{F}_{\varrho_{\mathbb{S}}}$. Setting $L = (\varrho_{\mathbb{S}}(N'))f^{-1}$, we obtain an epimorphism $g : L \to L'$, where $L \in \mathbb{F}_{\varrho_{\mathbb{S}}}$ and $L' = \varrho_{\mathbb{S}}(N') \in \mathbb{T}_{\varrho_{\mathbb{S}}}$. By assumption, $\mathrm{Res}_{\mathbb{S}}(L) + \mathrm{Ke}\, g \neq L$ and hence $X = L/(\mathrm{Res}_{\mathbb{S}}(L) + \mathrm{Ke}\, g)$ has to have a nonzero factor module in $\mathbb{S}$ (as factor module of $L/\mathrm{Res}_{\mathbb{S}}(L)$). On the other hand, $X$ is a factor module of $L' = L/\mathrm{Ke}\, g$, and thus $\mathrm{Hom}(L', S) = 0$ for all $S \in \mathbb{S}$ (since $L' \in \mathbb{T}_{\varrho_{\mathbb{S}}}$), a contradiction. It follows that the class $\mathbb{F}_{\varrho_{\mathbb{S}}}$ is cohereditary. $\square$

In the case of the dual concept of singular submodules, it is known that $Z_M^2(N) = Z_M(N/Z_M(N))$ is closed in $N$ and $Z_M^2(N) = Z_M^3(N)$. We give below some sufficient conditions for a module $N \in \sigma[M]$ to satisfy $\mathrm{Res}^2(N) = \mathrm{Res}^3(N)$.

**8.18. Cosmall submodules of $\mathrm{Res}_{\mathbb{S}}^{\alpha}(N)$ in $N$.** *Let $N \in \sigma[M]$ and $\alpha \geq 1$ be any ordinal. Then*

$$\mathrm{Res}_{\mathbb{S}}^{\alpha+1}(N) = \bigcap \left\{ X \mid X \xrightarrow[N]{cs} \mathrm{Res}_{\mathbb{S}}^{\alpha}(N) \right\}$$

*and the family of all cosmall submodules of $\mathrm{Res}_{\mathbb{S}}^{\alpha}(N)$ in $N$ is closed under finite intersections.*

*Proof.* It is clear that if $X \xrightarrow[N]{cs} \mathrm{Res}_{\mathbb{S}}^{\alpha}(N)$, then $\mathrm{Res}_{\mathbb{S}}^{\alpha+1}(N) \subseteq X$. Let

$$\mathrm{Res}_{\mathbb{S}}^{\alpha+1}(N) \subseteq X \subseteq \mathrm{Res}_{\mathbb{S}}^{\alpha}(N).$$

Then $\mathrm{Res}_{\mathbb{S}}^{\alpha}(N)/X$ is $M$-small. We claim that $\mathrm{Res}_{\mathbb{S}}^{\alpha}(N)/X \ll N/X$ and hence $X \xrightarrow[N]{cs} \mathrm{Res}_{\mathbb{S}}^{\alpha}(N)$. Suppose that $N/X = \mathrm{Res}_{\mathbb{S}}^{\alpha}(N)/X + L/X$. We show that $N/L$ is both a fully non-$M$-small and an $M$-small module which will imply $N = L$. For this let $f : N \to N/L$ be the natural map. Then

$$N/L = (\mathrm{Res}_{\mathbb{S}}^{\alpha}(N) + L)/L \subseteq (\mathrm{Res}_{\mathbb{S}}(N) + L)/L = (\mathrm{Res}_{\mathbb{S}}(N))f \subseteq \mathrm{Res}_{\mathbb{S}}(N/L)$$

and so $N/L$ is fully non-$M$-small. Next, if $g : N/X \to N/L$ is the natural map, then $(\mathrm{Res}_{\mathbb{S}}^{\alpha}(N)/X)g = N/L$ and so $N/L$ is $M$-small. Therefore $L = N$ and so $\mathrm{Res}_{\mathbb{S}}^{\alpha}(N)/X \ll N/X$. Hence

$$\mathrm{Res}_{\mathbb{S}}^{\alpha+1}(N) = \bigcap \{ X \mid X \xrightarrow[N]{cs} \mathrm{Res}_{\mathbb{S}}^{\alpha}(N) \}.$$

Let $X$, $Y$ be cosmall submodules of $\mathrm{Re_S}^\alpha(N)$ in $N$. Then $\mathrm{Re_S}^\alpha(N)/X$ and $\mathrm{Re_S}^\alpha(N)/Y$ are $M$-small modules. Since $\mathrm{Re_S}^\alpha(N)/(X \cap Y)$ is isomorphic to a submodule of $\mathrm{Re_S}^\alpha(N)/X \oplus \mathrm{Re_S}^\alpha(N)/Y$ and, by 8.2 (4), the class of $M$-small modules is closed under finite direct sums and submodules, $\mathrm{Re_S}^\alpha(N)/(X \cap Y)$ is an $M$-small module. Thus $(X \cap Y) \overset{cs}{\underset{N}{\hookrightarrow}} \mathrm{Re_S}^\alpha(N)$. $\qquad\square$

**8.19. Unique coclosure of $\mathrm{Re_S}(N)$.** *Let $N \in \sigma[M]$. If $\mathrm{Re_S}(N)$ has a coclosure in $N$, then $\mathrm{Re_S}^2(N)$ is the unique coclosure of $\mathrm{Re_S}(N)$. Hence $\mathrm{Re_S}^2(N)$ is the largest fully non-$M$-small submodule of $N$ and $\mathrm{Re_S}^\alpha(N) = \mathrm{Re_S}^2(N)$ for all ordinals $\alpha$ at least 2.*

*Proof.* Let $S$ and $T$ both be coclosures of $\mathrm{Re_S}(N)$ in $N$. Then, by 8.18, we have $S \cap T \overset{cs}{\underset{N}{\hookrightarrow}} \mathrm{Re_S}(N)$. Since $S$ and $T$ are both coclosed, $S = T$. This proves uniqueness. Moreover, if $X \overset{cs}{\underset{N}{\hookrightarrow}} \mathrm{Re_S}(N)$, then, by 8.18, $(S \cap X) \overset{cs}{\underset{N}{\hookrightarrow}} \mathrm{Re_S}(N)$ and so, since $S$ is coclosed in $N$, we get $S \subseteq X$. By 8.18, $S \subseteq \mathrm{Re_S}^2(N)$. Thus $\mathrm{Re_S}^2(N)$ is the unique coclosure of $\mathrm{Re_S}(N)$ in $N$. Now if $X \overset{cs}{\underset{N}{\hookrightarrow}} \mathrm{Re_S}^2(N)$, then since $\mathrm{Re_S}^2(N) \overset{cs}{\underset{N}{\hookrightarrow}} \mathrm{Re_S}(N)$, we have $X \overset{cs}{\underset{N}{\hookrightarrow}} \mathrm{Re_S}(N)$. Consequently, again by 8.18, we get $\mathrm{Re_S}^3(N) = \mathrm{Re_S}^2(N)$ and so $\mathrm{Re_S}^2(N)$ is the largest fully non-$M$-small submodule of $N$. $\qquad\square$

**8.20. Remark.** Notice that in general there may be modules in $\sigma[M]$ which are not $M$-small-cogenerated. In other words, there may be modules $N \in \sigma[M]$ such that $\mathrm{Re_S}(N) \neq 0$. For example, if $\widehat{S}$ is the $M$-injective hull of a simple module $S \in \sigma[M]$, then $S \subseteq \mathrm{Re_S}(\widehat{S})$. More generally, injective modules with essential socle are not $M$-small-cogenerated. However, there are modules $M$ such that for every $N \in \sigma[M]$, $\mathrm{Re_S}^2(N) = 0$. For example, consider $R = \mathbb{Z}$ and $M = \mathbb{Z}/8\mathbb{Z}$.

**8.21. Splitting of the torsion theory cogenerated by $\mathbb{S}$.** *Let $M$ be an $R$-module. Then the torsion theory $(\mathbb{S}^\circ, \mathbb{S}^{\circ\bullet})$ splits in the following cases.*

(i) *$M$ is a cosemisimple module. In this case $\mathbb{T}_\mathbb{S} = \sigma[M]$.*

(ii) *Every $M$-small-cogenerated module in $\sigma[M]$ is projective in $\sigma[M]$.*

*Proof.* (i) If $M$ is cosemisimple, then $\mathbb{S}^\circ = \sigma[M]$.

(ii) Let $N \in \sigma[M]$. Since $N/\mathrm{Re_S}(N)$ and $\mathrm{Re_S}(N)/\mathrm{Re_S}^2(N)$ are $M$-small-cogenerated, they are projective in $\sigma[M]$. Hence

$$N = \mathrm{Re_S}(N) \oplus K \text{ and } \mathrm{Re_S}(N) = \mathrm{Re_S}^2(N) \oplus T$$

where $K$ and $T$ are $M$-small-cogenerated. Therefore $N/\mathrm{Re_S}^2(N)$ is $M$-small-cogenerated. By 8.10, $\mathrm{Re_S}(N) = \mathrm{Re_S}^2(N)$ and so is the largest fully non-$M$-small submodule of $N$. Thus $N$ is a direct sum of a fully non-$M$-small and an $M$-small-cogenerated module. $\qquad\square$

Another instance of when $(\mathbb{S}^{\circ}, \mathbb{S}^{\circ \bullet})$ splits will be considered in 28.6.

**8.22. Exercises.**

(1) Let $R$ be a left hereditary ring with no simple injective modules. Show that $R$ is small in $R$-Mod ([291]).

(2) Let $R$ be a left artinian ring and let $Q$ be a minimal injective cogenerator in $R$-Mod. Prove that for an $R$-module $N$ the following are equivalent ([294]):

    (a) $N$ is small in $R$-Mod;

    (b) for every morphism $f : N \to Q$, $\operatorname{Im} f \ll Q$;

    (c) $\operatorname{An}_R(\operatorname{Jac}(R))N = 0$, where $\operatorname{An}_R$ denotes the left annihilator in $R$.

(3) Prove that the following are equivalent for a ring $R$ ([294]):

    (a) $R$ is a QF ring;

    (b) every module is a direct sum of a projective module and a small module;

    (c) every module is a direct sum of an injective module and a singular module.

(4) Assume that $M$ has a projective cover $P \to M$ in $\sigma[M]$. Prove:

    (i) Any module $N \in \sigma[M]$ with $\operatorname{Hom}(P, N) = 0$ is $M$-small.

    (ii) If $M$ is fully non-$M$-small, then for the class $\mathbb{S}$ of small modules in $\sigma[M]$,

        ($\alpha$) $\mathbb{S} = \{N \in \sigma[M] \mid \operatorname{Hom}(P, N) = 0\}$,

        ($\beta$) $\mathbb{S}$ is closed under extensions, direct sums and products in $\sigma[M]$, and

        ($\gamma$) for $N \in \sigma[M]$, $\operatorname{Re}_{\mathbb{S}}(N)$ is fully non-$M$-small and $N/\operatorname{Re}_{\mathbb{S}}(N)$ is $M$-small.

(5) Let $M$ be pseudo-injective in $\sigma[M]$ (see 7.5) with $\nabla(M) = 0$. Prove that for all submodules $N$ with $M/N \in \mathbb{S}$, $\operatorname{Hom}(M/N, M) = 0$.

(6) A module $Q$ is called *coretractable* if $\operatorname{Hom}(Q/N, Q) \neq 0$ for all proper submodules $N$ of $Q$. Let $Q \in \sigma[M]$ be coretractable and $T = \operatorname{End}(Q)$. Denote by $Z(M^*)$ be the singular submodule of $M^* := \operatorname{Hom}(M, Q)$ as right $T$-module. Prove:

    (i) $Z(M^*) \subseteq \nabla(M, Q)$.

    (ii) If $Q$ is pseudo-injective in $\sigma[M]$, then $Z(M^*) = \nabla(M, Q)$.

Note that the proof of (ii) also applies if $Q$ is semi-injective.

(7) Let $M$ be coretractable and pseudo-injective in $\sigma[M]$. Prove that the following statements are equivalent:

    (a) $M$ is fully non-$M$-small;

    (b) $\nabla(M) = 0$;

    (c) $\operatorname{End}(M)$ is a right non-singular ring.

Modules with these properties are *copolyform* (see 9.17).

**References.** Ganesan and Vanaja [114]; Generalov [119]; Keskin [199]; Keskin and Tribak [204]; Leonard [220]; Lomp [226, 228]; Mbuntum and Varadarajan [230]; Özcan [278]; Ramamurthi [291]; Rayar [294]; Stenström [323]; Talebi and Vanaja [329, 330].

# 9   Corational modules

Again let $M$ denote a left $R$-module. Recall that an *M-rational module* $X \in \sigma[M]$ is characterised by the property

$$\mathrm{Hom}(Y, M) = 0 \text{ for any submodule } Y \subseteq X.$$

This class of modules is used to build up the maximal ring of quotients of a ring and more generally to construct quotients for modules. Maximal rational extensions of a module always exist and can be identified with some submodule of the (self-) injective hull. In this section we are concerned with the dual concept. Because of the possible lack of projective covers and of supplements (the duals to complements), the resulting theory is not completely dual to the rational case.

**9.1. Definition.** An $R$-module $X \in \sigma[M]$ is called *M-corational* if

$$\mathrm{Hom}(M, X/Y) = 0 \text{ for any submodule } Y \subset X.$$

With the notation from 7.6, the class of all *M*-corational modules in $\sigma[M]$ is

$$\{M\}^{\blacktriangleright} = \{X \in \sigma[M] \mid \mathrm{Hom}(M, X/Y) = 0 \mid \text{ for all } Y \subset X\} \subseteq \{M\}^{\bullet}.$$

An epimorphism $f : M \to N$ is said to be an *M-corational cover* if $\mathrm{Ke}\, f$ is an *M*-corational module. In this case $\mathrm{Ke}\, f \ll M$ (see 9.4).

**9.2. Example.** As a $\mathbb{Z}$-homomorphism, the projection $\mathbb{Q} \to \mathbb{Q}/\mathbb{Z}$ is a $\mathbb{Q}$-corational cover since $\mathrm{Hom}(\mathbb{Q}, \mathbb{Z}/\mathbb{Z}n)$ is clearly zero for all $n \in \mathbb{N}$. (Here $\mathbb{Q}$ is not projective.)

Notice that *M*-corational is the same as $M^{(\Lambda)}$-corational for any index set $\Lambda$. As noted in 7.6, $\{M\}^{\blacktriangleright}$ is closed under submodules, factor modules and extension. If $P$ is a pseudo-projective module in $\sigma[M]$, then $\{P\}^{\blacktriangleright} = \{P\}^{\bullet}$ (see 7.7).

As remarked in 7.10, *M*-rational modules $X$ are those with $\mathrm{Hom}(X, \widehat{M}) = 0$ where $\widehat{M}$ is the *M*-injective hull of *M*. In particular, the class of *M*-rational modules is a hereditary pretorsion class.

A dual characterisation of *M*-corational modules is possible provided *M* has a (pseudo-)projective cover in $\sigma[M]$ (see 7.7). Otherwise $\{M\}^{\blacktriangleright}$ need not be closed under products. For example, for $R = \mathbb{Z}$ and $M = \mathbb{Q}$, $\{\mathbb{Q}\}^{\blacktriangleright}$ is not closed under products and direct sums in $\mathbb{Z}$-Mod, since $\mathbb{Z} \in \{\mathbb{Q}\}^{\blacktriangleright}$ but $\mathbb{Q} \simeq \mathbb{Z}^{(\mathbb{N})}/U$ for some submodule $U \subset \mathbb{Z}^{(\mathbb{N})}$.

**9.3. Corational torsion theory.** *For a module M the following are equivalent:*

(a) $\{M\}^{\blacktriangleright}$ *is closed under products in* $\sigma[M]$;

(b) $\{M\}^{\blacktriangleright} = \{P\}^{\bullet}$ *for some pseudo-projective module* $P \in \sigma[M]$.

*These conditions are satisfied provided there is a small epimorphism* $P \to M$ *where* $P \in \sigma[M]$ *is pseudo-projective.*

*Proof.* (a)⇒(b). Since $\{M\}^{\blacktriangleright}$ is always closed under factor modules, submodules and extensions (see 7.6), the condition in (a) implies that it is a torsion free class. By 7.3, this gives $\{M\}^{\blacktriangleright} = (\{M\}^{\blacktriangleright})^{\circ\bullet}$ and we have the cohereditary torsion theory $((\{M\}^{\blacktriangleright})^{\circ}, \{M\}^{\blacktriangleright})$. Then, by 6.13, $(\{M\}^{\blacktriangleright})^{\circ} = \text{Gen}(P)$ for some pseudo-projective module $P \in \sigma[M]$ and

$$\{M\}^{\blacktriangleright} = (\{M\}^{\blacktriangleright})^{\circ\bullet} = \text{Gen}(P)^{\bullet} = \{P\}^{\bullet}.$$

(b)⇒(a). The equality implies that the class under consideration is closed under products.

As shown in 7.7, if $P \to M$ is a pseudo-projective cover in $\sigma[M]$, $\{M\}^{\blacktriangleright} = \{P\}^{\bullet}$, and in this case $M$-corational is the same as $P$-corational. $\square$

We point out that if $M$ is a generator in $\sigma[M]$, there are no nonzero $M$-corational modules in $\sigma[M]$. In particular, any $R$-corational module has to be zero.

**9.4. Corational submodules are small.** *Let $N$ be an $M$-generated module. Then every $M$-corational submodule $K \subset N$ is small in $N$.*

*Proof.* Let $K \subset N$ be an $M$-corational submodule and $N = K + L$ for some submodule $L \subseteq N$. Then $\text{Hom}(M, N/L) \simeq \text{Hom}(M, K/(K \cap L)) = 0$. Since $N/L$ is $M$-generated this implies $L = N$, that is, $K$ is small in $N$. $\square$

**9.5. A corational cover.** *Let $f : P \to M$ be a pseudo-projective cover in $\sigma[M]$. Then $P/\text{Tr}(P, \text{Ke}(f)) \to M$ is a corational cover.*

*Proof.* It follows from 9.3 that the $M$-corational modules induce a cohereditary preradical for which $\text{Ke}\,f/\text{Tr}(P, \text{Ke}\,f)$ is $M$-corational. This is the kernel of the map $P/\text{Tr}(P, \text{Ke}(f)) \to M$. $\square$

The converse of 9.5 need not be true since, for any ring $R$, the small epimorphism $R \to R/\text{Jac}(R)$ is not corational unless $\text{Jac}(R) = 0$.

**9.6. Corational submodules.** Suppose $A \subseteq B \subseteq M$. We call $A$ a *corational submodule of $B$ in $M$*, if the natural epimorphism $M/A \to M/B$ is an $M/A$-corational cover, that is, $B/A$ is $M/A$-corational. We denote this by $A \xrightarrow[M]{cr} B$. Thus $A \xrightarrow[M]{cr} B$ if and only if $\text{Hom}(M/A, B/C) = 0$ for all $A \subseteq C \subseteq B$.

In particular, 0 is a corational submodule of $B$ in $M$, that is, $0 \xrightarrow[M]{cr} B$, if and only if $B$ is an $M$-corational module.

For the next result we recall that $A \xrightarrow[M]{cs} B$ denotes the property $B/A \ll M/A$.

**9.7. Properties of corational submodules.** *Let $M$ be an $R$-module.*

(1) *If $A \xrightarrow[M]{cr} B$, then $A \xrightarrow[M]{cs} B$. In general the converse is not true.*

(2) *Suppose $A \subseteq B \subseteq C \subseteq M$. $A \xrightarrow[M]{cr} C$ if and only if $A \xrightarrow[M]{cr} B$ and $B \xrightarrow[M]{cr} C$.*

(3) *Suppose $A \subseteq A' \subseteq B' \subseteq B \subseteq M$. If $A \xrightarrow[M]{cr} B$, then $A' \xrightarrow[M]{cr} B'$.*

(4) *If $A \xrightarrow[M]{cr} B$ and $C \xrightarrow[M]{cr} D$, then $(A+C) \xrightarrow[M]{cr} (C+D)$.*

(5) *Suppose $X \subseteq A \subseteq B \subseteq M$. Then $A/X \xrightarrow[M/X]{cr} B/X$ if and only if $A \xrightarrow[M]{cr} B$.*

(6) *Suppose $A \subseteq B \subseteq M$ and $f : M \to L$ is an epimorphism. If $A \xrightarrow[M]{cr} B$, then*
$(A)f \xrightarrow[L]{cr} (B)f$.

**Proof.** (1) Suppose that $A \xrightarrow[M]{cr} B$ and $B/A + Y/A = M/A$ for some $A \subseteq Y \subseteq M$. Then there is an epimorphism

$$f : M/A \to (B/A + Y/A)\, /\, Y/A \simeq B/(B \cap Y).$$

Hence $B = B \cap Y$ and so $Y = M$. Thus $A \xrightarrow[M]{cs} B$.

For the converse consider $R = \mathbb{Z}$ and $M = \mathbb{Z}/4\mathbb{Z}$. Then $0 \xrightarrow[M]{cs} 2\mathbb{Z}/4\mathbb{Z}$ but $0 \xrightarrow[M]{cr}\!\!\!\!\!/\;\; 2\mathbb{Z}/4\mathbb{Z}$.

(2) Suppose $A \xrightarrow[M]{cr} C$. Then $\mathrm{Hom}(M/A, C/X) = 0$ for all $A \subseteq X \subseteq C$. Hence $\mathrm{Hom}(M/A, B/Y) = 0$ for all $A \subseteq Y \subseteq B$ and so $A \xrightarrow[M]{cr} B$. Suppose $B \subseteq Z \subseteq C$ and $f : M/B \to C/Z$. Then $\eta f = 0$, where $\eta : M/A \to M/B$ is the natural map, and hence $f = 0$. Thus $B \xrightarrow[M]{cr} C$.

Let $A \xrightarrow[M]{cr} B$ and $B \xrightarrow[M]{cr} C$. Suppose $f : M/A \to C/X$, where $A \subseteq X \subseteq C$ is a nonzero map. Let $\mathrm{Im}\, f = Y/X$. Suppose $Y + B = X + B$. Then $Y/X \subseteq (X+B)/X \simeq B/(B \cap X)$, and so, since $A \subseteq (B \cap X) \subseteq B$ and $A \xrightarrow[M]{cr} B$, we get $f = 0$, a contradiction. Hence $Y + B \neq X + B$. Let $g : C/X \to C/(X+B)$ be the natural map. Put $D := X + B$ and $L := Y + B$. The map $h = fg : M/A \to C/D$ is a nonzero map with $\mathrm{Im}\, h = L/D$. Suppose $(B/A)h = T/D$. As $B/A \ll M/A$ we get $T/D \ll L/D$. Therefore $h$ induces a nonzero map $\overline{h} : M/B \to C/T$. Now $B \subseteq T \subseteq C$ and $B \xrightarrow[M]{cr} C$ imply $\overline{h} = 0$, a contradiction. Thus $A \xrightarrow[M]{cr} C$.

(3) This follows easily from (2).

(4) By (2), it is enough to prove that if $A \xrightarrow[M]{cr} B$, then $(A+X) \xrightarrow[M]{cr} (B+X)$ for any $X \subseteq M$. Let $f : M/(A+X) \to (B+X)/Y$ where $(A+X) \subseteq Y \subseteq (B+X)$. There exists $T$ such that $A \subseteq T \subseteq B$ and $Y = T + X$. Let $g : M/A \to M/(A+X)$ and $h : (B+X)/Y = (B+T+X)/(T+X) \to B/(B \cap (T+X))$ be the obvious maps.

Since $A \overset{cr}{\underset{M}{\hookrightarrow}} B$, $gfh = 0$. Then, since $h$ is an isomorphism and $g$ is an epimorphism, we get $f = 0$.

(5) is easy to prove, and (6) follows from (5) and (4). $\qquad \square$

**9.8. Minimal corational submodule.** Let $A$ be a submodule of an $R$-module $M$. We call a submodule $L$ of $A$ a *minimal corational submodule* of $A$ in $M$ if it is minimal among all corational submodules of $A$ in $M$ under set inclusion. If the set of all corational submodules of $A$ in $M$ has only one minimal element, then we refer to this submodule as the *unique minimal corational submodule* or simply as the *smallest corational submodule* of $A$ in $M$.

**9.9. Lemma.** *Let $M$ be any module and $B$ a small submodule of $M$. Then we have $\mathrm{Tr}(M,B) \subseteq A$ for any corational submodule $A$ of $B$ in $M$. If $M$ is pseudo-projective in $\sigma[M]$, then $\mathrm{Tr}(M,B)$ is the unique minimal corational submodule of $B$ in $M$, that is,*

$$\mathrm{Tr}(M,B) = \bigcap \{A \subseteq B \mid A \overset{cr}{\underset{M}{\hookrightarrow}} B\} \overset{cr}{\underset{M}{\hookrightarrow}} B.$$

*Proof.* Let $B \ll M$ and $A \overset{cr}{\underset{M}{\hookrightarrow}} B$. For any $f \in \mathrm{Hom}(M,B)$, let $C = A + (A)f$ and $\overline{f} : M/A \to B/C$ be the induced map $(m + A)\overline{f} := (m)f + C$. Since $A \overset{cr}{\underset{M}{\hookrightarrow}} B$, we have $\mathrm{Hom}(M/A, B/C) = 0$ and so $\overline{f} = 0$. This gives $(M)f = C = A + (A)f$. However, since $A \ll M$, we also have $(A)f \ll (M)f$. Thus $(M)f = A$, that is, $\mathrm{Im}\, f \subseteq A$. This shows $\mathrm{Tr}(M,B) \subseteq A$.

If $M$ is pseudo-projective in $\sigma[M]$, then $\mathrm{Tr}(M,B) \overset{cr}{\underset{M}{\hookrightarrow}} B$ by 9.5. Since any corational submodule $A$ of $B$ in $M$ contains $\mathrm{Tr}(M,B)$, this is the unique minimal corational submodule of $B$ in $M$. $\qquad \square$

**9.10. Corollary.** *If $f : P \to M$ is a pseudo-projective cover in $\sigma[M]$, then $\mathrm{Tr}(P, \mathrm{Ke}\, f)$ is the unique minimal corational submodule of $\mathrm{Ke}\, f$ in $P$.*

An example of a module $M$ and $B \subseteq M$ such that $\{A \subseteq B \mid A \overset{cr}{\underset{M}{\hookrightarrow}} B\}$ has no minimal element under set inclusion is given in 9.13.

**9.11. Minimal corational submodules in fully non-$M$-small modules.** *Let $N$ be a fully non-$M$-small module with submodule $K$. Then the set of corational submodules of $K$ in $N$ is closed under finite intersection. In particular, any minimal corational submodule of $K$ in $N$ is the smallest corational submodule of $K$ in $N$.*

*Proof.* If $X$ and $Y$ are two corational submodules of $K$ in $N$, then

$$K/(X \cap Y) \subseteq K/X \oplus K/Y \ll N/X \oplus N/Y$$

is $M$-small. By 8.13, $X \cap Y \overset{cs}{\underset{N}{\hookrightarrow}} K$ and, since $N$ is a fully non-$M$-small module,

$X \cap Y \overset{cr}{\underset{N}{\hookrightarrow}} K.$ $\qquad \square$

**9.12. Corollary.** *Let $\{K_i\}_I$ be a family of nonzero corational submodules of a submodule $K$ in a fully non-$M$-small module $N$. If $\bigcap_I K_i = 0$, then either $K$ has no minimal corational submodule in $N$ or $K \ll N$.*

*Proof.* Suppose that $K$ contains a minimal corational submodule $X$ in $N$. Then, by 9.11, $X \cap K_i \overset{cr}{\underset{N}{\hookrightarrow}} K$ for any $i \in I$. Since $X$ is minimal, $X = X \cap K_i$ and so $X \subseteq K_i$. Thus $X \subseteq \bigcap_I K_i = 0$, that is, $0 \overset{cr}{\underset{N}{\hookrightarrow}} K$. Thus $K \ll N$. $\qquad\square$

**9.13. Example.** Take $R = M = \mathbb{Z}$. Let $A_i = \mathbb{Z}/2^i\mathbb{Z}$, for all $i \in \mathbb{N}$, and put $E = E(A_1)$, the injective hull of $A_1$. Consider $N = \bigoplus_{\mathbb{N}} E_i$, where $E_i = E$, for all $i$. Take $B = \bigoplus_{\mathbb{N}} A_i$ and $B_n = \bigoplus_{k>n} A_k$. Since $N$, and hence every factor module of $N$, is an injective $\mathbb{Z}$-module, $N$ is fully non-$\mathbb{Z}$-small. As there exists an onto map from $B$ onto the injective $\mathbb{Z}$-module $E$, $B$ is non-$\mathbb{Z}$-small. For $n \in \mathbb{N}$, $B/B_n$ is $\mathbb{Z}$-small. Since $N/B_n$ is a fully non-$\mathbb{Z}$-small module, by 8.12, $B/B_n \ll N/B_n$ and so $B_n \overset{cr}{\underset{N}{\hookrightarrow}} B$. By 9.12, $B$ has no minimal corational submodule in $N$.

**9.14. Example.** Consider the ring $R = \begin{pmatrix} K & K \\ 0 & K \end{pmatrix}$, where $K$ is a field. For any $x \in K$, define $D_x = R(e_{11} + xe_{12})$, where $e_{ij}$ denote the matrix units. Then $D_x$ is a direct summand of $R$. For each $x \in K$, $D_x$ is a coclosure of $\mathrm{Soc}\,(R) = Re_{11} \oplus Re_{12}$, $D_x$ is a minimal corational submodule of $\mathrm{Soc}\,(R)$ and the natural map $f_x : R/D_x \to R/\mathrm{Soc}\,(R)$ is a corational cover. These covers are isomorphic.

**9.15. "Uniqueness" of minimal corational submodules.** *Suppose $M$ is an $R$-module and $f : P \to M$ is an epimorphism with $P$ self-projective and $\mathrm{Ke}\,f = K$. Let $A$ and $B$ be two minimal corational submodules of $K$ in $P$. If $h : P/A \to P/B$ is a homomorphism such that $h\eta_B = \eta_A$ where $\eta_A : P/A \to P/K$ and $\eta_B : P/B \to P/K$ are the natural maps, then $h$ is an isomorphism.*

*Proof.* Since $\eta_A$ and $\eta_B$ are small epimorphisms, $h$ is a small epimorphism. Hence there exists $g : P \to P/A$ such that $gh = \eta$ where $\eta : P \to P/B$ is the canonical map. As $h$ is small, $g$ is an epimorphism. Then $\mathrm{Ke}\,g \subseteq \mathrm{Ke}\,\eta = B$ and $P/A \simeq P/\mathrm{Ke}\,g$. It can be easily verified that $\mathrm{Ke}\,g \overset{cr}{\underset{P}{\hookrightarrow}} B$. By the choice of $B$ we have $\mathrm{Ke}\,g = B$. Since $(\mathrm{Ke}\,h)g^{-1} = \mathrm{Ke}\,\eta = B = \mathrm{Ke}\,g$, $\mathrm{Ke}\,h = 0$ and so $h$ is an isomorphism. $\quad\square$

**9.16. $\mathrm{Tr}(P, -)$-colocalisation of $M$.** *Let $p : P \to M$ be a projective cover in $\sigma[M]$. Then $\mathrm{Tr}(P, -)$ induces the $M$-corational torsion theory, $P/\mathrm{Tr}(P, \mathrm{Ke}\,p)$ is a $\{P\}^{\bullet}$-projective module, and $P/\mathrm{Tr}(P, \mathrm{Ke}\,p) \to M$ is the $\mathrm{Tr}(P, -)$-colocalisation of the module $M$.*

*Proof.* Under the given conditions, the $M$-corational modules form a cohereditary torsion free class (see 9.3) and the assertions follow from 6.24. $\qquad\square$

We know that all corational covers are small covers. We now study those modules that admit a converse to this fact.

**9.17. Copolyform Modules.** A module $M$ is called *copolyform* if any small submodule $K \ll M$ is $M$-corational.

**9.18. Characterisation of copolyform modules.** *For $M$ the following are equivalent:*

(a) *$M$ is copolyform;*

(b) $0 \overset{cr}{\underset{M}{\to}} K$ *for all small submodules $K$ of $M$;*

(c) $\nabla(M, M/K) = 0$ *for all $K \ll M$;*

(d) *for any $k \in \mathbb{N}$, $M^k$ is copolyform.*

*Proof.* (a)⇔(b) is just the definition.

(a)⇒(c). Let $f \in \nabla(M, M/K)$ for some $K \ll M$. Then $\operatorname{Im} f = N/K \ll M/K$ and hence, by 2.2 (3), $N \ll M$. By (a), this means $f \in \operatorname{Hom}(M, N/K) = 0$.

(c)⇒(a). Let $K \ll M$. For any submodule $L \subseteq K$, $K/L \ll M/L$ and $\operatorname{Hom}(M, K/L) \subseteq \nabla(M, M/L) = 0$.

(c)⇔(d). This follows from 2.5. □

As another consequence of 2.5 we observe:

**9.19. Direct sums of copolyform modules.** *A direct sum $M_1 \oplus \cdots \oplus M_n$ of finitely many modules is copolyform if and only if*

(i) *$M_i$ is copolyform for all $i \leq n$ and*

(ii) $\nabla(M_i, M_j/U) = 0$ *for any $i \neq j$ and $U \ll M_j$.*

For epi-projective modules we conclude from 9.18 and 4.25:

**9.20. Epi-projective copolyform modules.** *Let $M$ be an epi-projective module. Then $M$ is copolyform if and only if $\operatorname{Jac}(\operatorname{End}(M)) = 0$.*

An immediate consequence of 9.20 is that a ring $R$ is copolyform as left $R$-module if and only if $\operatorname{Jac}(R) = 0$.

Factor modules of a copolyform module $M$ need not be copolyform unless $M$ is weakly supplemented (see 17.19). Hence this is a property of its own interest.

**9.21. Definition.** A module $M$ is called *strongly copolyform* if every factor module of $M$ is copolyform.

**9.22. Characterisation of strongly copolyform modules.** *The following statements are equivalent for an $R$-module $M$:*

(a) *$M$ is strongly copolyform;*

(b) *every factor module of $M$ is (strongly) copolyform;*

(c) $\operatorname{Hom}(M/K, S/U) = 0$, *for any $K \subset M$ and $U \subset S \ll M/K$;*

(d) $L \overset{cr}{\underset{M}{\hookrightarrow}} N$ whenever $L \overset{cs}{\underset{M}{\hookrightarrow}} N$ for all $L \subset N \subset M$;

(e) $\nabla(M/L, M/N) = 0$ for all $L \overset{cs}{\underset{M}{\hookrightarrow}} N$.

*Proof.* (a)$\Leftrightarrow$(b)$\Rightarrow$(c) and (a)$\Leftrightarrow$(d) are obvious from the definitions.

(c)$\Rightarrow$(a). Let $A \overset{cs}{\underset{M}{\hookrightarrow}} B$ and $A \subseteq X \subseteq B$. Then $B/A \ll M/A$ and so, by (c) we have $\mathrm{Hom}(M/A, (B/A)/(X/A)) = 0$, that is, (a) holds.

(b)$\Leftrightarrow$(e) follows from 9.18 applied to the factor modules of $M$.                              $\square$

**9.23. Remark.** We note that, by 8.11 (f), for any module $M$, any fully non-$M$-small module $N$ is strongly copolyform.

**9.24. Small covers of strongly copolyform modules.** *Let $f : N \to M$ be a small epimorphism in $\sigma[M]$. Then the following are equivalent:*

(a) *$N$ is strongly copolyform;*

(b) *$M$ is strongly copolyform and $0 \overset{cr}{\underset{N}{\hookrightarrow}} \mathrm{Ke}\, f$.*

*Proof.* (a)$\Rightarrow$(b). As $N$ is strongly copolyform, by 9.22, $M$ is strongly copolyform. Since $K \ll N$, we have $0 \overset{cs}{\underset{N}{\hookrightarrow}} K$ in $N$. Now $N$ is copolyform implies that $0 \overset{cr}{\underset{N}{\hookrightarrow}} K$.

(b)$\Rightarrow$(a). We can identify $M$ with $N/K$. Suppose $A \overset{cs}{\underset{N}{\hookrightarrow}} B$. It is easy to see that $(A)f \overset{cs}{\underset{M}{\hookrightarrow}} (B)f$. As $M$ is strongly copolyform, $(A)f \overset{cr}{\underset{M}{\hookrightarrow}} (B)f$. Now $(A)f = (A+K)f$ and $(B)f = (B+K)f$. By 9.7(5), $(A+K) \overset{cr}{\underset{N}{\hookrightarrow}} (B+K)$. Since $0 \overset{cr}{\underset{N}{\hookrightarrow}} K$ and $A \overset{cr}{\underset{N}{\hookrightarrow}} A$, $A \overset{cr}{\underset{N}{\hookrightarrow}} (A+K)$ (see 9.7 (4)). Now $(A+K) \overset{cr}{\underset{N}{\hookrightarrow}} (B+K)$ implies $A \overset{cr}{\underset{N}{\hookrightarrow}} (B+K)$ (see 9.7 (2)). Again by 9.7 (2), $A \overset{cr}{\underset{N}{\hookrightarrow}} B$. Thus $N$ is strongly copolyform.          $\square$

A ring is called *left fully idempotent* if every left ideal is idempotent.

**9.25. Projective strongly copolyform modules.** *Let $P$ be a finitely generated self-projective module and $S = \mathrm{End}(P)$. If $P$ is strongly copolyform, then $S$ is left fully idempotent.*

*Proof.* Consider a nonzero $f \in S$. Since $P$ is finitely generated and self-projective, by 4.18, $\mathrm{Hom}(P, Pf) = Sf$ and $Pf\mathrm{Hom}(P, Pf) = PfSf$. As $P$ is $\pi$-projective, $PfSf$ is a cosmall submodule of $Pf$ in $P$ by 4.16. Now $P$ is strongly copolyform implies that $\mathrm{Hom}(P, Pf/PfSf) = 0$ and hence $Pf = PfSf$. Thus

$$Sf = \mathrm{Hom}(P, Pf) = \mathrm{Hom}(P, PfSf) = \mathrm{Hom}(P, P(Sf)^2) = (Sf)^2,$$

where the last equality follows by 4.18 (c). Hence cyclic left ideals of $S$ are idempotent and this implies that every left ideal is idempotent.                     $\square$

It now follows that a ring $R$ which is strongly copolyform as a left $R$-module must be left fully idempotent. Hence any ring $R$ with $\text{Jac}\,R = 0$ which is not left fully idempotent gives an example of a module which is copolyform but not strongly copolyform.

**9.26. The projective cover as colocalisation.** *Let* $p : P \to M$ *be a projective cover of* $M$ *in* $\sigma[M]$. *The following statements are equivalent:*

(a) $M$ *is copolyform;*

(b) $\nabla(P, M) = 0$;

(c) $P$ *is copolyform;*

(d) $\text{Jac}\,(\text{End}(P)) = 0$.

*In this case,* $p : P \to M$ *is a* $\text{Tr}(P, -)$-*colocalisation and endomorphisms of* $M$ *lift uniquely to endomorphisms of* $P$, *that is,* $\text{End}(M)$ *is a subring of* $\text{End}(P)$. *Moreover, small surjective endomorphisms of* $M$ *lift to isomorphisms of* $P$.

*Proof.* Note that, since $P$ is projective, we have $\nabla(P) = \text{Jac}\,(\text{End}(P))$ (see 4.25).

(a)$\Rightarrow$(b). $M$ copolyform implies that every small submodule $L \ll M$ is in $\{M\}^{\blacktriangleright}$. By 7.7, it follows that $\{M\}^{\blacktriangleright} = \{P\}^{\bullet}$. Hence $\text{Hom}(P, L) = 0$ and so $\nabla(P, M) = 0$.

(b)$\Rightarrow$(d). By Lemma 2.6, $\nabla(P) = 0$ since $P$ is $M$-cogenerated (projective). Hence our claim follows.

(d)$\Rightarrow$(a). Let $f : P \to K$ be a nonzero homomorphism with $K \ll M$ and consider the diagram

$$
\begin{array}{ccc}
P & \xrightarrow{\ f\ } & K \\
{\scriptstyle g}\downarrow & & \downarrow{\scriptstyle i} \\
P & \xrightarrow{\ p\ } & M \longrightarrow 0
\end{array}
$$

where $i$ is the inclusion map. Since $P$ is projective, there is a nonzero $g \in \text{End}(P)$ with $f = gp$ and $\text{Im}\,f = \text{Im}\,gp \ll M$. Since $p$ is a small epimorphism, $\text{Im}\,g \ll P$, and so $g \in \nabla(P) = 0$. Hence $f = 0$, a contradiction. Consequently, $\text{Hom}(P, K) = 0$ for every $K \ll M$, that is, $K \in \{P\}^{\bullet} = \{M\}^{\blacktriangleright}$ and hence all small covers $M \to N$ are corational.

(c)$\Leftrightarrow$(d) follows similarly setting $M = P$.

Now suppose that each (or any) of the statements holds. Since $K := \text{Ke}\,p \ll P$ and $\text{Hom}(P, K) = 0$, we have $\text{Tr}(P, K) = 0$, and so $P$ is the $\text{Tr}(P, -)$-colocalisation of $M$ by 9.16. For every $f \in \text{End}(M)$, the diagram

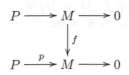

can be extended commutatively by some $\bar{f} \in \mathrm{End}(P)$. Assume there are two endomorphisms $g, h \in \mathrm{End}(P)$ that extend $f$. Then $\mathrm{Im}\,(g - h) \in \mathrm{Hom}(P, K) = 0$, that is, $g = h$. Hence we can regard $\mathrm{End}(M)$ as a subring of $\mathrm{End}(P)$.

If $f$ is a small epimorphism, then $\bar{f}$ is a small epimorphism and hence an isomorphism, by the projectivity of $P$.                                              □

Some finiteness conditions can force $\mathrm{End}(P)$ to be semisimple artinian.

**9.27. Semisimple endomorphism rings.** *Let $P \to M$ be a projective cover in $\sigma[M]$. The following statements are equivalent:*

(a) *$M$ is copolyform with finite hollow dimension;*

(b) *$\mathrm{End}(P)$ is semisimple artinian.*

*In this case any epimorphism of $M$ can be lifted to an isomorphism of $P$.*

*Proof.* We know that $\mathrm{h.dim}(M) = \mathrm{h.dim}(P) = \mathrm{h.dim}(\mathrm{End}(P))$ in case one of the dimensions is finite. This means that $M$ has finite hollow dimension if and only if $\mathrm{End}(P)$ is semilocal. We have also shown that $M$ is copolyform if and only if $\mathrm{Jac}\,(\mathrm{End}(P)) = 0$. Thus the statement follows.

Any epimorphism of $M$ lifts to an isomorphism of $P$ since any surjective endomorphism of a module with finite hollow dimension has small kernel.    □

A module $M$ is called *monoform* if it is a uniform polyform module. These modules have been studied in conjunction with the structure theory of noetherian rings as well as in the context of a general Jacobson density theorem. They are characterised by the property that any nonzero partial endomorphism (from a submodule $N$ to $M$) must be a monomorphism. We now consider their dual.

**9.28. Epiform modules.** A module $M$ is called *epiform* if, given any proper factor module $F$ of $M$, any nonzero homomorphism $M \to F$ is an epimorphism.

**9.29. Characterisation of epiform modules.** *For $M$ the following are equivalent.*

(a) *$M$ is epiform;*

(b) *$M$ is a hollow copolyform module;*

(c) *every proper submodule of $M$ is $M$-corational.*

*In this case all endomorphisms of $M$ are epimorphisms and $\mathrm{End}(M)$ is a domain.*

*Proof.* (a)$\Rightarrow$(b). For any proper submodule $N$ of $M$, $\mathrm{Hom}(M, N/K) = 0$ for all $K \subseteq N$ if $M$ is epiform. Hence $M$ is copolyform and any endomorphism of $M$ has to be surjective. Again let $N$ be a proper submodule of $M$. For any $K \subseteq M$ with $M = N + K$, consider the map $f : M \to M/N \simeq K/(N \cap K) \to M/(N \cap K)$. By hypothesis, $f$ is surjective and so $K = M$. Hence $N$ is small in $M$ and so $M$ is hollow.

(b)$\Rightarrow$(c) is trivial.

(c)⇒(a). Let $N$ be a proper submodule of $M$ and $f : M \to M/N$ be a nonzero homomorphism. Write $\operatorname{Im} f = K/N$ with $K \subset M$. Then, by (c), if $K \neq M$ we have $f \in \operatorname{Hom}(M, K/N) = 0$, a contradiction. Thus $K = M$ and so $f$ is surjective, as required. □

**9.30. Corollary.** *A module $M$ with projective cover $P$ in $\sigma[M]$ is epiform if and only if $\operatorname{End}(P)$ is a division ring.*

*Proof.* This follows from 9.27 since in the epiform case $M$ and $\operatorname{End}(P)$ have hollow dimension 1. □

**9.31. Exercises.**

(1) Let $R$ be a ring with $_R R$ strongly copolyform. Prove that, for any ideal $I \subset R$, $\operatorname{Jac} R/I = 0$.

(2) Recall that for $R$-modules $M$, $N$ we write $M \rightsquigarrow N$ if $\nabla(M, N') = 0$ for all factor modules $N'$ of $N$ (see 4.45).

   (i) Prove that if $f : M \to N$ is a corational cover, then $M \rightsquigarrow \operatorname{Ke} f$.

   (ii) Show that the following statements are equivalent:

      (a) $M \rightsquigarrow N$;

      (b) $\operatorname{Im} f$ is coclosed in $N'$, for any factor module $N'$ of $N$ and nonzero $f : M \to N'$;

      (c) $K \ll (M)f$ whenever $K \ll \operatorname{Im} f$, for any epimorphism $f : M \oplus N \to L$ and submodule $K \subset (M)f$.

**References.** Bican [40]; Courter [76]; Golan [123]; Güngöroğlu [138]; Güngöroğlu and Keskin [139]; Jambor-Malcolmson-Shapiro [181], Lomp [226]; McMaster [232]; Talebi-Vanaja [330].

# 10  Proper classes and $\tau$-supplements

With the intention of formalising the theory of Ext functors depending on a specific choice of monomorphisms, Buchsbaum introduced in [47] certain conditions on a class of monomorphisms which are needed to study *relative homological algebra*. Since in an abelian category any monomorphism induces an epimorphism and a short exact sequence, these conditions lead to the notion of *proper classes* of such sequences. Well-known examples of such classes are the *pure exact sequences* which can be defined by choosing a class $\mathcal{P}$ of (finitely presented) modules and considering those sequences on which $\mathrm{Hom}(P, -)$ is exact for each $P \in \mathcal{P}$. These techniques are outlined in, for example, Mishina and Skornjakov [234], Sklyarenko [316] and [363]. It was observed in abelian group theory that complement submodules induce a proper class of short exact sequences and this motivated the investigation of this phenomenon for modules over arbitrary rings. First results in this direction were obtained, for example, by Stenström [321] and Generalov [117, 118]. For a more comprehensive presentation of results and their sources we refer to E. Mermut's Ph. D. thesis [233].

Our interest in this approach to the structure theory of modules is based on an observation mentioned in Stenström [321], namely that, over any ring, supplement submodules induce a proper class of short exact sequences (see 20.7). This makes the theory presented in this section of interest for the study of lifting modules. It may also help to bring some order to the recent variations and generalisations of supplemented and lifting modules involving properties from torsion theory. This opens new ways to study some interdependencies neglected thus far, but we will not be able to fully cover the rich theory provided by this approach. We encourage the reader to examine these problems further.

**10.1. Proper classes.** Let $\mathbb{E}$ be a class of short exact sequences in $\sigma[M]$. If the exact sequence

$$0 \longrightarrow K \xrightarrow{\;f\;} L \xrightarrow{\;g\;} N \longrightarrow 0$$

belongs to $\mathbb{E}$, then $f$ is called an $\mathbb{E}$-*mono* and $g$ is said to be an $\mathbb{E}$-*epi*. The class $\mathbb{E}$ is called *proper* if it satisfies each of the following conditions:

P1. $\mathbb{E}$ is closed under isomorphisms;

P2. $\mathbb{E}$ contains all splitting short exact sequences in $\sigma[M]$;

P3. the class of $\mathbb{E}$-monomorphisms is closed under composition;
    if $f', f$ are monomorph and $f \diamond f'$ is an $\mathbb{E}$-mono, then $f$ is an $\mathbb{E}$-mono;

P4. the class of $\mathbb{E}$-epimorphisms is closed under composition;
    if $g, g'$ are epimorph and $g' \diamond g$ is an $\mathbb{E}$-epi, then $g$ is an $\mathbb{E}$-epi.

The class of all splitting short exact sequences in $\sigma[M]$ is an example of a proper class.

**10.2. Purities.** Let $\mathcal{P}$ be a class of modules in $\sigma[M]$. Let $\mathbb{E}^{\mathcal{P}}$ denote the class of all short exact sequences in $\sigma[M]$ on which $\mathrm{Hom}(P, -)$ is exact for each $P \in \mathcal{P}$. It is straightforward to prove that $\mathbb{E}^{\mathcal{P}}$ is a proper class. This type of class is called *projectively generated* and its elements are called $\mathcal{P}$-*pure sequences*.

The "classical" purity is obtained by taking for $\mathcal{P}$ all finitely presented modules in $\sigma[M]$; for $M = R$ this is the so-called *Cohn purity* (e.g., [363, § 33]).

**10.3. Copurities.** Let $\mathcal{Q}$ be a class of modules in $\sigma[M]$. Denote by $\mathbb{E}_{\mathcal{Q}}$ the class of all short exact sequences in $\sigma[M]$ on which $\mathrm{Hom}(-, Q)$ is exact for each $Q \in \mathcal{Q}$. Then $\mathbb{E}_{\mathcal{Q}}$ is a proper class. This type of class is called *injectively generated* and its elements are called $\mathcal{Q}$-*copure sequences* (see [363, § 38]).

**10.4. Relative injectivity and projectivity.** Let $\mathbb{E}$ be a proper class of short exact sequences in $\sigma[M]$.

A module $P \in \sigma[M]$ is called $\mathbb{E}$-*projective* if $\mathrm{Hom}(P, -)$ is exact on all short exact sequences in $\mathbb{E}$.

Dually, a module $Q \in \sigma[M]$ is called $\mathbb{E}$-*injective* if $\mathrm{Hom}(-, Q)$ is exact on all short exact sequences in $\mathbb{E}$.

It follows from standard arguments that the class of all $\mathbb{E}$-projective modules is closed under direct sums and direct summands, and the class of all $\mathbb{E}$-injective modules is closed under direct products and direct summands.

**10.5. Class of complement submodules.** *Let $\mathbb{E}_c$ be the class of short exact sequences* $0 \to K \to L \to N \to 0$ *in $\sigma[M]$ such that $K$ is a complement (closed) submodule of $L$. Then:*

(1) $\mathbb{E}_c$ *is a proper class in $\sigma[M]$ and*

(2) *every (semi-)simple module (in $\sigma[M]$) is $\mathbb{E}_c$-projective.*

*Proof.* (1) The conditions P1 and P2 are easily verified.

P3. Let $K \subset L$ and $L \subset N$ be closed submodules. Choosing $M$-injective hulls $\widehat{K} \subset \widehat{L} \subset \widehat{N}$, we have

$$K = \widehat{K} \cap L = \widehat{K} \cap \widehat{L} \cap N = \widehat{K} \cap N,$$

proving that $K$ is closed in $N$ (see 1.10).

Consider $K \subset L \subset N$. If $K$ is closed in $N$, then $K = \widehat{K} \cap N = \widehat{K} \cap L$ and so $K$ is closed in $L$.

P4. The composition of two epimorphisms $g$ and $h$ can be presented by the following commutative diagram with exact rows and columns,

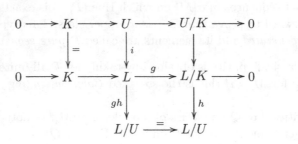

where $U/K = \operatorname{Ke} h$, for some $K \subset U \subset L$.

Assume $g$ and $h$ to have closed kernels. Suppose first that $U = \operatorname{Ke} gh$ is not closed in $L$ and let $\overline{U}$ be an essential closure of $U$ in $L$. Then $K$ is closed in $\overline{U}$ and, by 1.11, $\operatorname{Ke} h = U/K \trianglelefteq \overline{U}/K$, contradicting the assumption that $h$ has a closed kernel.

Now assume that $U = \operatorname{Ke} gh$ is closed. Suppose that $\operatorname{Ke} h$ has a proper essential extension $V$ in $L/K$. Then $U = (\operatorname{Ke} h)g^{-1} \trianglelefteq (V)g^{-1}$, contradicting our condition on $\operatorname{Ke} h$. This shows that $\operatorname{Ke} h$ is a closed submodule of $L/K$.

(2) Let $K \subset L$ be a closed submodule and $S \subseteq L/K$ be any simple submodule. Then $S \simeq N/K$ for some submodule $K \subset N \subseteq L$. By assumption, $K$ is a maximal submodule and is not essential in $N$. Hence the map $N \to N/K \simeq S$ splits showing that $\operatorname{Hom}(S, L) \to \operatorname{Hom}(S, L/K)$ is surjective and that the module $S$ is $\mathbb{E}_c$-projective.

Since direct sums of $\mathbb{E}_c$-projectives are again $\mathbb{E}_c$-projective, every semisimple module is $\mathbb{E}_c$-projective.                                                                                 □

**10.6. $\tau$-complement submodules.** *Let $\tau$ be an idempotent preradical for $\sigma[M]$ with associated classes $\mathbb{T}_\tau$ and $\mathbb{F}_\tau$. Then for a submodule $K \subset L$ where $L \in \sigma[M]$, the following statements are equivalent.*

(a) *Every $N \in \mathbb{T}_\tau$ is projective with respect to the projection $L \to L/K$;*

(b) *there exists a submodule $U \subseteq L$ such that*

$$K \cap U = 0 \quad \text{and} \quad \tau(L/K) = (U + K)/K \simeq U;$$

(c) *there exists a submodule $U \subseteq L$ such that*

$$K \cap U = 0 \quad \text{and} \quad \tau(L/K) \subseteq (U + K)/K \simeq U.$$

*If these conditions are satisfied, then $K$ is called a $\tau$-complement in $L$.*

*Proof.* (a)⇒(b). A pullback construction yields the following commutative diagram with exact rows, where $i, i_K$ are the inclusion maps and $g$ is the projection.

$$
\begin{array}{ccccccccc}
0 & \longrightarrow & K & \xrightarrow{i_K} & \widetilde{K} & \longrightarrow & \tau(L/K) & \longrightarrow & 0 \\
 & & \Big\| & & \Big\downarrow & & \Big\downarrow{\scriptstyle i} & & \\
0 & \longrightarrow & K & \longrightarrow & L & \xrightarrow{g} & L/K & \longrightarrow & 0.
\end{array}
$$

Since $\tau(L/K) \in \mathbb{T}_\tau$, there exists a morphism $h : \tau(L/K) \to L$ with $i = h \diamond g$. By the Homotopy Lemma (e.g., [363, 7.16]), this implies that the top row splits, that is, $\widetilde{K} = K \oplus U$ for some $U \subseteq \widetilde{K}$, and $(U + K)/K = \tau(L/K)$.

(b)⇒(a). Let $N \in \mathbb{T}_\tau$ and $f \in \mathrm{Hom}(N, L/K)$. Then $\mathrm{Im}\, f \subset \tau(L/K)$ and it can be seen from the diagram above that there is a morphism $h : N \to L$ with $f = h \diamond g$.

(b)⇒(c) is obvious.

(c)⇒(b). Assume we have $\tau(L/K) \subseteq (U + K)/K$ and $U \cap K = 0$. Then $\tau(L/K) = (V + K)/K$ for some submodule $V \subset U$, and $V \cap K = 0$. □

**10.7. Corollary.** *Let $\tau$ be a preradical for $\sigma[M]$ and $K \subset L$ where $L \in \sigma[M]$.*

(1) *If $K$ is a $\tau$-complement in $L$ and $L/K \in \mathbb{T}_\tau$, then $K$ is a direct summand.*

(2) *If $L \in \mathbb{T}_\tau$, then every $\tau$-complement submodule of $L$ is a direct summand.*

For the preradical induced by the class of all (semi-)simple modules we find an interesting relationship associated with the complement (closed) submodules.

**10.8. Neat submodules.** A monomorphism $f : K \to L$ is called *neat* if any simple module $S$ is projective relative to the projection $L \to L/\mathrm{Im}\, f$, that is, the Hom sequence $\mathrm{Hom}(S, L) \to \mathrm{Hom}(S, L/\mathrm{Im}\, f) \to 0$ is exact. Short exact sequences with neat monomorphisms are said to be *neat*. These sequences form a projectively generated class in the sense of 10.2.

It follows from 10.5 that all sequences in $\mathbb{E}_c$ are neat.

Recall that a module $M$ is said to be *semi-artinian* if every nonzero factor module of $M$ has nonzero socle (see [85, 3.12]).

**10.9. When neat submodules are closed in $\sigma[M]$.** *The following statements are equivalent for a module $M$.*

(a) *Every neat submodule of $M$ is closed;*

(b) *a submodule of $M$ is closed if and only if it is neat;*

(c) *for every $L \in \sigma[M]$, neat submodules of $L$ are closed;*

(d) *for every essential submodule $U \subset M$, $\mathrm{Soc}\, M/U \neq 0$;*

(e) *every $M$-singular module is semi-artinian.*

*Proof.* (a)⇔(b) is clear since closed submodules are neat.

(c)⇒(a) is obvious.

(a)⇒(d). Let $U \trianglelefteq M$ be a proper submodule. Then $U$ is not closed and hence not neat in $M$. Thus there is a morphism $g : S \to M/U$ where $S$ is simple, that can not be extended to a morphism $S \to M$. In particular, this implies that $\operatorname{Im} g \neq 0$ and so $\operatorname{Soc} M/U \neq 0$.

(d)⇒(e). Let $U \subset V \subset M$. If $U \trianglelefteq M$, then $V \trianglelefteq M$ and hence (d) implies that every factor module of $M/U$ has nonzero socle, that is, $M/U$ is semi-artinian.

By [85, Proposition 4.3], the set $\{M/U \,|\, U \trianglelefteq M\}$ is a generating set for all the $M$-singular $M$-generated modules and every $M$-singular module is a submodule of an $M$-generated $M$-singular module. Thus, since every proper factor $M/U$ has nonzero socle, this is also true for all $M$-singular modules.

(e)⇒(c). Let $K \subset L$ be a neat submodule and assume that it has a proper essential extension $\overline{K} \subset L$. Then $\overline{K}/K$ is an $M$-singular module and hence, by assumption, contains a simple submodule $S = N/K$ where $K \trianglelefteq N \subset \overline{K}$. Now neatness of $K \subset L$ implies that the map $N \to N/K = S$ splits. This contradicts $K \trianglelefteq N$, proving that $K$ is closed in $L$. ☐

For $M = R$ we obtain the following corollary which was (partly) proved in [117, Theorem 5].

**10.10. When neat submodules are closed in $R$-Mod.** *The following statements are equivalent for a ring $R$.*

(a) *Every neat left ideal of $R$ is closed;*

(b) *a left ideal of $R$ is closed if and only if it is neat;*

(c) *for every left $R$-module, closed submodules are neat;*

(d) *for every essential left ideal $I \subset R$, $\operatorname{Soc} R/I \neq 0$;*

(e) *every singular module is semi-artinian.*

Rings with these properties are called *C-rings* (in [296]).

Dualising the notions considered above yields the following.

**10.11. $\tau$-supplement submodules.** *Let $\tau$ be a radical for $\sigma[M]$ with associated classes $\mathbb{T}_\tau$ and $\mathbb{F}_\tau$. Then for a submodule $K \subset L$ where $L \in \sigma[M]$, the following statements are equivalent.*

(a) *Every $N \in \mathbb{F}_\tau$ is injective with respect to the inclusion $K \to L$;*

(b) *there exists a submodule $U \subseteq L$ such that*

$$K + U = L \ \text{ and } \ U \cap K = \tau(K);$$

(c) *there exists a submodule $U \subseteq L$ such that*

$$K + U = L \ \text{ and } \ U \cap K \subseteq \tau(K).$$

If these conditions are satisfied, then $K$ is called a $\tau$-*supplement in $L$*.

*Proof.* (a)$\Rightarrow$(b). Consider the commutative diagram with exact rows, where $i$ is the inclusion and $p$ is the projection map,

$$
\begin{array}{ccccccccc}
0 & \longrightarrow & K & \overset{i}{\longrightarrow} & L & \longrightarrow & L/K & \longrightarrow & 0 \\
& & \downarrow{\scriptstyle p} & & \downarrow & & \downarrow{\scriptstyle =} & & \\
0 & \longrightarrow & K/\tau(K) & \longrightarrow & L/\tau(K) & \longrightarrow & L/K & \longrightarrow & 0.
\end{array}
$$

Since $K/\tau(K) \in \mathbb{F}_\tau$, there exists $h : L \to K/\tau(K)$ with $p = i \diamond h$. By the Homotopy Lemma (e.g., [363, 7.16]), this implies that the bottom row splits, that is,

$$L/\tau(K) = K/\tau(K) \oplus U/\tau(K), \text{ for some } \tau(K) \subseteq U \subseteq L.$$

This gives the required $L = K + U$ and $U \cap K = \tau(K)$.

(b)$\Rightarrow$(a). By the given data, $L/\tau(K) = K/\tau(K) \oplus U/\tau(K)$. Let $N \in \mathbb{F}_\tau$ and $f \in \mathrm{Hom}(K, N)$. Then $\tau(K) \subseteq \mathrm{Ke}\, f$ and we have the commutative diagram (with $i, p$ as before)

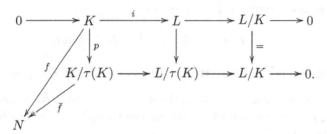

Since the second row splits we obtain a morphism $h : L \to N$ with $f = i \diamond h$.

(b)$\Rightarrow$(c) is trivial.

(c)$\Rightarrow$(b). Assume $K + U = L$ and $K \cap U \subseteq \tau(K)$. Putting $U' = U + \tau(K)$, we have $K + U' = L$ and $K \cap U' = K \cap U + \tau(K) = \tau(K)$. $\qquad\square$

**10.12. Corollary.** *Let $\tau$ be a radical for $\sigma[M]$ and $K \subset L$ where $L \in \sigma[M]$.*

(1) *If $K$ is a $\tau$-supplement in $L$ and $K \in \mathbb{F}_\tau$, then $K$ is a direct summand.*

(2) *If $L \in \mathbb{F}_\tau$, then every $\tau$-supplement submodule of $L$ is a direct summand.*

(3) *If $K$ is a $\tau$-supplement in $L$ and $X \subset L$, then $K/X$ is a $\tau$-supplement in $L/X$.*

*Proof.* (1) and (2) follow from the preceding observations.

(3) Let $U \subset L$ be such that $K + U = L$ and $K \cap U \subseteq \tau(K)$. Then $L/X = K/X + (U + X)/X$ and

$$K/X \cap (U + X)/X = (K \cap U + K)/X \subseteq (\tau(K) + X)/X \subset \tau(L/X).$$

$\qquad\square$

As a special case we consider the radical (for $\sigma[M]$) cogenerated by the simple modules.

**10.13. Co-neat submodules.** A monomorphism $f : K \to L$ is called *co-neat* if any module $Q$ with $\operatorname{Rad} Q = 0$ is injective relative to it, that is, the Hom sequence $\operatorname{Hom}(L, Q) \to \operatorname{Hom}(\operatorname{Im} f, Q) \to 0$ is exact.

Short exact sequences with co-neat monomorphisms are said to be *co-neat*. They form an injectively generated class in the sense of 10.3.

The following is a consequence of 10.11.

**10.14. Characterisation of co-neat submodules.** *For a submodule $K \subset L$, the following are equivalent:*

(a) *$K \subset L$ is a co-neat submodule;*

(b) *there exists a submodule $U \subseteq L$ such that*

$$K + U = L \quad \text{and} \quad U \cap K = \operatorname{Rad} K;$$

(c) *there exists a submodule $U \subseteq L$ such that*

$$K + U = L \quad \text{and} \quad U \cap K \subseteq \operatorname{Rad} K.$$

If these conditions are satisfied, then $K$ is also called a Rad-*supplement in $L$*.

Dual to the relationship between the neat and complement submodules, there is a relationship between co-neat submodules and *supplement submodules*. This will be investigated in 20.7.

**10.15. Exercises.**

(1) Prove that a submodule $K \subset L$ is neat if and only if for any maximal left ideal $I \subset R$ and $l \in L$ with $Il = 0$, there exists $k \in K$ such that $I(l - k) = 0$ ([233, Corollary 3.2.5]).

(2) Let $R$ be a left noetherian ring. Prove that $R$ is a C-ring (see 10.10) if and only if, for every essential left ideal $I \subset R$, $R/I$ has finite length ([296]).

(3) Let $R$ be a two-sided hereditary noetherian ring. Prove that $R$ is a C-ring ([231, Proposition 5.4]).

(4) Let $R$ be a commutative noetherian ring in which every nonzero prime ideal is maximal. Prove that $R$ is a C-ring ([233, Proposition 3.3.6]).

(5) Let $K \subset L$ be $\mathbb{Z}$-modules (see, e.g., [233]).

    (i) Show that $K$ is a supplement if and only if it is a neat submodule;

    (ii) If $K$ is a supplement, show that $K$ is also a complement submodule.

    (iii) If $K$ is finite, show that $K$ is a supplement if and only if it is a complement.

(iv) For a prime $p \in \mathbb{N}$, consider the exact sequence of $\mathbb{Z}$-modules

$$0 \longrightarrow \mathbb{Z} \xrightarrow{\ f\ } \mathbb{Z}/p\mathbb{Z} \oplus \mathbb{Z} \xrightarrow{\ g\ } \mathbb{Z}/p^2\mathbb{Z} \longrightarrow 0 \ ,$$

where
$$f(a) \quad = \quad a \cdot (-1 + p\mathbb{Z}, p) \text{ and}$$
$$g(a + p\mathbb{Z}, b) \quad = \quad (ap + b) + p^2\mathbb{Z}, \quad \text{for } a, b \in \mathbb{Z}.$$
Prove that $\mathbb{Z}$ is a complement but not a supplement submodule .

**References.** Generalov [117, 118]; Al-Takhman, Lomp and Wisbauer [9]; Mermut [233]; Mishina and Skornjakov [234]; Renault [296, 297]; Sklyarenko [316]; Stenström [321]; Wisbauer [363].

# Chapter 3

# Decompositions of modules

## 11   The exchange property

The exchange property appears in rudimentary form as the Steinitz Exchange Lemma for vector spaces (see, for example, [73, Lemma 4.3.4]) and, as a slightly more general forerunner, in the following well-known property of semisimple modules (see [363, 20.1]).

**11.1. Steinitz' Exchange Lemma for semisimple modules.** *Let* $M = \bigoplus_I M_i$ *where each* $M_i$ *is a simple module. If* $K$ *is a submodule of* $M$, *then there is a subset* $I_K$ *of* $I$ *for which*

$$M = K \oplus \left( \bigoplus_{i \in I_K} M_i \right).$$

In module theory, the exchange property plays a prominent rôle in the study of isomorphic refinements of direct sum decompositions of modules. We first record two elementary lemmas on submodules of direct sums which will be frequently useful.

**11.2. Lemma.** *Let* $M_1, M_2$ *and* $N$ *be modules with* $M_1 \subseteq N \subseteq M_1 \oplus M_2$. *Then* $N = M_1 \oplus N_1$ *where* $N_1 = N \cap M_2$.

*Proof.* Since $M_1 \subseteq N$, the modularity property gives

$$N \cap (M_1 + M_2) = M_1 + (N \cap M_2)$$

and so $N = M_1 + N_1$. Moreover, $M_1 \cap N_1 \subseteq M_1 \cap M_2 = 0$. □

**11.3. Lemma.** *Let* $M = M_1 \oplus M_2$, *let* $\pi_1 : M \to M_1$ *be the projection map, and let* $N$ *be a submodule of* $M$. *Then*

$$M = N \oplus M_2$$

*if and only if the restriction* $\pi_1|_N : N \to M_1$ *is an isomorphism. In this case, the projection map* $\pi_N : M \to N$ *associated with the direct sum* $M = N \oplus M_2$ *is given by* $\pi_1 (\pi_1|_N)^{-1}$.

*Proof.* $\operatorname{Ke}(\pi_1|_N) = N \cap \operatorname{Ke}\pi_1 = N \cap M_2$ and so $\pi_1|_N$ is monic if and only if $N \cap M_2 = 0$. Moreover, since $\operatorname{Ke}\pi_1 = M_2$ and $\pi_1|_{M_1} = 1_{M_1}$, we have

$$
\begin{aligned}
(N)\pi_1 &= (N + M_2)\pi_1 = ((N + M_2) \cap (M_1 + M_2))\pi_1 \\
&= (((N + M_2) \cap M_1) + M_2)\pi_1 = ((N + M_2) \cap M_1)\pi_1 \\
&= (N + M_2) \cap M_1
\end{aligned}
$$

and so $(N)\pi_1 = M_1$ if and only if $M_1 \subseteq N + M_2$, that is, $N + M_2 = M$.

Finally, if $M = N \oplus M_2$, then, for any $m \in M$, we have $m = (m)\pi_N + m_2$ for some $m_2 \in M_2$. This gives $(m)\pi_1 = ((m)\pi_N)\pi_1 + (m_2)\pi_1 = ((m)\pi_N)\pi_1$, and so $(m)\pi_N = (m)\pi_1 (\pi_1|_N)^{-1}$. $\qquad\square$

The following definition is due to Crawley and Jónnson [77] who introduced it in the wider context of general algebra.

**11.4. The exchange property.** Let $c$ be a cardinal number. A module $M$ is said to have the *c-exchange property* if, for any module $A$ and any internal direct sum decompositions of $A$ given by

$$
A = M' \oplus N = \bigoplus_I A_i
$$

for modules $M', N, A_i$ where $M' \simeq M$ and $\operatorname{card}(I) \leq c$, there always exist submodules $B_i \leq A_i$ for each $i \in I$ such that

$$
A = M' \oplus \left( \bigoplus_I B_i \right).
$$

If $M$ has the $n$-exchange property for every positive integer $n$, then $M$ is said to have the *finite exchange property*.

If $M$ has the $c$-exchange property for every cardinal number $c$, then $M$ is said to have the (*full* or *unrestricted*) *exchange property*.

**11.5. Remarks.** We emphasise that the direct sums in the definition are *internal* direct sums. In contrast to isomorphisms involving external direct sums, this ensures that direct summands with common complements are isomorphic. Thus, for example, the submodule $N$ above is isomorphic to $\bigoplus_I B_i$ since they share $M'$ as a complementary summand.

It is also important to note that the submodules $B_i$ of the definition are actually direct summands of the $A_i$, since applying Lemma 11.2 to the inclusions

$$
B_i \subseteq A_i \subseteq B_i \oplus \left( M' \oplus \left( \bigoplus_{j \neq i} B_j \right) \right)
$$

gives $A_i = B_i \oplus C_i$ where $C_i = A_i \cap \left( M' \oplus \left( \bigoplus_{j \neq i} B_j \right) \right)$.

It is clear from the definition that a finitely generated module with the finite exchange property has the full exchange property. However, it is particularly intriguing that, as of this writing, it is unknown whether *every* module with the finite exchange property also has the full exchange property.

We also note that, trivially, every module satisfies the 1-exchange property.

The corollary of the next lemma will be useful later in induction arguments.

**11.6. Lemma.** *Let $\{A_i\}_I$ be a family of modules with submodules $B_i \subseteq A_i$ for each $i \in I$. If $A, M, N$ and $L$ are modules such that*

$$
\begin{aligned}
A &= M \oplus N \oplus L = (\bigoplus_I A_i) \oplus L \quad and \\
A/L &= ((M+L)/L) \oplus (\bigoplus_I (B_i + L)/L),
\end{aligned}
$$

*then*

$$
A = M \oplus \left(\bigoplus_I B_i\right) \oplus L.
$$

*Proof.* The hypothesis for $A/L$ gives $A = M + (\sum_I B_i) + L$ so it suffices to show that this sum is direct. To this end, consider $m + \sum_I b_i + l = 0$ where $m \in M, l \in L$ and $b_i \in B_i$ are almost all zero. Then $(m + L) + \sum_I (b_i + L) = 0$ in $A/L$ and so, by hypothesis, $m \in L$ and $b_i \in L$ for all $i$. But then the first given equation shows that $m \in M \cap L = 0$ and $b_i \in B_i \cap L \subseteq A_i \cap L = 0$. This in turn gives $l = 0$, as required. $\qquad\square$

**11.7. Corollary.** *Let $c$ be a cardinal number and $I$ be an index set of cardinality at most $c$. Let $A, M, N, L,$ and $A_i$, $i \in I$, be modules such that*

$$
A = M \oplus N \oplus L = (\bigoplus_I A_i) \oplus L.
$$

*If $M$ has the $c$-exchange property, then there exist submodules $B_i \subseteq A_i$ such that*

$$
A = M \oplus (\bigoplus_I B_i) \oplus L.
$$

*Proof.* This follows from the Lemma after factoring out $L$ and applying the $c$-exchange property. $\qquad\square$

**11.8. Direct sums and the $c$-exchange property.** *Let $c$ be a cardinal number and let $M = X \oplus Y$. Then $M$ has the $c$-exchange property if and only if both $X$ and $Y$ have the $c$-exchange property.*

*Proof.* Assume that both $X$ and $Y$ have the $c$-exchange property and let $A = M \oplus N = \bigoplus_I A_i$, where $\mathrm{card}(I) \leq c$. Then, by our assumption on $Y$,

$$
A = X \oplus Y \oplus N = Y \oplus (\bigoplus_I B_i) \text{ with } B_i \subseteq A_i.
$$

By our assumption on $X$ we get from Corollary 11.7 that $A = X \oplus Y \oplus (\bigoplus_I C_i)$ with $C_i \subseteq B_i$. Thus $M$ has the $c$-exchange property.

Conversely, assume $M = X \oplus Y$ has the $c$-exchange property and consider

$$A = X' \oplus N = \bigoplus_I A_i$$

where $X' \simeq X$ and $\mathrm{card}(I) \leq c$. Then, setting $B = A \oplus Y$, we have

$$B = M' \oplus N = Y \oplus (\bigoplus_I A_i) \text{ where } M' = X' \oplus Y \simeq M.$$

Now fix $j \in I$. Then $B = M' \oplus N = (Y \oplus A_j) \oplus (\bigoplus_{i \neq j} A_i)$ and so, by the $c$-exchange property of $M$, there are submodules $C \subseteq Y \oplus A_j$ and $B_i \subseteq A_i$ for each $i \neq j$ for which

$$B = M' \oplus C \oplus (\bigoplus_{i \neq j} B_i).$$

Now, since $Y \subseteq Y \oplus C \subseteq Y \oplus A_j$, taking $M_1, N$ and $M_2$ as $Y, Y \oplus C$ and $A_j$ in Lemma 11.2 gives $Y \oplus C = Y \oplus B_j$ where $B_j = (Y \oplus C) \cap A_j$. Hence $M' \oplus C = (X' \oplus Y) \oplus C = X' \oplus Y \oplus B_j$. Substituting into the above then gives

$$B = X' \oplus Y \oplus B_j \oplus (\bigoplus_{i \neq j} B_i) = X' \oplus Y \oplus (\bigoplus_I B_i).$$

Since $X' \oplus (\bigoplus_I B_i) \subseteq A$, we may now apply the modular identity to give

$$A \cap (Y + (X' \oplus (\bigoplus_I B_i))) = (A \cap Y) + (X' \oplus (\bigoplus_I B_i)).$$

This implies that $A = X' \oplus (\bigoplus_I B_i)$. Hence $X$ has the $c$-exchange property.  $\square$

As a corollary we obtain:

**11.9. Finite direct sums and the exchange properties.** *Let $M = M_1 \oplus \cdots \oplus M_k$. Then $M$ has the (finite) exchange property if and only if $M_i$ has the (finite) exchange property for each $i = 1, \ldots, k$.*

Example 12.15 will show that 11.9 does not generalise to infinite direct sums.

**11.10. Proposition.** *Let $M = L \oplus K = N \oplus T$. If $T$ has the finite exchange property and $N \subseteq L$, then $K$ has the finite exchange property.*

*Proof.* Since $N \subseteq L$, $L = N \oplus (L \cap T)$. Hence

$$M = L \oplus K = N \oplus (L \cap T) \oplus K \text{ and } T = (L \cap T) \oplus ((N \oplus K) \cap T)$$

and so $M = N \oplus T = N \oplus (L \cap T) \oplus ((N \oplus K) \cap T) = L \oplus ((N \oplus K) \cap T)$. Thus $K \simeq M/L \simeq (N \oplus K) \cap T$, a direct summand of $T$. Hence $K$ has the finite exchange property.  $\square$

**11.11. The 2-exchange property.** *If a module M has the 2-exchange property, then it has the finite exchange property.*

*Proof.* Suppose $M$ has the 2-exchange property. We prove by induction that $M$ has the $n$-exchange property, for every $n > 2$. Assume that $M$ has the $(n-1)$-exchange property and let $N$ and $A_1, \ldots, A_n$ be modules with $M \oplus N = \bigoplus_{i=1}^{n} A_i$. Set $A = \bigoplus_{i=1}^{n} A_i$ and $B = \bigoplus_{i=2}^{n} A_i$, so that $A = A_1 \oplus B$. By the 2-exchange property of $M$, we have

$$A = M \oplus A_1' \oplus B' \text{ where}$$

$$A_1 = A_1' \oplus A_1'', \quad B = B' \oplus B'' \text{ and } B'' = B \cap (M \oplus A_1').$$

Now $M \simeq A_1'' \oplus B''$ has, by the induction hypothesis, the $(n-1)$-exchange property and so, if we write

$$A = (A_1 \oplus A_2) \oplus A_3 \oplus A_4 \oplus \cdots \oplus A_n,$$

then

$$A = A_1'' \oplus B'' \oplus (A_1 \oplus A_2)' \oplus \left( \bigoplus_{k=3}^{n} A_k' \right),$$

where $(A_1 \oplus A_2)' \subseteq A_1 \oplus A_2$ and $A_k' \subseteq A_k$, for $3 \leq k \leq n$. We have

$$B'' \oplus \left( \bigoplus_{k=3}^{n} A_k' \right) \subseteq B$$

and so

$$B = B'' \oplus \left( \bigoplus_{k=3}^{n} A_k' \oplus L \right),$$

where $L = B \cap (A_1'' \oplus (A_1 \oplus A_2)') \subseteq A_2$. Renaming $L$ as $A_2'$, we get

$$B = B'' \oplus \left( \bigoplus_{k=2}^{n} A_k' \right), \quad \text{where } A_k' \subseteq A_k \text{ for } 2 \leq k \leq n.$$

Now $B''$ is a direct summand of $B \oplus A_1'$, say $B \oplus A_1' = B'' \oplus T$. Then

$$\begin{aligned}
A &= M \oplus A_1' \oplus B' = B'' \oplus T \oplus B' \\
&= T \oplus B = T \oplus B'' \oplus \left( \bigoplus_{k=2}^{n} A_k' \right) \\
&= M \oplus A_1' \oplus A_2' \oplus A_3' \oplus \cdots \oplus A_n'
\end{aligned}$$

where $A_k' \subseteq A_k$, for $1 \leq k \leq n$. Hence $M$ has the $n$-exchange property and so, by induction, the finite exchange property. $\square$

We now introduce a concept which will often provide a more tractable method of handling the exchange properties.

**11.12. Summable homomorphisms.** Let $M$ and $N$ be $R$-modules. We say that a family $\{f_i : M \to N\}_I$ of homomorphisms is *summable* if, given any $m \in M$, then $(m)f_i = 0$ for all but a finite number of indices $i \in I$. In this case we can define the sum $\sum_I f_i$ of the family, in the obvious way.

Clearly, if $\{f_i : M \to N\}_I$ is summable, so too is $\{f_i \diamond g_i : M \to K\}_{i \in I}$ for any $R$-module $K$ and any family of homomorphisms $\{g_i : N \to K\}_I$.

The following result shows that for an $R$-module $M$ it suffices to check the exchange property in the case when $M$ is isomorphic to a direct sum of clones of itself. Moreover, it also characterises the exchange property using summable families in $M$'s endomorphism ring.

**11.13. The exchange property and summable families.** *Let $M$ be an $R$-module, $S = \text{End}(M)$, and $c$ be any cardinal. Then the following statements are equivalent.*

(a) *$M$ has the $c$-exchange property;*

(b) *for any index set $I$ with $\text{card}(I) \leq c$, if there are modules $M'$ and $N$ with $M' \simeq M$ and*

$$M' \oplus N = \bigoplus_I A_i \text{ where } A_i \simeq M \text{ for each } i \in I,$$

*then there are submodules $B_i \leq A_i$ for each $i \in I$ such that*

$$M' \oplus \left( \bigoplus_I B_i \right) = \bigoplus_I A_i;$$

(c) *for any summable family $\{f_i\}_I$ of homomorphisms in $S$ with $\text{card}(I) \leq c$ and $\sum_{i \in I} f_i = 1_M$, there is a set of orthogonal idempotents $\{e_i \in f_i S : i \in I\}$ for which $\sum_I e_i = 1_M$.*

*Proof.* (a)$\Rightarrow$(b) is automatic.

(b)$\Rightarrow$(c). Assume (b) holds and that $\{f_i\}_I$ is a summable family in $S$ with $\text{card}(I) \leq c$ and $\sum_I f_i = 1_M$. Let $A = \bigoplus_I A_i$ where $A_i = M$ for each $i \in I$. Next, define $M^\star = \{((m)f_i)_I \mid m \in M\} \subseteq A$. Then, by the properties of the $f_i$, we have $M \simeq M^\star$ under the correspondence $f : m \mapsto ((m)f_i)_I$. Next let $i_{M^\star} : M^\star \to A$ be the inclusion map and define $g : A \to M$ by $((m_i)_I)g = \sum_I m_i$. Then $i_{M^\star}$ splits since, for all $m \in M$ and all $((m)f_i)_I \in M^\star$, we have

$$(((m)f_i)_I)i_{M^\star}gf = (((m)f_i)_I)gf = (\textstyle\sum_I(m)f_i)f = (m)f = ((m)f_i)_I.$$

Thus $M^\star$ is a direct summand of $A$ and so, by (b), for each $i \in I$ there are $B_i, C_i \leq A_i$ for which $A_i = B_i \oplus C_i$ and

$$A = M^\star \oplus \left( \bigoplus_I B_i \right) = \left( \bigoplus_I B_i \right) \oplus \left( \bigoplus_I C_i \right).$$

Let $\pi_C : A \to \bigoplus_I C_i$ be the projection map induced by the second of these direct sums. Then, by Lemma 11.3, its restriction $\tau = \pi_C|_{M^\star} : M^\star \to \bigoplus_I C_i$ is an isomorphism. Next, for each $i \in I$, let $\pi_i : \bigoplus_I C_i \to C_i$ and $\varepsilon_i : C_i \to \bigoplus_I C_i$ be the natural projection and injection respectively and define $e_i \in S$ by

$$e_i = f\tau\pi_i\varepsilon_i\tau^{-1}i_{M^*}g.$$

Then, a simple analysis shows that $\{e_i\}_I$ is a set of orthogonal idempotents in $S$. Now, for each $i \in I$, let $p_i$ be the projection of $A_i = B_i \oplus C_i$ onto $B_i$. Then $f_ip_i = f\tau\pi_i$ and this gives $e_i = f_ip_i\varepsilon_i\tau^{-1}i_{M^*}g$. Consequently, $e_i \in f_iS$ and so $\{e_i\}_I$ is summable. We finalise (3) by readily verifying that $\sum_I e_i = 1_M$.

(c)$\Rightarrow$(a). Assume (c) holds and we have

$$A = M \oplus N = \bigoplus_I A_i$$

for modules $N$ and $A_i$ for each $i \in I$, where $\text{card}(I) \leq c$. Let $\pi : A \to M$ and, for each $i$, $\pi_i : A \to A_i$ be the natural projections. Then, defining $f_i = \pi_i|_M \pi \in S$, the family $\{f_i\}_I$ is summable with $\sum_I f_i = 1_M$. Thus, by (c), there is a family of orthogonal idempotents $\{e_i\}_I$ in $S$ with $\sum_I e_i = 1_M$ and $e_i = f_is_i$ for some $s_i \in S$, for each $i \in I$.

Now take $\tilde{e}_i : A \to M$ to be $\pi_i\pi s_ie_i$. We complete the proof by showing that

$$A = M \oplus \left( \bigoplus_I (A_i \cap \text{Ke}(\tilde{e}_i)) \right).$$

Since $\{\tilde{e}_i\}_I$ is also summable we may let $\phi = \sum_I \tilde{e}_i$. Moreover, since $\tilde{e}_i|_M = \pi_i|_M \pi s_i e_i = f_is_ie_i = e_ie_i = e_i$, we have $\tilde{e}_i\tilde{e}_j = \delta_{ij}\tilde{e}_i$ and $\phi|_M = 1_M$. Hence $\phi$ provides a splitting map for the inclusion of $M$ into $A$ and so it suffices to show that $\text{Ke}(\phi) = \bigoplus_I (A_i \cap \text{Ke}(\tilde{e}_i))$. We leave this routine verification to the reader. $\qquad\square$

We may now give a neat characterisation of modules with the finite exchange property.

**11.14. Corollary.** *Let $M$ be any $R$-module and let $S = \text{End}(M)$. The following statements are equivalent.*

(a) *$M$ has the finite exchange property;*

(b) *for any finite subset $\{f_1, \ldots, f_n\} \subseteq S$ with $\sum_{i=1}^{n} f_i = 1$, there are orthogonal idempotents $e_1, \ldots, e_n \in S$ with $e_i \in f_iS$ for each $i$ and $\sum_{i=1}^{n} e_i = 1$;*

(c) *for any $f \in S$ there is an idempotent $e \in fS$ with $1 - e \in (1 - f)S$.*

*Proof.* This is immediate from 11.13 and 11.11. $\qquad\square$

**11.15. Exchange rings.** The ring $R$ is called a *left exchange ring* if the module $_RR$ has the (finite) exchange property.

Note that, by 11.8, this is equivalent to saying that every finitely generated projective left $R$-module has the exchange property.

Right exchange rings are defined similarly. However the definitions are in fact equivalent, as we'll now prove.

**11.16. Characterisations I.** *For any ring $R$ the following are equivalent.*

(a) *$R$ is a left exchange ring;*

(b) *for any $r_1, \ldots, r_n \in R$ for which $\sum_{i=1}^{n} r_i = 1$, there are orthogonal idempotents $e_1, \ldots, e_n$ in $R$ with $e_i \in r_i R$ and $\sum_{i=1}^{n} e_i = 1$;*

(c) *for any $r \in R$, there is an idempotent $e$ in $R$ with $e \in rR$ and $1 - e \in (1-r)R$;*

(d) *for any $r \in R$, there is an idempotent $f$ in $R$ with $f \in Rr$ and $1 - f \in R(1-r)$;*

(e) *for any $r_1, \ldots, r_n \in R$ for which $\sum_{i=1}^{n} r_i = 1$, there are orthogonal idempotents $f_1, \ldots, f_n$ in $R$ with $f_i \in Rr_i$ and $\sum_{i=1}^{n} f_i = 1$;*

(f) *$R$ is a right exchange ring.*

*Proof.* The equivalence of (a), (b), and (c) and that of (d), (e), and (f) follow from Corollary 11.14 and its right-handed version, taking $M = R$. Thus, to complete the proof, we need only show the equivalence of (c) and (d) and, indeed, by symmetry, (c)$\Rightarrow$(d) will suffice.

(c)$\Rightarrow$(d). Clearly (c) may be restated as: if $a, b \in R$ with $a + b = 1$, there are $e = e^2 \in aR$ and $f = f^2 \in bR$ with $e + f = 1$.

In this case, let $r, s \in R$ with $e = ar$ and $f = bs$. Then $e = e^2 = a(re)$ and $f = f^2 = b(sf)$ and so we may replace $r, s$ by $r = re, s = sf$ respectively. It is straightforward to verify that

$$rar = r, \quad sbs = s, \quad rbs = 0, \quad sar = 0. \qquad (\star)$$

Now define
$$\bar{r} = 1 - sb + rb \quad \text{and} \quad \bar{s} = 1 - ra + sa.$$

Then $\bar{r}s = 0 = \bar{s}r$ by $(\star)$ and so, since $ar + bs = e + f = 1$, we get

$$a\bar{r} = a(1 - sb) + arb = a(1 - sb) + (1 - bs)b = a(1 - sb) + b(1 - sb) = 1 - sb.$$

Hence $\bar{r}a\bar{r} = \bar{r}(1 - sb) = \bar{r} - \bar{r}sb = \bar{r}$. A similar argument gives $b\bar{s} = 1 - ra$ and $\bar{s}b\bar{s} = \bar{s}$. Thus the elements $\bar{e} = \bar{r}a$ and $\bar{f} = \bar{s}b$ are both idempotents, in $Ra$ and $Rb$, respectively. Moreover, since $ab = b - b^2 = ba$, we get

$$\begin{aligned} \bar{e} + \bar{f} \;&=\; \bar{r}a + \bar{s}b \;=\; (1 - sb + rb)a + (1 - ra + sa)b \\ &=\; a - sba + rba + b - rab + sab \;=\; a + b \;=\; 1. \end{aligned}$$

This verifies (d).                                                                                    □

Of course, Corollary 11.14 can now be rephrased as follows:

**11.17. Corollary.** *A module $M$ has the finite exchange property precisely when $\mathrm{End}(M)$ is a (left or right) exchange ring.*

As a simple consequence of 11.16 we have

**11.18. Corollary.** *Any factor ring of an exchange ring is an exchange ring.*

Next we give another characterisation of exchange rings, using their Jacobson radical. Recall that if $I$ is a left (or right) ideal of $R$, we say that *idempotents lift modulo $I$* if, given any $a \in R$ with $a^2 - a \in I$, there exists $e^2 = e \in R$ with $e - a \in I$.

**11.19. Characterisations II.** *For any ring $R$, with $J = \mathrm{Jac}\,(R)$, the following statements are equivalent.*

 (a) *$R$ is an exchange ring;*

 (b) *idempotents lift modulo $L$ for every left ideal $L$ of $R$;*

 (c) *$R/J$ is an exchange ring and idempotents lift modulo $J$.*

*Proof.* (a)⇒(b). Let $L$ be a left ideal of $R$ and $a \in R$ such that $a^2 - a \in L$. Then, by 11.16 (d), there is an idempotent $e \in R$ with $e \in Ra$ and $1 - e \in R(1 - a)$. Hence

$$e - a = e - ea + ea - a = e(1 - a) - (1 - e)a \in Ra(1 - a) + R(1 - a)a \subseteq L,$$

as required.

(b)⇒(a). Let $r \in R$ and set $L = R(r^2 - r)$. Then our hypothesis produces an idempotent $e^2 = e \in R$ such that $e - r \in L$. Since $L \subseteq Rr \cap R(1 - r)$, it quickly follows that $e \in Rr$ and $1 - e \in R(1 - r)$. Thus, by 11.16 (e), $R$ is an exchange ring.

(a)+(b)⇒(c). $R/J$ is an exchange ring by Corollary 11.18 while the idempotent lifting property is immediate from (b).

(c)⇒(a). We show that $R$ is an exchange ring by verifying 11.16 (d). Thus let $r \in R$. Then, since $R/J$ is an exchange ring, by 11.16 (d) there are elements $a, b \in R$ such that $(a + J)^2 = a + J \in (R/J)(r + J)$ and $1 - a + J = b(1 - r + J)$ in $R/J$. Moreover, since idempotents lift modulo $J$, there is an idempotent $f \in R$ for which $f - a \in J$. Then the element $u = 1 - f + a$ is a unit in $R$, say with inverse $v$. Moreover, $fu = fa \in Rr$ and $1 - u = f - a \in J$ and this infers that $1 - v = v(u - 1) \in J$.

Now set $e = vfu = vfa$. Then $e^2 = e \in Rr$ and $f - e = f - vf + vf - e = (1 - v)f + vf(1 - u) \in J$. This gives

$$(1 - e) - b(1 - r) = (a - f) + (f - e) + (1 - a) - b(1 - r) \in J$$

and so $R(1 - e) \subseteq R(1 - r) + J$. From this we infer that $R = Re + R(1 - e) = Re + R(1 - r) + J$ and so, since $J$ is small in $R$, we have $R = Re + R(1 - r)$. Set $1 = se + t(1 - r)$, where $s, t \in R$, and $g = e + (1 - e)sc$. Then $g^2 = g \in Re \subseteq Rr$ and $1 - g = 1 - e - (1 - e)se = (1 - e)(1 - se) = (1 - e)t(1 - r) \in R(1 - r)$, as required.                                                                    □

The preceding characterisation allows us to identify a large class of exchange rings, which we now define.

**11.20. Semiregular rings.** A ring $R$ with Jacobson radical $J$ is called *semiregular* or *f-semiperfect* if its factor ring $R/J$ is (von Neumann) regular and every idempotent of $R/J$ can be lifted to an idempotent of $R$ (e.g., [363, 42.11]).

**11.21. Corollary.** *Every semiregular ring $R$ is an exchange ring.*

*Proof.* By the definition and 11.19, it is sufficient to show that if $R$ is a regular ring, then $R$ is an exchange ring. Thus let $r \in R$. Then, since $R$ is regular, we have $r = rar$ for some $a \in R$. It is then readily checked that $e = ra + r(1 - ra) \in rR$ is an idempotent and $1 - e = 1 - ra - r(1 - ra) = (1 - r)(1 - ra) \in (1 - r)R$. Hence $R$ is an exchange ring by 11.16 (d).                                           □

We note that the converse to 11.21 is false in general, that is, there are exchange rings $R$ for which $R/J$ is not regular. We refer the reader to Exercise 11.42 (3) for an example.

We now wish to show that the lifting property of 11.19 (b) which characterises exchange rings can in fact be strengthened. First we need a definition and a lemma.

**11.22. Strongly lifting ideals.** We say that a left (or right) ideal $T$ of the ring $R$ is *strongly lifting*, or that idempotents *lift strongly* modulo $T$, if, given any $a \in R$ for which $a^2 - a \in T$, there exists $e^2 = e \in Ra$ such that $e - a \in T$.

This concept has some symmetry, as the following lemma shows.

**11.23. Lemma.** *The following statements are equivalent for a left ideal $L$ of $R$.*

(a) *$L$ is strongly lifting;*

(b) *given any $a \in R$ for which $a^2 - a \in L$, there exists $e^2 = e \in aRa$ such that $e - a \in L$;*

(c) *given any $a \in R$ for which $a^2 - a \in L$, there exists $e^2 = e \in aR$ such that $e - a \in L$.*

*Proof.* Given $r, s \in R$, we write $r \equiv s$ to mean $r - s \in L$. Note that if $r \equiv s$, then $xr \equiv xs$ for all $x \in R$.

(a)$\Rightarrow$(b). If $a^2 \equiv a$, then $(a^2)^2 - a^2 = (a^2 + a)(a^2 - a) \in L$ and so, by (a), there exists an $e^2 = e \in Ra^2$ such that $e \equiv a^2 \equiv a$. Setting $e = xa^2$ we may assume that $ex = x$. Now define $f = axa \in aRa$. Then $f^2 = a(xa^2)xa = aexa = axa = f$ and $f = (ax)a \equiv (ax)a^2 = ae \equiv a^2 \equiv a$, proving (b).

(b)$\Rightarrow$(c) is immediate.

(c)$\Rightarrow$(a). Given $a^2 \equiv a$, by (c) there is an $e^2 = e \in aR$ such that $e \equiv a$. Let $e = ay$, where we may suppose that $ye = y$. Then $yay = ye = y$ and so, setting $f = ya$ we have $f^2 = f \in Ra$ and $af = aya = ea$. Moreover, $fa = ya^2 \equiv ya = f$ and so

$$af \equiv a(fa) = (af)a = (ea)a = ea^2 \equiv ea \equiv e^2 = e \equiv a.$$

Thus if we define $g = f + (1 - f)af$, then $g^2 = g \in Rf \subseteq Ra$ and

$$g = f + af - f(af) \equiv f + a - fa \equiv f + a - f = a.$$

This proves (a). $\qquad\qquad\qquad\qquad\qquad\qquad\qquad\qquad\qquad\qquad\qquad\qquad\square$

Strong lifting of idempotents is indeed stronger than ordinary lifting, as the following simple example shows.

**11.24. Example.** Let $I$ be a proper two-sided ideal of the ring $R$ such that 0 and 1 are the only idempotents in both $R$ and $R/I$. Then, if $I \nsubseteq J(R)$, idempotents lift modulo $I$ but do not lift strongly. (For example, take $R = \mathbb{Z}$ and $I = p^k\mathbb{Z}$ where $p$ is a prime and $k \in \mathbb{N}$.)

To see this first note that idempotents lift modulo $I$ because the only idempotents in $R$ and $R/I$ are trivial. Next, since $I \nsubseteq J(R)$, there is a $b \in I$ for which $1 - b$ has no left inverse. Setting $a = 1 - b$ gives $a^2 - a = b^2 - b \in I$. However 0 doesn't lift $a$ since otherwise $I = R$. Thus 1 must lift $a$ modulo $I$, but not strongly since $1 \notin Ra$.

We now strengthen the equivalence of (a) and (b) in 11.19.

**11.25. Strongly lifting and the exchange property.** *The ring $R$ is an exchange ring if and only if every left (respectively right) ideal of $R$ is strongly lifting.*

*Proof.* Let $L$ be a left ideal of the exchange ring $R$ and let $a \in R$ with $a^2 - a \in L$. Then, by 11.16, there exists $e^2 = e \in R$ such that $e \in Ra$ and $1 - e \in R(1 - a)$. Then $e - a = c$   $ca \mid ca - a = e(1 - a) - (1 - e)a \in Ru(1 - a) = R(a - a^2)$ and so $e - a \in L$. Hence $L$ is strongly lifting as required.

The converse follows from 11.19 since strongly lifting implies lifting. $\qquad\square$

We next investigate the conversion of sums of submodules into direct sum decompositions and show how this is intimately related to the finite exchange property.

**11.26. Refinable modules and rings.** An $R$-module $M$ is called *refinable* (or *suitable*) if, for any submodules $U, V \subset M$ with $U + V = M$, there exists a direct summand $U'$ of $M$ with $U' \subset U$ and $U' + V = M$.

If moreover there always also exists a direct summand $V' \subset V$ with $M = U' \oplus V'$, then $M$ is said to be *strongly refinable*.

The ring $R$ is called *left refinable* if $_RR$ is a (strongly) refinable module (see 11.28).

Note that a finitely generated module $M$ is (strongly) refinable if the defining conditions are satisfied for finitely generated submodules $U, V$. Thus, for example, a finitely generated module $M$ in which every finitely generated submodule is a direct summand is refinable. (Such an $M$ is sometimes called *regular*.)

We say that *direct summands lift modulo a submodule* $K \subset M$ if, under the canonical projection $\pi : M \to M/K$, every direct summand of $M/K$ is an image of a direct summand of $M$. Similarly, we say that *(finite) decompositions lift modulo* $K$ if, whenever $M/K$ is expressed as a (finite) direct sum of submodules $M_i/K$, then $M = \bigoplus_I N_i$ where $(N_i)\pi = M_i/K$ for each $i \in I$.

**11.27. Refinability and summands lifting.** *Let $M$ be a module.*

(1) *$M$ is refinable if and only if direct summands lift modulo every submodule of $M$.*

(2) *$M$ is strongly refinable if and only if finite decompositions lift modulo every submodule of $M$.*

(3) *If $M$ is refinable and $\pi$-projective, then $M$ is strongly refinable.*

*Proof.* (1) and (2) are shown in [364, 8.4].

 (3) follows from 4.13 (1).                                                  □

The following result is proved as part of [364, 18.7] and its proof is omitted.

**11.28. Self-projective finitely generated refinable modules.** *Let $M$ be a self-projective finitely generated module with $S = \mathrm{End}(M)$. Then the following statements are equivalent.*

(a) *$M$ is (strongly) refinable;*

(b) *$M/\mathrm{Rad}\,M$ is refinable and direct summands lift modulo $\mathrm{Rad}\,M$;*

(c) *$S$ is left refinable.*

As a consequence we have

**11.29. Corollary.** *A ring $R$ is left refinable if and only if $R/J(R)$ is left refinable and idempotents lift modulo $J(R)$.*

We now use this to show that refinable rings are just exchange rings in disguise.

**11.30. Refinable endomorphism rings.** *For a left $R$-module $M$ with $S = \mathrm{End}(M)$ the following are equivalent:*

(a) *$M$ has the finite exchange property;*

(b) *$S$ is left refinable;*

(c) *$S$ is right refinable;*

(d) *$S$ is a left (right) exchange ring.*

*Proof.* (a)⇔(d) is simply a restatement of Corollary 11.14. Thus, by the left-right symmetry of 11.16, it suffices to show the equivalence of (b) and (d) for any ring $S$. Furthermore, by 11.19 and 11.29, it suffices to assume that $J(S) = 0$.

 (d)⇒(b). Suppose that $S$ is a left exchange ring. Let $K, L$ be proper ideals of $S$ such that $K + L = S$. Then $1 = k + l$ where $k$ and $l$ are nonzero elements of $K$ and $L$ respectively. Then, by our assumption, there is an idempotent $e \in Sk$ such that $1 - e \in S(1 - k) = Sl \subseteq L$. This infers that $Se$ is a direct summand of $S$ with $Se \subseteq K$ and $Se + L = S$. Hence $_S S$ is refinable.

(b)$\Rightarrow$(d). Let $r \in S$. We will show that $S$ is a left exchange ring by verifying that condition (d) of 11.16 is satisfied. Firstly, since $S$ is left refinable and $S = Sr + S(1-r)$, there is an idempotent $e \in S$ for which $Se \subseteq Sr$ and $S = Se + S(1-r)$. Then there are $s, t \in S$ such that $1 = se + t(1 - r)$. Now set $f = e + (1 - e)se$. Then $f^2 = f$ and $f \in Se$ so $f \in Sr$. Moreover, $1 - f = (1 - e) - (1 - e)se = (1 - e)(1 - se) = (1 - e)t(1 - r) \in S(1 - r)$, as required. $\qquad\square$

**11.31. Self-projective exchange modules.** *For a self-projective R-module $M$ the following are equivalent:*

  (a) *$M$ is refinable;*

  (b) *$M$ is strongly refinable;*

  (c) *$M$ has the finite exchange property.*

*Proof.* (a)$\Rightarrow$(b). This follows from 11.27 (3).

(b)$\Rightarrow$(c). Put $S = \mathrm{End}(M)$. Let $f, g \in \mathrm{End}(M)$ with $f + g = 1_S$. Then $M = Mf + Mg$ and hence there exist idempotents $e_1, e_2 \in S$ with $Me_1 \subseteq Mf$ and $Me_2 \subseteq Mg$ and $e_1 + e_2 = 1_S$. Self-projectivity of $M$ implies $e_1 \in Sf$ and $e_2 \in Sg$ showing that $S$ is a left refinable ring. By 11.30 this implies that $M$ has the finite exchange property.

(c)$\Rightarrow$(a). Let $U, V$ be submodules of $M$ with $U + V = M$. Since $M$ is self-projective, by 4.10 there are $f, g \in S = \mathrm{End}(M)$ such that $(M)f \subseteq U, (M)g \subseteq V$ and $f + g = 1_M$. Then, since $M$ has the finite exchange property, $S$ is an exchange ring by 11.17 and so, by 11.16, there is an idempotent $e \in S$ for which $e \in Sf$ and $1_M - e \in S(1_M - f) = Sg$. From this we get $(M)e \subseteq (M)f \subseteq U$ and $(M)(1_M - e) \subseteq (M)g \subseteq V$. Since $(M)e$ is a summand of $M$, it follows that $M$ is refinable as required. $\qquad\square$

We now turn our attention to a generalisation of the exchange property.

**11.32. Exchange decompositions.** We say that $M = \bigoplus_I M_i$ is an *exchange decomposition* for the module $M$ or that the decomposition is *exchangeable* if, for any direct summand $X$ of $M$, there exist $N_i \subseteq M_i$ for all $i \in I$ such that

$$M = X \oplus \left( \bigoplus_I N_i \right).$$

In this case $N_i$ is a direct summand of $M_i$ for each $i \in I$, as can be seen by applying Lemma 11.2 to the inclusions

$$N_i \subseteq M_i \subseteq N_i \oplus \left( X \oplus \left( \bigoplus_{j \neq i} N_j \right) \right).$$

**11.33. Direct summands of exchange decompositions.** *Let $M$ be a module with exchange decomposition $M = \bigoplus_I M_i$. If $M_i = M_i' \oplus M_i''$, for every $i \in I$, then $N = \bigoplus_I M_i'$ is an exchange decomposition.*

*Proof.* Let $K$ be a direct summand of $N$. As $M = \bigoplus_I M_i$ is an exchange decomposition, we get $M = K \oplus \bigoplus_I L_i$, where each $L_i \subseteq M_i$. Then

$$N = K \oplus (N \cap (\bigoplus_I L_i)) = K \oplus (\bigoplus_I (L_i \cap M_i')).$$

$\square$

**11.34. The internal exchange property.** Let $c$ be a cardinal number. The module $M$ is said to have the *c-internal exchange property* if every decomposition $M = \bigoplus_I M_i$ with card$(I) \leq c$ is exchangeable.

We say that the module $M$ has the *(finite) internal exchange property* if it has the $c$-internal exchange property for every (finite) cardinal $c$.

Recall that, by 11.9, any direct summand of a module with the (finite) exchange property has the (finite) exchange property. Using this, a routine argument gives

**11.35. Exchange implies internal exchange.** *If $M$ has the (finite) exchange property, then $M$ has the (finite) internal exchange property.*

The converse of 11.35 is false in general: any indecomposable module $M$ has the internal exchange property (trivially) but, as we will see in 12.2, it has the exchange property only if its endomorphism ring is local.

**11.36. 2-internal exchange property of direct summands.** *The 2-internal exchange property is inherited by direct summands.*

*Proof.* Suppose that the module $M$ has the 2-internal exchange property and that $M = K \oplus L$. Let $K = K_1 \oplus K_2$ and let $X$ be a direct summand of $K$. The 2-internal exchange property for $M$ applied to the decomposition $M = K_1 \oplus (K_2 \oplus L)$ produces $K_1' \subseteq K_1$ and $(K_2 \oplus L)' \subseteq K_2 \oplus L$ such that

$$M = (X \oplus L) \oplus K_1' \oplus (K_2 \oplus L)' = X \oplus K_1' \oplus (K_2 \oplus L)' \oplus L.$$

Let $p : M \to K$ be the projection of $M$ along $L$ and set $K_2' = (K_2 \oplus L)'p$. Then $K_2' \subseteq (K_2 \oplus L)p = K_2$ and so $K_2' \oplus L = (K_2 \oplus L)' \oplus L$. Hence $M = X \oplus K_1' \oplus K_2' \oplus L$ and therefore $K = X \oplus K_1' \oplus K_2'$, with $K_1' \subseteq K_1$ and $K_2' \subseteq K_2$. $\square$

In parallel with 11.11, the 2-internal exchange property implies the finite internal exchange property. To prove this, we first establish the following lemma.

**11.37. Lemma.** *Let $M$ be an $R$-module and $K \subseteq M$ with an exchange decomposition $K = \bigoplus_I K_i$. If $M = N \oplus K'$, where $K' \subseteq K$, then $M = N \oplus (\bigoplus_I K_i')$ with $K_i' \subseteq K_i$, for all $i \in I$.*

*Proof.* Since $K' \subseteq K$, the modular law gives $K = (N \cap K) \oplus K'$. From the assumption on $K$ we get $K = (N \cap K) \oplus (\bigoplus_I K_i')$ with $K_i' \subseteq K_i$, for all $i \in I$. Put $L = \bigoplus_I K_i'$. Then

$$M = N \oplus K' = N + K = N + (N \cap K) + L = N + L,$$

and $N \cap L = N \cap (K \cap L) = (N \cap K) \cap L = 0.$ $\qquad\qquad\square$

**11.38. 2-internal exchange implies finite internal exchange.** *An R-module M satisfying the 2-internal exchange property has the finite internal exchange property.*

*Proof.* Suppose $M$ has the 2-internal exchange property. We prove by induction that $M$ has the $n$-internal exchange property for every $n > 2$. Assume that $M$ has the $(n-1)$-internal exchange property. Suppose $M = \bigoplus_{i=1}^n M_i$ and $X$ is a direct summand of $M$. Put $K = \bigoplus_{i=2}^n M_i$. Then $M = M_1 \oplus K$. By the 2-internal exchange property of $M$ we have $M = X \oplus M_1' \oplus K'$, with $M_1' \subseteq M_1$ and $K' \subseteq K$. As $K$ is a direct summand of $M$, $K$ has the 2-internal exchange property by 11.36 and hence has the $(n-1)$-internal exchange property by the induction hypothesis. It follows from Lemma 11.37 that $M = X \oplus M_1' \oplus M_2' \oplus \cdots \oplus M_n'$, with $M_i' \subseteq M_i$, for $i = 1, 2, \ldots, n$. $\qquad\qquad\square$

From 11.36 and 11.38 we get

**11.39. Corollary.** *The finite internal exchange property is inherited by direct summands.*

The following gives a necessary and sufficient condition for a finite direct sum of modules to have the finite internal exchange property.

**11.40. Finite direct sums and the finite internal exchange property.** *Suppose $M = \bigoplus_{i=1}^n M_i$. Then M has the finite internal exchange property if and only if each $M_i$ has the finite internal exchange property and $M = \bigoplus_{i=1}^n M_i$ is an exchange decomposition.*

*Proof.* The "only if part" follows from 11.39. For the converse, suppose each $M_i$ has the finite internal exchange property and the given decomposition of $M$ is exchangeable. By 11.38 it suffices to show that $M$ has the 2-internal exchange property.

Let $M = K \oplus L$ and $C$ be a direct summand of $M$. Then $M = L \oplus (\bigoplus_{i=1}^n M_i')$ where each $M_i'$ is a summand of $M_i$, say $M_i = M_i' \oplus M_i''$, for $i = 1, 2, \ldots, n$. Then $K \simeq \bigoplus_{i=1}^n M_i'$ and $L \simeq \bigoplus_{i=1}^n M_i''$. Write $K = \bigoplus_{i=1}^n K_i$ and $L = \bigoplus_{i=1}^n L_i$, where $K_i \simeq M_i'$ and $L_i \simeq M_i''$ for $i = 1, 2, \ldots, n$.

For each $i$, define $X_i = K_i \oplus L_i$. Then $X_i \simeq M_i' \oplus M_i'' = M_i$ and hence the decomposition $M = \bigoplus_{i=1}^n X_i$ is exchangeable. Thus $M = C \oplus X_1' \oplus X_2' \cdots \oplus X_n'$, where $X_i' \subseteq X_i$ for $i = 1, 2, \ldots, n$. Since each $X_i$ has the finite internal exchange property, applying Lemma 11.37, we get

$$M = C \oplus K_1' \oplus L_1' \oplus K_2' \oplus L_2' \cdots \oplus K_n' \oplus L_n',$$

where $K_i' \subseteq K_i$ and $L_i' \subseteq L_i$ for $i = 1, 2, \ldots, n$.                    $\square$

**11.41. Generalised projectivity and the internal exchange property.** *Let $K, L$ be modules where $K$ is generalised $L$-projective and has the internal finite exchange property. Then, for any direct summands $K^* \subset K$ and $L^* \subset L$, $K^*$ is generalised $L^*$-projective.*

*Proof.* Let $f : K^* \to X$ be a homomorphism, let $g : L \to X$ be an epimorphism and let $p : K = K^* \oplus K^{**} \to K^*$ be the projection. Since $K$ is generalised $L$-projective, there exist decompositions $K = K_1 \oplus K_2$, $L = L_1 \oplus L_2$, a morphism $\varphi_1 : K_1 \to L_1$ and an epimorphism $\varphi_2 : L_2 \to K_2$ such that $p|_{K_1} \diamond f = \varphi_1 \diamond g$ and $g|_{L_2} = \varphi_2 \diamond p \diamond f$. Since $K$ has the finite internal exchange property and $K^{**}$ is a direct summand of $K$, there exist decompositions $K_i = \overline{K_i} \oplus \widetilde{K_i}$, $i = 1, 2$, such that $K = K^{**} \oplus \overline{K_1} \oplus \overline{K_2}$. So there is an isomorphism

$$p|_{\overline{K_1} \oplus \overline{K_2}} : \overline{K_1} \oplus \overline{K_2} \simeq K^*.$$

Putting $K_i^* = (\overline{K_i})p$, $i = 1, 2$, then

$$K^* = (\overline{K_1})p \oplus (\overline{K_2})p = K_1^* \oplus K_2^*.$$

Let $\pi_{K_1^*} : K^* = K_1^* \oplus K_2^* \to K_1^*$ be the projection. Put

$$\alpha = p|_{\widetilde{K_2}} \diamond \pi_{K_1^*} \diamond p^{-1} : \widetilde{K_2} \to \overline{K_1}.$$

Write $U = \{\tilde{a}_2 - (\tilde{a}_2)\alpha \mid \tilde{a}_2 \in \widetilde{K_2}\}$. Then $\widetilde{K_2} \oplus \overline{K_1} = \widetilde{K_2} \oplus U$. Also

$$K = \overline{K_1} \oplus \widetilde{K_1} \oplus \overline{K_2} \oplus U \quad \text{and} \quad (U)p \subseteq K_2^*.$$

Now define
$$\alpha^* : K_2 = \overline{K_2} \oplus \widetilde{K_2} \to \overline{K_1}, \quad \overline{a_2} + \tilde{a}_2 \mapsto (\tilde{a}_2)\alpha.$$

Put $\beta = \varphi_2 \diamond \alpha^* \diamond \varphi_1 : L_2 \to L_1$ and $V = \{b_2 - (b_2)\beta \mid b_2 \in L_2\}$. Then $L = L_2 \oplus V$. Define isomorphisms

$$\tau_1 : K_2 \to \overline{K_2} \oplus U, \quad \overline{a_2} + \tilde{a}_2 \mapsto \overline{a_2} + (\tilde{a}_2 - (\tilde{a}_2)\alpha),$$
$$\tau_2 : V \to L_2, \quad b_2 - (b_2)\beta \mapsto b_2,$$

and put

$$\psi_1 = p^{-1} \diamond \varphi_1 : K_1^* \to L_1 \quad \text{and} \quad \psi_2 = \tau_2 \diamond \varphi_2 \diamond \tau_1 \diamond p : V \to K_2^*.$$

Then $\psi_2$ is an epimorphism. Given $a_1^* = (\overline{a_1})p \in K_1^*$, then

$$(a_1^*)f = ((\overline{a_1})p)f = (\overline{a_1})\varphi_1 g = (a_1^*)p^{-1}\varphi_1 g = (a_1^*)\psi_1 g.$$

Hence $f|_{K_1^*} = \psi_1 \diamond g$. Now given $x = b_2 - (b_2)\beta \in V$ put $(b_2)\varphi_2 = \overline{a_2} + \tilde{a}_2$. Then

$$
\begin{aligned}
(x)g &= (b_2)g - (b_2)\beta g = (b_2)\varphi_2 pf - \alpha^*((b_2)\varphi_2 \alpha^* \varphi_1)g \\
&= (\overline{a_2} + \tilde{a}_2)pf - ((\overline{a_2} + \tilde{a}_2)\alpha^* \varphi_1)g = (\overline{a_2})pf + (\tilde{a}_2)pf - (\tilde{a}_2)\alpha\varphi_1 g \\
&= (\overline{a_2})pf + (\tilde{a}_2)pf - (\tilde{a}_2)\alpha pf.
\end{aligned}
$$

On the other hand,

$$
\begin{aligned}
(x)\psi_2 f &= (b_2 - (b_2)\beta)\tau_2\varphi_2\tau_1 pf = (b_2)\varphi_2\tau_1 pf = (\overline{a_2} + \tilde{a}_2)\tau_1 pf \\
&= (\overline{a_2} + (\tilde{a}_2 - (\tilde{a}_2)\alpha))pf \\
&= (\overline{a_2})pf + (\tilde{a}_2)pf - (\tilde{a}_2)\alpha pf.
\end{aligned}
$$

Thus $g|_V = \psi_2 \diamond f$. Therefore $K^*$ is generalised $L$-projective and hence, by 4.43, $K^*$ is generalised $L^*$-projective. $\qquad\square$

### 11.42. Exercises.

(1) Let $M$ be an $R$-module with $M = B \oplus C$ and let $A$ be a submodule of $M$. Following Leron [221], we say that $A$ *exchanges* $B$ if $M = A \oplus C$ and that $A$ *subexchanges* $B$ if there is a decomposition $B = B' \oplus B''$ such that $M = A \oplus B'' \oplus C$.

Let $f : M \to N$ be a monomorphism, $A$ be a submodule of $M$ with inclusion map $i_A : A \to M$, and $N = B \oplus C$ with projection map $\pi_B : N \to B$. Show that $(A)f$ exchanges (subexchanges) $B$ if and only if the homomorphism $i_A f \pi_B$ from $A$ to $B$ is an isomorphism (has a right inverse). (The "exchanges" part in the special case of $M = N$ and $f = 1_M$ is given by 11.3.)

(2) Prove directly, without using 12.2, that $\mathbb{Z}$ is not an exchange ring as follows. Consider the $\mathbb{Z}$-modules $\mathbb{Z} \oplus \mathbb{Z} = A \oplus B = A' \oplus C$ where $A = \mathbb{Z}(1,0), B = \mathbb{Z}(0,1), A' = \mathbb{Z}(7,3), C = \mathbb{Z}(5,2)$ (all isomorphic to $\mathbb{Z}$). Show, using the indecomposability of $\mathbb{Z}$, that there is no decomposition $\mathbb{Z} \oplus \mathbb{Z} = A \oplus X \oplus Y$ where $X \subseteq A'$ and $Y \subseteq C$. (See Lam [217].)

(3) Let $S$ denote the field of rational numbers and $T$ denote its subring $\mathbb{Z}_{(2)}$, the integers localised at the prime ideal $2\mathbb{Z}$. Let $R$ be the ring of eventually constant sequences $r = (s_1, s_2, \ldots, s_n, t, t, t, \ldots)$ where $s_1, s_2, \ldots, s_n \in S$, the eventual constant $t$ is in $T$, and $n$ depends on $r$. Show that $J(R) = 0$, $R$ is an exchange ring, but $R$ is not von Neumann regular.

(4) Let $R$ be the ring of eventually constant sequences $r = (s_1, s_2, \ldots, s_n, t, t, t, \ldots)$ where $s_1, s_2, \ldots, s_n \in \mathbb{Q}$, the eventual constant $t$ is from $\mathbb{Z}$, and $n$ depends on $r$. Show that, in contrast to the previous exercise, $R$ is not an exchange ring. (See [194, § II.5].)

(5) Show that if $M$ is a (strongly) refinable module and $N$ is a fully invariant submodule of $M$, then $M/N$ is (strongly) refinable. ([364, 8.5 (1)].)

(6) Show that a $\pi$-projective module is strongly refinable. ([364, 8.5 (2)].)

(7) Let $M = K \oplus L$. Suppose $K = \bigoplus_{i=1}^{s} K_i$ and $L = \bigoplus_{j=1}^{t} L_j$ are exchange decompositions. Show that, if $M = K \oplus L$ is an exchange decomposition, then $M = K_1 \oplus \cdots \oplus K_s \oplus L_1 \oplus \cdots \oplus L_t$ is also an exchange decomposition.

(8) The module $M$ is said to have the $D2$-exchange property if, for every nonzero direct summand $M_0$ of $M$, any module $A$ and any decompositions

$$A = M_0 \oplus N = A_1 \oplus A_2,$$

there always exist a nonzero submodule $M_0'$ of $M_0$ and submodules $B_i \leq A_i$ for $i = 1, 2$ such that

$$A = M_0' \oplus B_1 \oplus B_2.$$

(Here "D" signifies "direct (summand)".)

  (i) Show that any module $M$ with the 2-exchange property has the $D2$-exchange property.

  (ii) Prove that the following conditions are equivalent for a module $M$.

    (a) $M$ has the $D2$-exchange property;

    (b) every nonzero direct summand of $M$ has the $D2$-exchange property;

    (c) every nonzero direct summand of $M$ contains a nonzero direct summand which has the $D2$-exchange property.

  (iii) Let $R$ be the ring of eventually constant sequences $r = (s_1, \ldots, s_n, t, t, \ldots)$ where $s_1, \ldots, s_n \in \mathbb{Q}$, the eventual constant $t$ is from $\mathbb{Z}$, and $n$ depends on $r$. Show that ${}_R R$ has the $D2$-exchange property but not the 2-exchange property.

  (iv) Show that if $M$ is an indecomposable module with the $D2$-exchange property, then it has the 2-exchange property.

  (See [195], [194, Chap. II, §1], and [311].)

(9) Let $I$ be a ring without unit and let $R$ be a unital ring containing $I$ as a two-sided ideal. Given $x \in I$, show that the following statements are equivalent.

  (a) There exists $e^2 = e \in I$ with $e - x \in R(x - x^2)$.

  (b) There exist $e^2 = e \in Ix$ and $c \in R$ with $(1 - e) - c(1 - x) \in J(R)$.

  (c) There exists $e^2 = e \in Ix$ such that $R = Ie + R(1 - x)$.

  (d) There exists $e^2 = e \in Ix$ with $1 - e \in R(1 - x)$.

  (e) There exist $r, s \in I$ and $e^2 = e \in I$ such that $e = rx = s + x - sx$.

(This exercise is the starting point in Ara [18] for his extension of the exchange property to non-unital rings, with Nicholson's unital ring version [256, Proposition 1.1] as forerunner.)

(10) Let $N \subseteq I$ be two-sided ideals of the ring $R$ and set $\overline{R} = R/N$ and $\overline{I} = I/N$.

  (i) Show that if ${}_R R = I \oplus L$ where $L$ is a left ideal, then $I$ is strongly lifting.

  (ii) Show that if $N$ and $\overline{I}$ are strongly lifting in $R$ and $\overline{R}$ respectively, then $I$ is strongly lifting in $R$. (See [262].)

(11) Let $M$ be an $R$-module with $S = \text{End}(M)$ and $\Delta = \{f \in S \,|\, \text{Ke}\, f \trianglelefteq M\}$. Let $\{f_i\}_I$ and $\{g_i\}_I$ be two families of homomorphisms from $S$ with common index set $I$.

   (i) Show that if $\{g_i\}_I$ is summable and $\{(m)f_i \,|\, i \in I\}$ is finite for each $m \in M$, then $\{f_i \diamond g_i\}_I$ is summable.

   (ii) Show that if $\{f_i\}_I$ and $\{g_i\}_I$ are both summable and $f_i - g_i \in \Delta$ for each $i \in I$, then $\sum_I f_i - \sum_I g_i \in \Delta$. (See [240, Lemma 4].)

(12) A ring $R$ is called $\pi$-*regular* if, for every $a \in R$, there exists an $x \in R$ and an $n \in \mathbb{N}$ (depending on $a$) such that $a^n = a^n x a^n$. Using 11.16, show that a $\pi$-regular ring is an exchange ring. (See [324, Example 2.3].)

(13) Show that the directed union of a family of exchange rings is an exchange ring.

**11.43. Comments.** As already indicated, the exchange property was introduced by Crawley and Jónnson in their pioneering paper [77]. The fundamentals of the property described in the first few pages of this section, up to 11.11, appear in [77] in a general algebra setting.

Warfield [357] proved that every injective module has the (full) exchange property. This was generalised to quasi-injectives by Fuchs [107], then to continuous modules by Mohamed and Müller [240]. In fact the exchange property for quasi-injectives also follows from a later result by Zimmermann-Huisgen and Zimmermann [385] showing that the property is satisfied by any submodule $M$ of an algebraically compact module $X$ for which $(M)f \subseteq M$ for all homomorphisms $f : M \to X$. (This also implies that artinian modules over a commutative ring have the exchange property.)

Crawley and Jónnson also raised the still-unresolved questions of whether the finite exchange property always implies the full exchange property and, for given infinite cardinals $m$ and $n$ with $1 < m < n$, the $m$-exchange property always implies the $n$-exchange property. As we will see in the next section (see 12.14), Zimmermann-Huisgen and Zimmermann [385] have shown that finite exchange does imply full exchange for any module with an indecomposable decomposition. An affirmative answer has also been provided for quasi-continuous modules by Oshiro and Rizvi [276] and then simplified by Mohamed and Müller [242] (although quasi-continuous modules need not enjoy the finite exchange property, as evidenced by the $\mathbb{Z}$-module $\mathbb{Z}$ in Exercise (1) above). Yu [375] shows that if $M$ has the finite exchange property and idempotents in $\text{End}(M)$ are all central (equivalently, all direct summands of $M$ are fully invariant), then $M$ has the $\aleph_0$-exchange property and this has been recently extended to the full exchange property by Nielsen [263].

It was noted by Crawley and Jónnson in [77] that, in the equations $A = M' \oplus N = \bigoplus_I A_i$ used in definition 11.4 of the $c$-exchange property for $M \simeq M'$, it suffices to take the $A_i$ to be isomorphic to submodules of $M$. The significant improvement on this in 11.13, reducing the $\bigoplus_{i \in I} A_i$ to a clone of $M$ and proved using summable families of homomorphisms, is due to Zimmermann-Huisgen and Zimmermann [385, Proposition 3]. The equivalence of (1) and (3) of its Corollary 11.14 was first established by Nicholson [256, Theorem 2.1], with Monk [249] as a precursor.

The left-right symmetry of exchange rings was first noted by Warfield in [359, Corollary 2] while their characterisation in 11.16 using idempotents was obtained independently by Goodearl [134, p. 167] and Nicholson [256, Proposition 1.1 and Theorem 2.1]. Our proof of 11.16 takes advantage of Nicholson's much later [258]. Warfield [359, Theorem 2] also proved that a module $M$ has the finite exchange property precisely when End($M$) is an exchange ring, while Nicholson [256, Proposition 1.5] provided the characterisation of exchange rings in 11.19. Apropos condition (b) of 11.19, Khurana and Gupta [209] show that, given an ideal $I$ of a ring $R$, idempotents lift modulo every left ideal of $R$ contained in $I$ precisely when they lift modulo every right ideal contained in $I$. Taking $I = R$ again yields the left-right symmetry of exchange rings.

Oberst and Schneider [264] refer to semiregular rings as *F-semiperfect* rings and characterise them as the rings for which every finitely presented module has a projective cover. This terminology is used by Azumaya [27] and others (see also [363, §42]).

Nicholson [256] calls a ring $R$ *clean* if every element of $R$ is the sum of a unit and an idempotent in $R$ and he proves that every clean ring is an exchange ring. That the converse is false in general is established by Camillo and Yu [52] through an example of Bergman in [145], but they also show that if $R$ contains no infinite set of orthogonal idempotents, then $R$ is clean precisely when $R$ is an exchange ring and precisely when $R$ is semiperfect. An alternative proof of the consequential characterisation of semiperfect rings, bypassing exchange property theory, is given by Nicholson in [255].

Adapting the characterisation of exchange rings given in 11.16, Ara [18] has defined a ring $I$ without unit to be an exchange ring if, for each $x \in I$, there exist an idempotent $e \in I$ and $r, s \in I$ for which $e = xr = x + s - xs$. (See Exercise (9) above.) As in the unital ring case, this definition is left-right symmetric. He extends 11.19 by proving in [18, Theorem 2.2] that if $I$ is an ideal of a (possibly non-unital) ring $R$, then $R$ is an exchange ring if and only if $I$ and $R/I$ are both exchange rings and idempotents can be lifted modulo $I$. Some applications of this are surveyed by Ara in [19]. Moreover, Baccella [31, Theorem 1.4] uses Ara's characterisation to show that every semi-artinian ring is an exchange ring. (Recall that a ring $R$ is (left) *semi-artinian* if each nonzero left $R$-module contains a simple submodule.)

The strongly lifting results 11.23–11.25 are from Nicholson and Zhou's recent [262], although the terminology is taken from Mohamed and Müller's [242]. Among other interesting results, Nicholson and Zhou show in [262, Theorem 10, Lemma 5] that, for any ring $R$, each term in the left or right socle series of $R$ is strongly lifting and, if idempotents lift modulo $J(R)$, then every one-sided ideal of $R$ contained in $J(R)$ is strongly lifting.

Birkenmeier [42] contains 11.31 and this also later appears in Tuganbaev [334]. (Much earlier, Nicholson [256, Proposition 2.9] proved the result for projective modules.)

The internal exchange property is used by Hanada, Kuratomi and Oshiro in [143] and [144] and by Mohammed and Müller in [244] in their investigations into the thorny question of when a direct sum of extending modules is extending. The basic features of the property given in 11.36 to 11.40 appear in [244] and [245] and these and exchange decompositions are used significantly in Chapter 4.

For more on the exchange property, we refer the reader to Facchini's text [92] (in particular for its connection with Azumaya's Theorem (see 12.6)) and the survey articles by Lam [217] and Tuganbaev [335], [337]. Chapters 6 and 7 of Tuganbaev's text [338] include recent work on exchange rings due to Ara, Goodearl, O'Meara, Raphael, and Pardo (see, for example, [20]). Ara, O'Meara, and Perera [21] have also recently established a fresh source of exchange rings, including one found a little earlier by O'Meara [267], namely the ring of $\omega \times \omega$ row-and-column-finite matrices over any regular ring.

**References.** Ara [18, 19]; Ara, Goodearl, O'Meara, and Pardo [20]; Ara, O'Meara, and Perera [21]; Azumaya [27, 24]; Baccella [31]; Birkenmeier [42]; Camillo and Yu [52]; Crawley and Jónnson [77]; Facchini [92]; Goodearl [134]; Hanada, Kuratomi, and Oshiro [143, 144]; Khurana and Gupta [209]; Kuratomi [214]; Lam [217]; Mohamed and Müller [240, 242, 244]; Monk [249]; Nicholson [256, 255]; Nicholson and Zhou [262]; Oberst and Schneider [264]; O'Meara [267]; Oshiro and Rizvi [276] Tuganbaev [334, 335, 337, 338]; Warfield [357, 359]; Yu [375]; Zimmermann-Huisgen and Zimmermann [385].

# 12    LE-modules and local semi-T-nilpotency

**12.1. LE-modules and LE-decompositions.** An $R$-module $M$ is said to be an *LE-module* if its endomorphism ring $\mathrm{End}(M)$ is local.

(Harada [158] uses the term *completely indecomposable* instead, while other authors, for example Lam [217], use *strongly indecomposable* — of course, every LE-module is indecomposable.)

A decomposition $M = \bigoplus_I M_i$ is called an *LE-decomposition* of $M$ if each $M_i$ is an LE-module.

The next result produces another plentiful supply of modules with the exchange property (and a plentiful supply of those that lack the property).

**12.2. Indecomposable exchange modules.** *Let $M$ be an indecomposable $R$-module. Then the following conditions are equivalent.*

(a) *$M$ is an LE-module;*

(b) *$M$ has the finite exchange property;*

(c) *$M$ has the exchange property.*

*Proof.* Note first that, since $M$ is indecomposable, the only idempotents in $S = \mathrm{End}(M)$ are the identity map $1_M$ and the zero map $0_M$ on $M$. Moreover, any one-sided invertible element in $S$ is a unit since, if $f, g \in S$ with $fg = 1_M$, then the nonzero idempotent $gf$ must also be $1_M$.

(a)$\Rightarrow$(b). Suppose that $S = \mathrm{End}(M)$ is a local ring and let $f \in S$. Then, since $S$ is local, either $f$ or $1 - f$ is a unit in $S$, say $f$. Then $1_M^2 = 1_M \in fS$ and $0_M^2 = 0_M \in (1-f)S$. It now follows from Corollary 11.14 (c) that $M$ has the finite exchange property.

(b)$\Rightarrow$(c). Assume that the indecomposable module $M$ has the finite exchange property and let $I$ be an infinite index set. Suppose that $A, N, A_i$ (for each $i \in I$) are modules such that

$$A = M \oplus N = \bigoplus_I A_i.$$

Let $m$ be a nonzero element of $M$. Then there is a finite subset $J$ of $I$ for which $m \in \bigoplus_J A_j$. Set $K = I \setminus J$ and $A_K = \bigoplus_{k \in K} A_k$. Then, since

$$A = M \oplus N = A_K \oplus \left( \bigoplus_J A_j \right)$$

and $M$ has the finite exchange property, we have decompositions $A_K = B_K \oplus C_K$ and $A_j = B_j \oplus C_j$ for each $j \in J$ for which

$$A = M \oplus B_K \oplus \left( \bigoplus_J B_j \right) = A_K \oplus \left( \bigoplus_J A_j \right) = (B_K \oplus C_K) \oplus \left( \bigoplus_J (B_j \oplus C_j) \right).$$

Factoring the submodule $B_K \oplus (\bigoplus_J B_j)$ from the second and fourth terms in these last equations shows that $M$ is isomorphic to $C_K \oplus (\bigoplus_J C_j)$. Since $M$ is indecomposable, it follows that either there is an index $j_0 \in J$ for which we have $C_K \oplus \left( \bigoplus_{j \in J, j \neq j_0} C_j \right) = 0$ or $\bigoplus_J C_j = 0$. Thus we have either

$$A = M \oplus A_K \oplus B_{j_0} \oplus \left( \bigoplus_{j \in J, j \neq j_0} A_j \right) \text{ or } A = M \oplus B_K \oplus \left( \bigoplus_J A_j \right).$$

However, the latter is impossible since $x$ is a nonzero element of $M \cap (\bigoplus_J A_j)$. Consequently we have the desired decomposition, given by

$$A = M \oplus A_K \oplus B_{j_0} \oplus \left( \bigoplus_{j \in J, j \neq j_0} A_j \right) = M \oplus B_{j_0} \oplus \left( \bigoplus_{i \in I, i \neq j_0} A_i \right).$$

(c)$\Rightarrow$(a). Here our proof is by contradiction. Thus suppose that $S$ is not local. Then there is an $f \in S$ for which both $f$ and $1 - f$ are non-isomorphisms of $M$. Since $M$ has the exchange property, it follows from Corollary 11.14 (c) that there is an idempotent $e \in S$ for which $e \in fS$ and $1 - e \in (1 - f)S$. Since $e$ is either $1_M$ or $0_M$, it follows that either $f$ or $1 - f$ has a right inverse and so, by our introductory remarks, we have our contradiction. $\qquad\square$

Applying 11.9 we get

**12.3. Corollary.** *A finite direct sum of LE-modules has the exchange property.*

We now recall some useful decomposition properties. This then enables us to record an important classical result on LE-decompositions due to Azumaya [24] which will be indispensable in later proofs.

**12.4. Complementing and equivalent decompositions.** Let $M = \bigoplus_I M_i$ be a decomposition of the module $M$ into nonzero summands $M_i$.

(1) This decomposition is said to *complement direct summands* if, whenever $A$ is a direct summand of $M$, there is a subset $J$ of $I$ for which $M = (\bigoplus_J M_j) \oplus A$. Note that, in this case, each $M_i$ must be indecomposable. From this it follows that an indecomposable decomposition complements direct summands if and only if it is exchangeable (see 11.32).

(2) The decomposition is said to *complement maximal direct summands* if, whenever $A_1, A_2$ are submodules of $M$ for which $M = A_1 \oplus A_2$ and $A_1$ is indecomposable, then $M = M_i \oplus A_2$ for some $i \in I$.

(3) Given a second decomposition $M = \bigoplus_J N_j$ of $M$, the two decompositions are said to be *equivalent* or *isomorphic* if there is a bijection $\sigma : I \to J$ such that $M_i \simeq N_{\sigma(i)}$ for each $i \in I$.

The following theorem, due to Anderson and Fuller [12], collects together some useful properties of indecomposable decompositions which complement maximal direct summands. We refer the reader to their textbook for its proof ([13, §12]).

**12.5. Complementing indecomposable decompositions.** *Let $M$ be a module with an indecomposable decomposition $M = \bigoplus_I M_i$ that complements (maximal) direct summands. Then*

(1) *all indecomposable decompositions of $M$ are equivalent;*

(2) *every indecomposable decomposition of $M$ complements (maximal) direct summands;*

(3) *if the summand $M_i$ appears at least twice in the decomposition, that is, if there are distinct indices $i, j \in I$ for which $M_i \simeq M_j$, then $M_i$ is an LE-module;*

(4) *if $N_1, N_2, \ldots, N_t$ are a finite number of indecomposable submodules of $M$ for which*

$$M = N_1 \oplus N_2 \oplus \cdots \oplus N_t \oplus D,$$

*then there are indices $i_1, i_2, \ldots, i_t \in I$ for which $N_k \simeq M_{i_k}$ for $1 \le k \le t$ and, for each $1 \le s \le t$,*

$$M = M_{i_1} \oplus \cdots \oplus M_{i_s} \oplus N_{s+1} \oplus \cdots \oplus N_t \oplus D$$

*and, consequently, $D$ has an indecomposable decomposition;*

(5) *for any subset $J$ of $I$, $M' = \bigoplus_J M_j$ is a decomposition of $M'$ which complements (maximal) direct summands.*

We now state Azumaya's generalisation of the Krull–Schmidt Theorem. Again we refer to [13, § 12] for a proof.

**12.6. Azumaya's Theorem.** *Let $M = \bigoplus_I M_i$ be an LE-decomposition. Then*

(1) *every nonzero direct summand of $M$ has an indecomposable direct summand,*

(2) *the decomposition $M = \bigoplus_I M_i$ complements maximal direct summands,*

(3) *and consequently the decomposition is equivalent to every indecomposable decomposition of $M$.*

While LE-decompositions guarantee very favourable decompositions, the following example (from [99]) shows that the local endomorphism ring property is not essential for a decomposition to complement summands.

**12.7. Example.** Given any index set $I$, there is a family $\{H_i\}_I$ of hollow modules which are not LE-modules such that the decomposition $H = \bigoplus_I H_i$ is exchangeable or, equivalently, complements direct summands.

To see this, let $\mathbb{Z}_{(p)}$ and $\mathbb{Z}_{(q)}$ be the localisations of $\mathbb{Z}$ at two distinct primes $p$ and $q$ respectively. Consider the ring

$$R = \begin{bmatrix} \mathbb{Z}_{(p)} & \mathbb{Q} \\ 0 & \mathbb{Z}_{(q)} \end{bmatrix}$$

and its left ideal

$$L = \begin{bmatrix} \mathbb{Z}_{(p)} & \mathbb{Z}_{(q)} \\ 0 & 0 \end{bmatrix}.$$

Then $R/L$ is a local module but $\mathrm{End}(R/L) \simeq \mathbb{Z}_{(p)} \cap \mathbb{Z}_{(q)}$ is a semilocal ring with exactly two maximal ideals. Now consider the ring $S = \prod_I R_i$, where each $R_i \simeq R$ for each index $i$. For each $i \in I$, let $M_i$ be an $R_i$-module isomorphic to $R/L$ and consider the $S$-module $M = \bigoplus_I M_i$. Then the decomposition $M = \bigoplus_I M_i$ complements direct summands, where each $M_i$ is a local module whose endomorphism ring is not local.

The next lemma, due to Dung [83], establishes an important replacement property of indecomposable decompositions which complement maximal direct summands.

**12.8. Lemma.** *Let $M = \bigoplus_I M_i$ be an indecomposable decomposition which complements maximal direct summands. Suppose that*

$$M = N_1 \oplus N_2 \oplus \cdots \oplus N_t \oplus D$$

*where $N_1, N_2, \ldots, N_t$ are a finite number of indecomposable submodules of $M$. Then there are indices $i_1, i_2, \ldots, i_t \in I$ for which $N_k \simeq M_{i_k}$ for $1 \leq k \leq t$ and*

$$M = N_1 \oplus N_2 \oplus \cdots \oplus N_t \oplus \left( \bigoplus_{i \in I \setminus \{i_1, i_2, \ldots, i_t\}} M_i \right).$$

*Proof.* We use induction on $t$. For $t = 1$ we have $M = N_1 \oplus D$ and so, since $D$ is then a maximal direct summand of $M$, by hypothesis there is an index $i_1 \in I$ for which $M = M_{i_1} \oplus D$. Clearly $D \simeq \bigoplus_{i \in I \setminus \{i_1\}} M_i$ and so $D = \bigoplus_{I'} D_i$ where $D_i \simeq M_i$ for each $i \in I' = I \setminus \{i_1\}$. Moreover, by 12.5 (2), the indecomposable decomposition $M = N_1 \oplus (\bigoplus_{I'} D_i)$ also complements maximal summands. Thus, since $\bigoplus_{I'} M_i$ is a maximal direct summand of $M$, we must have either $M = (\bigoplus_{I'} M_i) \oplus N_1$ just as required or, for some index $j \in I'$ we have $M = (\bigoplus_{I'} M_i) \oplus D_j$. In this latter case we have $D_j \simeq M_{i_1}$ and so, since $M = N_1 \oplus D = M_{i_1} \oplus D$, we also have $D_j \simeq N_1$. Thus, since $N_1$ appears more than once in the decomposition $M = N_1 \oplus (\bigoplus_{I'} D_i)$, it follows from 12.5 (3) that $N_1$ is an LE-module. Consequently, by 12.2, $N_1$ has the exchange property and so, for each $i \in I$, there is a summand $B_i$ of $M_i$ such that $M = N_1 \oplus (\bigoplus_I B_i)$. Since each $M_i$ is indecomposable, either $B_i = 0$ or $B_i = M_i$ and, since $N_1$ is indecomposable, from this it follows that $M = N_1 \oplus \left( \bigoplus_{i \neq k} B_i \right)$ for some $k \in I$, as required.

Now suppose that $t \geq 1$, that $M = N_1 \oplus \cdots \oplus N_t \oplus N_{t+1} \oplus D$ where $N_1, \ldots, N_t, N_{t+1}$ are indecomposable submodules of $M$, and that the Lemma is true for all collections of at most $t$ indecomposables. Then, by this inductive assumption, there are $i_1, \ldots, i_t$ in $I$ for which $N_k \simeq M_{i_k}$ for $1 \leq k \leq t$ and

$$M = N_1 \oplus N_2 \oplus \cdots \oplus N_t \oplus \left( \bigoplus_{i \in I_0} M_i \right) \qquad (\dagger)$$

where $I_0 = I \setminus \{i_1, i_2, \ldots, i_t\}$. Since $(\dagger)$ is another indecomposable decomposition of $M$, by 12.5 (2) it also complements maximal summands and so, applying the case for $t = 1$ proved above to the summand $N_{t+1}$ and (†), we have either

$$M = N_1 \oplus N_2 \oplus \cdots \oplus N_t \oplus N_{t+1} \oplus \left( \bigoplus_{i \in I_0 \setminus \{j\}} M_i \right)$$

for some $i_{t+1} \in I_0$, in which case we're finished since then clearly $N_{t+1} \simeq M_{i_{t+1}}$, or there is a $k \in \mathbb{N}$ with $1 \le k \le t$ for which

$$M = N_1 \oplus \cdots \oplus N_{k-1} \oplus N_{k+1} \oplus \cdots \oplus N_t \oplus N_{t+1} \oplus \left( \bigoplus_{i \in I_0} M_i \right).$$

If the second case holds, clearly $N_{t+1} \simeq N_k$. Now, from the initial decomposition $M = N_1 \oplus N_2 \oplus \cdots \oplus N_{t+1} \oplus D$, it follows from 12.5 (4) that $D$ also has an indecomposable decomposition. Thus the initial decomposition provides one which complements maximal summands in which $N_{t+1}$ occurs at least twice. Hence, again by 12.5 (3) and 12.2, $N_{t+1}$ is an LE-module and so has the exchange property. Then Lemma 11.6 and the indecomposability of the $M_i$ produce a subset $J$ of $I$ for which

$$M = N_1 \oplus N_2 \oplus \cdots \oplus N_t \oplus N_{t+1} \oplus \left( \bigoplus_{i \in J} M_i \right). \qquad (\ddagger)$$

Comparison of (†) with (‡) then shows that $J = I_0 \setminus \{i_{t+1}\}$ for some $i_{t+1} \in I_0$ and $N_{t+1} \simeq M_{i_{t+1}}$ and this completes our induction.                                    $\square$

Next we use Corollary 12.3 and Azumaya's Theorem to establish a technical lemma which will prove useful in handling arbitrary LE-decompositions.

**12.9. Lemma.** *Let $M = \bigoplus_I M_i$ be an LE-decomposition and let $P, Q$ be submodules of $M$ such that $M = P \oplus Q$. If $J$ is a finite subset of $I$ for which $P \cap (\bigoplus_J M_j) \ne 0$, there is an index $j_0 \in J$ and a submodule $P_0$ of $P$ for which*

$$M = M_{j_0} \oplus P_0 \oplus Q.$$

*Proof.* Let $M_J = \bigoplus_J M_j$. Then, by Corollary 12.3, $M_J$ has the exchange property and so we may write $P = P_1 \oplus P_2$ and $Q = Q_1 \oplus Q_2$ with

$$M = M_J \oplus P_1 \oplus Q_1.$$

Then, since this gives $M_J \cap P_1 = 0$, it follows from our hypothesis on $J$ that $M_J \cap P_2 \ne 0$ and so $P_2 \ne 0$. Note also that $M = (P_1 \oplus Q_1) \oplus (P_2 \oplus Q_2)$ and so, if $\pi : M \to P_2 \oplus Q_2$ is the projection map, it follows from Lemma 11.3 that its restriction

$$\pi|_{M_J} : M_J \to P_2 \oplus Q_2$$

is an isomorphism. Then, applying Azumaya's Theorem to the LE-decomposition $\bigoplus_J (M_j) \pi|_{M_J}$ of $P_2 \oplus Q_2$, we may write $P_2 = P_{21} \oplus P_{22}$ where $P_{21}$ is indecomposable and there is an index $j_0 \in J$ such that

$$P_2 \oplus Q_2 = P_{21} \oplus P_{22} \oplus Q_2 = (M_{j_0}) \pi_{j_0} \oplus P_{22} \oplus Q_2$$

where $\pi_{j_0}$ is the restriction of $\pi|_{M_J}$ to $M_{j_0}$. Then using Lemma 11.3 again (with the modules $M_1, M_2$ and $N$ there taken here to be $P_{21}, P_{22} \oplus Q_2$ and $(M_{j_0})\pi_{j_0}$ respectively), it follows that if $\pi_{21} : P_2 \oplus Q_2 \to P_{21}$ is the natural projection, then the composition

$$M_{j_0} \xrightarrow{\pi_{j_0}} P_2 \oplus Q_2 \xrightarrow{\pi_{21}} P_{21}$$

is an isomorphism. However this composition is simply the restriction to $M_{j_0}$ of the natural projection $\pi : M \to P_{21}$. Thus, using Lemma 11.3 once more, we get

$$M = P_{21} \oplus (P_{22} \oplus P_1 \oplus Q) = M_{j_0} \oplus (P_{22} \oplus P_1 \oplus Q)$$

so taking $P_0 = P_{22} \oplus P_1$ we are finished. □

We now introduce a condition on a family of modules which, as we will see, has a strong affinity with the exchange property. Moreover, while appearing somewhat technical at first glance, it often provides tangible evidence of exchange and decomposition.

**12.10. Local semi-T-nilpotency.** A family of modules $\{M_i\}_I$ is called *locally semi-T-nilpotent* if, for any countably infinite set $\{f_n : M_{i_n} \to M_{i_{n+1}}\}_{n \in \mathbb{N}}$ of non-isomorphisms where all the $i_n$ are distinct indices from $I$, for any $x \in M_{i_1}$ there is a $k \in \mathbb{N}$ (depending on $x$) such that $(x)f_1 \cdots f_k = 0$.
In the literature, *locally semi-T-nilpotent* is often abbreviated to *lsTn*.

**12.11. Remarks.** (1) In order to check that the family $\{M_i\}_I$ is locally semi-T-nilpotent, it suffices to assume that either
   (i) all the non-isomorphisms $f_n : M_{i_n} \to M_{i_{n+1}}$ are non-monomorphisms or
   (ii) for each $n \geq 2$, $f_n$ is a monomorphism and so a non-epimorphism.
To see this, suppose that there are an infinite number of the $f_n$'s which are not monic, say $f_{n_1}, f_{n_2}, f_{n_3}, \ldots$ where $n_1 < n_2 < n_3 < \cdots$. For each $t \in \mathbb{N}$, define $g_t = f_{n_t} \diamond \cdots \diamond f_{n_{t+1}-1}$. Then each $g_t$ is a non-monomorphism and if $(x)g_1 \cdots g_k = 0$ for some $k \geq 1$, then $(x)f_1 \cdots f_{m_k} = 0$.

On the other hand, if there are only finitely many of the $f_n$'s which are not monomorphisms, let $t$ be the largest integer for which $f_t$ is not monic. Then here we may take $g_1$ to be $f_1 \diamond \cdots \diamond f_t$ and $g_n = f_{t+n-1}$ for each $n \geq 2$.

(2) Let $\{M_i\}_I$ be an infinite family of modules which are pairwise isomorphic. If the family is locally semi-T-nilpotent, then a straightforward argument shows that each $M_i$ must be indecomposable.

The next two results illustrate the influence of local semi-T-nilpotency on the transfer of the exchange property to direct sums.

**12.12. Local semi-T-nilpotency and the exchange property I.** *Let $\{M_j\}_J$ be a locally semi-T-nilpotent family of R-modules, each with the exchange property, and such that $M_i$ and $M_j$ have no nontrivial isomorphic direct summand if $i \neq j$. Then $M = \bigoplus_J M_j$ has the exchange property.*

*Proof.* By well-ordering the index set $J$ we may assume that, for some fixed ordinal $\rho$, we have $J = \{\alpha : \alpha \text{ is an ordinal}, \alpha \leq \rho\}$. To establish the exchange property for $M = \bigoplus_{\alpha \leq \rho} M_\alpha$, by 11.13 we need only test its compatibility with $I$-clones of $M$ and so it suffices to consider any set $I$ and

$$A = \left(\bigoplus_{\alpha \leq \rho} M_\alpha\right) \oplus N = \bigoplus_{\substack{i \in I \\ \alpha \leq \rho}} A_{\alpha i}$$

where $A_{\alpha i} \simeq M_\alpha$ for each $i \in I$. The essence of the proof is showing that each $M_\alpha$ in turn, in the first direct sum, can be slotted into the second by replacing some of the $A_{\alpha i}$'s. We formalise this process as follows.

For each $\beta \leq \rho$, we construct by transfinite induction two families of submodules of $A$, $\{F_\alpha : \alpha \leq \beta\}$ and $\{G_\alpha : \alpha \leq \beta\}$, where $F_\alpha = \bigoplus_I B_{\alpha i}$ and $G_\alpha = \bigoplus_I C_{\alpha i}$ for each $\alpha$, such that

(1) $A_{\alpha i} = B_{\alpha i} \oplus C_{\alpha i}$ for each $\alpha \leq \beta$ and each $i \in I$, and

(2) $A = \left(\bigoplus_{\alpha \leq \beta} M_\alpha\right) \oplus \left(\bigoplus_{\alpha \leq \beta} F_\alpha\right) \oplus \left(\bigoplus_{i \in I, \alpha > \beta} A_{\alpha i}\right)$

$\quad = \left(\bigoplus_{\alpha < \beta} M_\alpha\right) \oplus \left(\bigoplus_{\alpha < \beta} F_\alpha\right) \oplus \left(\bigoplus_{i \in I, \alpha \geq \beta} A_{\alpha i}\right).$

The final stage in the construction, at $\beta = \rho$, then finishes the proof.

We use transfinite induction for the construction. Thus assume that for some ordinal $\gamma < \rho$ we have families $\{F_\alpha : \alpha < \gamma\}$ and $\{G_\alpha : \alpha < \gamma\}$ satisfying the above conditions.

We first deal with the case where $\gamma$ is a successor ordinal, say $\gamma = \beta + 1$. Then, using the exchange property of $M_\gamma = M_{\beta+1}$ and Corollary 11.7 (taking $M, N, L$ and $\bigoplus_I A_i$ there respectively as $M_{\beta+1}, \left(\bigoplus_{\alpha > \beta+1} M_\alpha\right) \oplus N, \bigoplus_{\alpha \leq \beta} M_\alpha$, and $(\bigoplus_{\alpha \leq \beta, i \in I} B_{\alpha i}) \oplus (\bigoplus_{\alpha > \beta, i \in I} A_{\alpha i}))$, the decompositions

$$\begin{aligned} A &= \left(\bigoplus_{\alpha \leq \beta} M_\alpha\right) \oplus M_{\beta+1} \oplus \left(\bigoplus_{\alpha > \beta+1} M_\alpha\right) \oplus N \\ &= \left(\bigoplus_{\alpha \leq \beta} M_\alpha\right) \oplus \left(\bigoplus_{\alpha \leq \beta, i \in I} B_{\alpha i}\right) \oplus \left(\bigoplus_{i \in I, \alpha > \beta} A_{\alpha i}\right) \end{aligned}$$

induce decompositions $B_{\alpha i} = K_{\alpha i} \oplus L_{\alpha i}$ for $\alpha \leq \beta$ and $A_{\alpha i} = K_{\alpha i} \oplus L_{\alpha i}$ for $\alpha > \beta$ such that

$$A = \left(\bigoplus_{\alpha \leq \beta} M_\alpha\right) \oplus M_{\beta+1} \oplus \left(\bigoplus_{i \in I, \alpha \leq \rho} K_{\alpha i}\right).$$

From this it follows that

$$M_{\beta+1} \simeq \bigoplus_{i \in I, \alpha \leq \rho} L_{\alpha i}$$

and so, by the direct summand hypothesis on the $M_j$'s, we must have $L_{\alpha i} = 0$ for all $\alpha \neq \beta + 1$ and all $i$. Now set $B_{\beta+1,i} = K_{\beta+1,i}$ and $C_{\beta+1,i} = L_{\beta+1,i}$ for each $i$

and then define $F_{\beta+1} = \bigoplus_I B_{\beta+1,i}$ and $G_{\beta+1} = \bigoplus_I C_{\beta+1,i}$. This gives

$$
\begin{aligned}
A &= \left(\bigoplus_{\alpha\leq\beta} M_\alpha\right) \oplus \left(\bigoplus_{\alpha\leq\beta} F_\alpha\right) \oplus \left(\bigoplus_{i\in I, \alpha>\beta} A_{\alpha i}\right) \\
&= \left(\bigoplus_{\alpha\leq\beta} M_\alpha\right) \oplus \left(\bigoplus_{\alpha\leq\beta} F_\alpha\right) \oplus \left(\bigoplus_I A_{\beta+1,i}\right) \oplus \left(\bigoplus_{i\in I, \alpha>\beta+1} A_{\alpha i}\right) \\
&= \left(\bigoplus_{\alpha\leq\beta} M_\alpha\right) \oplus \left(\bigoplus_{\alpha\leq\beta} F_\alpha\right) \oplus \left(\bigoplus_I (K_{\beta+1,i} \oplus L_{\beta+1,i})\right) \oplus \left(\bigoplus_{i\in I, \alpha>\beta+1} A_{\alpha i}\right) \\
&= \left(\bigoplus_{\alpha\leq\beta} M_\alpha\right) \oplus \left(\bigoplus_I L_{\beta+1,i} t\right) \oplus \left(\bigoplus_{\alpha\leq\beta} F_\alpha\right) \oplus \left(\bigoplus_I K_{\beta+1,i}\right) \oplus \left(\bigoplus_{i\in I, \alpha>\beta+1} A_{\alpha i}\right) \\
&= \left(\bigoplus_{\alpha\leq\beta} M_\alpha\right) \oplus M_{\beta+1} \oplus \left(\bigoplus_{\alpha\leq\beta} F_\alpha\right) \oplus F_{\beta+1} \oplus \left(\bigoplus_{i\in I, \alpha>\beta+1} A_{\alpha i}\right) \\
&= \left(\bigoplus_{\alpha\leq\beta+1} M_\alpha\right) \oplus \left(\bigoplus_{\alpha\leq\beta+1} F_\alpha\right) \oplus \left(\bigoplus_{i\in I, \alpha>\beta+1} A_{\alpha i}\right)
\end{aligned}
$$

and this last decomposition is just what we want.

Now assume that $\gamma$ is a limit ordinal. Note first that

$$
\begin{aligned}
A &= \bigoplus_{i\in I, \alpha\leq\rho} A_{\alpha i} = \left(\bigoplus_{i\in I, \alpha<\gamma} A_{\alpha i}\right) \oplus \left(\bigoplus_{i\in I, \alpha\geq\gamma} A_{\alpha i}\right) \\
&= \left(\bigoplus_{i\in I, \alpha<\gamma} (B_{\alpha i} \oplus C_{\alpha i})\right) \oplus \left(\bigoplus_{i\in I, \alpha\geq\gamma} A_{\alpha i}\right) \\
&= \left(\bigoplus_{\alpha<\gamma} F_\alpha\right) \oplus \left(\bigoplus_{\alpha<\gamma} G_\alpha\right) \oplus \left(\bigoplus_{i\in I, \alpha\geq\gamma} A_{\alpha i}\right).
\end{aligned}
$$

Let $\pi_\gamma : A \to \bigoplus_{\alpha<\gamma} G_\alpha$ be the projection associated with this last decomposition and let $p_\gamma$ denote its restriction to $\bigoplus_{\alpha<\gamma} M_\alpha$. For each $\beta < \gamma$ there is a corresponding decomposition $A = \left(\bigoplus_{\alpha<\beta} F_\alpha\right) \oplus \left(\bigoplus_{\alpha<\beta} G_\alpha\right) \oplus \left(\bigoplus_{i\in I, \alpha\geq\beta} A_{\alpha i}\right)$ and so we can define the analogous maps $\pi_\beta$ and $p_\beta$. We shall show that

$$
A = \left(\bigoplus_{\alpha<\gamma} M_\alpha\right) \oplus \left(\bigoplus_{\alpha<\gamma} F_\alpha\right) \oplus \left(\bigoplus_{i\in I, \alpha\geq\gamma} A_{\alpha i}\right)
$$

and to do so it suffices, by Lemma 11.3, to show that $p_\gamma$ is an isomorphism.

Our induction hypothesis shows that $p_\beta$ is an isomorphism for all $\beta < \gamma$ and from this it follows that $p_\gamma$ is a monomorphism. We use the local semi-T-nilpotency of our family to show that $p_\gamma$ is an epimorphism.

Assume to the contrary that $p_\gamma$ is not onto. Then, for some $\beta_1 < \gamma$, there is an $x_1 \in G_{\beta_1}$ which is not in $\operatorname{Im}(p_\gamma)$. Since $p_{\beta_1+1} : \bigoplus_{\alpha<\beta_1+1} M_\alpha \to \bigoplus_{\alpha<\beta_1+1} G_\alpha$ is an isomorphism, we may define

$$
z_1 = (x_1) p_{\beta_1+1}^{-1} p_\gamma \in \bigoplus_{\alpha<\gamma} G_\alpha.
$$

We now show that there is an ordinal $\beta_2$ with $\beta_1 < \beta_2 < \gamma$ for which the $\beta_2$-component of $z_1$ in $\bigoplus_{\alpha<\gamma} G_\alpha$ is not in $\operatorname{Im}(p_\gamma)$. To see this, let $w = z_1 - (z_1)\pi_{\beta_1+1}$. Then $w \in \bigoplus_{\beta_1<\alpha<\gamma} G_\alpha$ and since $p_{\beta_1+1} = p_\gamma \circ \pi_{\beta_1+1}$ we get

$$
w = z_1 - (x_1) p_{\beta_1+1}^{-1} p_{\beta_1+1} = (x_1) p_{\beta_1+1}^{-1} p_\gamma - x_1.
$$

Thus since $x_1$ is not in $\operatorname{Im}(p_\gamma)$, neither is $w$. This produces the desired $\beta_2$.

We now take $x_2$ to be the $\beta_2$-component of $x_1$ and, since $x_2 \notin \mathrm{Im}\,(p_\gamma)$, we may repeat the foregoing process with $x_2$ replacing $x_1$. We proceed inductively as in the following diagram where $y_n = (x_n)\,p_{\beta_n+1}^{-1}$, $\pi_n$ is the obvious projection, $\beta_1 < \beta_2 < \beta_3 < \cdots$ is an increasing sequence of ordinals less than $\gamma$, and the sequence of elements $x_1, x_2, x_3, \ldots$ is constructed so that $x_n \in G_{\beta_n} \setminus \mathrm{Im}\,(p_\gamma)$ for each $n \in \mathbb{N}$. Moreover, if we set $f_n = p_{\beta_n+1}^{-1} \diamond p_\gamma \diamond \pi_{n+1}$, then $x_{n+1} = (x_n)f_n$ for each $n$.

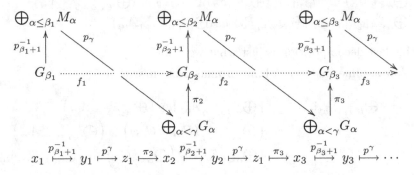

$$x_1 \overset{p_{\beta_1+1}^{-1}}{\longmapsto} y_1 \overset{p^\gamma}{\longmapsto} z_1 \overset{\pi_2}{\longmapsto} x_2 \overset{p_{\beta_2+1}^{-1}}{\longmapsto} y_2 \overset{p^\gamma}{\longmapsto} z_1 \overset{\pi_3}{\longmapsto} x_3 \overset{p_{\beta_3+1}^{-1}}{\longmapsto} y_3 \overset{p^\gamma}{\longmapsto} \cdots$$

Now, by our induction hypothesis, for each $n$ we have

$$
\begin{aligned}
A &= \left(\bigoplus_{\alpha \le \beta_n} M_\alpha\right) \oplus \left(\bigoplus_{\alpha \le \beta_n} F_\alpha\right) \oplus \left(\bigoplus_{i \in I, \alpha > \beta_n} A_{\alpha i}\right)\\
&= \left(\bigoplus_{\alpha < \beta_n} M_\alpha\right) \oplus \left(\bigoplus_{\alpha < \beta_n} F_\alpha\right) \oplus \left(\bigoplus_{i \in I, \alpha \ge \beta_n} A_{\alpha i}\right).
\end{aligned}
$$

Factoring out the common summand

$$\left(\bigoplus_{\alpha < \beta_n} M_\alpha\right) \oplus \left(\bigoplus_{\alpha < \beta_n} F_\alpha\right) \oplus \left(\bigoplus_{i \in I, \alpha < \beta_n} A_{\alpha i}\right)$$

gives $M_{\beta_n} \oplus F_{\beta_n} \simeq \bigoplus_I A_{\beta_n i}$. Then, since $\bigoplus_I A_{\beta_n i} = \bigoplus_I (B_{\beta_n i} \oplus C_{\beta_n i})$, $F_{\beta_n} = \bigoplus_I B_{\beta_n i}$ and $G_{\beta_n} = \bigoplus_I C_{\beta_n i}$, it follows that $M_{\beta_n} \simeq G_{\beta_n}$. Consequently, by our initial hypothesis on the family $\{M_j\}_J$, each $f_n : G_{\beta_n} \to G_{\beta_{n+1}}$ is a non-isomorphism.

Of course, since each $x_n$ is nonzero, this now provides a contradiction to the local semi-T-nilpotency hypothesis and so $p_\gamma : \bigoplus_{\alpha < \gamma} M_\alpha \to \bigoplus_{\alpha < \gamma} G_\alpha$ is indeed an isomorphism. This then allows us to proceed as in the successor ordinal case to produce modules $F_\gamma$ and $G_\gamma$ satisfying conditions (1) and (2), thereby completing the induction argument. $\qquad\square$

**12.13. Local semi-T-nilpotency and the exchange property II.** *Let $\{M_j\}_J$ be a locally semi-T-nilpotent family of R-modules, each with the exchange property. Then $M = \bigoplus_J M_j$ has the exchange property if $M_i \simeq M_j$ for all $i, j \in J$.*

*Proof.* As in the proof of 12.12, we need only consider a decomposition

$$A = M \oplus N = \bigoplus_I A_i$$

where $A_i \simeq M_j$ for each $i \in I$ and, by hypothesis, for each $j \in J$. Moreover, by 11.9, we may suppose that $J$ is infinite.

A simple Zorn's Lemma argument shows that there is a proper subset $L$ of $I$ maximal with respect to the two properties

(1) $M \cap (\bigoplus_L A_l) = 0$   and

(2) each finite subsum of $(\bigoplus_J M_j) \oplus (\bigoplus_L A_l)$ is a direct summand of $A$.

Then our goal is reached if we can show that $A = M \oplus (\bigoplus_L A_l)$. We suppose to the contrary and use this below to construct a sequence $(i_n)_{n \in \mathbb{N}}$ of indices from $I$, a corresponding sequence $(x_n)_{n \in \mathbb{N}}$ of nonzero elements $x_n \in A_{i_n}$, and a sequence of non-isomorphisms $f_n : A_{i_n} \to A_{i_{n+1}}$ such that $x_{n+1} = (x_1)f_1 \cdots f_n$. However, since $A_i \simeq M_j$ for each $i, j$, these sequences contradict the local semi-T-nilpotency hypothesis, noting that, since the $M_j$'s and $A_i$'s are all isomorphic to each other, the distinctness of the indices $i_n$ may be ignored.

If $A \neq M \oplus (\bigoplus_L A_l)$ there is an index $i_1 \in I$ and an element $x_1 \in A_{i_1}$ such that $x_1 \notin M \oplus (\bigoplus_L A_l)$. Then, by the maximality of $L$, there is a finite subsum

$$K = (\bigoplus_{j \in F} M_j) \oplus (\bigoplus_{i \in G} A_i) \quad \text{of} \quad (\bigoplus_J M_j) \oplus (\bigoplus_L A_l)$$

(where $F, G$ are finite subsets of $I, L$ respectively), such that

either $F \cap A_{i_1} \neq 0$ or $F + A_{i_1}$ is not a direct summand of $A$. $\quad (\star)$

Now, by 12.11 (2), each $M_j$ is indecomposable and so each $A_i$ is also. Moreover, by our hypothesis and 11.9, $F$ has the exchange property. Consequently, we have

$$A = F \oplus \left(\bigoplus_K A_k\right)$$

for some $K \subset I$. For each $k \in K$, let $\pi_k : A \to A_k$ be supplied by this decomposition. Then, since $x_1 \notin M \oplus (\bigoplus_L A_l)$, there is an $i_2 \in K$ for which $(x_1)\pi_{i_2} \neq 0$. Now let $f_1$ denote the restriction of $\pi_{i_2}$ to $A_{i_1}$. Then $f_1$ is a non-isomorphism since otherwise, by Lemma 11.3, we would have $A = F \oplus A_{i_1} \oplus \left(\bigoplus_{k \in K, k \neq i_2} A_k\right)$ contradicting the constraints on $F$ given by $(\star)$ above.

We now repeat the process with $x_2 = (x_1)f_1$ and then proceed inductively to complete the proof. $\qquad \square$

We now come to the main result of this section. It gives a striking characterisation of the exchange property for a module with an indecomposable decomposition.

**12.14. LE-decompositions, local semi-T-nilpotency and the exchange property.** *Let $\{M_i\}_I$ be a family of indecomposable R-modules and let $M = \bigoplus_I M_i$. Then the following statements are equivalent.*

(a) *$M = \bigoplus_I M_i$ is an LE-decomposition and the family $\{M_i\}_I$ is locally semi-T-nilpotent;*

(b) *M has the exchange property;*

(c) *M has the finite exchange property.*

*Proof.* (a)$\Rightarrow$(b). Let $\{M_k\}_K$ be a family of representatives of the isomorphism classes of the modules in $\{M_i\}_I$. For each such representative $M_k$, let $N_k$ be the direct sum of all those $M_i$'s which are isomorphic to $M_k$. Then, by 12.13, each $N_k$ has the exchange property. Moreover, using Azumaya's Theorem, one sees that $N_{k_1}$ and $N_{k_2}$ have no nontrivial isomorphic direct summand if $k_1, k_2$ are distinct indices in $K$.

Now let $k_1, k_2, \ldots$ be distinct indices in $K$ and $N_{k_1} \xrightarrow{f_1} N_{k_2} \xrightarrow{f_2} N_{k_3} \xrightarrow{f_3} \cdots$ be a sequence of non-isomorphisms. Suppose that for all $n \in \mathbb{N}$, $(x)f_1 f_2 \cdots f_n \neq 0$ for some $x \in N_{k_1}$. Then, for each $t \geq 2$, there is a module $M_{k_t}$ in our family isomorphic to a summand of $N_{k_t}$ such that, letting $\pi_t : N_{k_t} \to M_{k_t}$ and $\varepsilon_t : M_{k_t} \to N_{k_t}$ be the associated direct sum projection and injection, we have

$$(x)f_1 \pi_2 \varepsilon_2 f_2 \pi_3 \varepsilon_3 \cdots \varepsilon_n f_n \pi_{n+1} \neq 0$$

for all $n \in \mathbb{N}$. Thus, taking $g_t : M_{k_t} \to M_{k_{t+1}}$ to be the composition $\varepsilon_t f_t \pi_{t+1}$ for each $t \geq 2$, we get $((x)f_1 \pi_2)g_2 \cdots g_n \neq 0$ for each $n \in \mathbb{N}$. However this contradicts the local semi-T-nilpotency of $\{M_i\}_I$ and so it follows that the family $\{N_k\}_K$ is also locally semi-T-nilpotent. We may now apply 12.13 to conclude that $M = \bigoplus_K N_k$ has the exchange property.

(b)$\Rightarrow$(c) is automatic.

(c)$\Rightarrow$(a). Assume that $M$ has the finite exchange property. Then firstly, by 11.9, each $M_i$ inherits the property and so must be an LE-module by 12.2. This leaves us to establish the local semi-T-nilpotency of $\{M_i\}_I$.

To this end, let $\{f_n : M_{i_n} \to M_{i_{n+1}}\}_{n \in \mathbb{N}}$ be a countably infinite family of non-isomorphisms where all the $i_n$ are distinct indices from $I$. Clearly it suffices to take $I$ as $\mathbb{N}$ and so each $i_n$ as simply $n$. For each $n \in \mathbb{N}$, define

$$L_n = \{x + (x)f_n : x \in M_n\} \subseteq M_n + M_{n+1}.$$

Then it is easy to see that $L_n \simeq M_n$ and

$$M = L_1 \oplus M_2 \oplus L_3 \oplus M_4 \oplus \cdots = M_1 \oplus L_2 \oplus M_3 \oplus L_4 \oplus \cdots .$$

Moreover, by (c) and 11.8, $\bigoplus_{n \in \mathbb{N}} L_{2n}$ has the finite exchange property and this provides

$$M = \left(\bigoplus_{n \in \mathbb{N}} L_{2n}\right) \oplus A \oplus B \quad \text{where} \quad A \subseteq \bigoplus_{n \in \mathbb{N}} M_{2n} \quad \text{and} \quad B \subseteq \bigoplus_{n \in \mathbb{N}} L_{2n-1}.$$

Now we call on Remark 12.11 (1) to break the rest of the proof into two cases:

(i) all $f_n$ are non-monomorphisms and

(ii) for each $n \geq 2$, $f_n$ is a monomorphism and so a non-epimorphism.

Case (i). If the module $A$ above is nonzero, then there is a finite subset $J$ of $I$ for which $A \cap (\bigoplus_J M_j) \neq 0$. Then, taking $P = A$ and $Q = (\bigoplus_{n \in \mathbb{N}} L_{2n}) \oplus B$ in Lemma 12.9, there is a $k \in \mathbb{N}$ and a submodule $P_0$ of $P$ for which

$$M = M_{2k} \oplus P_0 \oplus \left(\bigoplus_{n \in \mathbb{N}} L_{2n}\right) \oplus B.$$

However this is impossible since $0 \neq \mathrm{Ke}\,(f_{2k}) \subseteq L_{2k} \cap M_{2k}$. Hence $A = 0$ and so $M = (\bigoplus_{n \in \mathbb{N}} L_{2n}) \oplus B$ where $B \subseteq \bigoplus_{n \in \mathbb{N}} L_{2n-1}$. Then, if $x \in M_1$, we have

$$x \in \left(\bigoplus_{n \in \mathbb{N}} L_{2n}\right) \oplus B \subseteq \bigoplus_{n \in \mathbb{N}} L_n,$$

say $x = \sum_{t=1}^p (x_t + (x_t) f_t)$ where $x_t \in M_t$ for $1 \leq t \leq p$. A simple argument now shows that $(x) f_1 f_2 \cdots f_p = (x_1) f_1 f_2 \cdots f_p = (x_2) f_2 \cdots f_p = \cdots = (x_p) f_p = 0$, and we are finished.

Case (ii). Let $X = (\bigoplus_{n \in \mathbb{N}} L_{2n}) \oplus A$. Our interim goal is to show that

$$M = X \oplus L_3 \oplus B_3$$

for some $B_3 \subseteq B$. From above,

$$M = X \oplus B = L_1 \oplus L_3 \oplus \left(\bigoplus_{n > 2} L_{2n-1}\right) \oplus \left(\bigoplus_{n \in \mathbb{N}} M_{2n}\right).$$

Let $\pi_X, \pi_B, \pi_3$ be the projections from these direct sums to $X, B, L_3$ respectively and let $\varepsilon_3 : L_3 \to M$ be the summand injection. Given $m \in M$ we may write (uniquely) $m = x + b$ where $x \in X, b \in B$ and hence $x = \sum_{n \in \mathbb{N}} l_{2n} + a$ uniquely where $l_{2n} \in L_{2n}, a \in A$. Then $l_{2n} = t_{2n} + (t_{2n}) f_{2n}$ for some $t_{2n} \in M_{2n}$, for each $n \in \mathbb{N}$. From this it follows that $(m) \pi_X \pi_3 = (x) \pi_3 = (t_2) f_2$. This shows that $\mathrm{Im}\,(\pi_X \pi_3) \subseteq \mathrm{Im}\,(f_2)$ and so, since $f_2$ is a non-epimorphism, so too is $\pi_X \pi_3$.

Consequently $\varepsilon_3 \pi_X \pi_3$ is not an automorphism of $L_3$. Since $\varepsilon_3 (\pi_X + \pi_B) \pi_3$ is the identity on $L_3$ and by assumption $L_3$ is an LE-module, it follows that $\varepsilon_3 \pi_B \pi_3$ is an automorphism on $L_3$. From this we see that the projection $\pi_B$ induces an isomorphism from $L_3$ onto a direct summand $B_3'$ of $B$, say $B = B_3 \oplus B_3'$. Then, taking $M_1 = X \oplus B_3$, $M_2 = B_3'$, and $N = L_3$ in Lemma 11.3, we reach our stated goal

$$M = X \oplus L_3 \oplus B_3 = \left(\bigoplus_{n \in \mathbb{N}} L_{2n}\right) \oplus A \oplus L_3 \oplus B_3.$$

We now repeat the above process on $L_5$ instead of $L_3$, with $B_3$ and $X \oplus L_3$ replacing $B$ and $X$ respectively, to give

$$M = X \oplus L_3 \oplus L_5 \oplus B_5 = \left(\bigoplus_{n \in \mathbb{N}} L_{2n}\right) \oplus A \oplus L_3 \oplus L_5 \oplus B_5$$

where $B_5 \subseteq B_3$. Continuing inductively, for each $t \in \mathbb{N}$ we produce a decomposition

$$M = \left(\bigoplus_{n \in \mathbb{N}} L_{2n}\right) \oplus A \oplus \left(\bigoplus_{n=1}^t L_{2n+1}\right) \oplus B_{2t+1}$$

where $B_{2t+1} \subseteq \bigoplus_{n \in \mathbb{N}} L_{2n-1}$.

If $A = 0$, then, as in Case (i), we can show that, for any $x \in M_1$, we have $(x)f_1 f_2 \cdots f_p = 0$ for some $p \in \mathbb{N}$. On the other hand, if $A \neq 0$, then there is a finite subset $J$ of $I = \mathbb{N}$ for which $A \cap \left( \bigoplus_J M_j \right) \neq 0$. Then, with $t \geq \max(J)$, taking $P = A$ and $Q = \left( \bigoplus_{n \in \mathbb{N}} L_{2n} \right) \oplus \left( \bigoplus_{n=1}^t L_{2n+1} \right) \oplus B_{2t+1}$ in Lemma 12.9, we find a $k_t \leq t$ and a submodule $A_t'$ of $A$ such that

$$M = M_{2k_t} \oplus A_t' \oplus \left( \bigoplus_{n \in \mathbb{N}} L_{2n} \right) \oplus \left( \bigoplus_{n=1}^t L_{2n+1} \right) \oplus B_{2t+1}.$$

Now if $A_t' \neq 0$ we can repeat the above argument to find an index $l_t$ different from $k_t$, both at most $t$, and a submodule $A_t''$ such that

$$M = M_{2k_t} \oplus M_{2l_t} \oplus A_t'' \oplus \left( \bigoplus_{n \in \mathbb{N}} L_{2n} \right) \oplus \left( \bigoplus_{n=1}^t L_{2n+1} \right) \oplus B_{2t+1}.$$

However, if $l_t < k_t$ we have the inclusion $M_{2l_t} \subseteq L_{2l_t} \oplus L_{2l_t+1} \oplus \cdots \oplus L_{2k_t-1} \oplus M_{2k_t}$, in contradiction to the preceding decomposition of $M$. Symmetry also gives a contradiction if $k_t < l_t$. Consequently $A_t' = 0$ and so we have

$$M = M_{2k_t} \oplus \left( \bigoplus_{n \in \mathbb{N}} L_{2n} \right) \oplus \left( \bigoplus_{n=1}^t L_{2n+1} \right) \oplus B_{2t+1}.$$

Now choose $u > k_t$. Since $f_{2u-1} : M_{2u-1} \to M_{2u}$ is a non-epimorphism, we may choose $m_{2u} \in M_{2u}$ with $m_{2u} \notin \operatorname{Im}(f_{2u-1})$. Moreover, since

$$M_{2u} \subseteq M_{2k_t} \oplus \left( \bigoplus_{n \in \mathbb{N}} L_{2n} \right) \oplus \left( \bigoplus_{n=1}^t L_{2n+1} \right) \oplus B_{2t+1},$$

we may write $m_{2u} = \sum_{s=1}^p (x_s + (x_s)f_s)$ for some $p \geq 2u - 1$ where $x_s \in M_s$ for each $s$. However, since $m_{2u} \notin \operatorname{Im}(f_{2u-1})$ and $f_n$ is a monomorphism for all $n \geq 2$, a straightforward argument shows that this is impossible. □

We can now readily give an example to show that the sufficiency part of Corollary 11.9 does not extend in general to infinite direct sums.

**12.15. Example.** Let $p$ be a fixed prime and, for each $k \in \mathbb{N}$, let $M_k$ be the cyclic $\mathbb{Z}$-module $\mathbb{Z}/\mathbb{Z}p^k$. Then, since $M_k$ is indecomposable and $\operatorname{End}(M_k)$ is local, $M_k$ has the exchange property by 12.2.

For each $k \in \mathbb{N}$, let $f_k : M_k \to M_{k+1}$ be the homomorphism defined by $(t + \mathbb{Z}p^k)f_k = pt + \mathbb{Z}p^{k+1}$ for all $t \in \mathbb{Z}$. Then each $f_k$ is a non-isomorphism and $(1 + \mathbb{Z}p)f_1 f_2 \cdots f_n \neq 0$ for all $n \in \mathbb{N}$. Thus the family $\{M_k\}_{k \in \mathbb{N}}$ is not locally semi-T-nilpotent. Consequently, by 12.14, $M = \bigoplus_{k \in \mathbb{N}} M_k$ does not have the exchange property.

We will make use of the following important result in the next section. It is due to F. Kasch and A. Zöllner (see [241, Theorem 2.26]).

**12.16.  Complementing summands implies local semi-T-nilpotency.** *Let $M = \bigoplus_I M_i$ be an indecomposable decomposition of the module $M$ which complements direct summands. Then $\{M_i\}_I$ is locally semi-T-nilpotent.*

*Proof.* The proof closely follows that of 12.14 (c)$\Rightarrow$(a). As there, it suffices to assume that $I = \mathbb{N}$ and so consider a family $\{f_n : M_n \to M_{n+1}\}_{n\in\mathbb{N}}$ of non-isomorphisms. Also as before, for each $n \in \mathbb{N}$, we define

$$L_n = \{x + (x)f_n \mid x \in M_n\} \subseteq M_n + M_{n+1}$$

and then we have $L_n \simeq M_n$. We will make use of the following extra properties of these $L_n$'s, the proofs of which are straightforward and so omitted.

(i)  $L_m \cap M_n = 0$ if $m \neq n$ and $L_n \cap M_n = 0$ if and only if $f_n$ is a monomorphism;

(ii)  the sum $\sum_{n\in\mathbb{N}} L_n$ is direct;

(iii)  for any $m \leq n$, $(\bigoplus_{i=m}^{n} L_i) \oplus M_{n+1} = \bigoplus_{i=m}^{n+1} M_i$.

In the first instance, we establish local semi-T-nilpotency if $M = \bigoplus_{n\in\mathbb{N}} L_n$. Here, given $x \in M_1$ we have, for some $p \in \mathbb{N}$ and $m_i \in M_i$ for each $i = 1, \ldots, p$,

$$x = m_1 + ((m_1)f_1 + m_2) + ((m_2)f_2 + m_3) + \cdots + ((m_{p-1})f_{p-1} + m_p) + (m_p)f_p$$

and from this we get $(x)f_1 f_2 \cdots f_p = 0$, as required.

Now suppose that $\bigoplus_{n\in\mathbb{N}} L_n \neq M$. Recall that for any subset $K \subset \mathbb{N}$, we set $M_K = \bigoplus_{k\in K} M_k$ and $L_K = \bigoplus_{k\in K} L_k$. Next let $D$ and $E$ denote the sets of odd and even natural numbers respectively. Then, as noted in 12.14, we have

$$M = M_D \oplus L_E = L_D \oplus M_E,$$

both indecomposable decompositions. Since our given decomposition complements direct summands, by 12.5 (2) so too does the second of these decompositions and so applying this to the first gives

$$M = L_E \oplus L_{D_0} \oplus M_{E_0} \qquad (*)$$

where $D_0 \subseteq D$ and $E_0 \subseteq E$. We now employ Remark 12.11 (1) to break the rest of the proof into two cases:

(a)  all $f_n$ are non-monomorphisms and

(b)  for each $n \geq 2$, $f_n$ is a monomorphism and so a non-epimorphism.

Case (a). By (i) above, for each $n \in \mathbb{N}$ we have $L_n \cap M_n \neq 0$ here, and so $E_0 = \emptyset$. Thus $M = L_E \oplus L_{D_0}$ and so $M = \bigoplus_{n\in\mathbb{N}} L_n$, giving local semi-T-nilpotency as shown above.

Case (b). To begin with, we show that $D_0 = D$. First note that, given a nonzero $m_1 \in M_1$, it is impossible to write $m_1$ as a sum of components from the summands in $(*)$ unless $1 \in D_0$. Next note that for each $k \in \mathbb{N}$, since $f_{2k}$ is a monomorphism, by hypothesis it is not an epimorphism and so there exists an $m_{2k+1} \in M_{2k+1} \setminus (M_{2k})f_{2k}$. Again, a careful analysis shows that we can not express $m_{2k+1}$ as a sum of components from the summands in $(*)$ unless $2k + 1 \in D_0$. Hence $D_0 = D$ and so $(*)$ gives

$$M = L_{\mathbb{N}} \oplus M_{E_0}.$$

Next we show that $E_0$ has at most one element. Suppose to the contrary that $2k, 2l \in E_0$ with $k < l$. Then, using (iii) above twice, we get

$$
\begin{aligned}
M &= L_{\mathbb{N}} \oplus M_{E_0} \\
&= \left( \bigoplus_{t=1}^{2k-1} L_t \right) \oplus M_{2k} \oplus \left( \bigoplus_{t \geq 2k} L_t \right) \oplus \left( \bigoplus_{\substack{s \in E_0 \\ s \neq 2k}} M_s \right) \\
&= \left( \bigoplus_{t=1}^{2k} M_t \right) \oplus \left( \bigoplus_{t \geq 2k} L_t \right) \oplus \left( \bigoplus_{\substack{s \in E_0 \\ s \neq 2k}} M_s \right) \\
&= \left( \bigoplus_{t=1}^{2k} M_t \right) \oplus \left( \bigoplus_{t=2k}^{2l-1} L_t \right) \oplus M_{2l} \oplus \left( \bigoplus_{t \geq 2l} L_t \right) \oplus \left( \bigoplus_{\substack{s \in E_0 \\ s \neq 2k, 2l}} M_s \right) \\
&= \left( \bigoplus_{t=1}^{2k} M_t \right) \oplus \left( \bigoplus_{t=2k}^{2l} M_t \right) \oplus \left( \bigoplus_{t \geq 2l} L_t \right) \oplus \left( \bigoplus_{\substack{s \in E_0 \\ s \neq 2k, 2l}} M_s \right)
\end{aligned}
$$

but the double appearance of $M_{2k}$ contradicts the sum's directness. Hence $E_0$ has at most one element and so, assuming $E_0 \neq \emptyset$, we have $M = L_{\mathbb{N}} \oplus M_{2k}$ for some $k \in \mathbb{N}$. Using (iii) again, we get

$$M = M_1 \oplus \cdots \oplus M_{2k} \oplus \left( \bigoplus_{i \geq k} L_i \right).$$

Thus, if $x \in M_{2k+1}$, there are elements $m_j \in M_j$ for $j = 1, \ldots, 2k$ and, for some $p \geq 2k$, elements $n_i \in M_i$ for $i = 2k, \ldots, p$ such that

$$x = m_1 + \cdots + m_{2k} + \sum_{i=2k}^{p} (n_i)(1 + f_i).$$

Then, rearranging this expression with respect to the original direct sum $M = \bigoplus_{\mathbb{N}} M_n$ and using the monomorphism property of the $f_i$'s, we deduce that $x = (n_{2k})f_{2k}$. However, this shows that $f_{2k}$ is onto and so an isomorphism. This long-awaited contradiction finishes the proof. $\qquad \square$

We finish this section with a quick application of 12.14 to direct sums of modules of finite length. First we record a well-known result due to Harada and Sai [162]. This was proved by them in the general setting of an abelian category, module-theoretic versions being available as [363, 54.1] and [22, Lemma VI.1.2, Corollary VI.1.3].

**12.17. The Harada–Sai Lemma.** *Let $\{M_i\}_{i \in \mathbb{N}}$ be a countably infinite family of indecomposable modules of finite length and assume there exists some $n \in \mathbb{N}$ such that* length $(M_i) \leq n$ *for each $i$. Then for any family $\{f_i : M_i \to M_{i+1}\}_{i \in \mathbb{N}}$ of non-isomorphisms we have $f_1 \cdots f_{2^n - 1} = 0$.*

**12.18. Corollary.** *Let $\{M_i\}_I$ be a family of modules of finite length for which there is an $n \in \mathbb{N}$ with* length $(M_i) \leq n$ *for each $i$. Then $\bigoplus_I M_i$ has the exchange property.*

*Proof.* Since each $M_i$ is a finite direct sum of LE-modules, we may assume that each $M_i$ is indecomposable. It then follows, by the Harada–Sai Lemma, that the family $\{M_i\}_I$ is locally semi-T-nilpotent and so $\bigoplus_I M_i$ has the exchange property by 12.14. $\qquad\square$

**12.19. Exercises.**

(1) Let $p, q$ be distinct primes, let $\mathbb{Z}_{(p)}, \mathbb{Z}_{(q)}$ be the localisations of $\mathbb{Z}$ at the prime ideals $(p), (q)$, and let $M$ denote the underlying abelian group of the ring $R = \mathbb{Z}_{(p)} \cap \mathbb{Z}_{(q)}$. Prove:

  (i) End$(M_{\mathbb{Z}}) = R$,

  (ii) $M_{\mathbb{Z}}$ is indecomposable, but

  (iii) $M_{\mathbb{Z}}$ does not have the 2-exchange property.

(2) Let $K$ be any field and let $R$ be the ring of sequences of elements from $K$, that is, $R = K^{\mathbb{N}}$. Prove that $_R R$ has no indecomposable decomposition.

(3) Let $M = \bigoplus_I M_i$ be a decomposition of $M$ which complements direct summands. Prove that each $M_i$ is indecomposable.

(4) This exercise has its origins in a 1912 paper of Steinitz on rings of algebraic integers [320]. In fact, for any Dedekind domain $R$ and invertible ideals $I_1, \ldots, I_m$ of $R$ there is an isomorphism $I_1 \oplus \cdots \oplus I_m \simeq R^{m-1} \oplus (I_1 \cdots I_m)$ (see [109, p. 165]).

  (i) Let $I$ and $J$ be left ideals of the ring $R$ such that $R = I + J$. By considering the natural epimorphism $I \oplus J \to R$, prove that the $R$-modules $I \oplus J$ and $R \oplus (I \cap J)$ are isomorphic.

  (ii) Let $R = \mathbb{Z}\left[\sqrt{-5}\right]$ and $I$ and $J$ be the ideals of $R$ generated by $\{3, 2 + \sqrt{-5}\}$ and $\{3, 2 - \sqrt{-5}\}$ respectively. By considering the standard (multiplicative) norm $N$ defined on $R$ by $N\left(a + b\sqrt{-5}\right) = a^2 + 5b^2$ for all $a, b \in \mathbb{Z}$, show that neither $I$ nor $J$ are principal ideals of $R$. Hence show that the $R$-module $M = I \oplus J$ has non-isomorphic indecomposable decompositions.

   (As noted in [183], this example shows that, in contrast to quasi-continuous modules, if an extending module $M$ is expressible as a direct sum of uniform modules, then this decomposition need not be unique up to isomorphism.)

(5) Let $\{M_i\}_I$ be a locally semi-T-nilpotent family of $R$-modules and $\{S_j\}_J$ be a family of simple $R$-modules. Prove that the disjoint union $\{M_i : i \in I\} \cup \{S_j : j \in J\}$ is also locally semi-T-nilpotent. (See [202].)

(6) Let $V$ be an infinite-dimensional left vector space over a division ring $D$ and let $R$ be the ring End$_D(V)$ of linear transformations on $V$. It is well-known that $R$ is a von Neumann regular left self-injective ring (see [128, Prop. 2.23]). Moreover, for each $n \in \mathbb{N}$, $_R R \simeq {}_R R^{(n)}$, the direct sum of $n$ copies of $R$ (see [13, Exercise 8.14]). Taking $M_n = R$ for each $n \in \mathbb{N}$, show that $\{M_n\}_{\mathbb{N}}$ is not locally semi-T-nilpotent. (See [70, Example 2].)

(7) A module $M$ is said to have the *mutual exchange property* if, for any module $A$ and any decompositions

$$A = M' \oplus N = \bigoplus_I A_i$$

for modules $M', N, A_i$ where $M' \simeq M$ and $\mathrm{card}(I)$ is arbitrary, there always exist decompositions $A_i = B_i \oplus C_i$ for each $i \in I$ such that

$$A = M' \oplus \left(\bigoplus_I B_i\right) = \left(\bigoplus_I C_i\right) \oplus N.$$

Using 11.3, prove that an LE-module $M$ has the mutual exchange property. (See [37]; there Beck raises the question as to whether the mutual exchange property is actually stronger than the exchange property and also whether the class of modules with the mutual exchange property is closed under finite direct sums.)

(8) Let $M$ be an indecomposable module with semiregular endomorphism ring. Show directly, without using Corollary 11.21 or 12.2, that $M$ is an LE-module.

**12.20. Comments.** Crawley and Jónnson [77] proved that any indecomposable module with the finite exchange property has the full exchange property, while Warfield [358, Proposition 1] completes 12.2 by showing that these are precisely the LE-modules. (In the proof we give here we take advantage of Corollary 11.14.) That a finite sum of LE-modules has the exchange property is also proved by Beck [37, Remark, p. 397] and appears as Lemma 26.4 in Anderson and Fuller's text [13].

As mentioned in the main text, the concept of complementing direct summands and 12.5 are due to Anderson and Fuller [12]. In [24], Azumaya established the important uniqueness of indecomposable decompositions for any module with an LE-decomposition (see property (3) of 12.6). This is also referred to as the Krull–Schmidt–Remak–Azumaya Theorem in deference to its earlier contributors — we refer the reader to Facchini's text [92] for some historical background as well as a detailed treatment of related direct sum decompositions.

We will use Dung's Lemma 12.8 (see [83, Lemma 3.1]) in the next section to look at when Azumaya's Theorem 12.6 may be extended to modules with indecomposable decompositions which complement maximal direct summands.

The local semi-T-nilpotency condition evolved in the 1970s in a series of papers, including [148], [149], [184], [150], [160], and [162], by M. Harada and his coauthors, particularly H. Kanbara and Y. Sai. They used it initially in their study of projective modules, in particular perfect and semiperfect modules, and were clearly motivated by the T-nilpotency condition introduced by Bass in his seminal study of perfect rings [34].

The centrepiece of this section, 12.14, appears as Corollary 6 in Zimmermann-Huisgen–Zimmermann's [385]. Here we have split up their preceding Theorem 5 into 12.12 and 12.13 and extracted 12.9 from the proof of their Corollary 6. The Theorem significantly extends and unifies earlier work by Harada, Ishii, Sai, and Yamagata. In particular, the implication (b)⇒(c) of 12.14 was established in [162] and [373] and (c)⇒(b) also in [373], while (c)⇒(a) was proved for the special cases in which all the $M_i$'s are injective in [371] and [372] or all isomorphic [159]. Moreover, in contrast to the categorical methods used by these forerunners, Zimmermann-Huisgen and Zimmermann's proof is completely module-theoretic. (They acknowledge the influence of arguments by Harada and Sai in [162] and Yamagata in [373] in their proof of (c)⇒(a) in 12.14, and of Stock (see [324]) in 12.13.)

Corollary 12.18 appears as Corollary 8 in [385]. Eisenbud and de la Peña [89] have sharpened the bound for the composition vanishing in the Harada–Sai Lemma in terms of the individual lengths of the modules.

A recent survey paper by Clark [69] looks at the impact of local semi-T-nilpotency on other areas of module theory. As a particular instance, we note that Huisgen–Zimmermann and Saorín [173] have shown that if $\{M_i\}_I$ is a family of LE-modules, then $\bigoplus_I M_i$ is noetherian over its endomorphism ring (i.e. endo-noetherian) and the family is locally semi-T-nilpotent if and only if there are only finitely many isomorphism classes of the $M_i$ and each $M_i$ is endo-noetherian.

**References.** Anderson and Fuller [13]; Azumaya [24]; Beck [37]; Crawley and Jónnson [77]; Dung [83]; Eisenbud and de la Peña [89]; Facchini [92]; Harada [148, 149, 150]; Harada and Kanbara [160]; Harada and Sai [162]; Huisgen–Zimmermann and Saorín [173]; Stock [324]; Kanbara [184]; Warfield [358]; Yamagata [371, 372]; Zimmermann-Huisgen–Zimmermann's [385].

# 13   Local direct summands

In this section we concentrate on yet another condition associated with indecomposable decompositions of a module $M$, namely when every local direct summand of $M$ is a direct summand of $M$. As we'll see, this condition is closely related to local semi-T-nilpotency. The material in the section is centred round an important paper by Dung [83], with the main theorem generalising much earlier work by Harada [158].

**13.1. Local direct summands.** An internal direct sum $\bigoplus_I A_i$ of submodules of a module $M$ is called a *local (direct) summand of $M$* if, given any finite subset $F$ of the index set $I$, the direct sum $\bigoplus_{i \in F} A_i$ is a direct summand of $M$.

If, moreover, the direct sum $\bigoplus_I A_i$ is itself a summand of $M$, then we say that the local direct summand $\bigoplus_I A_i$ is also a (direct) summand of $M$. This (somewhat verbose) phrasing is used to emphasise that local direct summands need not be summands in general and, indeed, the rest of this section will investigate modules $M$ for which every local direct summand of $M$ is also a direct summand of $M$.

Given any infinite family $\{M_i\}_I$ of $R$-modules, it is easy to see that $\bigoplus_I M_i$ is a local direct summand of $\prod_I M_i$.

It is also straightforward to show that, for any ring $R$, all the local direct summands of $_R R$ are summands if and only if $R$ has no infinite set of orthogonal idempotents.

The following two results are fundamental to the study of modules for which local summands are always summands.

**13.2. Local summands and chains of summands.** *For any module $M$, every local summand of $M$ is a summand if and only if the union of any chain of summands of $M$ (with respect to inclusion) is also a summand of $M$.*

*Proof.* First assume that every local summand of $M$ is a summand. Let the family $\{M_i\}_I$ be a chain of summands of $M$ and let $\mathcal{F}$ denote the family of all local summands $A = \bigoplus_I A_i$ for which $\bigoplus_I A_i = \bigcup_{i \in J} M_i$ for some subset $J = J(A)$ of $I$. Now partially order the nonempty $\mathcal{F}$ by inclusion $\subseteq$ and consider a chain $\mathcal{C} = \{A_\lambda : \lambda \in \Lambda\}$ in $\mathcal{F}$, where we take $A_\lambda$ to be the local summand $\bigoplus_{i_\lambda \in I_\lambda} A_{i_\lambda}$. Let $\bar{A} = \bigoplus_{I_\infty} A_i$ where $I_\infty = \bigcup_{\lambda \in \Lambda} I_\lambda$. Then, clearly, $\bar{A}$ is also a local summand. Moreover

$$\bigoplus_{I_\infty} A_i = \bigcup_{\lambda \in \Lambda} \left( \bigoplus_{i_\lambda \in I_\lambda} A_{i_\lambda} \right) = \bigcup_{\lambda \in \Lambda} \left( \bigcup_{i \in J(A_\lambda)} M_i \right) = \bigcup_{i \in \bigcup_{\lambda \in \Lambda} J(A_\lambda)} M_i$$

and so $\bar{A} \in \mathcal{F}$. This shows that every chain in $\mathcal{F}$ has an upper bound in $\mathcal{F}$.

Now invoke Zorn's Lemma to produce a maximal element $B = \bigoplus_K B_k$ in $\mathcal{F}$ and suppose that $B = \bigcup_{i \in J(B)} M_i$ where $J(B) \subseteq I$. Then, by our assumption, $B$ is a direct summand of $M$, say $M = B \oplus C$, and so our goal is achieved if we can show that $\bigcup_I M_i = B$. Suppose to the contrary that there is an index $k \in I$ for which $M_k \not\subseteq B$. Then, since $\{M_i\}_I$ is a chain, we must have $M_i \subsetneq M_k$ for all $i \in J(B)$ and so $B \subsetneq M_k$. As a consequence, $M_k = B \oplus D$, where $D = M_k \cap C \neq 0$. However this shows that $B \oplus D$ is also an element of $\mathcal{F}$, contradicting the maximality of $B$, as required.

Conversely, suppose that every chain of summands is a summand. Let $\bigoplus_I A_i$ be a local summand of $M$ and let $\mathcal{S}$ denote the family of all subsets $J$ of $I$ for which $\bigoplus_{i \in J \cup F} A_i$ is a summand of $M$ for every finite subset $F$ of $I$. Now partially order the nonempty $\mathcal{S}$ by inclusion $\subseteq$ and consider a chain $\mathcal{C} = \{J_\lambda : \lambda \in \Lambda\}$ in $\mathcal{S}$. Set $J_\infty = \bigcup_{\lambda \in \Lambda} J_\lambda$. Then, for any finite subset $F$ of $I$,

$$\bigoplus_{i \in J_\infty \cup F} A_i = \bigcup_{\lambda \in \Lambda} \left( \bigoplus_{i \in J_\lambda \cup F} A_i \right)$$

is a chain of summands of $M$ and so, by assumption, is also a summand of $M$. This shows that $J_\infty$ is an upper bound for the chain $\mathcal{C}$ in $\mathcal{S}$. Hence, by Zorn's Lemma, we can choose a maximal element $K$ of $\mathcal{S}$ and our proof is finished if we can show that $K = I$.

Assume to the contrary that there is an $i_0 \in I \setminus K$ and set $K_0 = K \cup \{i_0\}$. Then, for any finite subset $F$ of $I$, we have

$$\bigoplus_{i \in K_0 \cup F} A_i = \bigoplus_{i \in K \cup (F \cup \{i_0\})} A_i,$$

which is a direct summand of $M$. Consequently, $K_0 \in \mathcal{S}$, but this contradicts the maximality of $K$. $\square$

The next result was observed by Oshiro in [271].

**13.3. Local summands and decomposability.** *If every local summand of the module $M$ is a summand, then $M$ is a direct sum of indecomposable modules.*

*Proof.* A straightforward Zorn's Lemma argument shows that we can choose a local summand $A = \bigoplus_I A_i$ of $M$ in which each $A_i$ is indecomposable and $A$ is maximal with respect to inclusion and these properties. By assumption, we have $M = A \oplus B$ for some $B \subseteq M$ and of course we wish to show that $B = 0$.

Suppose by way of contradiction that there is a nonzero element $b$ in $B$. Then, by considering chains of summands of $B$ (and hence of $M$) which do not contain $b$, we infer from 13.2 that there is a summand $C$ of $B$ maximal with respect to the property that it omits $b$. Set $B = C \oplus D$, where $D \neq 0$. Then the maximal property of $A$ forces $D$ to be decomposable, say $D = D_1 \oplus D_2$ where both $D_i$ are nonzero. Now the maximal property of $C$ shows that $b$ is in both $C \oplus D_1$ and $C \oplus D_2$. However it follows easily from this that $b \in C$, giving our contradiction. $\square$

The following theorem gives us a first indication of the influence local semi-T-nilpotency has on local direct summands. The industrious reader will see how the ideas used in the proofs of 12.12, 12.13 and 12.14 have influenced the attack here.

**13.4. Local summands and local semi-T-nilpotency.** Let $M = \bigoplus_I M_i$ be an inde-composable decomposition which complements maximal direct summands. Suppose that $\{N_\lambda\}_\Lambda$ is a locally semi-T-nilpotent family of indecomposable submodules of $M$ for which $\bigoplus_\Lambda N_\lambda$ is a local direct summand. Then $M = (\bigoplus_\Lambda N_\lambda) \oplus (\bigoplus_{i \in H} M_i)$ for some subset $H$ of $I$.

*Proof.* If the index set $\Lambda$ is finite, then the required decomposition follows imme-diately from 12.8. Thus we henceforth assume that $\Lambda$ is infinite. The proof now breaks into two parts, the first assuming that all the $N_\lambda$ are pairwise isomorphic, while the second part uses the first to consider the general case.

*Part 1.* Assume that $N_\lambda \simeq N_\mu$ for all $\lambda, \mu \in \Lambda$. We use the subset $J$ of $I$ given by $J = \{j \in I : M_j \simeq N_\lambda$ for all $\lambda \in \Lambda\}$. Note that $J$ is infinite since, given any $n \in \mathbb{N}$ and distinct indices $\lambda_1, \ldots, \lambda_n \in \Lambda$, then by 12.8 there are indices $i_1, \ldots, i_n \in I$ for which $M_{i_k} \simeq N_{\lambda_k}$ for each $k = i, \ldots, n$. Moreover, since $\{N_\lambda\}_\Lambda$ is locally semi-T-nilpotent, $\{M_j\}_J$ is also locally semi-T-nilpotent.

Now set $K = I \setminus J$ and $P_{(K)} = (\bigoplus_{\lambda \in \Lambda} N_\lambda) \oplus (\bigoplus_{i \in K} M_i)$. We claim that the decomposition $P_{(K)}$ is a local direct summand of $M$. Indeed, given any finite subset $F = \{\lambda_1, \ldots, \lambda_n\} \subseteq \Lambda$, by 12.8 again there are indices $i_1, \ldots, i_n \in J$ for which $M_{i_k} \simeq N_{\lambda_k}$ for each $k = i, \ldots, n$ and

$$M = (N_{\lambda_1} \oplus \cdots \oplus N_{\lambda_n}) \oplus \left( \bigoplus_{i \in I \setminus \{i_1, \ldots, i_n\}} M_i \right).$$

This implies that $(N_{\lambda_1} \oplus \cdots \oplus N_{\lambda_n}) \oplus (\bigoplus_{i \in K} M_i)$ is a direct summand of $M$, verifying our claim. From this, a straightforward Zorn's Lemma argument shows that there is a subset $H$ of $I$ maximal with respect to inclusion and the properties that $K \subseteq H$ and the decomposition $P_{(H)} = (\bigoplus_\Lambda N_\lambda) \oplus (\bigoplus_{i \in H} M_i)$ is a local summand of $M$. We will reach our desired outcome by showing that $M = P_{(H)}$.

Suppose to the contrary that $M \neq P_{(H)}$. Then there is an index $i_1 \in I \setminus H$ and an element $x_1 \in M_{i_1}$ such that $x_1 \notin (\bigoplus_{\lambda \in \Lambda} N_\lambda) \oplus (\bigoplus_{i \in H} M_i)$. Since $K \subseteq H$, we have $i_1 \in J$. Now the maximality of $H$ shows that there are finite subsets $\Lambda_1 \subseteq \Lambda$ and $H_1 \subseteq H$ such that, if $Q$ denotes the finite subsum $(\bigoplus_{\lambda \in \Lambda_1} N_\lambda) \oplus (\bigoplus_{i \in H_1} M_i)$ of $P_{(H)}$, then either $M_{i_1} \cap Q \neq 0$ or $M_{i_1} + Q$ is not a summand of $M$. Now note that, since each of the (finitely many) summands in $Q$ are indecomposable, it follows from 12.8 that there is a subset $I' \subseteq I$ for which $M = Q \oplus (\bigoplus_{i \in I'} M_i)$.

For each $i \in I'$, let $\pi_i : M \to M_i$ be the natural projection corresponding to this last decomposition. Then, by our choice of $x_1$, there is an $i_2 \in I'$ for which $(x_1)\pi_{i_2} \notin P_{(H)}$. Next let $f_1 : M_{i_1} \to M_{i_2}$ be the restriction of $\pi_{i_2}$ on $M_{i_1}$. Now, if $f_1$ were an isomorphism, it would follow from Lemma 11.3 that $M = Q \oplus M_{i_1} \oplus (\bigoplus_{i \in I' \setminus \{i_2\}} M_i)$, which contradicts our choice of $Q$. Thus $f_1$ is not an isomorphism. Moreover, since $M_{i_2} \not\subseteq P_{(H)}$, we have $i_2 \in I \setminus H \subseteq J$.

We now repeat the preceding argument with $x_2 = (x_1)f_1$ replacing $x_1$ and continue inductively. Since $M_i \simeq M_j$ for all $i, j \in J$, this produces a sequence of non-isomorphisms $f_n : M_{i_n} \to M_{i_{n+1}}$, where $i_n \in J$ for each $n \in \mathbb{N}$, such that $(x_1)f_1 \cdots f_n \neq 0$ for all $n \geq 1$. However this contradicts the local semi-T-nilpotency of the family $\{M_j\}_J$. This contradiction shows that $M = P_{(H)}$, as required.

*Part 2.* We turn now to the general case. Thus let $\{N_\lambda\}_\Lambda$ be a locally semi-T-nilpotent family of indecomposable submodules of $M$ for which $\bigoplus_\Lambda N_\lambda$ is a local direct summand. Now select a family of representatives $\{N_\gamma\}_\Gamma$ of the isomorphism classes of the modules in $\{N_\lambda\}_\Lambda$. For each such representative $N_\gamma$, let $L_\gamma$ be the direct sum of all those $N_\lambda$'s which are isomorphic to $N_\gamma$. Then $\bigoplus_\Gamma L_\gamma = \bigoplus_\Lambda N_\lambda$.

For each $\gamma \in \Gamma$, clearly the family $\{N_\lambda : N_\lambda \simeq N_\gamma\}$ is also locally semi-T-nilpotent and its sum $L_\gamma$ is a local summand of $M$. Thus, by Part 1, there is a subset $J$ of $I$ for which $M = L_\gamma \oplus (\bigoplus_{i \in J} M_i)$. By 12.5 (2), (5), this decomposition of $M$ also complements maximal summands and so too does each indecomposable decomposition of $L_\gamma$. In particular, by 12.5 again, any indecomposable summand of $L_\gamma$ is isomorphic to $N_\gamma$. Thus, given distinct $\gamma_1, \gamma_2 \in \Gamma$, $L_{\gamma_1}$ and $L_{\gamma_2}$ have no isomorphic indecomposable summands in common.

We now let $\mathcal{P}$ denote the set of all pairs $(\Gamma', I')$ where $\Gamma' \subseteq \Gamma$ and $I' \subseteq I$ are such that $M = \left(\bigoplus_{\gamma \in \Gamma'} L_\gamma\right) \oplus \left(\bigoplus_{i \in I \setminus I'} M_i\right)$. Note that the previous paragraph shows that $\mathcal{P}$ is nonempty. We endow $\mathcal{P}$ with a partial ordering $\leq$ by simply setting $(\Gamma_1, I_1) \leq (\Gamma_2, I_2)$ provided $\Gamma_1 \subseteq \Gamma_2$ and $I_1 \subseteq I_2$.

Next, for the moment, assume that $(\mathcal{P}, \leq)$ has a maximal element $(\Gamma_0, I_0)$. Then we have

$$M = \left(\bigoplus_{\gamma \in \Gamma_0} L_\gamma\right) \oplus \left(\bigoplus_{i \in I \setminus I_0} M_i\right).$$

Rewriting this decomposition of $M$ by expanding each of the $L_\gamma$'s as direct sums of copies of $N_\gamma$, we get another indecomposable decomposition of $M$ which, by 12.5 (2), also complements maximal summands. Now suppose there is a $\gamma_1 \in \Gamma \setminus \Gamma_0$. Then, by Part 1 applied to $L_{\gamma_1}$ and this decomposition, remembering that $L_{\gamma_1}$ and $L_\gamma$ have no isomorphic indecomposable summands in common for any $\gamma \in \Gamma_0$, we find a subset $J_0$ of $I \setminus I_0$ for which

$$M = \left(\bigoplus_{\gamma \in \Gamma_0} L_\gamma\right) \oplus L_{\gamma_1} \oplus \left(\bigoplus_{i \in I \setminus (I_0 \cup J_0)} M_i\right).$$

Thus, setting $\Gamma_1 = \Gamma_0 \cup \{\gamma_1\}$ and $I_1 = I_0 \cup J_0$, we have $(\Gamma_1, I_1) \in \mathcal{P}$ with $(\Gamma_0, I_0) \lneq (\Gamma_1, I_1)$, contradicting the maximality of $(\Gamma_0, I_0)$. This contradiction shows that $\Gamma_0 = \Gamma$ and so

$$M = \left(\bigoplus_\Gamma L_\gamma\right) \oplus \left(\bigoplus_{i \in I \setminus I_0} M_i\right),$$

which is our desired outcome. Consequently, our proof will be complete if we can
show that $(\mathcal{P}, \leq)$ has a maximal element and so, by Zorn's Lemma, if we can show
that any chain in $(\mathcal{P}, \leq)$ has an upper bound in $(\mathcal{P}, \leq)$.

Let $\mathcal{C} = \{(\Gamma_k, I_k) : k \in K\}$ be a chain in $(\mathcal{P}, \leq)$ and set $\Gamma^* = \bigcup_{k \in K} \Gamma_k$ and
$I^* = \bigcup_{k \in K} I_k$. We complete the proof by showing that $(\Gamma^*, I^*) \in \mathcal{P}$, that is, that

$$M = \left( \bigoplus_{\gamma \in \Gamma^*} L_\gamma \right) \oplus \left( \bigoplus_{i \in I \setminus I^*} M_i \right). \qquad (*)$$

Letting $\pi : M = \left( \bigoplus_{i \in I^*} M_i \right) \oplus \left( \bigoplus_{i \in I \setminus I^*} M_i \right) \to \bigoplus_{i \in I^*} M_i$ be the natural projection map and then taking $f : \bigoplus_{\gamma \in \Gamma^*} L_\gamma \to \bigoplus_{i \in I^*} M_i$ to be the restriction of $\pi$ on
$\bigoplus_{\gamma \in \Gamma^*} L_\gamma$, it follows from Lemma 11.3 that $(*)$ is accomplished by showing that
$f$ is an isomorphism.

Given $x \in \operatorname{Ke} f$, we have $x \in \left( \bigoplus_{\gamma \in \Gamma^*} L_\gamma \right) \cap \left( \bigoplus_{i \in I \setminus I^*} M_i \right)$. Then there is an
index $k \in K$ with $x \in \bigoplus_{\gamma \in \Gamma_k} L_\gamma$. Since $M = \left( \bigoplus_{\gamma \in \Gamma_k} L_\gamma \right) \oplus \left( \bigoplus_{i \in I \setminus I_k} M_i \right)$, we
have

$$\left( \bigoplus_{\gamma \in \Gamma_k} L_\gamma \right) \cap \left( \bigoplus_{i \in I \setminus I^*} M_i \right) = 0,$$

and from this it follows that $x = 0$. Hence $f$ is a monomorphism.

This leaves us to show that $f$ is an epimorphism. Suppose to the contrary
that $\left( \bigoplus_{\gamma \in \Gamma^*} L_\gamma \right) f \neq \bigoplus_{i \in I^*} M_i$. Then, for some $k_1 \in K$, there is an $x_1 \in \bigoplus_{i \in I_{k_1}} M_i$
such that $x_1 \notin \left( \bigoplus_{\gamma \in \Gamma^*} L_\gamma \right) f$. Now, since $(\Gamma_{k_1}, I_{k_1}) \in \mathcal{P}$, we have

$$M = \left( \bigoplus_{\gamma \in \Gamma_{k_1}} L_\gamma \right) \oplus \left( \bigoplus_{i \in I \setminus I_{k_1}} M_i \right). \qquad (**)$$

Let $\pi_1 : M = \left( \bigoplus_{i \in I_{k_1}} M_i \right) \oplus \left( \bigoplus_{i \in I \setminus I_{k_1}} M_i \right) \to \bigoplus_{i \in I_{k_1}} M_i$ be the natural projection map and then take $f_1 : \bigoplus_{\gamma \in \Gamma_{k_1}} L_\gamma \to \bigoplus_{i \in I_{k_1}} M_i$ to be the restriction of $\pi_1$
on $\bigoplus_{\gamma \in \Gamma_{k_1}} L_\gamma$. Then, by 11.3 applied to $(**)$, $f_1$ is an isomorphism.

Set $y_1 = (x_1) f_1^{-1}$. Then, since $I$ is the disjoint union $I_k \cup (I^* \setminus I_{k_1}) \cup (I \setminus I^*)$
and $\bigoplus_{i \in I \setminus I_{k_1}} M_i \subseteq \operatorname{Ke} f$, we may write $y_1 = x_1 + x_2 + z_1$ where $x_2 \in \bigoplus_{i \in I^* \setminus I_{k_1}} M_i$
and $z_1 \in \bigoplus_{i \in I \setminus I^*} M_i$. From this we find a $k_2 \in K$ for which $I_{k_1} \subsetneq I_{k_2}$ and $x_2 \in$
$\bigoplus_{i \in I_{k_2} \setminus I_{k_1}} M_i$. Note also that $x_2$ is the image of $y_1$ under the natural projection

$$h_1 : M = \left( \bigoplus_{i \in I_{k_1}} M_i \right) \oplus \left( \bigoplus_{i \in I_{k_2} \setminus I_{k_1}} M_i \right) \oplus \left( \bigoplus_{i \in I \setminus I_{k_2}} M_i \right) \to \bigoplus_{i \in I_{k_2} \setminus I_{k_1}} M_i,$$

and so $x_2 = (x_1) f_1^{-1} h_1 = (x_1) g_1$ where $g_1 = f_1^{-1} \diamond h_1 : \bigoplus_{i \in I_{k_1}} M_i \to \bigoplus_{i \in I \setminus I_{k_2}} M_i$.
Also, since $x_1 + x_2 = (y_1) f$ and $x_1 \notin \left( \bigoplus_{\gamma \in \Gamma^*} L_\gamma \right) f$, it follows that $x_2 \notin \left( \bigoplus_{\gamma \in \Gamma^*} L_\gamma \right) f$.

Now we repeat the foregoing argument with $x_2$ replacing $x_1$ to produce an
index $k_3 \in K$ for which $I_{k_2} \subsetneq I_{k_3}$, an element $x_3 \in \bigoplus_{i \in I_{k_3} \setminus I_{k_2}} M_i$ such that

$x_3 \notin (\bigoplus_{\gamma \in \Gamma^*} L_\gamma) f$, and a homomorphism $g_2 : \bigoplus_{i \in I_{k_2} \setminus I_{k_1}} M_i \to \bigoplus_{i \in I_{k_3} \setminus I_{k_2}} M_i$ for which $x_3 = (x_2)g_2$. Then, proceeding inductively leads to an infinite sequence $(k_n)$ of indices from $K$ such that

$$I_{k_1} \subsetneqq I_{k_2} \subsetneqq \cdots \subsetneqq I_{k_n} \subsetneqq \cdots$$

with nonzero elements $x_n \in \bigoplus_{i \in I_{k_n} \setminus I_{k_{n-1}}} M_i$ for each $n \in \mathbb{N}$ (taking $I_{k_0} = \emptyset$), and homomorphisms

$$g_n : \bigoplus_{i \in I_{k_n} \setminus I_{k_{n-1}}} M_i \to \bigoplus_{i \in I_{k_{n+1}} \setminus I_{k_n}} M_i$$

for which $x_{n+1} = (x_1)g_1 g_2 \cdots g_n$ for each $n$.

Now note that, since

$$M = \left( \bigoplus_{\gamma \in \Gamma_{k_{n-1}}} L_\gamma \right) \oplus \left( \bigoplus_{i \in I \setminus I_{k_{n-1}}} M_i \right) = \left( \bigoplus_{\gamma \in \Gamma_{k_n}} L_\gamma \right) \oplus \left( \bigoplus_{i \in I \setminus I_{k_n}} M_i \right),$$

setting

$$\overline{L}_n = \bigoplus_{\gamma \in \Gamma_{k_n} \setminus \Gamma_{k_{n-1}}} L_\gamma \quad \text{and} \quad \overline{M}_n = \bigoplus_{i \in I_{k_n} \setminus I_{k_{n-1}}} M_i \qquad (***)$$

we have $\overline{L}_n \simeq \overline{M}_n$ for each $n \in \mathbb{N}$. Also recall that, for distinct $\gamma_1, \gamma_2 \in \Gamma$, $L_{\gamma_1}$ and $L_{\gamma_2}$ have no isomorphic indecomposable summands in common and so from $(***)$ we see that the $\overline{M}_n$ are pairwise non-isomorphic. In particular, the homomorphisms $g_n : \overline{M}_n \to \overline{M}_{n+1}$ are all non-isomorphisms.

Next note that, since the family $\{N_\lambda\}_\Lambda$ is locally semi-T-nilpotent, using König's Graph Theorem 31.1 one can show that the same is true for the family $\{\overline{L}_n\}_\mathbb{N}$. (We leave the details of this to the reader in 13.13 (11).) [1] Then, since $\overline{L}_n \simeq \overline{M}_n$ for each $n \in \mathbb{N}$, it follows that $\{\overline{M}_n\}_\mathbb{N}$ is also locally semi-T-nilpotent. However, as noted, each $g_n$ above is a non-isomorphism and so the corresponding sequence $(x_n)$ of nonzero elements contradicts this. This contradiction completes the proof. $\qquad \square$

We now give the last of our preparatory results needed for the proof of 13.6.

**13.5. Local summands inherit local semi-T-nilpotency.** *Let $M = \bigoplus_I M_i$ be an indecomposable decomposition which complements maximal direct summands and suppose that $\{M_i\}_I$ is locally semi-T-nilpotent. Then any family of indecomposable submodules $\{N_\lambda\}_\Lambda$ of $M$, for which $\bigoplus_\Lambda N_\lambda$ is a local summand, is also locally semi-T-nilpotent.*

---

[1] Although we have relegated this to the exercises, we note that a similar argument is used in the next section, specifically in Lemma 14.18 and the second half of the proof of 14.20.

*Proof.* We need only show that, if $\{N_{\lambda_k}\}_{k \in \mathbb{N}}$ is any countable subfamily of $\{N_\lambda\}_\Lambda$, then there are distinct indices $i_1, i_2, \ldots, i_k, \ldots$ for which $M_{i_k} \simeq N_{\lambda_k}$ for each $k \in \mathbb{N}$. Note first that, since $M = \bigoplus_I M_i$ complements maximal direct summands, each $N_\lambda$ is isomorphic to some $M_j$ for some $j \in I$. Thus it suffices to consider the case where $N_{\lambda_k} \simeq N_{\lambda_l}$ for all $k, l \in \mathbb{N}$. However, $\bigoplus_\Lambda N_\lambda$ is a local summand, it follows from Lemma 12.8 that there is an infinite number of indices $i \in I$ for which $M_i \simeq N_{\lambda_k}$, and so we are finished. $\qquad\square$

We now come to the major result of this section, due to Dung [83]. For an indecomposable decomposition, it pinpoints precisely the difference between complementing maximal direct summands and complementing all direct summands. By way of further introduction, note that if $P$ is an infinitely generated free left $R$-module over a semiperfect ring $R$ which is not left perfect, then $P$ has an indecomposable decomposition which complements maximal summands but does not complement every summand (see [12, Theorems 5 and 6] or [13, Theorems 27.11, 27.12, and 28.14]).

**13.6. Complementing decompositions, local direct summands and local semi-T-nilpotency.** *Let $M = \bigoplus_I M_i$ be an indecomposable decomposition which complements maximal direct summands. Then the following conditions are equivalent.*

(a) *The decomposition $M = \bigoplus_I M_i$ complements direct summands;*

(b) *the family $\{M_i\}_I$ is locally semi-T-nilpotent and every nonzero direct summand of $M$ contains an indecomposable direct summand;*

(c) *every local direct summand of $M$ is a direct summand.*

*Proof.* (a)$\Rightarrow$(b). If (a) is satisfied, it is straightforward to see that every nonzero summand of $M$ contains an indecomposable summand. The local semi-T-nilpotency of $\{M_i\}_I$ follows from 12.16.

(b)$\Rightarrow$(a). Suppose (b) holds and let $D$ be a nonzero direct summand of $M$. Then, by hypothesis, $D$ contains an indecomposable summand and so, as in the proof of 13.3, we may choose a local summand $N = \bigoplus_\Lambda N_\lambda$ of $D$ (and hence of $M$) in which each $N_\lambda$ is indecomposable and $N$ is maximal with respect to inclusion and these properties. By 13.5, $\{N_\lambda\}_\Lambda$ is also locally semi-T-nilpotent and so it follows from 13.4 that $M = (\bigoplus_\Lambda N_\lambda) \oplus (\bigoplus_{i \in H} M_i)$ for some subset $H$ of $I$. As a result, $N = \bigoplus_\Lambda N_\lambda$ is a direct summand of $D$, say $D = N \oplus L$, noting that $L$ is also a summand of $M$. Now, if $L$ were nonzero, then, by our assumption, $L$ would contain an indecomposable direct summand, $K$ say; but then $(\bigoplus_\Lambda N_\lambda) \oplus K$ would also be a local direct summand of indecomposables, in contradiction to the maximality of $N$. Hence $L = 0$ and so $D = N$, whence $M = D \oplus (\bigoplus_{i \in H} M_i)$. This shows that $M = \bigoplus_I M_i$ complements direct summands.

(b)$\Rightarrow$(c). Assume (b) holds and let $N = \bigoplus_\Lambda N_\lambda$ be a local summand of $M$. Then, since each $N_\lambda$ is a direct summand of $M$, it follows from (b)$\Rightarrow$(a) proved above that $N_\lambda$ also has an indecomposable decomposition, say $N_\lambda = \bigoplus_{j \in J_\lambda} N_{\lambda,j}$. We now see that $\bigoplus_{\lambda \in \Lambda, j \in J_\lambda} N_{\lambda,j}$ is also a local summand of $M$ and so, since each $N_{\lambda,j}$ is indecomposable and $\{M_i\}_I$ is locally semi-T-nilpotent, 13.5 shows that the family $\{N_{\lambda,j} \mid \lambda \in \Lambda, j \in J_\lambda\}$ is also locally semi-T-nilpotent. Hence, by 13.4, $N = \bigoplus_\Lambda N_\lambda = \bigoplus_{\lambda \in \Lambda, j \in J_\lambda} N_{\lambda,j}$ is a summand of $M$, as required.

(c)$\Rightarrow$(b). Assume that every local summand of $M$ is a summand. Then it is straightforward to see that every nonzero direct summand $D$ of $M$ also has this property and so, by 13.3, $D$ has an indecomposable decomposition and in particular an indecomposable summand.

To show that $\{M_i\}_I$ is locally semi-T-nilpotent, we may assume that $I = \mathbb{N}$ and so consider a sequence of non-isomorphisms

$$M_1 \xrightarrow{f_1} M_2 \xrightarrow{f_2} \cdots \xrightarrow{f_{n-1}} M_n \xrightarrow{f_n} \cdots .$$

Then, as in the proofs of (c)$\Rightarrow$(a) of 12.14 and 12.16, we define

$$L_n = \{x + (x)f_n \mid x \in M_n\} \subseteq M_n + M_{n+1},$$

for each $n \in \mathbb{N}$. As noted in the latter proof, we have

(i) $L_m \cap M_n = 0$ if $m \neq n$ and $L_n \cap M_n = 0$ if and only if $f_n$ is a monomorphism,

(ii) the sum $\sum_{n \in \mathbb{N}} L_n$ is direct, and

(iii) $(\bigoplus_{i=1}^n L_i) \oplus M_{n+1} = \bigoplus_{i=1}^{n+1} M_i$.

In particular it follows from (ii) and (iii) that $\bigoplus_\mathbb{N} L_n$ is a local summand of $M$ and so, by hypothesis, $L = \bigoplus_\mathbb{N} L_n$ is a summand of $M$, say $M = L \oplus K$. We show that $K = 0$.

Assume to the contrary that $K$ is nonzero. Then, as noted in our opening paragraph, $K$ contains an indecomposable summand $X$, say with $K = X \oplus Y$. Then $L \oplus Y$ is a maximal direct summand of $M$ and so our overriding hypothesis gives

$$M = L \oplus Y \oplus M_t$$

for some $t \in \mathbb{N}$. Similarly, if $Y \neq 0$, then $Y = U \oplus V$ for some indecomposable $U$ and summand $V$ of $Y$ and, since $L \oplus V \oplus M_t$ is then also a maximal summand of $M$, we have

$$M = L \oplus V \oplus M_t \oplus M_n$$

for some $n \in \mathbb{N}$. Without loss of generality we may assume that $t \leq n$. Then the preceding decomposition gives

$$\begin{aligned} M &= \left(\bigoplus_{i=1}^{n-1} L_i\right) \oplus M_n \oplus \left(\bigoplus_{i \geq n} L_i\right) \oplus M_t \oplus V \\ &= \left(\bigoplus_{i=1}^{n} M_i\right) \oplus \left(\bigoplus_{i \geq n} L_i\right) \oplus M_t \oplus V \end{aligned}$$

which contradicts the direct sum property since $t \leq n$. This forces $Y$ to be zero and so $M = L \oplus M_t$.

We now call on Remark 12.11 to break our argument into two cases:

($\alpha$) all $f_n$ are non-monomorphisms and

($\beta$) for each $n \geq 2$, $f_n$ is a monomorphism and so a non-epimorphism.

Case ($\alpha$). If all the $f_n$ are non-monomorphisms, then in particular $f_t$ is not a monomorphism and so, by (i) above, $L_t \cap M_t \neq 0$. This contradicts the direct sum $M = L \oplus M_t$.

Case ($\beta$). Suppose that $f_n$ is a monomorphism for all $n \geq 2$. From (iii) above we have

$$M = L \oplus M_t = \left( \left( \bigoplus_{i=1}^{t-1} L_i \right) \oplus M_t \right) \oplus \left( \bigoplus_{i \geq t} L_i \right) = \left( \bigoplus_{i=1}^{t} M_i \right) \oplus \left( \bigoplus_{i \geq t} L_i \right).$$

Then we may represent any $x \in M_t$ (uniquely) as

$$x = y + (x_t + (x_t)f_t) + \cdots + (x_m + (x_m)f_m) \qquad (*)$$

for some $m \geq t$, where $y \in \bigoplus_{i=1}^{t} M_i$ and $x_k \in M_k$ for $t \leq k \leq m$. Converting (*) to a representation in $M = \bigoplus_{\mathbb{N}} M_n$, if $t \geq 2$ it follows that, since $x$ was an arbitrary element of $M_t$, $f_t$ is onto and so an isomorphism. This contradicts the underlying assumptions for the $f_i$'s.

On the other hand if $t = 1$, again we get that $f_1$ is onto. Then, since $f_1$ is not an isomorphism, $f_1$ is not a monomorphism and so, by (i) above, $M_1 \cap L_1 \neq 0$. However, since $t = 1$, we have $M = L \oplus M_1$ and so in particular $M_1 \cap L = 0$, giving a contradiction again.

These contradictions show that $K = 0$. Thus $M = \bigoplus_{n \in \mathbb{N}} L_n$. Then, given any $x \in M_1$, for some $n \in \mathbb{N}$ we have

$$x = (x_1 + (x_1)f_1) + \cdots + (x_n + (x_n)f_n)$$

where $x_i \in M_i$ for $1 \leq i \leq n$. Rewriting this as a representation in the direct sum $M = \bigoplus_{k \in \mathbb{N}} M_k$ gives $x = x_1$, $x_{k+1} = -(x_k)f_k$ for $1 \leq k < n$, and $(x_n)f_n = 0$. From this it readily follows that $(x)f_1 \cdots f_n = 0$. Hence $\{M_i\}_I$ is locally semi-T-nilpotent. $\qquad\square$

Using 13.6, we now bring a result which will prove useful in considering the direct sum of quasi-discrete modules in 26.19.

**13.7. Summands complementing summands.** *Let $M = \bigoplus_I M_i$ be an exchange decomposition of the module $M$ where each $M_i$ has an indecomposable decomposition that complements direct summands. Then any indecomposable decomposition of $M$ also complements direct summands.*

*Proof.* For each $i \in I$, let $M_i = \bigoplus_{k_i \in K_i} M_{k_i}$ be an indecomposable decomposition of $M_i$ which complements direct summands. By 12.5 (2), it is enough to prove that the indecomposable decomposition $M = \bigoplus_I \bigoplus_{k_i \in K_i} M_{k_i}$ is a decomposition complementing direct summands. By 13.6, this will follow if we show that this decomposition complements maximal direct summands, every nonzero direct summand of $M$ contains an indecomposable direct summand and the family $\{M_{k_i} : k_i \in K_i, i \in I\}$ is lsTn.

Let $X$ be a maximal direct summand of $M$. By hypothesis, $M = X \oplus M_i'$ where $M_i'$ is an indecomposable direct summand of $M_i$ for some $i \in I$. Now $M_i = (X \cap M_i) \oplus M_i'$ and so, since the decomposition $M_i = \bigoplus_{k_i \in K_i} M_{k_i}$ complements direct summands, we have $M_i = (X \cap M_i) \oplus M_{k_i}$ for some $k_i \in K_i$. It is then easy to verify that $M = X \oplus M_{k_i}$. Thus our decomposition does indeed complement maximal direct summands.

Now let $A$ be a nonzero direct summand of $M$. Since the decomposition $M = \bigoplus_I M_i$ is exchangeable, there exists an $i \in I$ for which $A$ contains a nonzero direct summand $A_i$ such that $A_i \simeq B_i$ where $B_i$ is a direct summand $M_i$. Then, since $M_i$ has a decomposition that complements direct summands, $B_i$ and hence $A_i$ contains an indecomposable direct summand.

Now we prove that the family $\{M_{k_i} : k_i \in K_i, i \in I\}$ is lsTn. Let $\{B_n\}_\mathbb{N}$ be a countable subfamily of this family. Now, by 11.33 and 13.6, any family consisting of exactly one of the summands $M_{k_i}$ for each $i \in I$ is lsTn. This implies that the family $\{B_n\}_\mathbb{N}$ is lsTn if it contains only a finite subfamily of $\{M_{k_i}\}_{K_i}$, for each $i \in I$. On the other hand, if for some $i \in I$, the family $\{B_n\}_\mathbb{N}$ contains an infinite subfamily of $\{M_{k_i}\}_{K_i}$, then we are done since the family $\{M_{k_i}\}_{K_i}$ is lsTn (see 13.6). $\square$

Our next goal is to present a striking result of J. L. Gómez Pardo and P. A. Guil Asensio [125] which shows that if every local summand of every clone of an indecomposable module $M$ is a summand, then $M$ is an LE-module. The proof of this needs the following observation on direct limits.

**13.8. Lemma.** *Let $I$ be a directed set and let $(M_i, f_{i,j})$ be a direct system of modules and homomorphisms. Let $\pi : \bigoplus_I M_i \to \varinjlim M_i$ be the canonical homomorphism and, for each $i \in I$, let $\varepsilon_i : M_i \to \bigoplus_I M_i$ be the natural inclusion map. Given $i,j \in I$ with $i \leq j$, set $M_{i,j} = M_i$. Next let $\phi : \bigoplus_{i \leq j} M_{i,j} \to \bigoplus_I M_i$ be the homomorphism induced on $\bigoplus_{i \leq j} M_{i,j}$ by the family $\{f_{i,j} \diamond \varepsilon_j - \varepsilon_i : M_{i,j} \to \bigoplus_I M_i\}$ of homomorphisms. Then*

(1) *we have an exact sequence*

$$\bigoplus_{i \leq j} M_{i,j} \xrightarrow{\phi} \bigoplus_I M_i \xrightarrow{\pi} \varinjlim M_i \longrightarrow 0;$$

(2) *if $I$ is a totally ordered infinite set, then $\text{Im}(\phi) = \text{Ke}(\pi)$ is the union of a chain of direct summands $\{N_\lambda\}_\Lambda$ of $\bigoplus_I M_i$, where $\text{card}(\Lambda) = \text{card}(I)$.*

*Proof.* (1) is given by [363, 24.2].

(2). Since $\text{Im}\,(\phi) = \sum_{i \leq j} \text{Im}\,(f_{i,j} \diamond \varepsilon_j - \varepsilon_i)$ and $I$ is directed, it suffices to show that, for each $k \in I$,

$$\bigoplus_I M_i = \left( \sum_{i \leq j \leq k} \text{Im}\,(f_{i,j} \diamond \varepsilon_j - \varepsilon_i) \right) \oplus \left( \bigoplus_{k \leq l} M_l \right).$$

First note that if $x_i \in M_i$ for some $i \in I$ with $i \leq k$, in $\bigoplus_I M_i$ we have

$$(x_i)\varepsilon_i = -\left((x_i)f_{i,k}\varepsilon_k - (x_i)\varepsilon_i\right) + (x_i)f_{i,k}\varepsilon_k \in \text{Im}\,(f_{i,k}\varepsilon_k - \varepsilon_i) + M_k.$$

From this it follows that $\bigoplus_I M_i = \left( \sum_{i \leq j \leq k} \text{Im}\,(f_{i,j} \diamond \varepsilon_j - \varepsilon_i) \right) + \left( \bigoplus_{k \leq l} M_l \right)$ and so we're left to show that this sum is direct. To this end, let

$$x \in \left( \sum_{i \leq j \leq k} \text{Im}\,(f_{i,j} \diamond \varepsilon_j - \varepsilon_i) \right) \cap \left( \bigoplus_{k \leq l} M_l \right).$$

Then there is a finite set of pairs $(i_1, j_1), (i_2, j_2), \ldots, (i_t, j_t) \in I \times I$ with $i_u \leq j_u \leq k$ for $1 \leq u \leq t$ and elements $x_u \in M_{i_u}$ for which

$$x = \sum_{u=1}^{t} (x_u)(f_{i_u,j_u}\varepsilon_{j_u} - \varepsilon_{i_u}).$$

Now choose $r_1, \ldots, r_s \in I$ with $r_1 < \cdots < r_s \leq k$ such that $i_u, j_u \in \{r_1, \ldots, r_s\}$ for each $u = 1, \ldots, t$. If $i_u = r_m$ and $j_u = r_{m+n}$ we have

$$
\begin{aligned}
(x_u)(f_{i_u,j_u}\varepsilon_{j_u} - \varepsilon_{i_u}) &= (x_u)(f_{r_m,r_{m+n}}\varepsilon_{r_{m+n}} - \varepsilon_{r_m}) \\
&= \left[ (x_u)f_{r_m,r_{m+1}}\varepsilon_{r_{m+1}} - (x_u)\varepsilon_{r_m} \right] \\
&\quad + \left[ (x_u)f_{r_m,r_{m+1}})f_{r_{m+1},r_{m+2}}\varepsilon_{r_{m+2}} - (x_u)f_{r_m,r_{m+1}}\varepsilon_{r_{m+1}} \right] \\
&\quad + \cdots + \left[ (x_u)f_{r_m,r_{m+n-1}})f_{r_{m+n-1},r_{m+n}}\varepsilon_{r_{m+n}} - (x_u)f_{r_m,r_{m+n-1}}\varepsilon_{r_{m+n-1}} \right]
\end{aligned}
$$

and from this it follows that

$$x \in \sum_{p=1}^{s-1} \text{Im}\,(f_{r_p,r_{p+1}}\varepsilon_{r_{p+1}} - \varepsilon_{r_p}).$$

Then, since $x$ also belongs to $\bigoplus_{k \leq l} M_l$, a simple comparison shows that $x$ must be zero, as required. $\qquad\square$

We now state and prove Gómez Pardo and Guil Asensio's theorem.

**13.9. Indecomposables with well-behaved clones are LE-modules.** *Let $M$ be an indecomposable $R$-module. Suppose that, for each index set $I$, every local summand of $M^{(I)}$ is a summand. Then $M$ is an LE-module.*

*Proof.* Given $f \in \text{End}(M)$, we must show that either $f$ or $1-f$ is an isomorphism. Let $K = \bigcup_{n \in \mathbb{N}} \text{Ke}\,(f^n)$.

Now for the easy part. If $K = M$, then, given any $x \in M$, there is a $k \in \mathbb{N}$ for which $(x)f^k = 0$. This allows us to define $h \in \text{End}(M)$ by setting $(x)h = $

$x + \sum_{n \in \mathbb{N}} (x) f^n$. It is now straightforward to check that $h$ is an inverse for $1 - f$ and so we are done.

Next we consider the harder situation of $K \neq M$. For each ordinal $\alpha$, we set $N_\alpha = M$. We will now define, by transfinite induction, homomorphisms $h_{\alpha,\beta} : N_\alpha \to N_\beta$ for every pair of indices $\alpha \leq \beta$ so that the following two properties are satisfied.

(∗) For each ordinal $\gamma$, $h_{\gamma,\gamma+1} = f$ and the family

$$D_\gamma = \{ h_{\alpha,\beta} : N_\alpha \to N_\beta \mid \alpha \leq \beta \leq \gamma \}$$

is a direct system.

(∗∗) There is an element $x \in N_0$ for which $(x)h_{0,\alpha} \notin K$ for every ordinal $\alpha$.

We begin by setting $h_{\alpha,\alpha} = 1_{N_\alpha}$ and $h_{\alpha,\alpha+1} = f$ for every ordinal $\alpha$ (with $h_{0,1} = f$ in particular).

Now let $\gamma$ be a fixed ordinal and suppose that for each $\delta < \gamma$ we have already defined $h_{\alpha,\beta} : N_\alpha \to N_\beta$ for all $\alpha \leq \beta \leq \delta$ meeting the above requirements. Then, to complete our inductive construction, we must define $H_{\alpha,\gamma} : N_\alpha \to N_\gamma$ for each $\alpha < \gamma$ so that $(x)h_{0,\gamma} \neq 0$ and $D_\gamma = \{ h_{\alpha,\beta} : N_\alpha \to N_\beta \}_{\alpha \leq \beta \leq \gamma}$ is a direct system.

If $\gamma$ is not a limit ordinal, say $\gamma = \psi + 1$, then we simply define $h_{\alpha,\gamma} = h_{\alpha,\psi} \diamond f$ for each $\alpha \leq \psi$. Then it is easy to check that, since $D_\psi$ is a direct system, this creates a direct system $D_\gamma$. Moreover, since $(x)h_{0,\psi} \notin K$, it follows that $(x)h_{0,\gamma} \notin K$, and so both (∗) and (∗∗) hold.

Now suppose that $\gamma$ is a limit ordinal. Then, by our inductive hypothesis, $D_\delta$ is a direct system and $(x)h_{0,\delta} \notin K$ for all $\delta < \gamma$. From this one easily sees that $\{ h_{\alpha,\beta} : N_\alpha \to N_\beta \mid \alpha \leq \beta < \gamma \}$ is also a direct system and we denote its limit by $Y$. For each $\alpha < \gamma$, let $u_\alpha : N_\alpha \to Y$ denote the canonical map, so that $u_\alpha = u_\beta \diamond h_{\alpha,\beta}$ for all $\alpha \leq \beta < \gamma$.

By Lemma 13.8, we have an exact sequence

$$0 \to Z \to \bigoplus_{\alpha < \gamma} N_\alpha \xrightarrow{\pi} Y \to 0,$$

where $\pi$ is the canonical map and $Z = \mathrm{Ke}\,(\pi)$ is the union of a chain of summands of $\bigoplus_{\alpha < \gamma} N_\alpha$. Then, since by our initial hypothesis every local summand of $\bigoplus_{\alpha < \gamma} N_\alpha$ is a summand, we have by Lemma 13.2 that $Z$ is in fact a summand of $\bigoplus_{\alpha < \gamma} N_\alpha$ and so $\pi$ splits. Thus we have a homomorphism $\theta : Y \to \bigoplus_{\alpha < \gamma} N_\alpha$ for which $\theta \diamond \pi = 1_Y$. We show that $(x)u_0 \diamond \theta \notin \bigoplus_{\alpha < \gamma} K_\alpha \subseteq \bigoplus_{\alpha < \gamma} N_\alpha$ where we have set $K_\alpha = K$ for each $\alpha < \gamma$.

Note first that if $(x)u_0 \diamond \theta = 0$, then, since $\theta$ is a monomorphism, this would give $(x)u_0 = 0$ and so, by the properties of the direct limit, $(x)h_{0,\alpha} = 0$ for some $\alpha < \gamma$, in contradiction to our inductive hypothesis. Thus $(x)u_0 \diamond \theta \neq 0$. Then, since $\mathrm{Im}\,(u_0 \diamond \theta) \subseteq \mathrm{Im}\,(\theta)$ and $\mathrm{Im}\,(\theta) \cap Z = 0$, it follows that $(x)u_0 \diamond \theta \notin Z$. Thus, to establish the claim above, it is enough to show that $\bigoplus_{\alpha < \gamma} K_\alpha \subseteq Z$.

For this, let $y \in K$. Then, by definition, $(y)f^n = 0$ for some $n \in \mathbb{N}$ and so, since $h_{\alpha,\alpha+n} = h_{\alpha,\alpha+1}h_{\alpha+1,\alpha+2}\cdots h_{\alpha+n-1,\alpha+n} = f^n$, we have $(y)h_{\alpha,\alpha+n} = 0$. Thus, with $\varepsilon_\alpha : K_\alpha \to \bigoplus_{\alpha<\gamma}K_\alpha$ and $\epsilon_\alpha : N_\alpha \to \bigoplus_{\alpha<\gamma}N_\alpha$ denoting the natural injection maps for each $\alpha < \gamma$, we get

$$(y)\varepsilon_\alpha = (y)\epsilon_\alpha = (y)(\epsilon_\alpha - h_{\alpha,\alpha+n} \diamond \epsilon_{\alpha+n}) \in Z,$$

and this establishes $\bigoplus_{\alpha<\gamma}K_\alpha \subseteq Z$.

We have now reached our interim goal that $(x)u_0 \diamond \theta \notin \bigoplus_{\alpha<\gamma}K_\alpha$ and so, for some $\tau < \gamma$, there is a canonical projection $p_\tau : \bigoplus_{\alpha<\gamma}N_\alpha \to N_\tau$ for which $(x)u_0 \diamond \theta \diamond p_\tau \notin K$. Then, setting $h_{\alpha,\gamma} = u_0 \diamond \theta \diamond p_\tau : N_\alpha \to N_\tau = M = N_\gamma$, we have for any $\alpha \le \beta \le \gamma$, that

$$h_{\alpha,\gamma} \diamond h_{\beta,\gamma} = \begin{cases} h_{\alpha,\beta} \diamond u_\beta \diamond \theta \diamond p_\tau = u_\alpha \diamond \theta \diamond p_\tau = h_{\alpha,\gamma} & \text{if } \beta < \gamma, \\ h_{\alpha,\beta} \diamond 1_N = h_{\alpha,\gamma} & \text{if } \alpha < \beta = \gamma \end{cases}$$

and so the resulting $D_\gamma$ is a direct system. Moreover, by our definition of $h_{0,\gamma}$ we have $(x)h_{0,\gamma} \ne 0$ for all $\alpha \le \gamma$. This has completed our long desired construction satisfying $(*)$ and $(**)$.

We now establish that there is an ordinal $\mu$ for which, given any pair of ordinals $\alpha, \beta$ with $\mu \le \alpha \le \beta$, then there is an ordinal $\gamma \ge \beta$ such that $N_\beta = \operatorname{Im}(h_{\alpha,\beta}) + \operatorname{Ke}(h_{\beta,\gamma})$. We do this with a proof by contradiction.

Thus suppose there is no such $\mu$. Then there exist ordinals $\alpha_1 < \beta_1$ such that for every $\gamma \ge \beta_1$ we have $N_{\beta_1} \ne \operatorname{Im}(h_{\alpha_1,\beta_1}) + \operatorname{Ke}(h_{\beta_1,\gamma})$. Now, if given any $x \in M$ there exists an ordinal $\gamma$ such that $x \in \operatorname{Im}(h_{\alpha_1,\beta_1}) + \operatorname{Ke}(h_{\beta_1,\gamma})$, then taking $\gamma$ sufficiently large we would get $N_{\beta_1} = \operatorname{Im}(h_{\alpha_1,\beta_1}) + \operatorname{Ke}(h_{\beta_1,\gamma})$, contrary to our assumption. Thus there is an element $x_1 \in N_{\beta_1}$ such that

$$x_1 \notin \operatorname{Im}(h_{\alpha_1,\beta_1}) + \operatorname{Ke}(h_{\beta_1,\gamma})$$

for all $\gamma \ge \beta_1$. Then, again using our supposed nonexistence of $\mu$, we can choose ordinals $\alpha_2, \beta_2$ such that $\beta_1 < \alpha_2 < \beta_2$, and an element

$$x_2 \notin \operatorname{Im}(h_{\alpha_2,\beta_2}) + \operatorname{Ke}(h_{\beta_2,\gamma})$$

for all $\gamma \ge \beta_2$. Proceeding in this way, by transfinite induction we obtain, for each ordinal $\iota$, ordinals $\alpha_\iota < \beta_\iota$ and an element $x_\iota \in N_{\beta_\iota} = M$ such that the following two conditions hold:

(†)  $x_\iota \notin \operatorname{Im}(h_{\alpha_2,\beta_2}) + \operatorname{Ke}(h_{\beta_2,\gamma})$ for all $\gamma \ge \beta_\iota$ and
(‡)  $\alpha_\jmath < \beta_\jmath < \alpha_\iota < \beta_\iota$ for $\jmath < \iota$.

Now, for any ordinal $\xi$ and any $\jmath < \iota < \xi$, if $(x_\jmath)h_{\beta_\jmath,\beta_\xi} = (x_\iota)h_{\beta_\iota,\beta_\xi}$, then $0 = (x_\iota - (x_\jmath)h_{\beta_\jmath,\beta_\iota})h_{\beta_\iota,\beta_\xi}$. Then, since $\beta_\jmath < \alpha_\iota$ and so $\operatorname{Im}(h_{\beta_\jmath,\beta_\iota}) \subseteq \operatorname{Im}(h_{\alpha_\iota,\beta_\iota})$, this gives $x_\iota \in \operatorname{Im}(h_{\beta_\jmath,\beta_\iota}) + \operatorname{Ke}(h_{\beta_\iota,\beta_\xi}) \subseteq \operatorname{Im}(h_{\alpha_\iota,\beta_\iota}) + \operatorname{Ke}(h_{\beta_\iota,\beta_\xi})$, in contradiction to the definition of $x_\iota$. This contradiction shows that $(x_\jmath)h_{\beta_\jmath,\beta_\xi} \neq (x_\iota)h_{\beta_\iota,\beta_\xi}$ for any ordinals $\jmath < \iota < \xi$. Hence the set $\{(x_\iota)h_{\beta_\iota,\beta_\xi} : \iota < \xi\}$ has the same cardinality as $\xi$. However, since $\xi$ is arbitrary, this then implies that $M$ has unlimited cardinality. This contradiction has now established the existence of an ordinal $\mu$ with the properties we stated above, namely that for all ordinal pairs $\alpha, \beta$ with $\mu \leq \alpha < \beta$, there is an ordinal $\gamma \geq \beta$ for which $N_\beta = \text{Im}\,(h_{\alpha,\beta}) + \text{Ke}\,(h_{\beta,\gamma})$.

Now, since $Ke(h_{\mu,\alpha_1}) \subseteq \text{Ke}\,(h_{\mu,\alpha_2})$ whenever $\mu < \alpha_1 < \alpha_2$, these kernels form an increasing chain of subsets of $M$ and so there must be an ordinal $\kappa > \mu$ for which $\text{Ke}\,(h_{\mu,\lambda}) = \text{Ke}\,(h_{\mu,\kappa})$ for any ordinal $\lambda > \kappa$. Then, by definition of $\mu$, there is an ordinal $\gamma \geq \kappa + 1$ such that

$$N_{\kappa+1} = \text{Im}\,(h_{\mu,\kappa+1}) + \text{Ke}\,(h_{\kappa+1,\gamma}).$$

This sum is direct. To see this, suppose that $y \in \text{Im}\,(h_{\mu,\kappa+1})$ and $(y)h_{\kappa+1,\gamma} = 0$. Then $y = (z)h_{\mu,\kappa+1}$ for some $z \in N_\mu$ and so

$$0 = (y)h_{\kappa+1,\gamma} = (z)h_{\mu,\kappa+1} \diamond h_{\kappa+1,\gamma} = (z)h_{\mu,\gamma}$$

which gives $z \in \text{Ke}\,(h_{\mu,\gamma}) = \text{Ke}\,(h_{\mu,\kappa+1})$. Thus $y = (z)h_{\mu,\kappa+1} = 0$. This has shown $M = N_{\kappa+1} = \text{Im}\,(h_{\mu,\kappa+1}) \oplus \text{Ke}\,(h_{\kappa+1,\gamma})$. Then, since $0 \neq (x)h_{0,\kappa+1} \in \text{Im}\,(h_{0,\kappa+1}) \subseteq \text{Im}\,(h_{\mu,\kappa+1})$ and $M$ is indecomposable, our direct sum implies that $M = \text{Im}\,(h_{\mu,\kappa+1})$ and $\text{Ke}\,(h_{\kappa+1,\gamma}) = 0$. Since, by definition, $h_{\mu,\kappa+1} = h_{\mu,\kappa} \diamond f$, we have $\text{Im}\,(h_{\mu,\kappa+1}) \subseteq \text{Im}\,f$ and so $f$ is an epimorphism. Furthermore $h_{\kappa+1,\gamma} = h_{\kappa+1,\kappa+2} \diamond h_{\kappa+1,\gamma} = f \diamond h_{\kappa+2,\gamma}$ and so, since $\text{Ke}\,(h_{\kappa+1,\gamma}) = 0$, this shows that $f$ is a monomorphism. Hence $f$ is an isomorphism, completing the proof. $\square$

**13.10. Add $M$ and perfect decompositions.** For any $R$-module $M$, the class Add $M$ is defined to consist of all $R$-modules which are isomorphic to direct summands of direct sums of copies of $M$. Of course, Add $R$ is just the class of all projective $R$-modules.

Following Angeleri-Hügel and Saorín [15], if every module $A$ in Add $M$ has a decomposition that complements direct summands we say that $M$ has a *perfect decomposition*.

With these definitions, we are now able to give an interesting corollary to the previous theorem.

**13.11. Perfect decompositions.** *The following are equivalent for any module $M$.*

(a) *Every local summand of any module $N$ in* Add $M$ *is a summand;*

(b) *every nonzero module $N$ in* Add $M$ *has an LE-decomposition $N = \bigoplus_I N_i$ for which $\{N_i\}_I$ is locally semi-T-nilpotent;*

(c) *$M$ has a perfect decomposition;*

(d) *$M$ has an indecomposable decomposition and every module $N$ in* Add $M$ *has the (finite) exchange property.*

*Proof.* We first note that each of the properties expressed in statements (a)–(d) are inherited by direct summands. We detail this for (b). (See 11.8 and Exercises 13.13 (3), (7) for the others.)

Thus suppose that $N$ is a module with an LE-decomposition $N = \bigoplus_I N_i$ for which $\{N_i\}_I$ is locally semi-T-nilpotent. Let $N = A \oplus B$. We wish to show that $A$ has the same properties as $N$. By Azumaya's Theorem 12.6, the decomposition $N = \bigoplus_I N_i$ complements maximal direct summands and every nonzero summand of $N$ contains an indecomposable summand. Then, by Dung's result 13.6, the decomposition complements all summands and so from this it follows that $A$ has a decomposition of the form $A = \bigoplus_J A_j$ where $J \subseteq I$ and $A_j \simeq M_j$ for each $j \in J$. Clearly this gives an LE-decomposition of $A$ satisfying local semi-T-nilpotency, as required.

Since each of the properties is inherited by summands it now suffices to show that they are equivalent for all the clones $M^{(I)}$ of $M$ (for each index set $I$). Now if $M^{(I)}$ has an LE-decomposition, then (a), (b), and (c) are equivalent by Dung's result 13.6 and these are equivalent to (d) by the Zimmermann-Huisgen–Zimmermann result 12.14. Thus it suffices to show that, in each of the four cases, $M^{(I)}$ has an LE-decomposition.

Firstly, if every local summand of $M^{(I)}$ is a summand, then the same is true for $M$ itself and so, by 13.3, we may write $M = \bigoplus_\Lambda M_\lambda$ where each $M_\lambda$ is indecomposable. Then $M^{(I)} = \bigoplus_\Lambda M_\lambda^{(I)}$ for each $I$ and, moreover, every local summand of $M_\lambda^{(I)}$ is also a summand. Thus, by 13.9, each $M_\lambda$ is an LE-module, as required.

In the case of (b) there is nothing to prove.

If (c) holds, then $M^{(I)}$ has a decomposition which complements direct summands for each index set $I$ and so, in particular, $M$ also has such a decomposition, say $M = \bigoplus_\Lambda M_\lambda$. Then, because all indecomposable decompositions of $M^{(I)}$ are equivalent by 12.5 (1), it follows from 12.5 (3) applied to $M^{(I)} = \bigoplus_\Lambda M_\lambda^{(I)}$ that each $M_\lambda$ is an LE-module, as required.

Finally if (d) holds, then the summands in any indecomposable decomposition of $M^{(I)}$ must all have the (finite) exchange property by 11.8. Thus, by 12.2, each of these summands is an LE-module.                                          $\square$

Theorem 28.14 of [13] characterises left perfect rings $R$ as those for which every projective left $R$-module has a decomposition which complements direct summands. Taking $M = {}_R R$ in the above theorem we obtain the following expansion of this characterisation (and we have further support for the term *perfect decomposition* introduced above).

**13.12. Projective modules with perfect decompositions.** *The following statements for a ring $R$ are equivalent.*

(a) *$R$ is left perfect;*

(b) *every local summand of any projective left R-module P is a summand;*

(c) *every nonzero projective left R-module P has an LE-decomposition*
$P = \bigoplus_I P_i$ *for which* $\{P_i\}_I$ *is locally semi-T-nilpotent;*

(d) $_RR$ *has a perfect decomposition;*

(e) $_RR$ *has an indecomposable decomposition and every projective left R-module P has the (finite) exchange property.*

**13.13. Exercises.**

(1) Prove that any nonzero module $M$ contains an indecomposable submodule.

(2) A module is called *super-decomposable* if it has no indecomposable direct summands.

  (i) For any field $K$, show that the regular ring $K^{\mathbb{N}}/K^{(\mathbb{N})}$ is super-decomposable as a module over itself.

  (ii) Let $S$ denote the multiplicative monoid of non-negative rational numbers with product defined by $a \cdot b = \max(a, b)$ for all $a, b \in S$. Show that, for any field $K$, the monoid ring $K[S]$ is a super-decomposable module over itself by showing that every nonzero idempotent can be non-trivially decomposed into a sum of idempotents. (See [180].)

(3) Let $M$ be a module for which every local summand is a summand. Prove that any nonzero direct summand $D$ of $M$ also has this property.

(4) Let $R$ be the ring of eventually constant sequences $r = (s_1, s_2, \ldots, s_n, t, t, t, \ldots)$ where $s_1, s_2, \ldots, s_n \in \mathbb{Q}$, the eventual constant $t$ is from $\mathbb{Z}$, and $n$ depends on $r$. Show that the $R$-module $R$ has a local summand which is not a summand.

The next four exercises appear in §12 of [13].

(5) Let $f : M \to N$ be a module isomorphism. Prove that a decomposition $M = \bigoplus_I M_i$ complements (maximal) summands if and only if $N = \bigoplus_I (M_i)f$ complements (maximal) summands.

(6) A family of $R$-modules $\{M_i\}_I$ is said to be *homologically independent* provided $\operatorname{Hom}_R(M_i, M_j) = 0$ for any pair of distinct indices $i, j \in I$. Let $\{M_i\}_I$ be such a family with $M = \bigoplus_I M_i$.

  (i) Assume that each $M_i$ is indecomposable and let $\{e_i\}_I$ be the family of idempotents in $\operatorname{End}(M)$ given by the decomposition. Show that if $e$ is any idempotent in $\operatorname{End}(M)$, then $e_i e e_j = 0$ for any distinct $i, j \in I$.

  (ii) Show that if each $M_i$ is indecomposable, then for any nonzero summand $K$ of $M$ we have $K = \bigoplus_J M_j$ for some (not necessarily unique) subset $J \subseteq I$.

  (iii) Show that if each $M_i$ has an indecomposable decomposition that complements (maximal) summands, then so does $M$.

  (iv) Give an example of an indecomposable decomposition $M = \bigoplus_I M_i$ where $I$ is infinite which complements direct summands but no $M_i$ is an LE-module.

(7) Let $M$ be a module for which every direct summand has an indecomposable decomposition. Prove that if $M$ has a decomposition which complements maximal summands, then so does every summand of $M$.

(8) Use 12.5 (3) to show that the $\mathbb{Z}$-module $\mathbb{Z} \oplus \mathbb{Z}$ does not have an indecomposable decomposition which complements maximal direct summands. As an alternative, find a maximal summand of $\mathbb{Z} \oplus \mathbb{Z}$ which is not complemented by the decomposition $\mathbb{Z}(1,1) \oplus \mathbb{Z}(1,2)$.

(9) Show that a module $M$ satisfies the maximum condition on direct summands (with respect to inclusion) if and only if it satisfies the minimum condition on direct summands. (See [195].)

(10) Let $M$ be a module which satisfies the maximum condition on direct summands. Then (see 11.42 (8)) show that $M$ has the 2-exchange property if and only if it has the D2-exchange property. (See [195, Theorem 2.12].)

(11) Let $\{M_n\}_\mathbb{N}$ be a locally semi-T-nilpotent family. Let $\mathbb{N} = \bigcup_{k \in \mathbb{N}} J_k$ be a partition of $\mathbb{N}$ into infinitely countably many nonempty subsets $J_k$. For each $k \in \mathbb{N}$, set $N_k = \bigoplus_{j_k \in J_k} M_{j_k}$. Prove that $\{N_k\}_{k \in \mathbb{N}}$ is locally semi-T-nilpotent as follows. Consider a sequence of non-isomorphisms

$$N_1 \xrightarrow{f_1} N_2 \xrightarrow{f_2} \cdots \xrightarrow{f_{k-1}} N_k \xrightarrow{f_k} \cdots .$$

For each $k \in \mathbb{N}$ and each $j_k \in J_k$, let $\varepsilon_{j_k} : M_{j_k} \to N_k$ and $\pi_{j_k} : N_k \to M_{j_k}$ be the natural injection and projection maps and set $e_{j_k} = \pi_{j_k} \varepsilon_{j_k} : N_k \to N_k$. Given any $x \in N_1$, for any $n \in \mathbb{N}$ let $P_{x,n}$ be the set of all sequences $(j_1, j_2, \ldots, j_n) \in J_1 \times J_2 \times \cdots \times J_n$ such that

$$(x)e_{j_1} f_1 e_{j_2} \cdots e_{j_{n-1}} f_{n-1} e_{j_n} \neq 0.$$

Show that $(x)f_1 \cdots f_n = \sum \left\{ (x)e_{j_1} f_1 \cdots e_{j_{n-1}} f_{n-1} e_{j_n} \mid (j_1, \ldots, j_{n-1}, j_n) \in P_{x,n} \right\}$. Next let $G$ be the König graph $(P_1, P_2, \ldots, P_n, \ldots)$ where $P_n = P_{x,n}$ for each $n \in \mathbb{N}$. Then apply the König Graph Theorem (31.1) to $G$ and the local semi-T-nilpotency of $\{M_n\}_\mathbb{N}$ to show that $(x)f_1 \cdots f_n = 0$ for some $n \in \mathbb{N}$.
(See also Lemma 14.18 and the second half of the proof of 14.20.)

(12) Let $V$ be an infinite-dimensional left vector space over a division ring $D$ and let $R$ be the ring $\mathrm{End}_D(V)$ of linear transformations on $V$. It is well-known that the (right and left) socle $\mathrm{Soc}\, R$ of $R$ consists of those transformations with finite-dimensional image. Show that $\mathrm{Soc}\, R$ is a local direct summand of $R$ which is not a summand of $R$. (Cf. Exercise 12.19 (6).)

(13) Let $R$ be a ring with Jacobson radical $J$. (See [243, Lemmas 1, 2].)

　(i) If $e^2 = e \in R$ and $a \in R$ with $eae - e \in J$, show that there exists $f = f^2 \in aRe$ with $Re = Rf$. (Hint: first show that $x = eae$ has an inverse $y$ in the subring $eRe$ and then use $f = ay$.)

　(ii) Let $\{e_i\}_I$ be a family of idempotents in $R$ such that $e_j e_i \in J$ for $i < j$, for some total order on the index set $I$. Using (i), prove that $\sum_I Re_i$ is direct and a local direct summand of $R$. (Hint: for $I = \{1, 2\}$, show that there are idempotents $f_1, f_2$ with $f_1 \in (1 - e_2)Re_1$, $f_2 \in (1 - e_1)Re_2$, $Rf_i = Re_i$ and then that $R = Rf_1 \oplus Rf_2 \oplus R(1 - f_1 - f_2)$.)

(14) Azumaya [26] says that a submodule $N$ of a module $M$ is *locally split* if, for each $x \in N$, there is a homomorphism $h : M \to N$ such that $(x)h = x$. (Ramamurthi and Rangaswamy [292] say that $N$ is a *strongly pure* submodule.)

(i) Show that, if $x_1, \dots, x_n$ are elements of the locally split submodule $N$ of $M$, there is a homomorphism $h : M \to N$ such that $(x_i)h = x_i$ for $i = 1, \dots, n$.

(ii) Prove that every locally split submodule is a pure submodule.

(iii) Show that every local summand $N = \bigoplus_I N_i$ of $M$ is locally split.

(15) Paraphrasing Guil Asensio and Herzog [137], a pair $(aR, bR)$ of principal right ideals of a ring $R$ is said to be *comaximal* if $aR + bR = R$ and we may define a partial ordering $\leq$ on the set of all such pairs by setting

$$(a_1 R, b_1 R) \leq (a_2 R, b_2 R) \text{ if and only if } a_1 R \subseteq a_2 R \text{ and } b_1 R \subseteq b_2 R.$$

We also write

$$(a_1 R, b_1 R) < (a_2 R, b_2 R)$$
$$\text{if } (a_1 R, b_1 R) \leq (a_2 R, b_2 R) \text{ but not } (a_2 R, b_2 R) \leq (a_1 R, b_1 R).$$

A comaximal pair $(aR, bR)$ is called *minimal* if there are no comaximal pairs strictly less than $(aR, bR)$.

(i) Show that a pair $(aR, bR)$ of principal right ideals is comaximal if and only if the left $R$-module homomorphism $\mu_{a,b} : R \to R \oplus R$ given by $(t)\mu_{a,b} = (ta, tb)$ for all $t \in R$ is a split monomorphism.

(ii) Show that if $R = aR \oplus bR$, then $(aR, bR)$ is a minimal comaximal pair.

(iii) (Cf. 11.16 (c).) Given $a \in R$, show that the comaximal pair $(aR, (1 - a)R)$ is minimal if and only if $aR = eR$ for some idempotent $e \in R$.

(iv) Show that every minimal comaximal pair $(aR, bR)$ yields a decomposition $R = aR \oplus (1 - a)R$.

**13.14. Comments.** Local direct summands seem to have been considered first by Ishii [176] and the Zorn's Lemma argument used in the proof of 13.2 appears there (see also Lemma 2.16 of Mohamed and Müller's text [241]). Harada uses the concept extensively and effectively in proving his important precursor to Dung's result appearing as Theorems 7.3.15 and 8.2.1 in his text [158]. In [36] Beck approaches local direct summands through the more general concept of an *independence structure* of subsets of a set. Further details on both Harada and Beck's approach are given in a report by Zöllner [386].

The decomposability result 13.3 appears as Oshiro's [271, Lemma 2.4], but a Grothendieck category theory variant features earlier as Theorem 1 of Stenström [322]; here we have followed its presentation in Mohamed and Müller [241, Theorem 2.17]. Gomez Pardo and Guil Asensio use it in [126] to show that, given any module $M$, there is an infinite cardinal $\aleph$ such that if $M$ is $\aleph$-$\Sigma$-extending, then $M$ is in fact $\Sigma$-extending.

As already noted, given any infinite family $\{M_i\}_I$ of $R$-modules, $\bigoplus_I M_i$ is a local summand of $\prod_I M_i$, but it is seldom a summand. For instance, Zimmermann-Huisgen [384] shows that if $R$ is regular, then splitting occurs if and only if almost all the $M_i$ are semisimple with only finitely many isomorphism types of simple submodules in a cofinite summand. Camillo [51] gives a necessary condition for this splitting for general $R$ while Lenzing [219] shows that if $R^{(\mathbb{N})}$ is a direct summand of $R^{\mathbb{N}}$ as left $R$-modules, then $R$ is semiprimary with the ascending chain condition on left annihilators. Of course, any internal direct sum of injective submodules of a module $M$ is also a local direct summand of $M$.

Dung's result 13.6 and its essential ingredients 13.4 and 13.5 appear in [83]. The Theorem is important for two reasons. Firstly, it generalises part of the earlier theorem of Harada mentioned above, by replacing Harada's underlying LE-decomposition hypothesis to the weaker complementing maximal summands condition. Secondly, Dung's module-theoretic proof is quite different from Harada's, which uses his theory of factor categories. Using 13.6, Dung [83, Theorem 4.4] gives a complete characterisation of extending modules which have indecomposable decompositions that complement maximal direct summands.

Lemma 13.8 appears as Lemma 2.1 in Gomez Pardo and Guil Asensio's [125], (here modified slightly to become Lemma 1.2 of Angeleri Hügel and Saorín's Lemma 1.2 in [15]). Their Theorem 13.9 and its Corollary 13.11 are the main results in [125]. Corollary 13.12 has a long pedigree: using Bass's fundamental results in [34], Anderson and Fuller [12, Theorem 6] proved that $R$ is left perfect if and only if (in Angeleri Hügel and Saorín's terminology) $_RR$ has a perfect decomposition; Stenström [322, Lemma 2] proved that $R$ is left perfect if and only if the union of any chain of summands of any projective left $R$-module $P$ is also a summand of $P$; Harada and Ishii [159] proved that if $R$ is left perfect, then every projective left $R$-module has the exchange property and $\text{End}(P)/J(\text{End}(P))$ is von Neumann regular; independently, Yamagata [371] also showed that if $R$ is left perfect, then every projective left $R$-module has the exchange property and that the converse holds when the identity of $R$ is a finite sum of primitive orthogonal idempotents (in other words, $_RR$ has an indecomposable decomposition) (see also Theorem 8.5.12 in Harada's text [158]); the equivalence of conditions (a) and (e) of 13.12 is also a consequence of Zimmermann-Huisgen–Zimmermann's result 12.14. As noted in [385] and [125], if $_RR$ has an indecomposable decomposition, the exchange property for all projective $R$-modules follows from the property for the free $R$-module $R^{(\mathbb{N})}$.

On the other hand, if $_RR$ does not have an indecomposable decomposition, the equivalence of (a) and (e) in 13.12 is false in general. This was first shown by Kutami and Oshiro [215] who constructed a nonartinian Boolean ring all of whose projective modules have the exchange property. Oshiro [270] then showed that every projective module over a (von Neumann) regular ring has the *finite* exchange property. This was generalised by Stock [324] who proved that if $R/J(R)$ is regular and $J(R)$ is left T-nilpotent, then every projective $R$-module has the full exchange property, that is, in Stock's terminology, $R$ is a *P-exchange ring*. In fact, he obtains this from his [324, Theorem 3.1] which shows that if $M$ is a projective module for which every finitely generated submodule is a summand (in other words, a projective regular module), then $M$ has the exchange property.

Using Lemma 13.8, Guil Asensio has shown in [136, Theorem 3] that if $M$ is a uniform module for which every local summand of $M^{(\mathbb{N})}$ is a summand, then $M$ is an LE-module. This then implies that the following conditions are equivalent if $M$ is uniform: (a) $M$ is $\Sigma$-quasi-injective, (b) $M$ is $\Sigma$-extending, (c) $M$ is $\aleph_1$-$\Sigma$-extending, (d) $M$ is countably $\Sigma$-extending and every local summand of $M^{(\mathbb{N})}$ is a summand. Moreover, taking 13.9 also into account, he asks if $M$ is any indecomposable module for which every local summand of $M^{(\mathbb{N})}$ is a summand, must $M$ be an LE-module.

A ring $R$ is said to be *left cotorsion* if $\text{Ext}_R^1(F, {}_R R) = 0$ for every flat left $R$-module $F$. Using right comaximal pairs (see Exercise 15 above) and a transfinite induction argument similar to that used in the proof of 13.9, Guil Asensio and Herzog [137] show that if $R$ is a left cotorsion ring with Jacobson radical $J$, then $R$ is semiregular and the factor ring $R/J$ is left self-injective.

**References.** Anderson and Fuller [12]; Angeleri Hügel and Saorín [15]; Bass [34]; Beck [36]; Dung [83]; Gomez Pardo and Guil Asensio [125, 126]; Guil Asensio and Herzog [137]; Harada [158]; Harada and Ishii [159]; Ishii [176]; Kutami and Oshiro [215]; Mohamed and Müller [241]; Oshiro [271]; Stenström [322]; Stock [324]; Yamagata [371]; Zimmermann-Huisgen [384]; Zimmermann-Huisgen–Zimmermann [385]; Zöllner [386].

# 14   The total and LE-decompositions

Let $M = \bigoplus_I M_i$ be an LE-decomposition of the module $M$ and $S = \mathrm{End}(M)$. The main goal of this section is to show that $\{M_i\}_I$ is locally semi-T-nilpotent if and only $S$ is semiregular. This result was first proved by Harada (see [158]) using factor categories. Here our approach is due to Kasch using the total, a concept introduced by him in 1982.

We begin with a simple lemma which will lead to the definition of the total.

**14.1. Lemma.** *Given $R$-modules $M$ and $N$ and a homomorphism $f : M \to N$, the following statements are equivalent.*

(a) *There is a homomorphism $g : N \to M$ for which $e = fg$ is a nonzero idempotent endomorphism of $M$;*

(b) *there is a homomorphism $h : N \to M$ for which $d = hf$ is a nonzero idempotent endomorphism of $N$;*

(c) *there is a nonzero homomorphism $k : N \to M$ for which $kfk = k$;*

(d) *there are nonzero direct summands $A$ of $M$ and $B$ of $N$ for which the restriction $f|_A$ is an isomorphism from $A$ to $B$.*

*Proof.* (a)$\Rightarrow$(b). Since $e = fg$ is idempotent, $fge = e$ and so, if we set $d = gef$ we have $d^2 = (gef)(gef) = ge(fge)f = gef = d$ and $fdg = f(gef)g = e^3 \neq 0$. Thus we may take $h = ge$.

(b)$\Rightarrow$(c). Setting $k = dh$ gives $kfk = dhfdh = d^3h = dh = k$ and $k \neq 0$ since $kf = d^2 \neq 0$.

(c)$\Rightarrow$(d). Set $A = (M)fk$ and $B = (N)kf$. Then, since $fk$ and $kf$ are nonzero idempotents in $\mathrm{End}(M)$ and $\mathrm{End}(N)$ respectively, $A$ and $B$ are nonzero direct summands of $M$ and $N$ respectively. Now, given $a \in A$, say $a = (m)fk$ where $m \in M$, we have $(a)f = (m)fkf \in B$ and so $(A)f|_A \subseteq B$. Moreover, if $(a)f = 0$, then $a = (m)fk = (m)fkfk = (a)fk = 0$, so $f|_A$ is a monomorphism. Finally, given $b \in B$, we have $b = (n)kf$ for some $n \in N$ and so $b = (n)kfkf = (a)f$ where $a = (n)kfk \in (M)fk = A$, and this shows that $f_A : A \to B$ is onto.

(d)$\Rightarrow$(a). By hypothesis we have decompositions $M = A \oplus A_1$ and $N = B \oplus B_1$ where $A \neq 0$. Let $i_A : A \to M$ and $\pi_B : N \to B$ be the associated inclusion and projection maps and let $\phi = f|_A$ be the given isomorphism from $A$ to $B$, so that $\phi = i_A f \pi_B$. Then, setting $g = \pi_B \phi^{-1} i_A$, we have

$$(fg)^2 = f\pi_B \phi^{-1} i_A f \pi_B \phi^{-1} i_A = f\pi_B \phi^{-1} \phi \phi^{-1} i_A = f\pi_B \phi^{-1} i_a = fg$$

so that $fg$ is an idempotent endomorphism of $M$. Moreover, since

$$i_A fg = i_A f\pi_B \phi^{-1} i_A = \phi \phi^{-1} i_A = i_A,$$

we have $fg \neq 0$.                                                                                        $\square$

Applying the Lemma to the case where $M = N = R$ gives

**14.2. Corollary.** *The following statements are equivalent for an element $r$ of $R$.*

(a) *There is an element $s \in R$ for which $e = rs$ is a nonzero idempotent of $R$;*

(b) *there is an element $t \in R$ for which $d = tr$ is a nonzero idempotent of $R$;*

(c) *there is a nonzero element $u \in R$ for which $uru = u$;*

(d) *there are nonzero left ideals $L_1, L_2$ of $R$, both summands of $_RR$, for which the homomorphism $f : L_1 \to L_2$, $x \mapsto xr$ $(x \in L_1)$, is an isomorphism.*

**14.3. Partial invertibility and the total.** A module homomorphism $f : M \to N$ satisfying the equivalent conditions of Lemma 14.1 is said to be *partially invertible, pi* for short; the homomorphism $g$ of the lemma is then called a *right partial inverse* of $f$ while $f$ is called a *left partial inverse* of $g$.

For $R$-modules $M$ and $N$, the *total* of $M$ to $N$ is defined as

$$\mathrm{Tot}\,(M, N) = \{f \in \mathrm{Hom}(M, N) \mid f \text{ is not pi }\}.$$

An element $r \in R$ of Corollary 14.2 is said to be *partially invertible*, again *pi* for short; the element $s$ of the Corollary is called a *right partial inverse* of $r$ while $r$ is called a *left partial inverse* of $s$.

The *total* of $R$ is defined as

$$\mathrm{Tot}\,(R) = \{r \in R \mid r \text{ is not pi }\}.$$

We will be particularly interested in the relationship between the Jacobson radical of a ring $R$ and its total. While in general the two are sometimes distinct, we do have the following inclusion.

**14.4. The radical and the total.** *For any ring $R$, $J(R) \subseteq \mathrm{Tot}\,(R)$.*

*Proof.* Let $r$ be an element of $R$ which is not in $\mathrm{Tot}\,(R)$. Then $r$ is a pi element of $R$ and so $rs$ is a nonzero idempotent of $R$ for some $s \in R$. It follows that $rs \notin J(R)$ and so $r \notin J(R)$. $\qquad\square$

**14.5. Remarks.** (1) If $M$ and $N$ are both indecomposable, then Lemma 14.1(d) shows that $\mathrm{Tot}\,(M, N) = \{f \in \mathrm{Hom}(M, N) \mid f \text{ is not an isomorphism }\}$.

Consequently, if the ring $R$ is indecomposable, $\mathrm{Tot}\,(R)$ is the set of non-units of $R$. In particular we see that if $R$ is indecomposable, then $\mathrm{Tot}\,(R)$ is closed under addition if and only if $R$ is a local ring (and in this case $\mathrm{Tot}\,(R) = J(R)$).

(2) While $\mathrm{Tot}\,(R)$ may not be closed under addition, it follows from the next lemma that if $r \in R$ is not pi, then $rs$ and $sr$ are not pi for any $s \in R$. Thus, if $\mathrm{Tot}\,(R)$ *is* additively closed, then it will be a two-sided ideal of $R$.

**14.6. Lemma.** *Let $n \in \mathbb{N}$ and $\{f_i : M_i \to M_{i+i}\}_{i \le n}$ be a chain of $R$-module homomorphisms. Then $f = f_1 f_2 \cdots f_n$ is pi only if each $f_i$ is pi. Equivalently, if $f_i \in \mathrm{Tot}\,(M_i, M_{i+1})$ for some $i$, then $f \in \mathrm{Tot}\,(M_1, M_{n+1})$.*

*Proof.* We use induction on $n$. First, if $n = 2$, then $f = f_1 f_2 : M_1 \to M_3$ is pi and so, by Lemma 14.1 (a), (b), there are homomorphisms $g, h \in \text{Hom}(M_3, M_1)$ for which $fg$ and $hf$ are idempotent endomorphisms. Since $fg = f_1(f_2 g)$ and $hf = (h f_1) f_2$, it follows from Lemma 14.1 (a), (b) that both $f_1$ and $f_2$ are pi.

Now suppose that the result is true for some $n \geq 2$ and $f = f_1 f_2 \cdots f_n f_{n+1}$ is pi. Since $f = (f_1 f_2 \cdots f_n) f_{n+1}$, the $n = 2$ case shows that both $f_1 f_2 \cdots f_n$ and $f_{n+1}$ are pi and then our induction hypothesis completes the proof.         $\square$

We now give a simple but effective lemma on the factorisation of isomorphisms with an indecomposable intermediary. We omit its easy proof.

**14.7. Lemma.** *Let $f : A \to M$ and $g : M \to B$ be nonzero module homomorphisms for which $f \diamond g$ is an isomorphism. Then*

$$M = \text{Im}\,(f) \oplus \text{Ke}\,(g).$$

*Consequently if $M$ is indecomposable, then both $f$ and $g$ are isomorphisms.*

In our next result and its corollary, where we have decompositions $M = \bigoplus_I M_i$ and $N = \bigoplus_J N_j$, for each $i \in I$ and $j \in J$ we let $\pi_i : M \to M_i, \bar{\pi}_j : N \to N_j$, $\varepsilon_i : M_i \to M$, and $\bar{\varepsilon}_j : N_j \to N$ be the induced projection and inclusion maps. Furthermore we let $e_i$ and $\bar{e}_j$ denote the endomorphisms $\pi_i \varepsilon_i$ and $\bar{\pi}_i \bar{\varepsilon}_i$ of $M$ and $N$ respectively, called the *projectors* of the decompositions.

**14.8. Partial invertibility between LE-decompositions.** *Let*

$$M = \bigoplus_I M_i \ \text{and} \ N = \bigoplus_J N_j$$

*be LE-decompositions of the modules $M$ and $N$. Then $f \in \text{Hom}(M, N)$ is partially invertible if and only if, for some $i \in I$ and $j \in J$,*

$$\varepsilon_i\, f\, \bar{\pi}_j : M_i \to N_j$$

*is an isomorphism.*

*Proof.* If, for some $i \in I$ and $j \in J$, $\varepsilon_i\, f\, \bar{\pi}_j$ is an isomorphism, then it is pi and so, by Lemma 14.6, $f$ is also pi.

Conversely, assume that $f$ is pi. Then, by Lemma 14.1 (d), there are decompositions

$$M = A \oplus A_1 \ \text{and} \ N = B \oplus B_1$$

where $A \neq 0$ and $f|_A : A \to B$ is an isomorphism. From Lemma 12.9 it follows that there is an index $k \in I$ for which $M = M_k \oplus A_0 \oplus A_1$ where $A_0 \subseteq A$ respectively and so we have a decomposition

$$A = A_k \oplus A_0 \ \text{for which there is an isomorphism} \ g : A_k \simeq M_k$$

and this induces the decompositions

$$B = (A_k)f \oplus (A_0)f \quad \text{and} \quad N = (A_k)f \oplus (A_0)f \oplus B_1.$$

Let $\tilde{\pi}_k : N \to (A_k)f$ denote the induced projection and $\tilde{\varepsilon}_k : A_k \to M$ be the inclusion map. Then the homomorphism $g\tilde{\varepsilon}_k f\tilde{\pi}_k$, as shown next, is an isomorphism:

$$M_k \xrightarrow{g} A_k \xrightarrow{\tilde{\varepsilon}_k} M \xrightarrow{f} N \xrightarrow{\tilde{\pi}_k} (A_k)f.$$

Let $h$ denote the inverse of $g\tilde{\varepsilon}_k f\tilde{\pi}_k$. Next, for any nonzero $x$ in $M_k$, let $I_x$ and $J_x$ be the finite subsets of $I$ and $J$ given by $\{i \in I \mid (x)ge_i \neq 0\}$ and $\{j \in J \mid (x)gfe_j \neq 0\}$ respectively. Thus

$$(x)g = \sum_{i \in I_x}(x)ge_i \quad \text{and} \quad (x)gf = \sum_{j \in J_x}(x)gf\bar{e}_i. \qquad (\star)$$

For each $i \in I_x$ and $j \in J_x$ we define $s_{ij}$ and $t$ in $\text{End}(M_k)$ by

$$s_{ij} = g\tilde{\varepsilon}_k e_i f\bar{e}_j \tilde{\pi}_k h \quad \text{and} \quad t = 1_{M_k} - \sum_{i \in I_x}\sum_{j \in J_x} s_{ij}.$$

Now from $(\star)$ it follows easily that $(x)t = 0$ and so $t$ is not an isomorphism. Since $\text{End}(M_k)$ is local, this infers that $s_{i_0 j_0}$ must be an isomorphism for some $i_0 \in I_x, j_0 \in J_x$. Then, since

$$s_{i_0 j_0} = g\tilde{\varepsilon}_k e_{i_0} f\bar{e}_{j_0} \tilde{\pi}_k h = g\tilde{\varepsilon}_k \pi_{i_0} \diamond \varepsilon_{i_0} f\tilde{\pi}_{j_0} \bar{\varepsilon}_{j_0} \tilde{\pi}_k h = g\tilde{\varepsilon}_k \pi_{i_0} \diamond \varepsilon_{i_0} f\tilde{\pi}_{j_0} \diamond \bar{\varepsilon}_{j_0} \tilde{\pi}_k h,$$

Lemma 14.7 first implies that $\varepsilon_{i_0} f\tilde{\pi}_{j_0} \bar{\varepsilon}_{j_0} \tilde{\pi}_k h$ is an isomorphism (since $M_{i_0}$ is indecomposable) and then implies that $\varepsilon_{i_0} f\tilde{\pi}_{j_0}$ is an isomorphism (since $N_{j_0}$ is indecomposable), as required. $\qquad \square$

**14.9. The total and LE-decompositions.** *Let $M = \bigoplus_I M_i$ and $N = \bigoplus_J N_j$ be LE-decompositions of the modules $M$ and $N$ and let $S = \text{End}(M)$. Then $\text{Tot}\,(M,N)$ is additively closed and, consequently, $\text{Tot}\,(S) = \text{Tot}\,(M,M)$ is an ideal of $S$.*

*Proof.* Suppose to the contrary that there are $f, g \in \text{Tot}\,(M,N)$ for which $f + g$ is pi. Then, by 14.8, $\varepsilon_i\,(f+g)\,\tilde{\pi}_j : M_i \to N_j$ is an isomorphism for some $i \in I$ and $j \in J$, say with inverse $h$. Then

$$1_{M_i} = \varepsilon_i\,(f+g)\,\tilde{\pi}_j h = \varepsilon_i\,f\,\tilde{\pi}_j h + \varepsilon_i\,g\,\tilde{\pi}_j h$$

and so, since $\text{End}(M_i)$ is local, without loss of generality we may assume that $\varepsilon_i\,f\,\tilde{\pi}_j h$ is an isomorphism and hence pi. However, by Lemma 14.6, this implies that $f$ is pi, a contradiction. $\qquad \square$

While the Jacobson radical of the endomorphism ring of an LE-decomposition may be properly contained in its total, the following lemma shows that a significant part of the total is contained in the radical.

**14.10. Lemma.** *Let $M = \bigoplus_I M_i$ be an LE-decomposition, let $S = \mathrm{End}(M)$ and, for each $i \in I$, let $e_i = \pi_i \varepsilon_i$ be the $i$-th projector. Then*

$$e_i \,\mathrm{Tot}\,(S) + \mathrm{Tot}\,(S) e_i \subseteq J(S).$$

*Proof.* We first show that $1_M - e_i f e_i$ is invertible for any $f \in \mathrm{Tot}\,(S)$ and any $i \in I$.

Since $f \in \mathrm{Tot}\,(S)$, it follows from Lemma 14.6 that $\varepsilon_i f \pi_i \in \mathrm{Tot}\,(S_i)$ where $S_i = \mathrm{End}(M_i)$. However, since $M_i$ is an LE-module, $\mathrm{Tot}\,(S_i) = J(S_i)$ by Remark 14.5 and so $\varepsilon_i f \pi_i \in J(S_i)$. From this we infer that $e_i f e_i = \pi_i \varepsilon_i f \pi_i \varepsilon_i \in J(e_i S e_i)$ and so, noting that $e_i$ is the identity of $e_i S e_i$, it follows that $e_i - e_i f e_i$ is invertible in $e_i S e_i$. Thus, for some $g \in S$ we have

$$(e_i - e_i f e_i) e_i g e_i = e_i = e_i g e_i (e_i - e_i f e_i).$$

From this one easily verifies that $1_M - e_i f e_i$ is invertible in $S$ with inverse $h = 1_M - e_i + e_i g e_i$.

Since $e_i$ is an idempotent, it is now straightforward to show that $1_M - e_i f$ and $1_M - f e_i$ are both invertible, with inverses $1_M + e_i h e_i f$ and $1_M + f e_i h e_i$ respectively. Then, since $\mathrm{Tot}\,(S)$ is closed under right multiplication and left multiplication by elements from $S$ (see Remark 14.5 (2)), it follows that both $e_i \,\mathrm{Tot}\,(S)$ and $\mathrm{Tot}\,(S) e_i$ are contained in $J(S)$. The desired inclusion then follows since $\mathrm{Tot}\,(S)$ is additively closed by Corollary 14.9.                                                                      $\square$

Our next goal is to describe the endomorphism ring $S$ of an LE-decomposition modulo its total (which is an ideal of $S$ by Corollary 14.9). We first split the decomposition into homogeneous components and show that this splitting induces a factorisation of $S/\mathrm{Tot}\,(S)$. To describe this process we now introduce some notation.

Let $M = \bigoplus_I M_i$ be an LE-decomposition. Let $\{M_\lambda\}_\Lambda$ be a set of representatives of the isomorphism classes of the modules in $\{M_i\}_I$. For each $\lambda \in \Lambda$, let $I_\lambda = \{i \in I : M_i \simeq M_\lambda\}$ and set $H_\lambda = \bigoplus_{i \in I_\lambda} M_i$. We refer to the $H_\lambda$'s as the *homogeneous components* of the decomposition. If there is only one such component, the decomposition is said to be *homogeneous*. Now

$$M = \bigoplus_{\lambda \in \Lambda} H_\lambda$$

and, as earlier, for each $\lambda \in \Lambda$, we let $\varepsilon_\lambda : H_\lambda \to M$ and $\pi_\lambda : M \to H_\lambda$ be the associated injection and projection maps and let $e_\lambda$ be the projector $\varepsilon_\lambda \pi_\lambda \in \mathrm{End}(M)$.

Furthermore, for any $\lambda \in \Lambda$ and $k \in I_\lambda$, let $\varepsilon_k^\lambda : M_k \to H_\lambda$ and $\pi_k^\lambda : H_\lambda \to M_k$ be the inclusion and projection maps. Then, for any $k, l \in I_\lambda$,

$$\pi_\lambda \pi_k^\lambda = \pi_k, \qquad \varepsilon_k^\lambda \varepsilon_\lambda = \varepsilon_k, \qquad \varepsilon_l \, e_\lambda = \varepsilon_l \, \pi_\lambda \, \varepsilon_\lambda = \varepsilon_l. \qquad (\circledast)$$

Next set

$$S = \mathrm{End}(M), \quad \overline{S} = S/\mathrm{Tot}\,(S), \quad \text{and} \quad S_\lambda = \mathrm{End}(H_\lambda), \quad \overline{S_\lambda} = S_\lambda/\mathrm{Tot}\,(S_\lambda)$$

for each $\lambda \in \Lambda$. Also, given any $f$ in $S$ we let $\overline{f}$ denote the element $f + \mathrm{Tot}\,(S) \in \overline{S}$, and similarly for $f$ in $S_\lambda$.

**14.11. Factorisation of** $S/\mathrm{Tot}\,(S)$. *Let* $M = \bigoplus_I M_i$ *be an LE-decomposition. Then, using the notation above, there is a ring isomorphism*

$$\theta : \overline{S} \to \prod_{\lambda \in \Lambda} \overline{S_\lambda}$$

*given by* $\theta\left(\overline{f}\right) = \left(\overline{\varepsilon_\lambda f \pi_\lambda}\right)_{\lambda \in \Lambda}$ *for all* $f \in S$.

*Proof.* If $f, g$ are elements of $S$ for which $\overline{f} = \overline{g}$ in $\overline{S}$, then $f - g \in \mathrm{Tot}\,(S)$ and so $f - g$ is not pi. Then, by Lemma 14.6, $\varepsilon_\lambda f \pi_\lambda - \varepsilon_\lambda g \pi_\lambda = \varepsilon_\lambda (f - g) \pi_\lambda \in \mathrm{Tot}\,(S_\lambda)$ for each $\lambda$ and this shows that $\theta$ is well-defined.

It is easily seen that $\theta$ is additive and preserves the identity of $\overline{S}$.

Now suppose that $\theta(\overline{f}) = 0$. Then $\varepsilon_\lambda f \pi_\lambda \in \mathrm{Tot}\,(S_\lambda)$ for each $\lambda$ and so, by Lemma 14.6 and (⊛) above, for each $k, l \in I_\lambda$ we have

$$\varepsilon_k f \pi_l = \varepsilon_k^\lambda \varepsilon_\lambda f \pi_\lambda \pi_l^\lambda \in \mathrm{Tot}\,(M_k, M_l)$$

and so $\varepsilon_k f \pi_l$ is not an isomorphism. On the other hand, if $k \in I_{\lambda_1}$ and $l \in I_{\lambda_2}$ where $\lambda_1$ and $\lambda_2$ are distinct indices from $\Lambda$, then, since $M_k \not\cong M_l$, once again $\varepsilon_k f \pi_l$ is not an isomorphism. Thus, thanks to 14.8, $f \subset \mathrm{Tot}\,(S)$. This has shown that $\theta$ is injective.

Next let $(\overline{f_\lambda})_{\lambda \in \Lambda}$ be an arbitrary element of $\prod_{\lambda \in \Lambda} \overline{S_\lambda}$, where $f_\lambda \in S_\lambda$ for each $\lambda$. We define $f \in \overline{S}$ to be induced additively from $(x)f = (x)f_\lambda$ whenever $x \in H_\lambda$. Then $\varepsilon_\lambda f \pi_\lambda = f_\lambda$ for all $\lambda$ and so $\theta(\overline{f}) = (\overline{f_\lambda})_{\lambda \in \Lambda}$. This has shown that $\theta$ is surjective.

It remains to show that $\theta$ is multiplicative, that is, given $f, g \in S$, we have $\theta(\overline{fg}) = \theta(\overline{f})\theta(\overline{g})$. To show this, it suffices to show that, for any $\lambda \in \Lambda$,

$$\varepsilon_\lambda (fg) \pi_\lambda - \varepsilon_\lambda f \pi_\lambda \varepsilon_\lambda g \pi_\lambda \in \mathrm{Tot}\,(S_\lambda).$$

Since $\varepsilon_\lambda (fg) \pi_\lambda - \varepsilon_\lambda f \pi_\lambda \varepsilon_\lambda g \pi_\lambda = \varepsilon_\lambda (fg - f e_\lambda g) \pi_\lambda = \varepsilon_\lambda f (1_M - e_\lambda) g \pi_\lambda$, it follows from Lemma 14.6 that it suffices to show that $(1_M - e_\lambda) g \pi_\lambda : M \to H_\lambda$ is in $\mathrm{Tot}\,(M, H_\lambda)$.

If, on the contrary, $(1_M - e_\lambda) g \pi_\lambda : M \to H_\lambda$ is pi, then, by 14.8, there are indices $i \in I, k \in I_\lambda$ for which $\varepsilon_i (1_M - e_\lambda) g \pi_\lambda : M_i \to M_k$ is an isomorphism. Now if $i \in I_\lambda$, then, by (⊛), we have $\varepsilon_i (1_M - e_\lambda) = \varepsilon_i - \varepsilon_i = 0$ and it follows that $\varepsilon_i (1_M - e_\lambda) g \pi_\lambda$ can not be an isomorphism. On the other hand, if $i \notin I_\lambda$, then $M_i \not\cong M_k$ so once again we have a contradiction. Hence $\theta$ is multiplicative. $\qquad\square$

Our next result determines the structure of each of the factors in 14.11.

**14.12.  $S/\operatorname{Tot}(S)$ for a homogeneous LE-decomposition.** *Let $M = \bigoplus_I M_i$ be an LE-decomposition in which $M_i \simeq M_j$ for each $i, j \in I$. Then $\overline{S}$ is isomorphic to the endomorphism ring of a left vector space $V$ over a division ring $D$ where $\dim(_D V) = \operatorname{card}(I)$.*

*Proof.* Continuing with our notation above, here we have precisely one isomorphism class representative $M_\lambda$. For each $i \in I$, let $h_i : M_i \to M_\lambda$ be the given isomorphism, taking $h_\lambda = 1_{M_\lambda}$. Next let

$$D = \operatorname{End}(M_\lambda)/\operatorname{Tot}(\operatorname{End}(M_\lambda)) = \overline{S_\lambda}.$$

Then, since $M_\lambda$ is an LE-module, $D$ is a division ring by Remark 14.5 (1). The left vector space we seek is given by $V = D^{(I)}$. We let $B$ denote its standard basis $\{b_i : i \in I\}$, where the $j$-th component of $b_i$ is $\delta_{ij} 1_D$ for each $i, j \in I$, and use this to define the required isomorphism.

Now, for any $f \in S$ the family of homomorphisms $\{f\pi_j : M \to M_j\}_{j \in I}$ is summable and so, for any $i \in I$, the following defines a finite set $A_{f,i}$:

$$A_{f,i} = \{j \in I \mid \varepsilon_i f \pi_j : M_i \to M_j \text{ is an isomorphism}\}.$$

For each $i, j \in I$, we set

$$f_{ij} : M_i \to M_j \text{ to be} \quad \begin{cases} \varepsilon_i f \pi_j & \text{if } \varepsilon_i f \pi_j \text{ is an isomorphism;} \\ 0 & \text{otherwise.} \end{cases}$$

Note that $f_{ij} \neq 0$ if and only if $h_i^{-1} f_{ij} h_j \in \operatorname{End}(M_\lambda)$ is an automorphism.

We are now ready to define our isomorphism $\sigma : \overline{S} \to \operatorname{End}(_D V)$. Given $f \in S$, using the finiteness of $A_{f,i}$, we can define $\sigma(\overline{f}) \in \operatorname{End}(_D V)$ to be the endomorphism $\widehat{f}$ of $V$ given by

$$(b_i)\widehat{f} = \sum_{j \in I} \overline{h_i^{-1} f_{ij} h_j}\, b_j \qquad \text{for all } i \in I.$$

To see that $\sigma$ is well-defined, let $f \in S$ and $g \in \operatorname{Tot}(S)$ (so that $\overline{f} = \overline{f + g}$ in $\overline{S}$). Then, for any $i, j \in I$ we have, by Lemma 14.6, $h_i^{-1} g_{ij} h_j \in \operatorname{Tot}(S_\lambda)$ and so

$$\overline{h_i^{-1}(f + g)_{ij} h_j} = \overline{h_i^{-1} f_{ij} h_j} + \overline{h_i^{-1} g_{ij} h_j} = \overline{h_i^{-1} f_{ij} h_j},$$

as required. The first of these last two equations also shows that $\sigma$ is additive. Moreover, if $\sigma(\overline{f}) = 0$, then, by the definition of $\widehat{f}$, for all $i, j \in I$ we have $h_i^{-1} f_{ij} h_j \in \operatorname{Tot}(S_\lambda)$ and so $\varepsilon_i f \pi_j$ is a non-isomorphism. Thus, by 14.8, $f \in \operatorname{Tot}(S)$, that is, $\overline{f} = 0$. This has proved that $\sigma$ is a monomorphism.

We now show that $\sigma$ is onto. Thus suppose that $\theta \in \operatorname{End}(_D V)$. If $\theta = 0$, then clearly taking $f = 0 \in S$ gives $\sigma(\overline{f}) = \theta$. Thus suppose that $\theta$ is nonzero. For each $i \in I$, we let $S_i$ be (the possibly empty but always finite) subset of $I$ given by the support of $(b_i)\theta$ with respect to our basis $B$. Then, when $S_i$ is nonempty,

$$(b_i)\theta = \sum_{j \in S_i} d_{ij} b_j$$

where the $d_{ij}$ are (uniquely determined) nonzero elements of $D$. For each of these, choose $\vartheta_{ij} \in S_\lambda$ so that $\overline{\vartheta_{ij}} = d_{ij}$. Then, since $d_{ij} \neq 0$ and $S_\lambda$ is local, each $\vartheta_{ij}$ must be an automorphism of $M_\lambda$. Now set $\gamma_{ij} = h_i \vartheta_{ij} h_j^{-1} : M_i \to M_j$. Next define $f \in S$ by setting, for each $i \in I$ and each $x \in M_i$,

$$(x)f = \begin{cases} \sum_{j \in S_i} (x)\gamma_{ij} & \text{if } S_i \neq \emptyset, \\ 0 & \text{if } S_i = \emptyset \end{cases}$$

and extending this componentwise to $\bigoplus_I M_i$. We claim that $\widehat{f} = \sigma(\overline{f}) = \theta$.

To see this, let $y \in M_\lambda$ and set $x = (y)h_i^{-1}$. Then, for any $j \in A_{f,i}$, we have

$$\begin{aligned} (y)h_i^{-1} f_{ij} h_j &= (x)\varepsilon_i f \pi_j h_j = (x)f\pi_j h_j = \sum_{k \in S_i} (x)\gamma_{ik} \pi_j h_j \\ &= \sum_{k \in S_i} (x) h_i \vartheta_{ik} h_k^{-1} \pi_j h_j = (y)\vartheta_{ij} h_j^{-1} h_j = (y)\vartheta_{ij} \end{aligned}$$

and this produces what we want, namely

$$(b_i)\widehat{f} = \sum_{j \in I, \vartheta_{ij} \neq 0} \overline{\vartheta_{ij}} b_j = \sum_{j \in I, d_{ij} \neq 0} d_{ij} b_j = (b_i)\theta.$$

It remains to show that $\sigma$ is multiplicative. To this end, we just need to show that for any $i \in I$ and any $f, g \in S$, we have $(b_i)\widehat{fg} = ((b_i)\widehat{f})\widehat{g}$. Note first that

$$\begin{aligned} (b_i)\widehat{fg} &= \sum_{j \in A_{fg,i}} \overline{h_i^{-1}(fg)_{ij} h_j} \, b_j \text{ while} \\ (*) \qquad ((b_i)\widehat{f})\widehat{g} &= \left( \sum_{k \in A_{f,i}} \overline{h_i^{-1} f_{ik} h_k} \, b_k \right) \widehat{g} \\ &= \sum_{k \in A_{f,i}} \sum_{j \in A_{g,k}} \overline{h_i^{-1} f_{ik} h_k} \, \overline{h_k^{-1} g_{kj} h_j} \, b_j. \end{aligned}$$

To show that the coefficients in these two expressions in $(*)$ agree, we set

$$\alpha_{ij} = (fg)_{ij} - \sum_{k \in A_0} f_{ik} g_{kj} \in \text{Hom}(M_i, M_j)$$

for each $i, j \in I$, where

$$\begin{aligned} A_0 &= \{k \in I : f_{ik} \text{ and } g_{kj} \text{ are both isomorphisms}\} \\ &= \{k \in I : k \in A_{f,i} \text{ and } j \in A_{g,k}\}. \end{aligned}$$

Then, for the required agreement it suffices to show that $\alpha_{ij} \in \text{Tot}(M_i, M_j)$.

To prove this, first note that if $(M_i)f = 0$, then $\alpha_{ij} = 0 \in \text{Tot}(M_i, M_j)$. Thus suppose that $(a)f \neq 0$ for some $a \in M_i$ and let $A = \{k \in I \mid (a)f\pi_k \neq 0\}$, the support of $(a)f$ in our decomposition $\bigoplus_{k \in I} M_k$. Then, if $k \in I \setminus A$, we have $(a)f_{ik} = (a)f\pi_k = 0$ and so $k \notin A_0$. This shows that $A_0 \subseteq A$. Next we set

$$\beta_{ij} = -\sum_{k \in A \setminus A_0} f_{ik} g_{kj} \in \text{Hom}(M_i, M_j).$$

Then $\alpha_{ij} + \beta_{ij}$ is not an isomorphism since

$$(a)(\alpha_{ij} + \beta_{ij}) = (a)fg\pi_j - \sum_{k \in A}(a)f\pi_k\varepsilon_k g\pi_j = (a)fg\pi_j - (a)fg\pi_j = 0.$$

Thus, since $M_i, M_j$ are indecomposable, we have $\alpha_{ij} + \beta_{ij} \in \mathrm{Tot}\,(M_i, M_j)$ by Remark 14.5 (1). Moreover, if $k \notin A_0$, then either $f_{ik}$ or $g_{kj}$ is not an isomorphism and so, from Lemma 14.6, $f_{ik}g_{kj} \in \mathrm{Tot}\,(M_i, M_j)$. Consequently, since $\mathrm{Tot}\,(M_i, M_j)$ is additively closed by Corollary 14.9, $\beta_{ij} \in \mathrm{Tot}\,(M_i, M_j)$. Then, again since $\mathrm{Tot}\,(M_i, M_j)$ is additively closed, it follows that $\alpha_{ij} = (\alpha_{ij} + \beta_{ij}) - \beta_{ij} \in \mathrm{Tot}\,(M_i, M_j)$, which is just what we wanted.                                  $\square$

Combining the previous two results gives the following important result.

**14.13. Structure Theorem for $S/\mathrm{Tot}\,(S)$ for an LE-decomposition.** *Let* $M = \bigoplus_I M_i$ *be an LE-decomposition with* $S = \mathrm{End}(M)$. *Then there is a ring isomorphism* $\sigma$ *from* $S/\mathrm{Tot}\,(S)$ *to a direct product of endomorphism rings of left vector spaces over division rings, these endomorphism rings being in one-to-one correspondence with the homogeneous components of the LE-decomposition. Consequently,* $S/\mathrm{Tot}\,(S)$ *is a left self-injective regular ring.*

**14.14. Remark.** We now describe a particular feature of the isomorphism $\sigma$ of 14.13 which will prove useful later. We assume that the LE-decomposition $M = \bigoplus_I M_i$ is homogeneous. For any subset $J$ of $I$, let $e_J \in S$ denote the projector corresponding to the first summand of the decomposition

$$M = \left(\bigoplus_{i \in J} M_i\right) \oplus \left(\bigoplus_{i \in I \setminus J} M_i\right).$$

We determine $\sigma(\overline{e_J})$. Using our earlier notation, for any $i, j \in I$ we have

$$h_i^{-1}(e_J)_{ij}h_j \;=\; \begin{cases} 1_D & \text{if} \quad i = j \in J, \\ 0 & \text{if} \quad i = j \notin J, \\ 0 & \text{if} \quad i \neq j, \end{cases}$$

and from this we get

$$(b_i)\widehat{e_J} \;=\; \begin{cases} b_i & \text{if } i \in J, \\ 0 & \text{otherwise.} \end{cases}$$

This shows that $\sigma(\overline{e_J}) = \widehat{e_J}$ is the projector corresponding to the first summand of the vector space decomposition

$$V = \left(\bigoplus_{i \in J} Db_i\right) \oplus \left(\bigoplus_{i \in I \setminus J} Db_i\right).$$

While the above structure theorem illustrates that the ring $S/\mathrm{Tot}\,(S)$ is particularly accessible for the endomorphism ring $S$ of an LE-decomposition, it is still possible to have $\mathrm{Tot}\,(S) \neq J(S)$ here. Indeed, as we will see in 14.22, equality holds if and only if the LE-modules form a locally semi-T-nilpotent family. As a precursor to 14.22, we give the following illustrative example.

**14.15. Example.** Let $M = \bigoplus_{k \in \mathbb{N}} M_k$ be the LE-decomposition of Example 12.15. Thus, for each $k \in \mathbb{N}$, $M_k$ is the cyclic $\mathbb{Z}$-module $\mathbb{Z}/\mathbb{Z}p^k$ where $p$ is a fixed prime. We define $f \in S = \mathrm{End}(M)$ by setting

$$\left( (t_k + \mathbb{Z}p^k)_{k \in \mathbb{N}} \right) f = (0 + \mathbb{Z}p, pt_1 + \mathbb{Z}p^2, pt_2 + \mathbb{Z}p^3, pt_3 + \mathbb{Z}p^4, \dots),$$

where the $t_k \in \mathbb{Z}$ are almost all zero. Then it is straightforward to check that

$$\left( 1 + \mathbb{Z}p, 0 + \mathbb{Z}p^2, 0 + \mathbb{Z}p^3, 0 + \mathbb{Z}p^4, \dots \right) \notin \mathrm{Im}\,(1_M - f)$$

and so, $1_M - f$ is not an isomorphism. Consequently $f \notin J(S)$.

On the other hand, it is also easy to see that, for every natural injection $\varepsilon_k : M_k \to M$ and projection $\pi_l : M \to M_l$, the composition $\varepsilon_k f \pi_l : M_k \to M_l$ is not an isomorphism and so, by 14.8, $f \in \mathrm{Tot}\,(S)$.

To prepare for our investigation of the equality $\mathrm{Tot}\,(S) = J(S)$, we first give a useful lemma. Its proof is routine and so omitted.

**14.16. Lemma.** *Let $M$ be an $R$-module with $S = \mathrm{End}(M)$.*

(1) *If $_S S = K \oplus L$ is a decomposition of $S$ into left ideals $K$ and $L$, then*

$$M = (M)K \oplus (M)L.$$

(2) *If $e$ and $f$ are idempotents of $S$ for which $M = (M)e \oplus (M)f$, then*

$$_S S = Se \oplus Sf.$$

**14.17. LE-decompositions for which $J(S) = \mathrm{Tot}\,(S)$.** *Let $M = \bigoplus_I M_i$ be an LE-decomposition with $S = \mathrm{End}(M)$. If $J(S) = \mathrm{Tot}\,(S)$, then this decomposition of $M$ complements direct summands. Moreover, every indecomposable decomposition of $M$ is isomorphic to $M = \bigoplus_I M_i$ and complements direct summands.*

*Proof.* We first assume that the decomposition has just one homogeneous component. As in the proof of 14.12, let $D$ be the division ring $\mathrm{End}(M_\lambda)/\mathrm{Tot}\,M_\lambda$ for some fixed $\lambda \in I$, let $V$ be the left vector space over $D$ given by $V = D^{(I)}$, let $B$ denote its standard basis $\{b_i : i \in I\}$, and let $\sigma : S/\mathrm{Tot}\,(S) \to \mathrm{End}(_D V)$ be the resulting ring isomorphism.

Now assume that $M = A \oplus B$, let $\varepsilon_A : A \to M$ and $\pi_A : M \to A$ be the inclusion and projection maps, and let $e_A$ be the projector $\pi_A \varepsilon_A$. Then $e_A$ is an idempotent in $\mathrm{End}(M)$ and so $\sigma(\overline{e_A}) = \widehat{e_A}$ is an idempotent in $\mathrm{End}(_D V)$. Thus

$$V = (V)\widehat{e_A} \oplus (V)(1_V - \widehat{e_A}). \qquad (1)$$

Then, by Steinitz' Exchange Lemma 11.1, there is a subset $J$ of $I$ for which

$$V = (V)\widehat{e_A} \oplus \left( \bigoplus_J Db_j \right). \qquad (2)$$

We will achieve our goal by showing that

$$M = (M)e_A \oplus \left(\bigoplus_J M_j\right) = A \oplus \left(\bigoplus_J M_j\right). \qquad (3)$$

As in Remark 14.14, let $e_J$ be the projector corresponding to the first summand of the decomposition

$$M = \left(\bigoplus_{i \in J} M_i\right) \oplus \left(\bigoplus_{i \in I \setminus J} M_i\right).$$

Then, by (1) and 14.14, we have

$$V = \left(\bigoplus_{i \in J} Db_i\right) \oplus \left(\bigoplus_{i \in I \setminus J} Db_i\right) = (V)\widehat{e_J} \oplus \left(\bigoplus_{i \in I \setminus J} Db_i\right) = (V)\widehat{e_J} \oplus (V)\widehat{e_A}$$

and so, by Lemma 14.16 (2), $\mathrm{End}(_D V) = \mathrm{End}(_D V)\widehat{e_J} \oplus \mathrm{End}(_D V)\widehat{e_A}$. Consequently,

$$\overline{S} = \sigma^{-1}(\mathrm{End}(_D V)) = \sigma^{-1}(\mathrm{End}(_D V)\widehat{e_J} \oplus \mathrm{End}(_D V)\widehat{e_A}) = \overline{Se_J} \oplus \overline{Se_A}.$$

Now, since by our hypothesis $J(S) = \mathrm{Tot}\,(S)$, lifting this last decomposition modulo $\mathrm{Tot}\,(S)$ gives $S = Se_J + Se_a + J(S)$ and so, since $J(S) \ll S$, we have

$$S = Se_J + Se_A.$$

We now show that this sum is direct. First, since $S = S(1 - e_A) \oplus Se_A$ we have

$$S(1 - e_A) \simeq S/Se_A = (Se_J + Se_A)/Se_A \simeq Se_J/(Se_J \cap Se_A)$$

and so, since $S(1 - e_A)$ is a projective left $S$-module, so too is $Se_J/(Se_J \cap Se_A)$. Hence $Se_J \cap Se_A$ is a direct summand of $Se_J$ and so also of $_S S$. However, since $\overline{S} = \overline{Se_J} \oplus \overline{Se_A}$, we also have $Se_J \cap Se_A \leq J(S)$ and so the direct summand $Se_J \cap Se_A$ is also small in $_S S$. Hence $Se_J \cap Se_A = 0$, and so $S = Se_J \oplus Se_A$ as required. Our targeted decomposition (3) now follows from Lemma 14.16 (2).

Now assume that our decomposition is not necessarily homogeneous. Let $\theta : \overline{S} \to \prod_{\lambda \in \Lambda} \overline{S_\lambda}$ be the isomorphism of 14.11 and, for each $\lambda \in \Lambda$, let $V_\lambda$ be the vector space given by 14.12. Again let $e_A$ be the projector for the decomposition $M = A \oplus B$. Then, for each $\lambda$, the component $\varepsilon_\lambda \widehat{e_A \pi_\lambda}$ induced by $\theta(\overline{e_A})$ is an idempotent in $\mathrm{End}(V_\lambda)$ and so, as above, there is a subset $J_\lambda$ of $I_\lambda$ for which the analogue of (2) holds for $V_\lambda$ and $\varepsilon_\lambda \widehat{e_A \pi_\lambda}$. If we now set $J = \bigcup_{\lambda \in \Lambda} J_\lambda$, then we can show, by the same argument as before, that

$$M = A \oplus \left(\bigoplus_J M_j\right).$$

Thus our decomposition complements direct summands.

Finally, since $M$ has an LE-decomposition that complements direct summands, it follows from 12.5 (1), (2) that all indecomposable decompositions of $M$ are equivalent and each of these complements is a direct summand. $\qquad \square$

We now present two technical lemmas. Once again, $M = \bigoplus_I M_i$ is an indecomposable decomposition with $S = \mathrm{End}(M)$ and, for each $i \in I$, $e_i \in S$ is the projector $\pi_i \varepsilon_i$. Given any $x \in M$, we let

$$\sup(x) = \{i \in I \mid (x)e_i \neq 0\} \quad \text{and} \quad d_x = \textstyle\sum_{i \in \sup(x)} e_i,$$

noting that $\sup(x)$ is finite (and empty if and only if $x = 0$) and taking $d_x = 0$ if $\sup(x) = \emptyset$. Then $x = (x)d_x$.

**14.18. Lemma.** *Let $M = \bigoplus_I M_i$ and $S$ be as above. Given any $x \in M$ and $h_0, h_1, h_2, \ldots, h_n \in S$, where $h_0 = 1_M$, let $P_{x,n}$ be the set of all finite sequences $(i_0, i_1, \ldots, i_{n-1}, i_n)$ of indices from $I$ for which*

$$(x)h_0 e_{i_0} h_1 e_{i_1} h_2 \cdots e_{i_{n-1}} h_n e_{i_n} \neq 0.$$

*Then $(x)h_0 h_1 \cdots h_n$*

$$= \sum \left\{ (x)h_0 e_{i_0} h_1 e_{i_1} \cdots e_{i_{n-1}} h_n e_{i_n} \mid (i_0, i_1, \ldots, i_{n-1}, i_n) \in P_{x,n} \right\}. \quad (\dagger)$$

*Proof.* We use induction on $n$, noting first that if $n = 0$, then $P_{x,n}$ is simply $\sup(x)$ and so $x = (x)d_x = \sum\{(x)e_{i_0} \mid i_0 \in P_{x,0}\}$, as required.

Now suppose that $(\dagger)$ holds for some $n \geq 0$ and let $h_{n+1} \in S$. Applying $h_{n+1}$ to $(\dagger)$ gives $(x)h_0 h_1 \cdots h_n h_{n+1}$

$$= \sum \left\{ (x)h_0 e_{i_0} h_1 e_{i_1} \cdots e_{i_{n-1}} h_n e_{i_n} h_{n+1} \mid (i_0, \ldots, i_{n-1}, i_n) \in P_{x,n} \right\}.$$

Then the case for $n + 1$ follows by replacing $x$ in $(\dagger)$ by each of the summands $(x)h_0 e_{i_0} h_1 e_{i_1} \cdots e_{i_{n-1}} h_n e_{i_n} h_{n+1}$ and accumulating. $\qquad\square$

For the purposes of the second lemma, we define the following subset $A_n$ of $I$, for each $n \geq 0$. With terminology and hypothesis as above, let

$$A_n = \bigcup \{\{i_0, i_1, \ldots, i_n\} \mid (i_0, i_1, \ldots, i_n) \in P_{x,n}\}.$$

Note that $\sup((x)h_1 h_2 \cdots h_n) \subseteq A_n$ since if $(x)h_1 h_2 \cdots h_n e_i \neq 0$ for some $i \in I$, then, by Lemma 14.18, $i$ must be an index $i_n$ for which $(i_0, i_1, \ldots, i_n) \in P_{x,n}$ for some $i_0, i_1, \ldots, i_{n-1} \in I$. Also note that $A_0 = P_{x,0} = \sup(x)$.

**14.19. Lemma.** *We retain the hypotheses and the notation of Lemma 14.18 and further assume that $I$ is infinite. Then, for any nonzero $x \in M$ and any nonzero $f \in \mathrm{Tot}\,(S)$, there exist sequences*

$$h_1, h_2, h_3, \ldots \text{ in } \mathrm{Tot}\,(S) \quad \text{and} \quad g_1, g_2, g_3, \ldots \text{ in } S$$

*such that, for any $n \in \mathbb{N}$,*

$(\alpha)$ $(x)(1 - h_1 h_2 h_3 \cdots h_n) = (x)g_n(1 - f)$ *and*

($\beta$)  *if $\{i_0, i_1, \ldots, i_{n-1}\} \subseteq I$ is such that $y = (x)e_{i_0}h_1e_{i_1}\cdots e_{i_{n-1}}h_n \neq 0$, then*

$$\sup(y) \cap A_{n-1} = \emptyset,$$

*with $A_{n-1}$ as defined above.*

*Proof.* Let $d_x$ be as defined just before Lemma 14.18. Then, from there, $d_x$ is a nonzero idempotent in $S$ and, by Lemma 14.10, $fd_x \in J(S)$. Thus $1 - fd_x$ is invertible in $S$. Let $h_1 = g_1f(1 - d_x)$ where $g_1 = (1 - fd_x)^{-1}$. Then

$$1 - h_1 = 1 - g_1f(1 - d_x) = g_1((1 - fd_x) - f(1 - d_x)) = g_1(1 - f),$$

whence condition ($\alpha$) for $n = 1$. Moreover, if $i_0 \in P_{x,0}$, then $i_0 \in \sup(x)$ and so $(1 - d_x)e_{i_0} = e_{i_0} - e_{i_0} = 0$. This gives $h_1e_{i_0} = 0$ and so, in this case, condition ($\beta$) follows trivially.

With the first terms established, we now proceed to construct the required sequences $h_1, h_2, \ldots, h_n, \ldots$ in Tot $(S)$ and $g_1, g_2, \ldots, g_n, \ldots$ in $S$ using induction on $n$. Thus assume that ($\alpha$) and ($\beta$) are satisfied for some $n \geq 1$.

Next let $d_{n+1} = \sum_{i \in A_n} e_i$. Then, for any $i \in A_n$, we have $(1 - d_{n+1})e_i = 0$. This allows us to apply the first step of our induction argument above with $(x)h_1 \cdots h_n$ and $d_{n+1}$ assuming the rôles of $x$ and $d_x$ respectively, thereby producing $h_{n+1} \in S$ given by

$$h_{n+1} = k_{n+1}f(1 - d_{n+1}) \text{ where } k_{n+1} = (1 - fd_{n+1})^{-1}$$

such that

$$(x)h_1 \cdots h_n(1 - h_{n+1}) = (x)h_1 \cdots h_nk_{n+1}(1 - f).$$

Then, using our inductive assumption and setting $g_{n+1} = g_n + h_1 \cdots h_nk_{n+1} \in S$, we have

$$
\begin{aligned}
(x)(1 - h_1h_2\cdots h_nh_{n+1}) &= (x)(1 - h_1h_2\cdots h_n) + (x)(h_1h_2\cdots h_n(1 - h_{n+1})) \\
&= (x)(g_n(1 - f)) + (x)h_1h_2\cdots h_nk_{n+1}(1 - f) \\
&= (x)(g_{n+1}(1 - f)),
\end{aligned}
$$

so that property ($\alpha$) is satisfied. Moreover, by the definition of $A_n$ and $d_{n+1}$, it follows quickly that

$$\sup((x)e_{i_0}h_1e_{i_1}\cdots e_{i_{n-1}}h_ne_{i_n}h_{n+1}) \cap A_n = \emptyset$$

for any $(i_0, i_1, \ldots, i_n) \in P_{x,n}$, so property ($\beta$) is also fulfilled.            $\square$

We now bring back on stage the important local semi-T-nilpotency condition of the previous two sections. Recall from Remark 14.5 (1) that if $M_i$ and $M_j$ are indecomposable modules, then Tot $(M_i, M_j)$ is the set of all non-isomorphisms from $M_i$ to $M_j$.

**14.20. Tot $(S) = J(S)$ for a locally semi-T-nilpotent LE-decomposition.** *Let* $M = \bigoplus_I M_i$ *be an indecomposable decomposition with* $S = \mathrm{End}(M)$. *If* $\{M_i\}_I$ *is locally semi-T-nilpotent, then* $\mathrm{Tot}\,(S) = J(S)$ *and so*

$$J(S) = \{f \in S \mid \varepsilon_i f \pi_j \text{ is a non-isomorphism for all } i, j \in I\}.$$

*Proof.* By 14.4, $J(R) \subseteq \mathrm{Tot}\,(R)$ for any ring $R$. Thus we need only show that if $f \in \mathrm{Tot}\,(S)$, then $1 - f$ is invertible in $S$.

We prove first that $1 - f$ is a monomorphism. (This will not require local semi-T-nilpotency.) Recall that, for each $i \in I$, the projector $e_i \in S$ is defined as the composition $\pi_i \varepsilon_i$ where $\pi_i : M \to M_i$ and $\varepsilon_i : M_i \to M$ are the projection and injection maps. The subset $E = \{e_i : i \in I\}$ of $S$ is a set of nonzero orthogonal idempotents. Given any $x \in M$, we let

$$\sup(x) = \{i \in I \mid (x)e_i \neq 0\} \text{ and } d_x = \sum_{i \in \sup(x)} e_i,$$

noting that $\sup(x)$ is finite and taking $d_x = 0$ if $\sup(x) = \emptyset$. Then $d_x$ is an idempotent in $S$ and, since $x = (x)d_x$, we get

$$(x)(1 - f) = x - ((x)d_x)f = (x)(1 - d_x f).$$

Furthermore, by Lemma 14.10, $d_x f \in J(S)$ and so $1 - d_x f$ is invertible in $S$, say with inverse $h_x$. From this we get

$$x = (x)(1 - d_x f)h_x = (x)(1 - f)h_x.$$

Thus if $(x)(1 - f) = 0$, then $x = 0$, hence $1 - f$ is a monomorphism as required.

To complete the proof, that is, to show that $1 - f$ is an epimorphism, we use König's Graph Theorem (Appendix 1) and Lemmas 14.18 and 14.19. Let $x$ be a nonzero element of $M$. For this $x$ we define a König graph $G = (P_1, P_2, \ldots, P_n, \ldots)$ on $I$ by setting, for each $n \in \mathbb{N}$, $P_n = P_{x,n-1}$ as defined in Lemma 14.18. (It is straightforward to see that $G$ is indeed a König graph.) Now, if $G$ is infinite, that is, if $P_n \neq \emptyset$ for each $n \in \mathbb{N}$, then König's Graph Theorem 31.1 implies that $G$ has a path of infinite length, say $p = (I_0, i_1, i_2, \ldots, i_n, \ldots)$. In this case, for each $n \geq 0$ we have

$$(x)e_{i_0} h_1 e_{i_1} \cdots e_{i_{n-1}} h_n e_{i_n} \neq 0. \qquad (*)$$

Then each of these indices $i_k$ in $p$ are distinct since for any $k \in \mathbb{N}$ we have $i_0, i_1, \ldots, i_{k-1} \in A_{k-1}$ while $i_k \in \sup((x)e_{i_0} h_1 e_{i_1} \cdots e_{i_{n-1}} h_n)$ whence $i_k \notin A_{k-1}$ by property $(\beta)$ of Lemma 14.19.

Recalling that $e_i = \pi_i \varepsilon_i$ for each $i \in I$, we now define $f_k : M_{i_{k-1}} \to M_{i_k}$ for each $k \in \mathbb{N}$ by $f_k = \varepsilon_{i_{k-1}} h_k \pi_{i_k}$. Then $(*)$ gives

$$((x)\pi_{i_0})f_1 \cdots f_n \neq 0 \text{ for all } n \in \mathbb{N}.$$

However this contradicts the local semi-T-nilpotency of $\{M_i\}_I$. This contradiction shows that the König graph $G$ is not infinite and so $P_n = \emptyset$ for some $n \in \mathbb{N}$. Consequently, for such $n$ we have, by $P_n$'s definition, that

$$(x)e_{i_0}h_1e_{i_1}\cdots e_{i_{n-2}}h_{n-1}e_{i_{n-1}} = 0 \text{ for all } (i_0, i_1, \ldots, i_{n-1}) \in I^n$$

and so, from Lemma 14.18, we have $(x)h_1 \cdots h_{n-1} = 0$. Property $(\alpha)$ of Lemma 14.19 then gives

$$x = (x)(1 - h_1 \cdots h_{n-1}) = ((x)g_{n-1})(1 - f)$$

and so $1 - f$ is onto, as required.

The last statement of 14.20 is an immediate consequence of 14.8. $\qquad\square$

We record another important property of the total for an LE-decomposition. For the proof the interested reader can consult [386, Satz 2.34].

**14.21. Idempotents and the total for an LE-decomposition.** *Let $M = \bigoplus_I M_i$ be an indecomposable decomposition with $S = \mathrm{End}(M)$. Then idempotents lift modulo* Tot $S$.

We now bring together under one umbrella some of the more important results of this chapter thus far.

**14.22. Well-behaved LE-decompositions.** *Let $M = \bigoplus_I M_i$ be an LE-decomposition with $S = \mathrm{End}(M)$. Then the following statements are equivalent.*

(a) *$S$ is semiregular;*

(b) *$S$ is an exchange ring;*

(c) *$M$ has the exchange property;*

(d) *$M$ has the finite exchange property;*

(e) *$\{M_i\}_I$ is locally semi-T-nilpotent;*

(f) *every local summand of $M$ is a summand;*

(g) *the decomposition $M = \bigoplus_I M_i$ complements direct summands;*

(h) *Tot $(S) = J(S) = \{f \in S : \varepsilon_i f \pi_j$ is a non-isomorphism for all $i, j \in I\}$.*

*Proof.* The equivalence of (c), (d), and (e) comes from 12.14, while that of (e), (f), and (g) comes from 13.6 and Azumaya's Theorem (12.6).

Theorem 14.13 and 14.21 give (h)$\Rightarrow$(a), (a)$\Rightarrow$(b) is by Corollary 11.21, while (b)$\Rightarrow$(d) is from Corollary 11.17. Finally, (e)$\Rightarrow$(h) by 14.20. $\qquad\square$

We are also able to add to the characterisations given in 13.11 of modules with perfect decompositions.

**14.23. Semiregularity and perfect decompositions.** *A module $M$ has a perfect decomposition if and only if $M$ has an LE-decomposition and $\mathrm{End}(N)$ is semiregular for all nonzero $N \in \mathrm{Add}\, M$.*

*Proof.* If $M$ has a perfect decomposition, then by 13.11 every nonzero $N \in \text{Add } M$ has an LE-decomposition $N = \bigoplus_I N_i$ for which $\{N_i\}_I$ is locally semi-T-nilpotent. It then follows from 14.22 that $\text{End}(N)$ is semiregular, as required.

Conversely, suppose that $M$ has an LE-decomposition and $\text{End}(N)$ is semi-regular for all nonzero $N \in \text{Add } M$. Then, by 11.17, each such $N$ has the finite exchange property and so $M$ has a perfect decomposition by 13.11. $\qquad\square$

### 14.24. Exercises.

(1) Show that if $M$ and $N$ are semisimple modules, then $\text{Tot}(M, N) = 0$.

(2) Show that if $R$ is a (von Neumann) regular ring, then $\text{Tot}(R) = 0$.

(3) Let $R$ be the ring of countably infinite sequences of rational numbers which are eventually integers, that is, sequences of the form $q = (q_n)_{n \in \mathbb{N}}$ where $q_n \in \mathbb{Q}$ for each $n$ and for which there is an $n(q) \in \mathbb{N}$ such that $q_k \in \mathbb{Z}$ for all $k \geq q(n)$. Show that $R$ is not regular but $\text{Tot}(R) = 0$.

(4) Let $n \in \mathbb{N}$, $n \geq 2$. Show that $\text{Tot}(\mathbb{Z}_n) = 0$ if and only if $n$ is square-free. (See [192] for an explicit description of $\text{Tot}(\mathbb{Z}_n)$ for any $n$.)

(5) Given $R$-modules $M$ and $N$, define the *singular submodule* of $\text{Hom}_R(M, N)$ by (see [192] and [191])

$$\Delta(M, N) = \{f \in \text{Hom}(M, N) \mid \text{Ke } f \trianglelefteq M\}.$$

   (i) Prove that $\Delta(M, N) \subseteq \text{Tot}(M, N)$.

   (ii) Show that the following conditions are equivalent for the module $M$:

      (a) for every non-essential submodule $A$ of $M$, there is a nonzero injective submodule $Q$ of $M$ with $A \cap Q = 0$;

      (b) $\Delta(M, N) = \text{Tot}(M, N)$ for every $R$-module $N$.

      (In [191] such a module $M$ is called *locally injective*, although this term is used differently by other authors.)

(6) Given $R$-modules $M$ and $N$, recall from 2.4 that their *cosingular ideal* is defined by (see [192] and [191])

$$\nabla(M, N) = \{f \in \text{Hom}(M, N) \mid \text{Im } f \ll N\}.$$

   (i) Prove that $\nabla(M, N) \subseteq \text{Tot}(M, N)$.

   (ii) Show that the following conditions are equivalent for the module $N$:

      (a) for every nonsmall submodule $B$ of $N$, there is a nonzero projective direct summand $P$ of $N$ with $P \subseteq B$;

      (b) $\nabla(M, N) = \text{Tot}(M, N)$ for every $R$-module $M$.

      (In [191] such a module $N$ is called *locally projective*, although this term is used differently by other authors.)

(7) Let $e$ be an idempotent of the ring $R$.

   (i) Show that if $\text{Tot}(R)$ is an ideal of $R$, then $\text{Tot}(eRe)$ is an ideal of $eRe$.

(ii) Show that if $\text{Tot}\,(R) = J(R)$, then $\text{Tot}\,(eRe) = J(eRe)$.

(8) Let $M = \bigoplus_I M_i$ be an LE-decomposition of the module $M$ and let $f \in \text{Tot}\,(S)$ where $S = \text{End}(M)$. Note that, by the first part of the proof of 14.20, $1_M - f$ is a monomorphism. Prove that $(M)(1_m - f) = \bigoplus_I (M_i)(1_M - f)$ is a local direct summand of $M$. (See [311, Lemma 5.11].)

(9) Let $V$ be an infinite-dimensional left vector space over a division ring $D$ and let $R$ be the ring $\text{End}_D(V)$ of linear transformations on $V$. Using 14.22 show that the module $_RR$ does not have an LE-decomposition. (Cf. Exercises 8.18 (6) and 13.13 (12).)

**14.25. Comments.** Let $M = \bigoplus_I M_i$ be an LE-decomposition $M = \bigoplus_I M_i$ with $S = \text{End}(M)$. Early information on $S$ and $J(S)$ and the semiregular connection with local semi-T-nilpotency is given by Harada, Ishii, Kanbara, and Sai in [148], [159], [160], and [162], and by Yamagata in [372] and [373]. The equivalence of semiregularity of $S$ and lsTn of $\{M_i\}_I$ is usually cited as Theorem 7.3.15 of Harada's text [158]. As mentioned, Harada's proof uses his theory of factor categories.

Our approach to establishing Harada's result closely follows that given by Kasch and Mader in their text [194]. Here the central concept of the total was introduced by Kasch and developed by him and his coauthors and/or students Beidar, Schneider, and Zöllner in [190], [38], [194], [195], and [386]. However, most of its applications we use here appear elsewhere, expressed in other ways. For example, 14.9 and 14.12 appear respectively as a Corollary to Lemma 4 and Theorem 7 of Harada and Sai's [162], proved using category theory techniques, while 14.20 appears as part of Yamagata's [372, Proposition 1.1] and also in [160] and [184]. Leron [221] adopts a matrix approach to describe $\text{Hom}(M, N)$ where $M$ and $N$ are two modules with LE-decompositions while Beck [37], Lie [225], and Khurana and Gupta [208] all offer alternative proofs for the structure of $S = \text{End}(M)$ and $S/J(S)$ when $M$ has an LE-decomposition which complements direct summands. Zöllner [387] also provides a different approach to $J(S)$ when $M$ has an LE-decomposition forming a locally semi-T-nilpotent family.

A natural question stemming from 13.12 and 14.22 is the determination of the rings $R$ for which *every* $R$-module has the (finite) exchange property. In this regard, in 1974 Tachikawa [325] showed that if $R$ is of finite representation type (i.e., is left artinian and, up to isomorphism, has only finitely many indecomposable left $R$-modules), then every left $R$-module and every right $R$-module has a decomposition which complements direct summands. In 1975 Fuller and Reiten [112] established the converse to this, while Yamagata [373] proved that if the identity of $R$ is a sum of primitive orthogonal idempotents, then $R$ is of finite representation type if and only if every left and every right $R$-module has the finite exchange property. (The ring of eventually constant sequences over a field shows that the condition on the identity is necessary — see [373].) Shortly afterwards, Fuller [111] proved the one-sided result that every left module over the ring $R$ has a decomposition which complements direct summands if and only if all left $R$-modules are direct sums of finitely generated $R$-modules, that is, $R$ is a left pure-semisimple ring (see [363, §53]), while Zimmermann-Huisgen [383] then showed, using the exchange property for $\Sigma$-algebraically compact modules, that these rings are precisely those for which every left $R$-module has an indecomposable decomposition. Leszczyński and Simson [222] have also established some of these characterisations in a Grothendieck category setting.

More recently, Angeleri-Hügel [14] (also in her paper with Saorín [15]) has extended 13.12 and 14.22 to categories of finitely presented modules and shown how their conditions are related to covers and envelopes. Her results also imply earlier characterisations of perfect rings and pure semisimple rings of Azumaya and Facchini in [26] and [29] involving pure-projective modules and Mittag-Leffler modules. In particular, it is shown that a ring $R$ is left pure semisimple if and only if every pure-projective left $R$-module has a semiregular endomorphism ring if and only if every local summand of a pure-projective left $R$-module is a summand. Notice that these characterisations can be derived from the fact that a module $M$ with semiregular endomorphism ring which is projective in $\sigma[M]$ has small radical and is a direct sum of cyclic modules (see [363, 42.12], Nicholson [255]).

Also interestingly, using results of Herzog [169], Simson [313] has shown that every right pure semisimple ring $R$ is of finite representation type, that is, *the pure semisimplicity conjecture* is true, if and only if a particular strong form of local semi-T-nilpotency fails, namely, for any pair of division rings $F$ and $G$ and any simple $F$-$G$-bimodule $_F M_G$ for which $\dim M_G$ is finite and $\dim {_F M}$ is infinite, setting $R_M$ to be the hereditary right artinian ring

$$R_M = \begin{bmatrix} F & _F M_G \\ 0 & G \end{bmatrix},$$

one can construct a sequence $X_1 \xrightarrow{f_1} X_2 \xrightarrow{f_2} \cdots \xrightarrow{f_{n-1}} X_n \xrightarrow{f_n} X_{n+1} \xrightarrow{f_{n+1}} \cdots$ where each $X_i$ is an indecomposable right $R_M$-module of finite length, each $f_i$ is a non-isomorphism, and $f_n f_{n-1} \cdots f_2 f_1 \neq 0$ for all $n \in \mathbb{N}$.

In a recent paper [380], Zelmanowitz has given necessary and sufficient conditions for an element $s \in S = \mathrm{End}(M)$ to belong to $J(S)$ for any infinitely indexed decomposition $M = \bigoplus_I M_i$, where the $M_i$ are not necessarily indecomposable.

**References.** Angeleri-Hügel [14]; Angeleri-Hügel and Saorín [15]; Azumaya and Facchini in [26]; Beck [37]; Beidar and Kasch [38]; Fuller [111]; Fuller and Reiten [112]; Harada [148, 158] Harada and Ishii [159]; Harada and Kanbara [160]; Harada and Sai [162]; Herzog [169]; Kasch [190, 191] Kasch and Mader [194]; Kasch and Schneider [194, 195]; Khurana and Gupta [208]; Leron [221]; Lie [225]; Nicholson [255]; Simson [313] Tachikawa [325] Wisbauer [363]; Yamagata [372, 373]; Zelmanowitz [380]; Zimmermann-Huisgen [383]; Zöllner [386, 387].

# 15   Stable range 1 and cancellation

Given a ring $R$, we let $\mathrm{U}(R)$ denote the group of units of $R$.

**15.1. Stable range one.** A ring $R$ is said to have *stable range* 1, written $\mathrm{sr}(R) = 1$, if given $a, b \in R$ for which $aR + bR = R$, then $(a + bx)R = R$ for some $x \in R$.

The following result shows that we may strengthen the one-sided invertibility requirement of the definition.

**15.2. Elementary properties of** $\mathrm{sr}(R) = 1$. *Let $R$ be a ring.*

(1) *If $\mathrm{sr}(R) = 1$, then $R$ is directly finite, that is, if $uv = 1$ for $u, v \in R$, then $vu = 1$.*

(2) *The following conditions are equivalent:*

    (a) $\mathrm{sr}(R) = 1$;

    (b) *given $a, b \in R$ for which $aR + bR = R$, then $a + bx \in \mathrm{U}(R)$ for some $x \in R$;*

    (c) *given $a, b \in R$ for which $aR + bR = R$, then there are $k, m, n \in R$ for which $ak + bm = 1$ and $nk = 1$.*

*Proof.* (1) Suppose $uv = 1$ for $u, v \in R$. Then $vR + (1 - vu)R = R$ and so, since $\mathrm{sr}(R) = 1$, there exist $w, z \in R$ for which $(v + (1 - vu)w)z = 1$. Multiplying on the left by $u$ then gives $u((v + (1 - vu)w)z = u$, which yields $z = u$. Then $(v + (1 - vu)w)u = 1$ so the right invertible element $u$ is also left invertible. This gives $vu = 1$, as required.

(2) (a)$\Rightarrow$(b) follows from (1).

For (b)$\Rightarrow$(c), let $aR + bR = R$. Then there are $x, y \in R$ for which $(a + bx)y = 1$ and so $ay + b(xy) = 1$. Setting $k = y, m = xy$, and $n = a + bx$ gives (c).

For (c)$\Rightarrow$(a), let $aR + bR = R$ and choose $k, m, n \in R$ for which $ak + bm = 1$ and $nk = 1$. Setting $x = mn$ gives $(a + bx)k = 1$, whence (a).                                    $\square$

Taking $a = 2, b = 5$ in the above, it is straightforward to see that $\mathrm{sr}(\mathbb{Z}) \neq 1$.

However, the following elementary observations are useful for constructing rings with stable range 1.

**15.3. Stability of** $\mathrm{sr}(R) = 1$.

(1) *Let $R$ be a ring with Jacobson radical $J$. Then $\mathrm{sr}(R) = 1$ if and only if $\mathrm{sr}(R/J) = 1$.*

(2) *Let $R = R_1 \times \cdots \times R_n$ be a direct product of rings. Then $\mathrm{sr}(R) = 1$ if and only if $\mathrm{sr}(R_i) = 1$ for each $1 \leq i \leq n$.*

*Proof.* (1) For any $r \in R$, $r + J \in \mathrm{U}(R/J)$ if and only if $r \in \mathrm{U}(R)$ and so we may apply condition (b) of 15.2.

(2) The proof of this is straightforward.                                    $\square$

Our next result removes the one-sided bias from the definition of the stable range 1 condition.

**15.4. Symmetry of** $\mathrm{sr}(R) = 1$. *Let* $a, b, c, d$ *be elements of a ring* $R$ *for which* $ac + bd = 1$. *If there is an element* $x \in R$ *for which* $a + (bd)x \in \mathrm{U}(R)$, *then there is an element* $y \in R$ *for which* $c + y(bd) \in \mathrm{U}(R)$. *Consequently, the stable range 1 condition is right-left symmetric.*

*Proof.* Set $u = a + (bd)x \in \mathrm{U}(R)$. This allows us to define

$$v = c + (1 - cx)u^{-1}bd \quad \text{and} \quad w = a + x(1 - ca).$$

Then it suffices to show that $v \in \mathrm{U}(R)$ with $v^{-1} = w$. First note that

$$vx = cx + (1 - cx)u^{-1}bdx = cx + (1 - cx)u^{-1}(u - a) = 1 - (1 - cx)u^{-1}a$$

and so $vx(1 - ca) = (1 - ca) - (1 - cx)u^{-1}(1 - ac)a = 1 - ca - (1 - cx)u^{-1}bda$. Adding this to the equation $va = ca + (1 - cx)u^{-1}bda$ gives $vw = 1$. Next, we have

$$
\begin{aligned}
w(1 - cx) &= (a + x(1 - ca))(1 - cx) = a + x(1 - ca) - acx - xc(1 - ac)x \\
&= a + x(1 - ca) - acx - xcbdx = a + (1 - ac)x - xc(a + bdx) \\
&= a + bdx - xcu = (1 - xc)u
\end{aligned}
$$

and so $w(1 - cx)u^{-1}bd = (1 - xc)bd$. Adding this to the equation

$$wc = (a + x(1 - ca))c = ac + xc(1 - ac) = ac + xcbd$$

gives $wv = ac + xcbd + (1 - xc)bd = ac + bd = 1$, as required. $\qquad\qquad\square$

**15.5. The substitution property.** An $R$-module $M$ is said to have the *substitution property* or be *substitutable* if, given a module $A$ with internal decompositions

$$A = M_1 \oplus N_1 = M_2 \oplus N_2$$

where $M_1 \simeq M \simeq M_2$, then the summands $N_1$ and $N_2$ have a common complement $M_0$, necessarily isomorphic to $M$, that is, there is a submodule $M_0$ of $M$ for which

$$A = M_0 \oplus N_1 = M_0 \oplus N_2.$$

The next result is straightforward but fundamental.

**15.6. Substitutability and summands.** *Let* $K$ *and* $L$ *be modules. Then* $M = K \oplus L$ *is substitutable if and only if* $K$ *and* $L$ *are both substitutable.*

*Proof.* Suppose that $M$ is substitutable and that $B := K_1 \oplus N_1 = K_2 \oplus N_2$ where $K_1 \simeq K \simeq K_2$. Then $A := K_1 \oplus L \oplus N_1 = K_2 \oplus L \oplus N_2$ and so, since $K_1 \oplus L \simeq K \oplus L \simeq K_2 \oplus L$, we have $A = M_0 \oplus N_1 = M_0 \oplus N_2$ where $M_0 \simeq M$. Consequently, $B = (B \cap M_0) \oplus N_1 = (B \cap M_0) \oplus N_2$ and so $K$ is substitutable.

Conversely suppose that $K$ and $L$ are both substitutable. Let $A := C_1 \oplus N_1 = C_2 \oplus N_2$ where $C_1 \simeq K \oplus L \simeq C_2$. Then we have $C_1 = K_1 \oplus L_1$ and $C_2 = K_2 \oplus L_2$, where $K_1 \simeq K \simeq K_2$ and $L_1 \simeq L \simeq L_2$, and $A = K_1 \oplus L_1 \oplus N_1 = K_2 \oplus L_2 \oplus N_2$. Thus, since $K$ is substitutable, $L_1 \oplus N_1$ and $L_2 \oplus N_2$ have a common complement $K_0$ in $A$. Then, since $L$ is substitutable, $K_0 \oplus N_1$ and $K_0 \oplus N_2$ have a common complement $L_0$ in $A$. This gives $K_0 \oplus L_0$ as a common complement for $N_1$ and $N_2$ in $A$, as required.                                                                       $\square$

We now bond the substitution property to the stable range 1 condition.

**15.7. Modules with stable range 1 endomorphism rings.** *Let $M$ be an $R$-module and $S = \mathrm{End}(M)$. Then the following statements are equivalent:*

(a) *$M$ has the substitution property;*

(b) *given any module $N$, if $A = M \oplus N$, then for every splitting epimorphism $\tau : A \to M$ there is a splitting map $\phi : M \to A$ for which $A = M\phi \oplus N$;*

(c) *$S$ has stable range 1.*

*Proof.* (a)$\Rightarrow$(b). Let $\sigma : M \to A$ be a map with $\sigma\tau = 1_M$. Then $A = M\sigma \oplus \mathrm{Ke}\,\tau$ and $M\sigma \simeq M$. Thus, since $M$ is substitutable, we have $A = K \oplus N = K \oplus \mathrm{Ke}\,\tau$ for some $K \leq A$. It is easily seen that $\tau|_K : K \to M$ is an isomorphism. Then $\phi = (\tau|_K)^{-1}$ is a splitting for $\tau$ with $A = M\phi \oplus N$ as required.

(b)$\Rightarrow$(a). Let $A = M_1 \oplus N_1 = M_2 \oplus N_2$ where $M_1 \simeq M \simeq M_2$ and let $\pi : A \to M_2$ be the projection map and $\rho : M_2 \to M_1$ be the induced isomorphism. Then $\tau = \pi\rho$ is a split epimorphism from $M_1 \oplus N_1$ to $M_1$ and so, by (b), there is a splitting map $\phi : M_1 \to A$ for $\tau$ such that $A = M_1\phi \oplus N_1$. Clearly $M_1\phi \simeq M$ and it is easy to see that $A = M_1\phi \oplus N_2$.

(b)$\Rightarrow$(c). Assuming (b), we show that $\mathrm{sr}(S) = 1$ by establishing the left-hand version of condition (c) of 15.2 (ii). Let $f, g \in S$ such that $Sf + Sg = S$, say with $f_1 f + g_1 g = 1_M$ for $f_1, g_1 \in S$, and let $M'$ be a copy of $M$ with isomorphism $\gamma : M \to M'$. Next set $A = M \oplus M'$ with associated direct sum maps $\pi : A \to M$, $\pi' : A \to M'$, $\varepsilon : M \to A$, and $\varepsilon' : M' \to A$. Next define $\tau : A \to M$ and $\sigma : M \to A$ by setting, for all $m \in M, m' \in M'$,

$$\tau(m, m') = (m)f + (m')\gamma^{-1}g, \qquad \sigma(m) = ((m)f_1, (m)g_1\gamma).$$

Then $\sigma\tau = 1_M$ and so the epimorphism $\tau$ splits. Thus, by (b), there is a splitting map $\phi : M \to A$ with $A = M\phi \oplus M'$. Next let $f_2, g_2 \in S$ be given by $f_2 = \phi\pi, g_2 = \phi\pi'\gamma^{-1}$. Then it is easily checked that $f_2 f + g_2 g = 1_M$. Moreover, since $\phi$ is a splitting map with $A = M\phi \oplus M'$, $\phi$ has a right inverse $\alpha : A \to M$ with $M' = \mathrm{Ke}\,\alpha$. Then setting $h = \varepsilon\alpha \in S$ gives $f_2 h = \phi\pi\varepsilon\alpha = \phi\alpha = 1_M$, finishing this part of the proof.

(c)$\Rightarrow$(b). Let $N$ be any module and $\tau : M \oplus N \to M$ be a split epimorphism, with $\sigma : M \to M \oplus N$ giving $\sigma\tau = 1_M$. Let $\pi_M, \pi_N$ be the direct sum projections and $\varepsilon_M, \varepsilon_N$ be the injections. Then, defining $f_1 = \varepsilon_M \tau$, $f_2 = \varepsilon_N \tau$, $g_1 = \sigma \pi_M$, $g_2 = \sigma \pi_N$, we get $f_1, g_1$, and $g_2 f_2 \in S$ with $g_1 f_1 + g_2 f_2 = \sigma \tau = 1_M$. Thus, since $\mathrm{sr}(S) = 1$, there is a $t \in S$ for which $f_1 + t g_2 f_2$ is a unit in $S$, say with inverse $h$. Defining $\phi : M \to M \oplus N$ by $(m)\phi = ((m)h, (m)htg_2)$ for all $m \in M$, we get $\phi \tau = 1_M$. Moreover, defining $\gamma : M \oplus N \to M$ by $(m,n)\gamma = (m)h^{-1}$ for all $(m,n) \in M \oplus N$ gives $\mathrm{Ke}\, \gamma = N$ and also $\phi \gamma = 1_M$. Thus $M \oplus N = M\phi \oplus \mathrm{Ke}\, \gamma = M\phi \oplus N$, finishing the proof. $\qquad \square$

**15.8. Corollary.** *Let $R$ be a ring with stable range 1. Then $\mathrm{sr}(\mathrm{End}(P)) = 1$ for every finitely generated projective $R$-module.*

*Proof.* This is manifest from 15.6 and 15.7. $\qquad \square$

**15.9. Cancellation properties.** Given a collection $\mathcal{C}$ of left $R$-modules, a module $M \in \mathcal{C}$ is said to be *cancellable* in $\mathcal{C}$ if, whenever $M \oplus X \simeq M \oplus Y$ for $X, Y \in \mathcal{C}$, then $X \simeq Y$. In the case where $\mathcal{C}$ is the category of all left $R$-modules, we simply say that $M$ is *cancellable* or has the *cancellation property*.

**15.10. Substitution cancels.** *A substitutable module $M$ is cancellable.*

*Proof.* Let $A, M_1, M_2, N_1$, and $N_2$ be modules such that $A = M_1 \oplus N_1 \simeq M_2 \oplus N_2$, where $M_1 \simeq M \simeq M_2$. Then $A = M_1 \oplus N_1 = M_3 \oplus N_3$, where $M_3 \simeq M_2$ and $N_3 \simeq N_2$. Since $M$ is substitutable, this then gives $A = M_0 \oplus N_1 = M_0 \oplus N_3$ where $M_0 \simeq M$ and so $N_1 \simeq A/M_0 \simeq N_3 \simeq N_2$, as required. $\qquad \square$

As an immediate consequence of 15.7 and 15.10 we get the following result which appears in [90].

**15.11. Evans' Theorem.** *Let $M$ be a module whose endomorphism ring has stable range 1. Then $M$ is cancellable.*

**15.12. Example.** The $\mathbb{Z}$-module $\mathbb{Z}$ is cancellable but not substitutable. For the former see Cohn [72] or Walker [353]. We noted earlier that $\mathrm{sr}(\mathbb{Z}) \neq 1$ so 15.7 shows the latter.

**15.13. Summands and cancellation.** *Let $M$ and $N$ be modules in a collection $\mathcal{C}$ of left $R$-modules. Then $M \oplus N$ is cancellable in $\mathcal{C}$ if and only if $M$ and $N$ are both cancellable in $\mathcal{C}$.*

*Proof.* Suppose that both $M$ and $N$ are cancellable in $\mathcal{C}$ and there are modules $X$ and $Y$ in $\mathcal{C}$ with $M \oplus N \oplus X \simeq M \oplus N \oplus Y$. Then cancelling first $M$ and then $N$ gives $X \simeq Y$, as required.

Conversely, suppose that $M \oplus N$ is cancellable in $\mathcal{C}$ and $X$ and $Y$ are modules in $\mathcal{C}$ with $N \oplus X \simeq N \oplus Y$. Then $M \oplus N \oplus X \simeq M \oplus N \oplus Y$ and so cancelling $M \oplus N$ gives $X \simeq Y$. Hence $N$ is cancellable in $\mathcal{C}$ and symmetry completes the proof. $\qquad \square$

**15.14. Morphic endomorphisms.** An endomorphism $f$ of a module $M$ is called *morphic* if $M/Mf \simeq \operatorname{Ke} f$.

**15.15. Example.** We illustrate this concept with a simple example which will be useful later.

Let $V$ be a finite-dimensional vector space over a division ring $K$. Then, given any endomorphism $f$ of $V$, $\dim_K(V/Vf) = \dim_K(\operatorname{Ke} f)$ and so $V/Vf \simeq \operatorname{Ke} f$. Thus all endomorphisms of $V$ are morphic.

On the other hand, if $\dim_K(V)$ is infinite, then there is an endomorphism $g$ of $V$ which is onto but not mono and consequently $g$ is not morphic.

**15.16. Morphic endomorphisms characterised.** *Let $M$ be an $R$-module, $S = \operatorname{End}(M)$ and $f \in S$. Then the following statements are equivalent:*

(a) *$f$ is morphic;*

(b) *there is a $g \in S$ for which $\operatorname{Im} g = \operatorname{Ke} f$ and $\operatorname{Ke} g = \operatorname{Im} f$;*

(c) *there is a $g \in S$ for which $\operatorname{Im} g \simeq \operatorname{Ke} f$ and $\operatorname{Ke} g = \operatorname{Im} f$.*

*Proof.* Given (a), there is an isomorphism $h : M/\operatorname{Im} f \to \operatorname{Ke} f$. Define $g \in S$ by setting $(m)g = (m + \operatorname{Im} f)h$ for all $m \in M$. Then $\operatorname{Im} g = (M/\operatorname{Im} f)h = \operatorname{Ke} f$ and $\operatorname{Ke} g = \{m \in M \mid (m + \operatorname{Im} f)h = 0\} = \operatorname{Im} f$, hence (b).

(b)$\Rightarrow$(c) is clear.

If (c) holds, then $M/\operatorname{Im} f = M/\operatorname{Ke} g \simeq \operatorname{Im} g \simeq \operatorname{Ke} f$, hence (a).          $\square$

**15.17. Example.** In 15.16, we can not replace (a) or (b) by "there is a $g \in S$ for which $\operatorname{Im} g = \operatorname{Ke} f$ and $\operatorname{Ke} g \simeq \operatorname{Im} f$". To see this, consider the $\mathbb{Z}$-module $M = \mathbb{Z}_2 \oplus \mathbb{Z}_4$. Define $f, g \in S = \operatorname{End}(M)$ by $(x + 2\mathbb{Z}, y + 4\mathbb{Z})f = (0, 2x + 4\mathbb{Z})$ and $(x + 2\mathbb{Z}, y + 4\mathbb{Z})g = (0, y + 4\mathbb{Z})$ for all $x, y \in \mathbb{Z}$. Then $\operatorname{Im} g = \operatorname{Ke} f$ and $\operatorname{Ke} g \simeq \operatorname{Im} f$ but $M/\operatorname{Im} f \not\simeq \operatorname{Ke} f$, so $f$ is not morphic.

**15.18. Regularity, unit-regularity, and partial unit-regularity.** An element $a$ of the ring $R$ is called *(von Neumann) regular* if $axa = a$ for some $x \in R$. If there is such an element $x \in \operatorname{U}(R)$, then $a$ is called *unit-regular*.

We note that $a$ is unit-regular if and only if there is an idempotent $e \in R$ and $v \in \operatorname{U}(R)$ such that $a = ev$. (For the necessity, take $e = au$ and $v = u^{-1}$ where $u \in \operatorname{U}(R)$ is such that $aua = a$.)

The ring $R$ is called *partially unit-regular* if every regular element of $R$ is unit-regular, and is called *unit-regular* if all of its elements are unit-regular.

**15.19. Characterisation of regular and unit-regular endomorphisms.** *Let $M$ be an $R$-module and $S = \operatorname{End}(M)$. Then an element $f \in S$ is*

(1) *regular if and only if $\operatorname{Ke} f$ and $\operatorname{Im} f$ are both direct summands of $M$ and*

(2) *unit-regular if and only if it is both regular and morphic.*

*Proof.* (1) Let $i : \operatorname{Im} f \to M$ denote the inclusion map and first suppose that $f$ is regular, choosing $g \in S$ with $fgf = f$. Let $h : \operatorname{Im} f \to M$ denote the composition $i \diamond g$. Then, for all $x \in \operatorname{Im} f$, taking $x = (y)f$ where $y \in M$, we have $(x)h \diamond f = (y)f \diamond i \diamond g \diamond f = (y)f = x$ and so $i \diamond g \diamond f = h \diamond f = 1_{\operatorname{Im} f}$. This shows that $h$ splits the epimorphism $f : M \to \operatorname{Im} f$ and $g \diamond f$ splits the monomorphism $i : \operatorname{Im} f \to M$ and so $\operatorname{Ke} f$ and $\operatorname{Im} f$ are both direct summands of $M$, as required.

Conversely, suppose that $\operatorname{Ke} f$ and $\operatorname{Im} f$ are both direct summands of $M$. Then, by the first of these, there is a homomorphism $g' : \operatorname{Im} f \to M$ with $(y)g'f = y$ for all $y \in \operatorname{Im} f$. Since $\operatorname{Im} f$ is a summand, we may extend $g'$ to $g \in S$ by taking $g$ as zero on the complementary summand. Then, for any $x \in M$, $(x)fgf = (x)f$ and so $fgf = f$.

(2) First suppose that $f$ is unit-regular, with $fhf = f$ where $h \in U(S)$. Since $fh$ is an idempotent in $S$, we have $\operatorname{Im}(fh) = \operatorname{Ke}(1_M - fh)$ and $\operatorname{Ke}(fh) = \operatorname{Im}(1_M - fh)$. Then, setting $g = h \diamond (1_M - fh) \in S$, we have

$$\operatorname{Im} g = \operatorname{Im}(1_M - fh) = \operatorname{Ke}(fh) = \operatorname{Ke}(fhh^{-1}) = \operatorname{Ke} f \quad \text{and}$$

$$\operatorname{Ke} g = (\operatorname{Ke}(1_M - fh))h^{-1} = (\operatorname{Im}(fh))h^{-1} = \operatorname{Im}(fhh^{-1}) = \operatorname{Im} f$$

and so $f$ is morphic by 15.16.

Conversely suppose that $f$ is regular and morphic. Then, by (i), both $\operatorname{Im} f$ and $\operatorname{Ke} f$ are summands of $M$, say $M = \operatorname{Im} f \oplus K = N \oplus \operatorname{Ke} f$. Then $(N)f = \operatorname{Im} f$ and so $M = (N)f \oplus K$. Moreover, since $f$ is morphic, we have $K \simeq M/\operatorname{Im} f \simeq \operatorname{Ke} f$ and we let $h : K \to \operatorname{Ke} f$ be the isomorphism. This produces a well-defined $g \in S$ given by $((n)f + k)g = n + (k)h$ for all $n \in N, k \in K$. It is easily seen that $fgf$ and $f$ agree on $N$ and on $\operatorname{Ke} f$ and so, since $M = N \oplus \operatorname{Ke} f$, $fgf = f$. Finally, $(M)g = N + (K)h = N + \operatorname{Ke} f = M$ so that $g$ is onto, while $g$ is monic since $h$ is; thus $g \in U(S)$, as required. $\quad\square$

**15.20. Remark.** As an easy consequence of 15.19, we note that any simple artinian ring is unit-regular. In detail, if $V$ is a finite-dimensional vector space over a division ring $K$, then, as noted above, every $f \in \operatorname{End}_K(V)$ is morphic. Then, since $\operatorname{End}_K(V)$ is regular, 15.19 (2) applies.

**15.21. Internal cancellation.** An $R$-module $M$ is said to have *internal cancellation* or be *internally cancellable* (*IC* for short) if, given internal decompositions

$$M = A_1 \oplus B_1 = A_2 \oplus B_2$$

where $A_1 \simeq A_2$, then $B_1 \simeq B_2$. If $_R R$ has IC, $R$ is said to be a *left IC ring*.

**15.22. Lemma.** *A cancellable module $M$ is internally cancellable.*

*Proof.* Let $M = A_1 \oplus B_1 = A_2 \oplus B_2$, where $A_1 \simeq A_2$. Then, since $A_1$ is a summand of $M$, it is cancellable by 15.13. Thus, since $A_1 \oplus B_1 = A_2 \oplus B_2 \simeq A_1 \oplus B_2$, we have $B_1 \simeq B_2$, as required. $\quad\square$

**15.23. Lemma.** *Every direct summand $N$ of an IC module $M$ is IC.*

*Proof.* Let $M = N \oplus K$ and suppose that $N = A_1 \oplus B_1 = A_2 \oplus B_2$, where $A_1 \simeq A_2$. Then $M = (A_1 \oplus K) \oplus B_1 = (A_2 \oplus K) \oplus B_2$ where $A_1 \oplus K \simeq A_2 \oplus K$. The internal cancellability of $M$ then gives $B_1 \simeq B_2$, as required. $\qquad\qquad$ $\square$

**15.24. Remark.** In contrast to 15.13 and 15.23, the direct sum of two IC modules need not be IC. (See Lam [217].)

**15.25. Directly finite modules.** A module $M$ is called *directly finite* if $M$ is not isomorphic to a proper direct summand of itself.

Thus $M$ is directly finite if and only if $N = 0$ is its only submodule for which $M \oplus N \simeq M$. It is straightforward to show that $M$ is directly finite if and only if, for all $f, g \in \operatorname{End}(M)$ if $fg = 1_M$, then $gf = 1_M$ (see, e.g., [131, Lemma 5.1]). From this we see that the directly finite notion used for $R$ in 15.2 is the same.

**15.26. Direct finiteness and IC.** *A module $M$ with the internal cancellation property is directly finite. The converse is true if $M$ has a decomposition that complements direct summands.*

*Proof.* Let $M$ be IC and suppose that $A \simeq M$ where $M = A \oplus B$. Then $M = A \oplus B = M \oplus 0$ and so, by assumption, $B \simeq 0$. Thus $M$ is directly finite.

Conversely, assume that $M$ is directly finite, let $M = \bigoplus_I M_i$ be a decomposition that complements direct summands and suppose that

$$M = A_1 \oplus B_1 = A_2 \oplus B_2$$

where $A_1 \simeq A_2$. Then there exists $J_1, J_2 \subset I$ such that

$$A_1 \simeq \bigoplus_{j \in J_1} M_j, \ B_1 \simeq \bigoplus_{j \notin J_1} M_j, \ A_2 \simeq \bigoplus_{j \in J_2} M_j, \ B_2 \simeq \bigoplus_{j \notin J_2} M_j.$$

Since $M$ is directly finite it is not difficult to see that, for any $i \in I$, the set $\{j \in I : M_i \simeq M_j\}$ is finite. Thus if $A_1 \simeq A_2$, then $B_1 \simeq B_2$. $\qquad$ $\square$

The next result characterises internal cancellation via endomorphism rings.

**15.27. Partially unit-regular endomorphism rings.** *An $R$-module $M$ has internal cancellation if and only if $S = \operatorname{End}(M)$ is partially unit-regular.*

*Proof.* Suppose $M$ has IC and let $f$ be a regular element of $S$. Then, by 15.19 (1), $M = \operatorname{Im} f \oplus K = \operatorname{Ke} f \oplus N$ for some $K, N \leq M$. This gives $\operatorname{Im} f \simeq M/\operatorname{Ke} f \simeq N$ and so, since $M$ has IC, we get $K \simeq \operatorname{Ke} f$. Hence $M/\operatorname{Im} f \simeq \operatorname{Ke} f$ and so $f$ is morphic. Hence $f$ is unit-regular, by 15.19 (2). Thus $S$ is partially unit-regular.

Conversely, suppose that $S$ is partially unit-regular, we have internal decompositions $M = A_1 \oplus B_1 = A_2 \oplus B_2$ and there is an isomorphism $\gamma : A_1 \to A_2$. We wish to show that $B_1 \simeq B_2$. Define $f \in S$ by setting $(a_1 + b_1)f = (a_1)\gamma$ for all

$a_1 \in A_1, b_1 \in B_1$. Then $\operatorname{Im} f = (A_1)\gamma = A_2$ and $\operatorname{Ke} f = B_1$, so that $\operatorname{Im} f$ and $\operatorname{Ke} f$ are both summands of $M$. Hence, by 15.19 (1), $f$ is regular and so, by hypothesis, unit-regular. Then, by 15.19 (2), we get $B_2 \simeq M/A_2 = M/\operatorname{Im} f \simeq \operatorname{Ke} f = B_1$, as required. $\qquad \square$

**15.28. Corollary.** *Let $M$ be an $R$-module and $S = \operatorname{End}(M)$. Then $S$ is a unit-regular ring if and only if $S$ is regular and $M$ has IC.*

Taking $M = R$ gives

**15.29. Corollary.** *A ring $R$ is a unit-regular ring if and only if it is regular and ${}_R R$ (or equivalently $R_R$) has IC.*

It follows from 15.27 and the left-right symmetry of partial unit-regularity that any left IC ring is right IC and vice versa.

We may also characterise IC rings using isomorphic idempotents, as follows.

**15.30. Conjugate idempotents and IC rings.** *The following statements are equivalent for a ring $R$.*

(a) *$R$ is an IC ring;*

(b) *given idempotents $e, f \in R$, if $Re \simeq Rf$, then $R(1 - e) \simeq R(1 - f)$;*

(c) *given idempotents $e, f \in R$, if $Re \simeq Rf$, then $e$ and $f$ are conjugates, that is, $ueu^{-1} = f$ for some $u \in \operatorname{U}(R)$;*

(d) *the right analogues of (b) and (c).*

*Proof.* The equivalence of (a) and (b) is straightforward and (d) will comply once we have shown (b)$\Leftrightarrow$(c).

(b)$\Rightarrow$(c). Let $e, f$ be idempotents of $R$ for which there is an isomorphism $\theta : Re \to Rf$. Set $a = (e)\theta$ and $b = (f)\theta^{-1}$. Then it is easily checked that $a \in eRf, b \in fRe$ and $e = ba, f = ab$. Since (b) holds we also have an isomorphism $\phi : R(1 - e) \to R(1 - f)$ and so, similarly, there are elements $c \in (1 - e)R(1 - f)$, $d \in (1 - f)R(1 - e)$ such that $1 - e = dc, 1 - f = cd$. Then $ad = da = bc = cb = 0$. From this we get $(a + c)(b + d) = ab + cd = f + 1 - f = 1$ and $(b + d)(a + c) = ba + dc = e + 1 - e = 1$. Thus $u := a + c \in \operatorname{U}(R)$ with $u^{-1} = b + d$. Moreover

$$ueu^{-1} = (a + c)e(b + d) = (a + c)ba(b + d) = abab = f^2 = f,$$

as required.

(c)$\Rightarrow$(b). Let $e, f$ be idempotents of $R$ for which $Re \simeq Rf$. Then, by (c), $f = ueu^{-1}$ for some unit $u \in \operatorname{U}(R)$. This gives $1 - f = u(1 - e)u^{-1}$ and so induces an isomorphism $\phi : R(1 - e) \to R(1 - f)$ where $(r(1 - e))\phi = ru^{-1}(1 - f)$ for all $r \in R$. $\qquad \square$

**15.31. Abelian rings.** A ring $R$ is said to be *abelian* if all its idempotents are central elements.

Of course, any commutative ring is abelian, as are rings with 0 and 1 as their only idempotents, and reduced rings (rings which have no nonzero nilpotent elements).

We now utilise condition (c) of 15.30 to give a large class of IC rings.

**15.32. Abelian implies IC.** *Every abelian ring $R$ is IC.*

*Proof.* We show that if $e, f$ are idempotents in the abelian ring $R$ with $Re \simeq Rf$, then $e = f$. It then follows immediately using 15.30 (c) that $R$ is IC.

If $Re \simeq Rf$, then, by the proof of (b)$\Rightarrow$(c) in 15.30, there are $a, b \in R$ for which $e = ba, f = ab$. Then, using the abelian property of $R$, we have

$$f = f^2 = (ab)(ab) = aeb = abe = fe = fba = bfa = (ba)(ba) = e^2 = e,$$

as required.                                                                                     $\square$

The following result unites the various criteria introduced in this section for the $R$-module $R$, under the additional assumption that $R$ has the exchange property.

**15.33. Exchange rings, stable range 1 and cancellation.** *Let $R$ be an exchange ring. Then the following statements are equivalent.*

 (a) $\mathrm{sr}(R) = 1$;

 (b) *if $a, b \in R$ are such that $aR = bR$, then $b = au$ for some $u \in \mathrm{U}(R)$;*

 (c) *$R$ is partially unit-regular;*

 (d) *$_R R$ is internally cancellable;*

 (e) *if $a, e \in R$ are such that $e^2 = e$ and $aR + eR = R$, then $a + ex \in \mathrm{U}(R)$ for some $x \in R$;*

 (f) *for each $n \in \mathbb{N}$, the matrix ring $M_n(R)$ is partially unit-regular;*

 (g) *$_R R$ is cancellable in the category of all left $R$-modules;*

 (h) *$_R R$ is cancellable in the category $\mathcal{P}(R)$ of finitely generated projective left $R$-modules;*

 (i) *every module in $\mathcal{P}(R)$ is cancellable in $\mathcal{P}(R)$;*

 (j) *the right analogues of (b), (d), (e), (g), (h), and (i).*

*Proof.* (a)$\Rightarrow$(b). Suppose that $aR = bR$. Then $a = bc$ and $b = ad$ for some $c, d \in R$. Thus $a = adc$. Then, since $dR + (1 - dc)R = R$, (a) gives an $x \in R$ for which $u := d + (1 - dc)x \in \mathrm{U}(R)$. Thus $au = ad + (a - adc)x = ad + (a - a)x = ad = b$, as required.

(b)$\Rightarrow$(c). Suppose that $a, x \in R$ with $axa = a$. Then $aR = axR$ and so, by (b), $ax = au$ for some $u \in \mathrm{U}(R)$. This gives $a = axa = aua$ and so $R$ is partially unit-regular.

(c)⇔(d) is given by 15.27.

(c)⇒(e). Suppose that $a, e \in R$ are such that $e^2 = e$ and $aR + eR = R$. Then $ax + ey = 1$ for some $x, y \in R$ and so $1 - e = (1 - e)ax + (1 - e)ey = (1 - e)ax$. This gives

$$(1 - e)ax(1 - e)a = (1 - e)a.$$

Hence, by the partial unit-regularity of $R$, we have $(1 - e)au(1 - e)a = (1 - e)a$ for some $u \in U(R)$. Now set $f = u(1 - e)a$. Then $f^2 = u(1 - e)au(1 - e)a = u(1 - e)a = f$ and so

$$(1 + fxu^{-1}(1 - f))(1 - fxu^{-1}(1 - f)) = 1 - (fxu^{-1}(1 - f))^2 = 1,$$

giving $1 - fxu^{-1}(1 - f) \in U(R)$. Moreover

$$u(1 - e)axu^{-1} + ueu^{-1} = u(1 - e)u^{-1} + ueu^{-1} = uu^{-1} = 1$$

and so

$$fxu^{-1}(1 - f) + ueu^{-1}(1 - f) = 1 - f.$$

This gives

$$\begin{aligned} u(a + e(u^{-1}(1 - f) - a)) &= u((1 - e)a + eu^{-1}(1 - f)) \\ &= f + ueu^{-1}(1 - f) = 1 - fxu^{-1}(1 - f) \end{aligned}$$

and so $a + e(u^{-1}(1 - f) - a)) \in U(R)$, hence (e).

(e)⇒(a). Suppose that $aR + bR = R$ for some $a, b \in R$. Then $ax + by = 1$ for some $x, y \in R$. Since $R$ is an exchange ring, there is an idempotent $e \in R$ with $e \in byR$, say $e = byc$ where $c \in R$, and $1 - e \in axR$. Then

$$1 = ax + (1 - e)by + eby \in aR + eR$$

and so $aR + eR = R$. Thus, by (e), $a + ex \in U(R)$ for some $x \in R$ and so $a + bycx \in U(R)$. Hence $sr(R) = 1$.

(a)⇒(f). By 15.8, since $sr(R) = 1$ we have $sr(\text{End}(R^n)) = 1$. Then, by the implications (a)⇒(b)⇒(c) proved above but applied to $M_n(R)$, (f) follows. (Note that these implications do not rely on the exchange property.)

(a)⇒(g) follows from 15.7 and 15.10.

The implications (f)⇒(c) and (g)⇒(h) are automatic, while (h)⇒(i) follows quickly from 15.13.

(i)⇒(d). Suppose that $_R R = A_1 \oplus B_1 = A_2 \oplus B_2$ where $A_1 \simeq A_2$. Then, since $A_1, B_1, A_2, B_2 \in \mathcal{P}(R)$, it follows that $B_1 \simeq B_2$, as required.

The equivalence of (j) to the other conditions follows from the left-right symmetry of partial unit-regularity (or of the stable range one condition). □

**15.34. Remarks.** Note that the exchange property assumption was used in the proof of 15.33 only in showing (e)⇒(a). On the other hand the ring $\mathbb{Z}$ shows that (b)⇒(a) is not true in general. We now give an example, due to Kaplansky [185, p. 466] to show that (c)⇒(b) also fails in general. To see this, first note that, by 15.32, every commutative ring satisfies (c). However, setting

$$R = \{(n, f(x)) \in \mathbb{Z} \times \mathbb{Z}_5[x] : f(0) \equiv n \pmod 5\},$$

it is easily checked that $(0, x)$ and $(0, 2x)$ generate the same principal ideal in $R$ but are not associates.

**15.35. Corollary.** *A regular ring $R$ is unit-regular if and only if* $\mathrm{sr}(R) = 1$.

*Proof.* Since $R$ is regular, $R$ is an exchange ring by 11.21. Now apply the equivalence of (a) and (c) of 15.33.                                                                   □

We are now able to give a quick proof of a result first proved by Bass [35]. Recall that a ring $R$ with Jacobson radical $J$ is called *semilocal* if the factor ring $R/J$ is left semisimple (=semisimple artinian).

**15.36. Corollary (Bass).** *If $R$ is a semilocal ring, then* $\mathrm{sr}(R) = 1$.

*Proof.* By 15.3 (1), we may assume that $J(R) = 0$, that is, $R$ is semisimple artinian. Furthermore, by 15.3 (2), we may assume that $R$ is simple artinian. Then, as noted earlier, $R$ is unit-regular and so, by Corollary 15.35, $\mathrm{sr}(R) = 1$, as required.     □

Warfield [361, Theorem 2.4] shows that if a module $M$ has a semiregular endomorphism ring $S$ (in particular if $M$ is a self-injective module), then $\mathrm{sr}(S) = 1$ if and only if $M$ is cancellable. Since, by 11.21, semiregular rings are exchange rings, the following result generalises this (and provides a partial converse to Evans' theorem (15.11)).

**15.37. The finite exchange property and cancellation.** *Let $M$ be a module with the finite exchange property. Then the following statements are equivalent.*

(a)  $\mathrm{sr}(\mathrm{End}(M)) = 1$;

(b)  *$M$ is substitutable;*

(c)  *$M$ is cancellable;*

(d)  *$M$ is internally cancellable.*

*Proof.* (a)⇔(b), (b)⇒(c) and (c)⇒(d) follow from 15.7, 15.10 and Lemma 15.22, respectively.

(d)⇒(b) Now suppose that $M$ is internally cancellable. Then, by 15.27, the ring $S = \mathrm{End}(M)$ is partially unit-regular. Since $M$ has the finite exchange property, $S$ is also an exchange ring and so, by 15.33, $\mathrm{sr}(S) = 1$. Thus $M$ is substitutable, by 15.7.                                                         □

## 15.38. Exercises.

(1) Using 15.6 and 15.7, prove that the stable range 1 property is a Morita invariant. (See [363, §46] for a Morita invariant background, in particular 46.3 and 46.4 there.)

(2) Let $R$ be a ring with $\mathrm{sr}(R) = 1$ and let $e$ be an idempotent of $R$. Show that $\mathrm{sr}(eRe) = 1$. (Hint: use $\mathrm{End}(Re)$.) (See [351].)

(3) Let $R$ be a commutative ring in which every prime ideal is maximal. Show that $\mathrm{sr}(R) = 1$. (Hint: use 15.3 (1) to reduce to a reduced ring and then localise at every prime ideal to show that $R$ is regular.) (See [217, Prop. 2.12].)

(4) Let us say that two module epimorphisms $f, g : P \to M$ are *left equivalent* if there is an automorphism $h$ of $P$ for which $f = h \diamond g$. (See [57].)

   Prove that $\mathrm{sr}(R) = 1$ if and only if each pair of epimorphisms $f, g : P \to M$ from a finitely generated projective $R$-module $P$ to an $R$-module $M$ are left equivalent. (Hint: for the necessity note that $\mathrm{sr}(\mathrm{End}(P)) = 1$ for any finitely generated projective $R$-module $P$; for the sufficiency suppose that $a, b \in R$ with $Ra + Rb = R$ and take $P = R$ and $M = R/Rb$.)

(5) Let $p$ be a prime, at least 5, and choose $k \in \mathbb{Z}$ with $k \not\equiv 0, \pm 1 \pmod{p}$. Choose $m, n \in \mathbb{Z}$ satisfying $mk - np = 1$. Now consider the free $\mathbb{Z}$-module $F = \mathbb{Z} \oplus \mathbb{Z}$ with basis elements $x = (1, 0), y = (0, 1)$. Show that $F$ also has basis $\{x', y'\}$ where $x' = (m, n), y' = (p, k)$ and so $F = \mathbb{Z}x' \oplus \mathbb{Z}y'$. Now show that if $w \in F$ is such that $F = \mathbb{Z}w \oplus \mathbb{Z}y = \mathbb{Z}w \oplus \mathbb{Z}y'$, then $w = (\pm 1, l)$, where $(\pm 1)k - lp = \pm 1$, a contradiction (showing that $\mathbb{Z}$ fails to substitute; a proof of this fact without having recourse to 15.7.)

(6) Prove that a substitutable module is not isomorphic to any of its proper submodules.

(7) Give an example of a ring $R$ such that $_R R$ has the exchange property but is not substitutable.

(8) Let $f$ be a morphic endomorphism of the module $M$ (see [260]).

   (i) Show that $f$ is monic if and only if it is epic.

   (ii) Show that if $g$ is an automorphism of $M$, then both $fg$ and $gf$ are morphic.

(9) Following [260], a module $M$ is called *morphic* if every endomorphism of $M$ is morphic.

   (i) Show that $M$ is morphic if and only if, given any $K, N \subseteq M$ with $M/K \simeq N$, then $M/N \simeq K$.

   (ii) Let $M$ be morphic with $K \subseteq M$. Show that
   ($\alpha$) if $M \simeq K$, then $K = M$ and
   ($\beta$) if $M \simeq M/K$, then $K = 0$.

   (iii) Show that, for any $n \geq 2$, the $\mathbb{Z}$-module $\mathbb{Z}_n$ is morphic but neither $\mathbb{Z}$ nor, for any prime $p$, the Prüfer group $\mathbb{Z}_{p^\infty}$ are morphic $\mathbb{Z}$-modules.

   (iv) Show that the left $R$-module $R$ is morphic if and only if, for any $a \in R$, $R/\mathbf{l}(a) \simeq Ra$, where $\mathbf{l}(a)$ denotes the left annihilator of $a$. (In this case $R$ is said to be a *left morphic* ring.)

(v) A ring $R$ is called *right principally injective* if, for any $a \in R$, any homomorphism from $aR$ to $R$ can be extended to an endomorphism of $R_R$. Show that any left morphic ring $R$ (see (iv)) is right principally injective.

(10) Following [210], an element $a$ of the ring $R$ is said to have the *right unique generator property* (*right UG* for short) if, given any $b \in R$ for which $aR = bR$, then $b = au$ for some $u \in U(R)$. Left UG is defined accordingly. The equivalence of (a) and (b) in 15.33 says that if $R$ has the exchange property, then $\mathrm{sr}(R) = 1$ if and only if every $a \in R$ has the right (respectively, left) unique generator property (but note that (a)$\Rightarrow$(b) is true in general).

(i) Show that the following statements are equivalent. (See [58, Corollary 4.4].)

(a) Given any $x, y \in R$ for which $xR + \mathbf{r}(y) = R$, there is a $z \in \mathbf{r}(y)$ for which $x + z \in U(R)$ (where $\mathbf{r}(y)$ denotes the right annihilator of $y$);

(b) all elements of $R$ have the right UG property;

(c) given any $a, b \in R$ for which $aR = bR$, then $b = au$ for some right invertible element $u \in R$.

(ii) Let $R = \mathbb{Z}\langle x, y\rangle/(x^2, yx)$. Show that all elements of $R$ are right and left UG but $R$ does not have stable range 1. (See [58, Example 6.6].)

**15.39. Comments.** This section was strongly influenced by parts of Lam's survey [217] and the recent papers [210] by Khurana and Lam, [261] by Nicholson and Sánchez Campos, and [319] by Song et al.

Bass introduced the concept of stable range in his study [35] of the stability properties of the general linear group. Some of the basic properties of stable range 1 appear first in [35], in Vaserstein's [350], [351], and in notes by Kaplansky [187]. The proof of the symmetry of the condition in 15.4 is taken from Goodearl's [129] and, as he says, is adapted from the proof of Vaserstein's [350, Theorem 2].

As a variation on the regular ring theme, a ring $R$ is called *strongly $\pi$-regular* if for each $a \in R$ there is an $n \in \mathbb{N}$ and an $x \in R$ for which $a^n = a^{n+1}x$. Any strongly $\pi$-regular ring $R$ is $\pi$-regular and so, by Exercise (12) on page 121, is an exchange ring. Using this and an intricate calculation to show that elements of $R$ are unit-regular, Ara [17] deduces that every strongly $\pi$-regular ring has stable range 1.

Modules with the substitution property were first considered by Fuchs in [108]. There he proved 15.6, 15.10, and, under the assumption that the module is self-projective, 15.7. The general form of 15.7 was proved later by Warfield in [361]. The proof we give here is (slightly modified) from Fuchs and Salce [109, Proposition 8.1].

There has been considerable work done on the cancellation property for modules and Lam's exposition [217] gives details of and references for both positive and negative results.

Given two non-isomorphic $R$-modules $X$ and $Y$, if we set $M = X \oplus Y \oplus X \oplus Y \oplus \cdots$, then $M \oplus X \simeq M \oplus Y$, indeed both isomorphic to $M$, but $M$ is not cancellable. (This construction is known as *Eilenberg's trick* and gives some motivation to restricting the study of the cancellation property to finitely generated modules.)

Morphic endomorphisms, morphic modules and morphic rings (see the exercises for the latter two) are studied by Nicholson and Sánchez Campos in [260] and [261]. 15.16,

15.17, and 15.19 appear in [261], where part (1) of 15.19 is attributed to Ware [356] and, earlier, Azumaya [25]. The reader is also referred to two recent articles on these topics by Chen, Li and Zhou [67] and Nicholson [259].

The terminology *partially unit-regular* comes from Hall et al. [142] while *IC* is from [210]. The important characterisation given by 15.27, and its corollaries, are due to Ehrlich [86], but also appear in [261]. The conjugate idempotents characterisation given by 15.30 appears in various places and we have followed its treatment by Song et al in [319].

15.33 appears as Khurana and Lam's [210, Theorem 6.5]. As they say, it unifies the criteria for an exchange ring to have stable range 1 obtained by Yu and Camillo in [376, Theorem] and [53, Theorem 3], and by Chen in [62, Proposition 3.5], [63, Lemma 1.1], and [64, Theorem 6]. Earlier versions of it also appear in Canfell's [58] and Song et al. [319]. Corollary 15.35 was proved independently by Fuchs [108, Theorem 4, Corollary 1] and Kaplansky [187, Theorems 2, 3]. 15.37 appears as Yu's [376, Theorem 10].

It follows from 15.33 that every finitely generated projective module over a unit-regular ring $R$ is cancellable. However, Goodearl [131, Example 5.13] shows that, for any field $F$, if $R = \prod_{n \in \mathbb{N}} M_n(F)$, then $R$ is a unit-regular ring for which there is a cyclic module which is not cancellable.

**References.** Ara [17]; Azumaya [25]; Bass [35]; Chen, Li and Zhou [67]; Fuchs [108]; Goodearl [129, 131]; Kaplansky [187]; Khurana and Lam [210]; Lam [217]; Nicholson [259]; Nicholson and Sánchez Campos [260, 261]; Song, Chu, and Zhu [319]; Vaserstein [350, 351]; Ware [356]; Warfield [361]; Yu [376].

# 16   Decomposition uniqueness and biuniformity

Recall that if $M$ is an $R$-module and $\{M_i\}_I$ and $\{N_j\}_J$ are two families of submodules of $M$ such that $M$ can be decomposed as both $M = \bigoplus_I M_i$ and $M = \bigoplus_J N_j$, then these decompositions are said to be *equivalent* if there is a bijection $\sigma : I \to J$ such that $M_i \simeq N_{\sigma(i)}$ for all $i \in I$.

The classical Krull–Schmidt Theorem states that if $M$ is a module of finite composition length, then it decomposes as a direct sum of finitely many indecomposable modules and this decomposition is unique in the sense that any two such decompositions are equivalent.

In this section we will look at questions of uniqueness in the decomposition of a module into a direct sum of hollow or uniform modules. Although (as we will see in Example 16.18) the Krull–Schmidt Theorem is no longer true if we replace modules of finite length by finite direct sums of hollow or uniform modules (indeed even uniserial), such decompositions can still be "unique" in a more general sense. This leads to so-called *Weak Krull–Schmidt Theorems*, the study of which was initiated by Facchini (see, e.g., [81]). These are described using two weaker forms of equivalence and we now define the first of these.

**16.1.  Epigeny classes.** Two modules $M$ and $N$ are said to be *epi-equivalent*  or *belong to the same epigeny class* if there is an epimorphism from $M$ to $N$ and an epimorphism from $N$ to $M$. In this case we write $[M]_e = [N]_e$.

It is clear to see that if $[M]_e = [N]_e$, then the class of $M$-generated modules coincides with its $N$-generated counterpart.

**16.2.  Remark.** If $[M]_e = [N]_e$ with epimorphisms $f : M \to N$ and $g : N \to M$, then $fg : M \to M$ is an epimorphism. Then, if $M$ is also a noetherian module, $fg$ is in fact an isomorphism (see, e.g., [363, 31.13]) and so, in this case, $M \simeq N$.

We now use epi-equivalence to state and prove the following weak version of the Krull–Schmidt Theorem.

**16.3.  The Weak Krull–Schmidt Theorem for hollow modules.** *Let $A_1, A_2, \ldots, A_m$, $B_1, B_2, \ldots, B_n$ be hollow $R$-modules. Then*

$$[A_1 \oplus A_2 \oplus \cdots \oplus A_m]_e = [B_1 \oplus B_2 \oplus \cdots \oplus B_n]_e$$

*if and only if $m = n$ and there is a permutation $\sigma$ of $\{1, 2, \ldots, n\}$ such that*

$$[A_i]_e = [B_{\sigma(i)}]_e \text{ for each } i.$$

*Proof.* The sufficiency is clear.

For the necessity, we will use 30.7, the Krull–Schmidt Theorem for bipartite digraphs, proved in the Appendix.

Set $A = A_1 \oplus A_2 \oplus \cdots \oplus A_m$ and $B = B_1 \oplus B_2 \oplus \cdots \oplus B_n$. Then, by hypothesis, there are epimorphisms

$$\alpha : A \to B \text{ and } \beta : B \to A.$$

For each $i \in \{1, \ldots, m\}, j \in \{1, \ldots, n\}$, let

$$\varepsilon_i : A_i \to A, \ \pi_i : A \to A_i, \ \varepsilon'_j : B_j \to B, \ \pi'_j : B \to B_j$$

be the structural monomorphisms and epimorphisms and define

$$\varphi_{i,j} : A_i \to B_j \text{ and } \varphi'_{j,i} : B_j \to A_i$$

by $\varphi_{i,j} = \varepsilon_i \alpha \pi'_j$ and $\varphi'_{j,i} = \varepsilon'_j \beta \pi_i$.

Let $D = (X, Y, E)$ be the bipartite digraph in which $X = \{A_1, A_2, \ldots, A_m\}$, $Y = \{B_1, B_2, \ldots, B_n\}$, and there is a directed edge *from $A_i$ to $B_j$* if $\varphi'_{j,i} : B_j \to A_i$ is an epimorphism and one *from $B_j$ to $A_i$* if $\varphi_{i,j} : A_i \to B_j$ is an epimorphism.

Let $T \subseteq X$, set $t = |T|$ and $s = |\mathrm{N}^+(T)|$. We may relabel the vertices of $D$ so that

$$T = \{A_1, \ldots, A_t\} \text{ and } \mathrm{N}^+(T) = \{B_1, \ldots, B_s\}.$$

Then, for each $i = 1, \ldots, t$ and each $j = s + 1, \ldots, n$, the morphisms $\varphi'_{j,i}$ are not epimorphisms. Thus, since $A_i$ is hollow, for each $i = 1, \ldots, t$,

$$D_i := \sum_{j=s+1}^{n} (B_j)\varphi'_{j,i} \neq A_i.$$

Then, letting $\hat{\pi}_i : A_i \to A_i/D_i$ denote the natural epimorphism, the composite map

$$B_j \xrightarrow{\varepsilon'_j} B \xrightarrow{\beta} A \xrightarrow{\pi_i} A_i \xrightarrow{\hat{\pi}_i} A_i/D_i$$

is zero for each $i = 1, \ldots, t$ and $j = s + 1, \ldots, n$ since $(B_j)\varepsilon'_j \beta \pi_i = (B_j)\varphi'_{j,i} \subseteq D_j$. Thus, for each such $i$, it follows that $\bigoplus_{j=s+1}^{n} B_j = \sum_{j=s+1}^{n}(B_j)\varepsilon'_j \subseteq \mathrm{Ke}\,(\beta \pi_i \hat{\pi}_i)$ and so there is a homomorphism $\psi_i : \bigoplus_{j=1}^{s} B_j \to A_i/D_i$ for which the diagram

$$
\begin{array}{ccc}
\bigoplus_{j=1}^{n} B_j & \xrightarrow{\beta} & \bigoplus_{i=1}^{m} A_i \\
\downarrow & & \downarrow{\scriptstyle \pi_i \hat{\pi}_i} \\
\bigoplus_{j=1}^{s} B_j & \xrightarrow{\psi_i} & A_i/D_i
\end{array}
$$

is commutative (where the left vertical map is the natural epimorphism). Then, by the universal property of products, there is a homomorphism

$\gamma : \bigoplus_{j=1}^{s} B_j \to \bigoplus_{i=1}^{t} A_i/D_i$ such that the diagram

$$
\begin{array}{ccc}
\bigoplus_{j=1}^{n} B_j & \xrightarrow{\ \beta\ } & \bigoplus_{i=1}^{m} A_i \\
\downarrow & & \downarrow \\
\bigoplus_{j=1}^{s} B_j & \xrightarrow{\ \gamma\ } & \bigoplus_{i=1}^{t} A_i/D_i
\end{array}
$$

commutes, where the vertical maps are the natural epimorphisms. From this it follows that $\gamma$ is an epimorphism. Then, since each $A_i/D_i$ is a hollow module, it follows from 5.2 that $t \le s$, that is, $|T| \le |N^{+}(T)|$.

A similar argument shows that $|T| \le |N^{+}(T)|$ for all $T \subseteq Y$. Thus we may apply 30.7 to infer that $m = n$ and that a suitable reordering of the vertices of $D$ gives, for each $i = 1, \ldots, m$, a directed path from $A_i$ to $B_i$ and one from $B_i$ to $A_i$. These paths give epimorphisms from $A_i$ to $B_i$ and from $B_i$ to $A_i$ and so $[A_i]_e = [B_i]_e$ for each $i$, as required. $\qquad\square$

As noted above in 16.2, two noetherian modules are in the same epigeny class precisely when they are isomorphic. This immediately gives the following corollary.

**16.4. The Krull–Schmidt Theorem for hollow noetherian modules.**
Let $A_1, \ldots, A_m$, $B_1, \ldots, B_n$ be hollow noetherian $R$-modules. Then

$$
A_1 \oplus A_2 \oplus \cdots \oplus A_m \simeq B_1 \oplus B_2 \oplus \cdots \oplus B_n
$$

if and only if $m = n$ and there is a permutation $\sigma$ of $\{1, 2, \ldots, n\}$ such that $A_i \simeq B_{\sigma(i)}$ for each $i$.

We now give an example to show that 16.3 does not extend to infinite direct sums.

**16.5. Example.** Let $M$ be a module with submodule lattice isomorphic to $\mathbb{N} \cup \{\infty\}$, with the natural ordering in reverse:

$$
0 = M_\infty \subset \cdots \subset M_n \subset \cdots \subset M_2 \subset M_1 \subset M_0 = M.
$$

For example, if $R$ is the ring $\mathbb{Z}_p$ of integers localised at the prime $p$, we can take $M$ to be the $R$-module $R$ (with $M_n = \mathbb{Z}_p p^n$ for each $n \ge 1$).

For each $n \ge 1$, let $A_n = M/M_{2n+1}$ and $B_n = M/M_{2n}$. Then $A_n$ and $B_n$ are uniserial modules with composition series of lengths $2n + 1$ and $2n$ respectively. Moreover, for each $n$, there are natural epimorphisms $\pi_{2n+1} : A_n \to B_n$ and $\pi_{2n+2} : B_{n+1} \to A_n$. These induce an epimorphism from $\bigoplus_n A_n \to \bigoplus_n B_n$ and an epimorphism from $\bigoplus_n B_n \to \bigoplus_n A_n$, as displayed in the following diagram:

$$
\begin{array}{ccccccccccc}
A_1 & \oplus & A_2 & \oplus & A_3 & \oplus & \cdots & \oplus & A_n & \oplus & \cdots \\
\downarrow & \searrow & \downarrow & \searrow & \downarrow & \searrow & \cdots & \searrow & \downarrow & \searrow & \cdots \\
B_1 & \oplus & B_2 & \oplus & B_3 & \oplus & \cdots & \oplus & B_n & \oplus & \cdots
\end{array}
$$

Consequently, $[\bigoplus_n A_n]_e = [\bigoplus_n B_n]_e$. However, since all $A_m$ and $B_n$ have different composition lengths, no two of these belong to the same epigeny class.

We now consider the dual of epigeny.

**16.6. Monogeny classes.** Two modules $M$ and $N$ are said to be *mono-equivalent* or *belong to the same monogeny class* if there is a monomorphism from $M$ to $N$ and a monomorphism from $N$ to $M$. In this case we write $[M]_m = [N]_m$.

It is clear to see that if $[M]_m = [N]_m$, then the class of $M$-cogenerated modules coincides with its $N$-cogenerated counterpart. We note that a module $M$ is called *compressible* if it is mono-equivalent to each of its nonzero submodules.

**16.7. Remark.** Given $[M]_m = [N]_n$, with the monomorphisms $f : M \to N$ and $g : N \to M$, then $fg : M \to M$ is a monomorphism. Then, if $M$ is also an artinian module, $fg$ is in fact an isomorphism (see, e.g., [363, 31.13]) and so, in this case, $M \simeq N$.

We now state without proof the dual of 16.3. (Its proof is analogous to that of 16.3.)

**16.8. The Weak Krull–Schmidt Theorem for uniform modules.** *Let* $A_1, \ldots, A_k$, $B_1, \ldots, B_l$ *be uniform R-modules. Then*

$$[A_1 \oplus A_2 \oplus \cdots \oplus A_k]_m = [B_1 \oplus B_2 \oplus \cdots \oplus B_l]_m$$

*if and only if* $k = l$ *and there is a permutation* $\sigma$ *of* $\{1, 2, \ldots, k\}$ *such that*

$$[A_i]_m = [B_{\sigma(i)}]_m \text{ for each } i.$$

Dualising Example 16.5 shows that, as for 16.3, 16.8 also does not extend to infinite direct sums.

As a direct consequence of 16.8 and 16.7 we have the following analogue of 16.4.

**16.9. The Krull–Schmidt Theorem for uniform artinian modules.** *Let* $A_1, \ldots, A_m$, $B_1, \ldots, B_n$ *be uniform artinian R-modules. Then*

$$A_1 \oplus A_2 \oplus \cdots \oplus A_m \simeq B_1 \oplus B_2 \oplus \cdots \oplus B_n$$

*if and only if* $m = n$ *and there is a permutation* $\sigma$ *of* $\{1, 2, \ldots, n\}$ *such that* $A_i \simeq B_{\sigma(i)}$ *for each* $i$.

The rest of this section is mainly concerned with the following type of module.

**16.10. Biuniform modules.** A module $M$ which is both uniform and hollow is called *biuniform*.

The name is derived from the fact that hollow modules are also called *co-uniform* by some authors. Clearly uniserial modules are biuniform. The next example shows that biuniform modules need not be uniserial.

**16.11. Example.** Let $F$ be a field and $X = \{x_i \,|\, i \in I\}$ be a set of commuting indeterminates, with $|I| \geq 2$. Let $R$ be the polynomial ring $F[X]$ subject to the relations $x_i x_j = 0$ for any two distinct $i$ and $j$, $x_i^2 = x_j^2$ for all $i$ and $j$, and $x_i^3 = 0$ for all $i \in I$. Then $R$ is a local ring with radical generated by $\{x_i \,|\, i \in I\}$ and essential socle generated by $x_i^2$ for any $i \in I$. Consequently, the $R$-module $R$ is biuniform. However $R$ is not a uniserial module since, for example, the ideals $Rx_i$ and $Rx_j$ are incomparable for any distinct pair $i, j$.

Given the key rôle that local endomorphism rings have in Azumaya's generalisation of the Krull–Schmidt Theorem (12.6), in order to investigate uniqueness of decompositions into biuniform modules it is important to determine when these modules are LE-modules. In preparation for this, we now take a closer look at maps between uniform and hollow modules.

**16.12. Maps from uniform to hollow modules.** *Let $U$ be a uniform module and $H$ be a hollow module.*

(1) *If $f, g : U \to H$ are two homomorphisms such that $f$ is injective but not surjective and $g$ is surjective but not injective, then $f + g$ is an isomorphism.*

(2) *Let $f_1, \ldots, f_n : U \to H$ be $n$ homomorphisms such that $f_1 + \cdots + f_n$ is an isomorphism. Then either*

    (i) *one of the $f_i$'s is an isomorphism or*

    (ii) *there are distinct indices $i, j \in \{1, \ldots, n\}$ for which $f_i$ is injective but not surjective and $f_j$ is surjective but not injective.*

(3) *$U \simeq H$ if and only if $[U]_m = [H]_m$ and $[U]_e = [H]_e$.*

*Proof.* (1) We have $H = (U)g \subseteq (U)f + (U)(f + g)$. Thus, since $H$ is hollow and $f$ is not surjective, it follows that $f + g$ must be surjective. A dual argument shows that $f + g$ must also be injective, as required.

    (2) Since $\bigcap_{i=1}^n \mathrm{Ke}\, f_i \subseteq \mathrm{Ke}\, (f_1 + \cdots + f_n) = 0$ and $U$ is uniform, we have $\mathrm{Ke}\, f_i = 0$ for some $i$. Since $B = \mathrm{Im}\, f_1 + \cdots + f_n \subseteq \sum_{i=1}^n \mathrm{Im}\, f_i$ and $H$ is hollow, we have $\mathrm{Im}\, f_j = 0$ for some $j$. If either $f_i$ or $f_j$ is an isomorphism, then (i) holds; otherwise (ii) holds.

    (3) This follows easily from (1). $\qquad\qquad\qquad\qquad\qquad\qquad\qquad\qquad\square$

We now characterise biuniform LE-modules.

**16.13. Biuniform LE-modules.** *For a biuniform module $M$, the following statements are equivalent:*

(a) *$\mathrm{End}(M)$ is a local ring;*

(b) *in $\mathrm{End}(M)$ either all monomorphisms are isomorphisms or all epimorphisms are isomorphisms.*

*Proof.* (a)⇒(b). Assume there exists $f \in \text{End}(M)$ that is monomorph but not epimorph. Then, for any epimorphism $g \in \text{End}(M)$ which is not an isomorphism, the sum $f + g$ is an isomorphism (by 16.12). This contradicts the fact that in any local ring the sum of two non-units is a non-unit.

(b)⇒(a). For any $f \in \text{End}(M)$, $\text{Ke}\, f \cap \text{Ke}\,(1 - f) = 0$ and so the uniformity of $M$ implies that either $f$ or $1 - f$ is a monomorphism. Moreover, $\text{Im}\, f + \text{Im}\,(1 - f) = M$ and so, since $M$ is hollow, $\text{Im}\, f = M$ or $\text{Im}\,(1 - f) = M$. Now (b) implies that $f$ or $1 - f$ is an isomorphism which means that $\text{End}(M)$ is a local ring. $\qquad \square$

**16.14. Remarks.** Notice that $\mathbb{Z}_{p^\infty}$ is a uniserial module with local endomorphism ring for which epimorphisms need not be isomorphisms.

In general, the endomorphism ring of a biuniform module need not be local. Indeed the following example shows that, even over a uniserial domain $R$ of Krull dimension 2, a cyclic uniserial $R$-module $M$ need not have local endomorphism ring.

**16.15. Example. A biuniform module which is not an LE-module.** Let $\mathbb{Z}[i]$ denote the ring of Gaussian integers and $V$ be its localisation $\mathbb{Z}[i]_{(2+i)}$ at its prime ideal generated by $2 + i$. Then $V$ is a discrete valuation domain, that is, a commutative noetherian uniserial domain, with quotient field $\mathbb{Q}[i]$. Let $\sigma$ be the automorphism on $\mathbb{Q}[i]$ given by complex conjugation and $S$ be the skew power series ring $\mathbb{Q}[i][[x; \sigma]]$, writing the power series $a_0 + a_1 x + \cdots + a_n x^n + \cdots$ with coefficients on the left of the indeterminate $x$ and multiplication respecting the rule $xa - \sigma(a)x$ for all $a \in \mathbb{Q}[i]$.

Next let $R$ be the subring of $S$ consisting of those power series having constant term in $V$. Then $R$ is a left and right uniserial domain of Krull dimension 2 (see [287] for details). Let $M$ be the cyclic uniserial $R$-module $R/Rx$. Defining $f \in \text{End}_R(M)$ to be right multiplication by $2 + i$ gives a monomorphism $f$ which is not an epimorphism. Similarly, defining $g \in \text{End}_R(M)$ to be right multiplication by $2 - i$ gives an epimorphism $g$ which is not a monomorphism. By 16.13, the existence of such $f$ and $g$ imply that $\text{End}_R(M)$ is not local.

We also note that $Rx^2$ is a two-sided ideal of $R$. Then the factor ring $R/Rx^2$, although not a domain, is a uniserial ring of Krull dimension 1 and $M$ (as before) is a uniserial $R/Rx^2$-module, again with non-local endomorphism ring.

In fact, it follows from 19.6 and 18.10 that if $M$ is biuniform, then $S = \text{End}(M)$ is either local or semilocal with factor $S/\text{Jac}\, S$ of length 2.

Combining the Weak Krull–Schmidt Theorems for hollow and uniform modules 16.3 and 16.8, gives the necessity of the following result, also due to Facchini. We refer to [91, Theorem 1.9] or [92, Theorem 9.13] for the proof of the sufficiency.

**16.16. The Weak Krull–Schmidt Theorem for biuniform modules.** *Let* $A_1, \ldots, A_k$, $B_1, \ldots, B_l$ *be biuniform R-modules. Then*

$$A_1 \oplus A_2 \oplus \cdots \oplus A_k \simeq B_1 \oplus B_2 \oplus \cdots \oplus B_l$$

*if and only if* $k = l$ *and there are permutations* $\sigma, \tau$ *of* $\{1, 2, \ldots, k\}$ *such that*

$$[A_i]_m = [B_{\sigma(i)}]_m \text{ and } [A_i]_e = [B_{\tau(i)}]_e \text{ for each } i.$$

The next result prepares for the main example of this section.

**16.17. Lemma.** *Let* $M$ *be an R-module,* $U$ *be a uniform R-module, and* $H$ *be a hollow R-module such that* $[M]_m = [U]_m$ *and* $[M]_e = [H]_e$, *with monomorphisms* $f : M \to U$, $g : U \to M$ *and epimorphisms* $u : M \to H$, $v : H \to M$, *that is, we have the diagram*

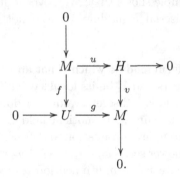

*Then* $M$ *is biuniform and* $M \oplus N \simeq U \oplus H$ *where* $N = (U)gv^{-1}$.

*Proof.* Since $f : M \to U$ is a monomorphism and $U$ is uniform, $M$ is also uniform. Since $v : H \to M$ is an epimorphism and $H$ is hollow, $M$ is also hollow and so biuniform.

If the monomorphism $fg : M \to M$ is also an epimorphism, then both $f$ and $g$ are isomorphisms. From this we get $N = (U)gv^{-1} = (M)v^{-1} = H$ and so $M \oplus N = M \oplus H \simeq U \oplus H$, as required. On the other hand, if the epimorphism $uv : M \to M$ is a monomorphism, then $M \oplus N = M \oplus (U)g \simeq H \oplus U$, as required. Thus we may suppose that $fg$ is not surjective and $uv$ is not injective. Then, by Lemma 16.12 (1), $\gamma := fg + uv : M \to M$ is an isomorphism. Next define $\alpha : M \to U \oplus H$ by $(m)\alpha = ((m)f, (m)u)$ for all $m \in M$ and $\beta : U \oplus H \to M$ by $(u, h)\beta = (u)g + (h)v$ for all $u \in U, h \in H$. Then it is straightforward to check that $\alpha$ is a monomorphism and that $\alpha\beta = \gamma$. Thus, since $\alpha(\beta\gamma^{-1}) = 1_M$, the monomorphism $\alpha : M \to U \oplus H$ splits giving $U \oplus H \simeq M \oplus \mathrm{Ke}\,\beta\gamma^{-1}$. It is easily seen that $\mathrm{Ke}\,\beta\gamma^{-1} = \mathrm{Ke}\,\beta \simeq (U)gv^{-1}$, completing the proof.  $\square$

The following example is due to Facchini, appearing in more generality as [91, Example 2.1], [92, Example 9.20], and [287, pp. 30–32]. It demonstrates that in general the monogeny-epigeny relations of 16.16 can not be strengthened to isomorphisms by exhibiting the failure of the Krull–Schmidt Theorem for serial modules; more specifically, it shows that over a serial ring $R$ the decomposition of a finitely presented $R$-module into uniserial summands may not be unique up to isomorphism, thereby answering in the negative a problem posed by Warfield [360]. (Recall that a module is called *serial* if it can be written as a direct sum of uniserial modules, while the ring $R$ is called *left serial* if $_RR$ is a serial module.)

**16.18. Example.** Let $M_2(\mathbb{Q})$ be the ring of $2 \times 2$ matrices with rational entries. Let $p$ and $q$ be two distinct primes and let $\Lambda_p$ and $\Lambda_q$ be the subrings of $M_2(\mathbb{Q})$ given by

$$\Lambda_p = \begin{bmatrix} \mathbb{Z}_p & p\mathbb{Z}_p \\ \mathbb{Z}_p & \mathbb{Z}_p \end{bmatrix} \text{ and } \Lambda_q = \begin{bmatrix} \mathbb{Z}_q & q\mathbb{Z}_q \\ \mathbb{Z}_q & \mathbb{Z}_q \end{bmatrix}$$

where $\mathbb{Z}_p$ and $Z_q$ are the localisations of $\mathbb{Z}$ at the primes $p$ and $q$ respectively. Next let $R$ be the subring of the ring $M_4(\mathbb{Q})$ of $4 \times 4$ matrices with rational entries given by

$$R = \begin{bmatrix} \Lambda_p & M_2(\mathbb{Q}) \\ 0 & \Lambda_q \end{bmatrix} = \begin{bmatrix} \mathbb{Z}_p & p\mathbb{Z}_p & \mathbb{Q} & \mathbb{Q} \\ \mathbb{Z}_p & \mathbb{Z}_p & \mathbb{Q} & \mathbb{Q} \\ 0 & 0 & \mathbb{Z}_q & q\mathbb{Z}_q \\ 0 & 0 & \mathbb{Z}_q & \mathbb{Z}_q \end{bmatrix}.$$

For each $i = 1, 2, 3, 4$, let $e_i$ be the diagonal matrix with 1 as the $i$-th diagonal entry. Then $\{e_1, \ldots, e_4\}$ is a complete set of primitive orthogonal idempotents for $R$ and the principal left and right ideals $Re_i$ and $e_iR$ are uniserial $R$-modules (with the submodule lattice of $Re_3$ detailed immediately below); thus $R$ is both a left and right serial ring (and so semiperfect). Moreover one can check that the $Re_i$'s are pairwise non-isomorphic and so, setting $S_i = Re_i/Je_i$ where $J$ is the Jacobson radical of $R$, $\{S_1, S_2, S_3, S_4\}$ is an irredundant set of representatives of the simple left $R$-modules (see, for example, [363, Exercise 21.17 (3)]).

Set $V = (\mathbb{Q}, \mathbb{Q}, \mathbb{Z}_q, \mathbb{Z}_q)$. Then, regarding it as a $4 \times 1$ column vector, $V$ is simply the left $R$-module $Re_3$. It is straightforward to check that the submodule lattice of $V$ is given by the chain

$$0 \subset \cdots \subset D_n \subset C_n \subset \cdots \subset D_1 \subset C_1 \subset W \subset \cdots$$
$$\cdots \subset B_n \subset A_n \subset \cdots \subset B_1 \subset A_1 \subset V$$

where, for each $i \in \mathbb{N}$,

$$A_i = (\mathbb{Q}, \mathbb{Q}, q^i\mathbb{Z}_q, q^{i-1}\mathbb{Z}_q), \quad B_i = (\mathbb{Q}, \mathbb{Q}, q^i\mathbb{Z}_q, q^i\mathbb{Z}_q), \quad W = (\mathbb{Q}, \mathbb{Q}, 0, 0),$$
$$C_i = (p^{i-1}\mathbb{Z}_p, p^{i-1}\mathbb{Z}_p, 0, 0), \quad D_i = (p^i\mathbb{Z}_p, p^{i-1}\mathbb{Z}_p, 0, 0).$$

For each $i \in \mathbb{N}$, the simple factors $A_i/B_i, B_i/A_{i+1}, C_i/D_i, D_i/C_{i+1}$ are isomorphic to $S_4, S_3, S_2, S_1$ respectively, while the remaining simple factors $V/A_1$ and $W/C_1$ are respectively isomorphic to $S_3$ and $S_1$.

Now let $K, L, M, N$ be the uniserial modules given by $V/C_1, A_1/C_1, V/D_1,$ $A_1/D_1$ respectively. We show that $M \oplus L \simeq N \oplus K$ using Lemma 16.17. To this end, define $f : M \to N$ to be multiplication by $q$ and $g : N \to M$ to be the inclusion map. Then both $f$ and $g$ are monomorphisms which are not epimorphisms. Next define $u : M \to K$ to be the natural epimorphism and $v : K \to M$ to be the epimorphism given by the composition of the natural epimorphism $\pi : K \to V/W$ and the isomorphism $V/W \simeq M$ induced by multiplication by $p$. Then neither $u$ nor $v$ are monomorphisms. Next note that $(N)gv^{-1} = A_1/C_1 = L$. Thus, by Lemma 16.17, we have $M \oplus L \simeq N \oplus K$ as claimed.

However the uniserial modules $K$, $L$, $M$, $N$ have socles isomorphic to $S_1$, $S_1, S_2, S_2$ respectively and (unique) simple factors isomorphic to $S_3$, $S_4$, $S_3$, $S_4$ respectively, showing that they are pairwise non-isomorphic.

We finish this section with two further propositions which show that biuniform modules provide some interesting splitting properties when they are involved in direct sum decompositions.

**16.19. Proposition.** *Let $A$ be a biuniform module and let $B, C_1, \ldots, C_n$ be modules such that $A \oplus B = C_1 \oplus \cdots \oplus C_n$, where $n \geq 2$. Then there are two distinct indices $i, j \in \{1, \ldots, n\}$ and modules $A', B'$ for which $A \simeq A'$, $A' \oplus B' \simeq C_i \oplus C_j$, and $B \simeq B' \oplus (\bigoplus_{k \neq i,j} C_k)$.*

*Proof.* Denote the canonical embeddings and projections with respect to the direct sums $A \oplus B$ and $C_1 \oplus \cdots \oplus C_n$ by $\varepsilon_A, \pi_A, \varepsilon_B, \pi_B$, and $\varepsilon_i, \pi_i$ for $i = 1, \ldots, n$. Then in the ring $E = \mathrm{End}(A)$ we have

$$1_E = \varepsilon_A \pi_A = \varepsilon_A \left( \sum_{i=1}^{n} \pi_i \varepsilon_i \right) \pi_A = \sum_{i=1}^{n} \varepsilon_A \pi_i \varepsilon_i \pi_A.$$

Since $A$ is biuniform, it follows from Lemma 16.12 (2) that either (i) one of the $\varepsilon_A \pi_i \varepsilon_i \pi_A$'s is an isomorphism or (ii) there are distinct indices $i, j \in \{1, \ldots, n\}$ for which $\varepsilon_A \pi_i \varepsilon_i \pi_A$ is injective but not surjective and $\varepsilon_A \pi_j \varepsilon_j \pi_A$ is surjective but not injective.

If (i), then taking 1 as the specific $i$, the monomorphism $\varepsilon_A \pi_1$ of $A$ into $C_1$ splits since $\varepsilon_A \pi_1 (\varepsilon_1 \pi_A (\varepsilon_A \pi_1 \varepsilon_1 \pi_A)^{-1}) = 1_A$. Thus there are modules $A', B^*$ with $A \simeq A'$ and $A' \oplus B^* = C_1$. This gives $A \oplus B = A' \oplus B^* \oplus C_2 \oplus (\bigoplus_{k \geq 2} C_k)$. However, since $A$ is biuniform, it cancels from direct sums (see 19.6, 15.36 and 15.11) so setting $B' = B^* \oplus C_2$ gives the required result.

If (ii), then for the specific indices $i, j$, it follows from Lemma 16.12(1) that $\alpha = \varepsilon_A \pi_i \varepsilon_i \pi_A + \varepsilon_A \pi_j \varepsilon_j \pi_A$ is an automorphism of $A$. Defining $\beta : A \to C_i \oplus C_j$ and $\gamma : C_i \oplus C_j \to A$ by

$$\beta : a \mapsto ((a)\varepsilon_A \pi_i, (a)\varepsilon_A \pi_j) \quad \text{and} \quad \gamma : (c_i, c_j) \mapsto (c_i)\varepsilon_i \pi_A \alpha^{-1} + (c_j)\varepsilon_j \pi_A \alpha^{-1},$$

it is easily checked that $\beta\gamma$ is the identity on $A$. Thus there are modules $A', B'$ with $A \simeq A'$ and $A' \oplus B' = C_i \oplus C_j$. Then, again since $A$ cancels from direct sums, the result follows. $\qquad\square$

**16.20. Proposition.** *Let $A$ and $B$ both be modules which are direct sums of finitely many biuniform modules. If $A$ is a direct summand of $B$, then $B/A$ is also a direct sum of (finitely many) biuniform modules.*

*Proof.* We use induction on $n = \text{u.dim}(B) = \text{h.dim}(B)$. If $n = 1$, then $B$ is biuniform and so either $A = B$ or $A = 0$ and so the result is trivially true.

Now suppose that $A \neq 0$ and $B = C_1 \oplus \cdots \oplus C_n$ where each $C_i$ is biuniform and $n \geq 2$. Then, by hypothesis, we may write $A = D \oplus A'$ where $D$ is biuniform and $B = A \oplus A^*$ for some submodule $A^*$ of $B$.

This gives $D \oplus (A' \oplus A^*) = C_1 \oplus \cdots \oplus C_n$ and so, by Proposition 16.19, there are distinct indices $i, j \in \{1, \ldots, n\}$ and modules $D', E'$ for which $D \simeq D'$, $D' \oplus E' \simeq C_i \oplus C_j$, and $A' \oplus A^* \simeq E' \oplus (\bigoplus_{k \neq i,j} C_k)$. Since u.dim and h.dim are additive on direct sums, it follows from $D' \oplus E' \simeq C_i \oplus C_j$ that $E'$ is also biuniform. Thus $E' \oplus (\bigoplus_{k \neq i,j} C_k)$ has uniform and hollow dimension $n - 1$. Applying the inductive hypothesis to the summand $A'$ of $A' \oplus A^*$ now shows that $A^*$ is a direct sum of biuniforms, as required. $\qquad\square$

**16.21. Exercises.**

(1) Let $M$ be a biuniform $R$-module and $S = \text{End}(M)$. Prove that the following are equivalent ([289, Theoerem 4.3]):

    (a) there exists an infinitely generated projective left $S$-module $\Gamma$ such that $P/\text{Rad}(P)$ is cyclic;

    (b) there exist non-isomorphisms $f, g \in S$ such that $f$ is epimorph, $g$ is mono-morph, and $fg = 0$;

    (c) there exists a cyclic flat left $S$-module which is not projective.

(2) An $R$-module $M$ is said to be *mono-correct* if, given any module $N$ for which $[M]_m = [N]_m$, then $N$ is in fact isomorphic to $M$.

    (i) Show that any artinian module $M$ is mono-correct (see [363, 31.13]).

    (ii) Show that if $M$ is uniserial and all injective endomorphisms of $M$ are automorphisms, then $M$ is mono-correct.

(3) An $R$-module $M$ is said to be *epi-correct* if, given any module $N$ for which $[M]_e = [N]_e$, then $N$ is in fact isomorphic to $M$.

    (i) Show that any noetherian module $M$ is epi-correct (see [363, 31.13]).

    (ii) Show that if $M$ is uniserial and all surjective endomorphisms of $M$ are automorphisms (i.e., $M$ is Hopfian), then $M$ is epi-correct.

(4) Show that the module $M$ is both mono-correct and epi-correct (see (2),(3)) in each of the following cases:

    (i) $M$ is noetherian and self-injective;

   (ii)  $M$ is artinian and self-projective;

   (iii)  $M$ has finite length;

   (iv)  $M$ is semisimple.

(5) Assume every projective left $R$-module to be mono-correct (see (2)). Show that $R$ is a left hereditary ring.

(6) Define a class $\mathcal{C}$ of $R$-modules to be

   *mono-correct* if, for any $N, M \in \mathcal{C}$, $[M]_m = [N]_m$ implies $M \simeq N$ and

   *epi-correct* if, for any $N, M \in \mathcal{C}$, $[M]_e = [N]_e$ implies $M \simeq N$.

   (a)  Prove that the following classes are mono-correct:

      (i)  the class of artinian modules;

      (ii)  the class of self-injective modules (see [48]);

      (iii)  the class of continuous modules (see [237, Corollary 3.18]);

      (iv)  the class of semisimple modules.

   (b)  Prove that the following classes are epi-correct:

      (i)  the class of noetherian modules;

      (ii)  the class of $\Sigma$-self-projective supplemented modules;
         (Hint: consider $M/\mathrm{Rad}\,(M)$ and $N/\mathrm{Rad}\,(N)$ and use the fact that these radicals are small (see [363, 42.3]).)

      (iii)  the class of semisimple modules.

(7) Prove that for a ring $R$ the following are equivalent (see [366]):

   (a)  The class of pure-injective left $R$-modules is mono-correct;

   (b)  every pure-injective left $R$-module is injective;

   (c)  every short exact sequence of left $R$-modules is pure;

   (d)  $R$ is a von Neumann regular ring.

(8) A module $M$ is called *pure mono-correct* if for any module $N$ such that $M$ and $N$ are isomorphic to pure submodules of each other, $M \simeq N$. Prove (see [301, 302, 366]):

   (i)  $R$ is a left noetherian ring if and only if all injective left $R$-modules are pure mono-correct.

   (ii)  $R$ is left pure semisimple (pure exact sequences in $R$-Mod split) if and only if every left $R$-module is pure mono-correct.


**16.22. Comments.** The Weak Krull–Schmidt Theorem for hollow modules, 16.3, appears as Diracca and Facchini's [81, Theorem 1.3] where it is stated in the more general setting of hollow objects in an abelian category. In fact there they prove in detail the Weak Krull–Schmidt Theorem for uniform objects and obtain the hollow result as an immediate corollary by transferring to the opposite category. Here we've chosen to illustrate the return journey. An alternative categorical approach for the uniform case is given by Krause in [212].

   Diracca and Facchini [81] refer to examples showing that the Weak Krull–Schmidt Theorem for hollow modules does not generalise to modules of finite hollow dimension.

These examples appear in Facchini's book [92, Chapter 8]. We note that while 16.9 shows that the Krull–Schmidt Theorem holds for uniform artinian modules, Facchini, Herbera, Levy and Vámos have shown in [96] that it fails for artinian modules in general.

Facchini [92, Definition 9.14] defines a biuniform module $B$ to be *Krull–Schmidt* if either it is the only biuniform module up to isomorphism in its monogeny class or if it is the only biuniform module up to isomorphism in its epigeny class. It follows immediately from 16.13 that this is the case if $\text{End}(B)$ is a local ring. He then shows [92, Corollary 9.15] that the biuniform module $B$ is Krull–Schmidt if and only if whenever $B$ is isomorphic to a direct summand of a direct sum of finitely many biuniform modules, then it is actually isomorphic to one of these biuniforms. As a consequence of this he shows that any biuniform module with local endomorphism ring is Krull–Schmidt, but he also gives an example of a Krull–Schmidt biuniform module with non-local endomorphism ring.

Dung and Facchini [84] show that if $\{U_i\}_I$ and $\{V_j\}_J$ are two locally semi-$T$-nilpotent families of nonzero uniserial modules, then their direct sums are isomorphic if and only if there are bijections $\sigma, \tau : I \to J$ such that $[U_i]_m = [V_{\sigma(i)}]_m$ and $[U_i]_e = [V_{\tau(i)}]_e$ for each $i \in I$. Their proof relies on Zimmermann-Huisgen and Zimmermann's result 12.14. They also define a module $M$ to be *quasismall* if, whenever $M$ is isomorphic to a direct summand of a direct sum $\oplus_{i \in I} N_i$ of arbitrary modules $N_i$, then there is a finite subset $J$ of $I$ for which $M$ is isomorphic to a direct summand of $\oplus_{j \in J} N_j$. Recently, Příhoda [285] has shown that the Krull–Schmidt Theorem holds for arbitrary families of uniserial modules that are not quasismall.

Example 16.15 initially came from Bessenrodt et al. [39, p. 66] and also appears, more accessibly, in Puninski's text [287, Example 2.12]. A further example of a cyclic uniserial module with non-local endomorphism ring is given by Facchini in [91, Example 2.4] and [92, Example 9.26].

16.19 and 16.20 appear in [91] and [82] respectively. Much more on the success and failure of the Krull–Schmidt Theorem in various scenarios can be found in Facchini's [91] – [95] and his coauthored works [33], [75], and [84].

The general notions of mono- and epi-correctness defined in the exercises were initiated in Rososhek's [301, 302] and developed further in [366]. As a consequence of their treatment, a ring $R$ is semisimple artinian if and only if the following equivalent conditions hold: (a) the class of all left $R$-modules is mono-correct (epi-correct); (b) the class of all (injective) left $R$-modules is mono-correct; (c) the class of all (projective) left $R$-modules is epi-correct.

**References.** Diracca and Facchini's [81, 82]; Dung and Facchini [84]; Facchini [91, 95, 92]; Facchini, Herbera, Levy and Vámos [96]; Krause [212]; Puninski[287, 289]; Rososhek [301, 302]; Wisbauer [366].

# Chapter 4

# Supplements in modules

The existence of complements of submodules $K$ in any module $M$ (i.e., submodules $L \subseteq M$ which are maximal with respect to $K \cap L = 0$) follows by Zorn's Lemma. Dually, submodules $L \subseteq M$ which are minimal with respect to $K + L = M$ need not exist in general, and so their existence has to be derived in special situations or stated as a condition. In this chapter we investigate various forms of such conditions and their consequences.

## 17  Semilocal and weakly supplemented modules

Recall that a ring $R$ is said to be *semilocal* provided $R/\operatorname{Jac} R$ is a left (right) semisimple ring. In this section we transfer this definition to modules $M$ and show how this is related to a weak form of supplements in $M$.

**17.1. Definition.** A module $M$ is called *semilocal* if $M/\operatorname{Rad} M$ is semisimple.

Thus a ring $R$ is semilocal if it is semilocal as a left (or right) $R$-module.

**17.2. Characterisation of semilocal modules.** *For a module $M$ the following statements are equivalent:*

(a) *$M$ is semilocal;*

(b) *for every $L \subseteq M$, there exists a submodule $K \subseteq M$ such that $L + K = M$ and $L \cap K \subseteq \operatorname{Rad} M$;*

(c) *there is a decomposition $M = M_1 \oplus M_2$ such that $M_1$ is semisimple and $M_2$ is semilocal with $\operatorname{Rad} M \trianglelefteq M_2$.*

*Proof.* (a)$\Rightarrow$(b) is clear, since $(L + \operatorname{Rad} M)/\operatorname{Rad} M$ is a summand of $M/\operatorname{Rad} M$.

(b)$\Rightarrow$(a). Let $L/\operatorname{Rad} M \subseteq M/\operatorname{Rad} M$. Then there is a submodule $K \subseteq M$ such that $L + K = M$ and $L \cap K \subseteq \operatorname{Rad} M$. Thus $L/\operatorname{Rad} M \oplus K/\operatorname{Rad} M = M/\operatorname{Rad} M$. Hence every submodule of $M/\operatorname{Rad} M$ is a direct summand.

(a)$\Rightarrow$(c). Let $M_1$ be a complement of $\operatorname{Rad} M$ in $M$. Then $M_1 \oplus \operatorname{Rad} M$ is essential in $M$. Since $M/\operatorname{Rad} M$ is semisimple, $M_1 = (M_1 \oplus \operatorname{Rad} M)/\operatorname{Rad} M$ is a direct summand in $M/\operatorname{Rad} M$, hence semisimple, and there is a semisimple submodule $M_2/\operatorname{Rad} M$ such that

$$(M_1 \oplus \operatorname{Rad} M)/\operatorname{Rad} M \oplus M_2/\operatorname{Rad} M = M/\operatorname{Rad} M.$$

Hence $M = M_1 + M_2$ and $M_1 \cap M_2 \subseteq M_1 \cap \operatorname{Rad} M = 0$. Thus $M = M_1 \oplus M_2$. Because $M_1$ is a complement, $\operatorname{Rad} M$ is essential in $M_2$.

(c)$\Rightarrow$(a) is clear, since $M/\operatorname{Rad} M \simeq M_1 \oplus (M_2/\operatorname{Rad} M)$.                □

**17.3. Properties of semilocal modules.** *The class of semilocal modules is closed under factor modules, coclosed submodules, direct sums and small covers.*

*Proof.* Suppose $f : M \to Q$ is an epimorphism. As $(\operatorname{Rad} M)f \subseteq \operatorname{Rad} Q$, $f$ induces an epimorphism $\overline{f} : M/\operatorname{Rad} M \to Q/\operatorname{Rad} Q$. This shows that the class of semilocal modules is closed under factor modules.

Similarly one shows that the class is closed under direct sums.

Now let $f : M \to Q$ be a small cover. Then $\operatorname{Ke} f \subseteq \operatorname{Rad} M$ and $(\operatorname{Rad} M)f = \operatorname{Rad} Q$. Hence $M/\operatorname{Rad} M \subseteq Q/\operatorname{Rad} Q$. Thus if $Q$ is semilocal, then so is $M$. Hence the class is closed under small covers.

Finally let $N$ be a coclosed submodule of a semilocal module $M$. Then $\operatorname{Rad} N = N \cap \operatorname{Rad} M$ (see 3.7) and so $N/\operatorname{Rad} N \simeq (N + \operatorname{Rad} M)/\operatorname{Rad} M \subseteq M/\operatorname{Rad} M$ is semisimple. Hence $N$ is semilocal.                □

**17.4. Definition.** A module $M$ is called *homogeneous semilocal* if $M/\operatorname{Rad} M$ is homogeneous semisimple.

Obviously any local module is homogeneous semilocal. The preceding proof also yields

**17.5. Class of homogeneous semilocal modules.** *The class of homogeneous semilocal modules is closed under factor modules, coclosed submodules, direct sums and small covers.*

**17.6. Properties of homogeneous semilocal modules.** *Let $M$ be an $R$-module.*

(1) *If $\operatorname{Rad} M \ll M$ and $M$ is homogeneous semilocal, then $\operatorname{Rad} M$ is the largest fully invariant proper submodule of $M$.*

(2) *If $M$ is semilocal and has a proper largest fully invariant submodule, then $M$ is homogeneous semilocal.*

(3) *Let $M$ be a finitely generated self-projective homogeneous semilocal module.*

  (i) *There exists a unique (up to isomorphism) $M$-cyclic indecomposable projective module $P$ in $\sigma[M]$.*

  (ii) *Every finitely $M$-generated $M$-projective module $Q$ is isomorphic to $P^k$, for some $k \in \mathbb{N}$.*

  (iii) *$P^{(\Lambda)} \simeq P^{(\Lambda')}$ if and only if the index sets $\Lambda$ and $\Lambda'$ have the same cardinality.*

*Proof.* We first note that a semisimple module $N$ is homogeneous semisimple if and only if $0$ and $N$ are its only fully invariant submodules. Also if $N$ is a semisimple module which is not homogeneous, then $N$ has no largest fully invariant proper submodule.

(1) As noted in 2.8, $\operatorname{Rad} M$ is a fully invariant submodule of $M$. Let $U \subset M$ be a proper fully invariant submodule of $M$. Then $(U + \operatorname{Rad} M)/\operatorname{Rad} M$ is a fully invariant submodule of the homogeneous semisimple module $M/\operatorname{Rad} M$, and so either $U + \operatorname{Rad} M = M$ or $U \subseteq \operatorname{Rad} M$. Now $U + \operatorname{Rad} M = M$ implies $U = M$, a contradiction. Hence $U \subseteq \operatorname{Rad} M$ as claimed.

(2) Conversely, let $U \subset M$ be the largest fully invariant proper submodule of $M$. Then $\operatorname{Rad} M \subseteq U$ and $U/\operatorname{Rad} M$ is the largest fully invariant proper submodule of the semisimple module $M/\operatorname{Rad} M$. This clearly implies that $M/\operatorname{Rad} M$ is homogeneous semilocal.

(3) Let $E \subset M/\operatorname{Rad} M$ be a simple submodule (unique up to isomorphism). Then for every finitely $M$-generated module $Q$, $Q/\operatorname{Rad} Q$ is generated by $M/\operatorname{Rad} M$ and thus $Q/\operatorname{Rad} Q \simeq E^{n_Q}$, for some $n_Q \in \mathbb{N}$.

(i) Choose a finitely $M$-generated $M$-projective module $P$ with $n_P$ minimal. Then for any finitely $M$-generated $M$-projective module $Q$ there exists an epimorphism $Q \to P/\operatorname{Rad} P$ which lifts to an epimorphism $Q \to P$ (since $Q$ is $P$-projective). By $M$-projectivity of $P$ this map splits and hence $P$ is isomorphic to a direct summand of $Q$. In particular $P$ is $M$-cyclic and claim (i) holds.

(ii) Choose an epimorphism $P^k \to Q/\operatorname{Rad} Q$ with $k \in \mathbb{N}$ minimal. Therefore $(k-1)n_P < n_Q \le k n_P$. As $P^k$ is $Q$-projective, there exists an epimorphism (since $\operatorname{Rad} Q \ll Q$) $f : P^k \to Q$. Since $Q$ is $P^k$-projective we have $P^k \simeq Q \oplus Q'$. Therefore $k n_P = n_Q + n'_Q$. Applying (i) we get $Q' = 0$. Hence $P^k \simeq Q$.

(iii) Let $P^{(\Lambda)} \simeq P^{(\Lambda')}$. Since $P \simeq E^{n_P}$ this implies $(E^{n_P})^{(\Lambda)} \simeq (E^{n_P})^{(\Lambda')}$. By the decomposition properties of semisimple modules (e.g., [363, 20.5]) this means $n_P|\Lambda| = n_P|\Lambda'|$ and so $|\Lambda| = |\Lambda'|$. $\qquad\square$

The ring $R$ is called *left homogeneous semilocal* provided $R$ is homogeneous semilocal as a left $R$-module, that is, $R/\operatorname{Jac} R$ is a simple left artinian ring.

From 17.6 we derive:

## 17.7. Homogeneous semilocal rings.

(1) *The following statements are equivalent for the ring $R$:*

    (a) *$R$ is left homogeneous semilocal;*

    (b) *$R$ is (left) semilocal and $\operatorname{Jac} R$ is a maximal ideal in $R$;*

    (c) *$R$ is right homogeneous semilocal.*

(2) *Let $R$ be a homogeneous semilocal ring.*

    (i) *There exists a unique (up to isomorphism) cyclic indecomposable projective $R$-module $P$.*

    (ii) *Every finitely generated projective $R$-module is isomorphic to $P^k$, for some $k \in \mathbb{N}$.*

(iii) $P^{(\Lambda)} \simeq P^{(\Lambda')}$ *if and only if the index sets $\Lambda$ and $\Lambda'$ have the same cardinality.*

We note that (2) (ii) holds more generally for projective modules which are not necessarily finitely generated (e.g., [75, Proposition 2.3]).

**17.8. Definition.** A submodule $N \subset M$ is called a *weak supplement* of a submodule $L$ of $M$ if $N+L = M$ and $N \cap L \ll M$. The module $M$ is called *weakly supplemented* if every submodule $N$ of $M$ has a weak supplement.

**17.9. Properties of weak supplements.** *Let $K \subseteq N \subseteq M$ be submodules.*

(1) *If $K$ and $N$ have the same weak supplement in $M$, then $K \subseteq N$ is cosmall in $M$.*

(2) *If $N$ is a weak supplement of $L$ in $M$, then*

$$\mathrm{h.dim}(M) = \mathrm{h.dim}(M/L) + \mathrm{h.dim}(M/N).$$

(3) *Every weakly supplemented module $M$ is semilocal and so has a decomposition $M = M_1 \oplus M_2$ with $M_1$ semisimple and $\mathrm{Rad}\, M \trianglelefteq M_2$.*

(4) *A module $M$ with small radical is weakly supplemented if and only if $M$ is semilocal.*

(5) *If $K \overset{cs}{\underset{M}{\longrightarrow}} N$ and $N$ has a weak supplement in $M$, then $N = K + S$, where $S \ll M$.*

(6) *Suppose $M$ is weakly supplemented and $M = A + B$. Then $A$ has a weak supplement $C$ in $M$ such that $C \subseteq B$.*

*Proof.* (1) Let $L$ be a weak supplement of both $K$ and $N$. Then, since $M = K+L$, we have $N = K + (N \cap L)$. Now let $X$ be a submodule of $M$ with $M = N + X$. Then $M = K + (N \cap L) + X = K + X$ since $N \cap L \ll M$. Thus $K \subseteq N$ is cosmall in $M$ by 3.2 (1).

(2) Let $N$ be a weak supplement of $L$ in $M$. Then, since $L \cap N \ll M$,

$$
\begin{aligned}
\mathrm{h.dim}(M) &= \mathrm{h.dim}(M/(L \cap N)) \text{ (see 5.4 (2))} \\
&= \mathrm{h.dim}(L/(L \cap N) \oplus N/(L \cap N)) \\
&= \mathrm{h.dim}(L/(L \cap N)) + \mathrm{h.dim}(N/(L \cap N)) \text{ (see 5.4 (1))} \\
&= \mathrm{h.dim}(M/L) + \mathrm{h.dim}(M/N).
\end{aligned}
$$

(3) Suppose $M$ is a weakly supplemented module and $T/\mathrm{Rad}\, M \subseteq M/\mathrm{Rad}\, M$. Let $L$ be a weak supplement of $T$ in $M$. Then it is easy check that

$$M/\mathrm{Rad} M = T/\mathrm{Rad}\, M \oplus (L + \mathrm{Rad}\, M)/\mathrm{Rad}\, M.$$

Thus $M$ is a semilocal module. Now (3) follows from 17.2.

(4) By (3), a weakly supplemented module is semilocal. Conversely, assume $M$ is semilocal and $\operatorname{Rad} M \ll M$. By 17.2, for every submodule $N$ of $M$ there exists a submodule $L$ of $M$ such that $N + L = M$ and $N \cap L \subseteq \operatorname{Rad} M \ll M$. Hence $L$ is a weak supplement of $N$.

(5) Suppose $L$ is a weak supplement of $N$ in $M$. Then $N + L = M$ and $N \cap L \ll M$. Since $K \overset{cs}{\underset{M}{\hookrightarrow}} N$, we have $K + L = M$. Thus $N = K + (N \cap L)$ and $N \cap L \ll M$.

(6) Suppose $M = A + B$. As $M$ is weakly supplemented, there exists $K \subseteq M$ such that $M = (A \cap B) + K$ and $(A \cap B) \cap K \ll M$. Now $B = (A \cap B) + (B \cap K)$. Thus $B \cap K$ is a weak supplement of $A$. $\qquad \square$

It follows from 17.9 (4) and 2.8 (6) that a ring $R$ is weakly supplemented as a left $R$-module if and only if $R$ is semilocal. However, we now give an example of a weakly supplemented module $M$ with $\operatorname{Rad} M = M$:

**17.10. Example.** $\mathbb{Q}/\mathbb{Z}$ *is a weakly supplemented $\mathbb{Z}$-module.*

*Proof.* First write $M := \mathbb{Q}/\mathbb{Z} = \bigoplus_p M_p$ as the direct sum of its (prime) $p$-components $M_p := \mathbb{Z}_{p^\infty}$. Every submodule $N$ of $M$ is of the form $N = \bigoplus N_p$ where $N_p = N \cap M_p \subseteq M_p$ are the $p$-components of $N$. Since $M_p$ is hollow, either $N_p = M_p$ or $N_p \ll M_p$. Thus $N \ll M$ if and only if $N_p \neq M_p$ for all $p$. If $N$ is not small in $M$, set $\Lambda = \{p \mid N_p \neq M_p\}$ and $L := \bigoplus_{p \in \Lambda} M_p$. Then $N + L = M$ and $N \cap L = \bigoplus_\Lambda N_p \ll M$. Hence $L$ is a weak supplement of $N$ in $M$. Finally, since $M$ has no maximal submodules, $\operatorname{Rad} M = M$. $\qquad \square$

To show (closure) properties of weakly supplemented modules we need the following technical observation.

**17.11. Lemma.** *Let $N \subset M$ be a submodule such that $N/L$ is weakly supplemented for some submodule $L \subseteq N$ with $L \ll M$. If $K \subset M$ is a submodule such that $N + K$ has a weak supplement in $M$, then $K$ also has a weak supplement in $M$.*

*Proof.* By assumption $N + K$ has a weak supplement $X \subseteq M$. Then $(N + K) + X = M$ and $(N + K) \cap X \ll M$. We consider two cases. Firstly suppose that $N \cap (K + X) \subseteq L \ll M$. Then

$$(N + X) \cap K \subseteq (N \cap (K + X)) + (X \cap (N + K)) \ll M$$

and so, since $(N + X) + K = M$, $N + X$ is a weak supplement of $K$ in $M$.

On the other hand, if $N \cap (K + X) \not\subseteq L$, we may choose a weak supplement $Y/L$ of $(L + (N \cap (K + X)))/L$ in $N/L$. Hence $N = Y + (N \cap (K + X))$ and

$$[Y \cap (L + (N \cap (K + X)))]/L \ll N/L \subseteq M/L, \text{ and so}$$

$$Y \cap (K + X) \subseteq Y \cap ((K + X) + L) = Y \cap (L + (N \cap (K + X))) \ll M$$

by 2.2 (3) since $L \ll M$. Hence $M = N + X + K = Y + X + K$ and

$$(Y + X) \cap K \quad \subseteq (Y \cap (K + X)) + (X \cap (Y + K))$$
$$\subseteq (Y \cap (K + X)) + (X \cap (K + N)) \ll M$$

by 2.2 (4), showing that $Y + X$ is a weak supplement of $K$ in $M$. $\qquad \square$

As a consequence we get

**17.12. Corollary.** *If $M = M_1 + M_2$, with $M_1$ and $M_2$ weakly supplemented, then $M$ is weakly supplemented.*

*Proof.* For every submodule $N \subseteq M$, $M_1 + (M_2 + N)$ has a trivial weak supplement and by the lemma $M_2 + N$ also has a weak supplement in $M$. Applying the lemma again we get a weak supplement for $N$. $\qquad \square$

**17.13. The class of weakly supplemented modules.** *The class of weakly supplemented modules is closed under coclosed submodules, homomorphic images, finite direct sums and small covers.*

*Proof.* Let $L$ be a coclosed submodule of a weakly supplemented module $M$. Suppose $T \subseteq L$ and $N$ is a weak supplement of $T$ in $M$. Then $L = T + (N \cap L)$ and $T \cap (N \cap L) \subseteq N \cap T \ll M$. As $L$ is coclosed in $M$, $T \cap (N \cap L) \ll L$ (see 3.7) showing that $L$ is weakly supplemented.

Let $N$ be a submodule of the weakly supplemented module $M$. Given a submodule $K/N$ of $M/N$, let $L$ be a weak supplement of $K$ in $M$, so that $M = K + L$ and $K \cap L \ll M$. Thus $K/N + (L + N)/N = M/N$ and, by 2.2(5), $(K/N) \cap ((L + N)/N) \ll M/N$. Thus $(L + N)/N$ is a weak supplement of $K/N$ in $M/N$. Hence $M/N$ is weakly supplemented.

17.12 shows that the (direct) sum of finitely many weakly supplemented modules is also weakly supplemented.

Let $M$ be a small cover of a weakly supplemented module $N$. Then $N \simeq M/K$ for some $K \ll M$. Take a submodule $L$ of $M$ and a weak supplement $X/K$ of $(L + K)/K$ in $M/K$. Since $K \ll M$, by 2.2 (4), we get $(X \cap L) + K = X \cap (L + K) \ll M$. Hence $X \cap L \ll M$ and $X$ is a weak supplement of $L$ in $M$. Thus $M$ is weakly supplemented. $\qquad \square$

As announced earlier (see 4.32), for some classes of modules $M$ the Rad-$M$-projective modules and small $M$-projective modules are $M$-projective.

**17.14. Small projectivity.** *Let $M$ and $L$ be $R$-modules.*

(1) *If $M$ is a weakly supplemented module, then:*

    (i) *$L$ is $M$-projective if and only if $L$ is small $M$-projective;*

    (ii) *$L$ is im-summand $M$-projective if and only if $L$ is im-summand small $M$-projective;*

> (iii) *L is im-coclosed M-projective if and only if L is im-coclosed small M-projective.*

(2) *If M is semilocal, the following are equivalent:*

> (a) *L is M-projective;*
>
> (b) *L is Rad-M-projective.*

*Proof.* We only prove (1)(i) since the proofs of the other parts follow the same pattern. One direction is obvious.

Let $L$ be a small $M$-projective module and $f : L \to M/N$ be any morphism. Let $\eta : M \to M/N$ be the natural map. As $M$ is weakly supplemented, there exists $T \subseteq M$ such that $T + N = M$ and $T \cap N \ll M$. Let $\eta_1 : N \to N/(N \cap T)$ be the natural map. Then the map

$$g : M \to M/N \oplus N/(N \cap T), \quad t + n \mapsto ((t)\eta, (n)\eta_1), \text{ for } t \in T, \; n \in N$$

is a well-defined small epimorphism. Let $i : M/N \to M/N \oplus N/(N \cap T)$ be the inclusion map. By assumption, in the diagram

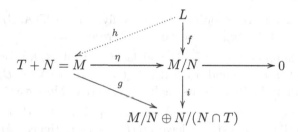

the map $f \diamond i$ can be lifted to a map $h : L \to M$ such that $h \diamond g = f \diamond i$. Then $h \diamond \eta = f$ thus proving that $L$ is $M$-projective. $\qquad\square$

**17.15. Example.** $\mathbb{Q}$ *is a weakly supplemented $\mathbb{Z}$-module.*

*Proof.* Since $\mathbb{Q}$ is a small cover of the weakly supplemented module $\mathbb{Q}/\mathbb{Z}$ (see 17.10), we now see that $\mathbb{Q}$ is also a weakly supplemented module. $\qquad\square$

Weak supplements provide more results on copolyform modules. As a prelude to this, we exhibit a relationship between the im-small projectivity condition and the small ideal $\nabla$ for weakly supplemented modules.

**17.16. Lemma.** *Let $M$ and $N$ be modules where $N$ is weakly supplemented. Then the following statements are equivalent:*

(a) *$M$ is im-small $N$-projective;*

(b) *$\nabla(M, N/K) = \nabla(M, N)\pi_K$ for all $K \subseteq N$ where $\pi_K : N \to N/K$ denotes the canonical projection.*

*Proof.* (a)⇒(b). Let $K \subseteq N$ and choose a weak supplement $L$ of $K$ in $N$. Then $N/(K \cap L) \simeq N/K \oplus N/L$. Let $j : N/K \to N/(K \cap L)$ denote the inclusion map and $p : N/(K \cap L) \to N/K$ be the projection map. For any $0 \neq f \in \nabla(M, N/K)$ we have $fj \in \nabla(N, N/(K \cap L))$. If $M$ is im-small $N$-projective, then there is a map $g : M \to N$ such that $g\pi_{K \cap L} = fj$ where $\pi_{K \cap L}$ denotes the canonical projection $N \to N/(K \cap L)$. Since $\pi_{K \cap L}$ is a small epimorphism and $\operatorname{Im} g\pi_{K \cap L} = \operatorname{Im} fj$ is small, we have $\operatorname{Im} g \ll N$. Hence $g \in \nabla(M, N)$ and $g\pi_K = g\pi_{K \cap L}p = fjp = f$. Thus $\nabla(M, N/K) \subseteq \nabla(M, N)\pi_K$. Since any homomorphism in $\nabla(M, N)\pi_K$ has a small image in $N/K$, equality holds.

(b)⇒(a) is trivial.                                                                                 □

While a direct summand of a module is always coclosed, it need not be strongly coclosed. We now look at a special case when it is strongly coclosed.

**17.17. Direct summands that are strongly coclosed.** *The following statements are equivalent for a module* $M = A \oplus B$ *such that* $B$ *is weakly supplemented:*

(a)  $A$ *is strongly coclosed in* $M$;

(b)  $A$ *is im-small* $B$-*projective and* $\nabla(A, B) = 0$.

*Proof.* By Lemma 17.16, condition (b) is equivalent to $\nabla(A, B/K) = 0$ for all submodules $K \subseteq B$. Let $\pi_A$ be the canonical projection from $M$ onto $A$.

(a)⇒(b). Suppose that $A \overset{scc}{\hookrightarrow} M$ and let $f \in \nabla(A, B/K)$ for some $K \subseteq B$. Then $\pi_A f \in \nabla(M, B/K)$ and, by 3.12 (1), $f = 0$. Hence $\nabla(A, B/K) = 0$ for all $K \subseteq B$. By Lemma 17.16, $A$ is im-small $B$-projective. Moreover $\nabla(A, B) = 0$.

(b)⇒(a). Let $f : M \to X$ be a homomorphism such that $(A)f \ll (M)f$. Since $(A)f + (B)f = (M)f$, we have $(B)f = (M)f$. Hence $(M)f \simeq B/K$ for $K = B \cap \operatorname{Ke} f$ and $f \in \nabla(A, B/K) = 0$, that is, $(A)f = 0$. Then, by 3.12, $A$ is strongly coclosed in $M$.                                                                      □

**17.18. Example.** Consider the ring $R$ of $2 \times 2$ upper triangle matrices defined in 9.14. Let $D_1 = Re_{11}$ and $M = Re_{22}$. Then $R = D_1 \oplus M$ and $M$ is a projective supplemented copolyform module with $\nabla(M, D_1) = 0$ and hence strongly coclosed in $R$. On the other hand, $D_1$ is a simple copolyform module that is $M$-projective but, since $\nabla(D_1, M) \neq 0$, $D_1$ is not strongly coclosed in $R$. Note also that the direct sum $R = D_1 \oplus M$ of these two copolyform modules (which are relatively projective) is not copolyform, since $\nabla(R) \simeq \operatorname{Jac}(R) \neq 0$.

**17.19. Copolyform weakly supplemented modules.** *For $M$ the following hold.*

(1)  *If $M$ is im-small $M$-projective and $\nabla(M) = 0$, then $M$ is copolyform.*

(2)  *If $M$ is weakly supplemented, the following statements are equivalent:*

(a)  $M$ *is copolyform;*

(b)  $M$ *is im-small $M$-projective and $\nabla(M) = 0$;*

(c)  $\nabla(M, M/K) = 0$ *for all* $K \subseteq M$;

(d)  $\nabla(M/L, M/K) = 0$ *for all* $K \subseteq M$ *and* $L \subseteq M$;

(e)  *every factor module of* $M$ *is copolyform;*

(f)  $M$ *is strongly coclosed in* $M \oplus M$.

*Proof.* (1) Let $\pi : M \to F$ be a small cover with $N = \mathrm{Ke}\,\pi \ll M$ and $f \in \mathrm{Hom}(M, N/K)$ for some $K \subseteq N$. Since $N \ll M$, $N/K \ll M/K$. As $M$ is im-small $M$-projective, there exists an endomorphism $g : M \to M$ such that $\mathrm{Im}\,f = (\mathrm{Im}\,g + K)/K$. Since $K \ll M$ and $\mathrm{Im}\,f \ll M/K$, we have $\mathrm{Im}\,g \ll M$, that is, $g \in \nabla(M) = 0$. Hence $f = 0$ and so $N$ is $M$-corational. Hence $M$ is copolyform.

(2) Suppose $M$ is weakly supplemented.

(a)$\Rightarrow$(b). Let $f \in \nabla(M, M/K)$ for some submodule $K$ of $M$. Let $L$ be any weak supplement of $K$ in $M$ and

$$ j : M/K \to M/K \oplus M/L \simeq M/(K \cap L) $$

be the inclusion map. Then $fj \in \nabla(M, M/(K \cap L)) = 0$, by 9.18. Hence $f = 0$, that is, $\nabla(M, M/K) = 0$. In particular $\nabla(M) = 0$ and, by 17.16, $M$ is im-small $M$-projective.

(b)$\Rightarrow$(c)$\Rightarrow$(d)$\Rightarrow$(e)$\Rightarrow$(a) are obvious.

(f)$\Leftrightarrow$(b) follows from 17.17. $\qquad\qquad\square$

The module $M$ is said to be $\Sigma$-*weakly supplemented* if any direct sum of copies of $M$ is weakly supplemented, and $M$ is called $\Sigma$-*copolyform* if any direct sum of copies of $M$ is copolyform.

**17.20. Copolyform and fully non-$M$-small modules.**

(1) *If* $M$ *is projective in* $\sigma[M]$, *then* $M$ *is* $\Sigma$-*copolyform if and only if* $M$ *is copolyform.*

(2) *If* $M$ *is* $\Sigma$-*weakly supplemented, then* $M$ *is* $\Sigma$-*copolyform if and only if* $M$ *is fully non-$M$-small.*

*Proof.* (1) Let $M$ be copolyform and projective in $\sigma[M]$. Then for any set $\Lambda$, $\nabla(M^{(\Lambda)}) = 0$ by 2.5(1), as $\nabla(M) = 0$. Moreover, $M^{(\Lambda)}$ is self-projective. Hence $M^{(\Lambda)}$ is copolyform by 17.19 (1).

(2) Let $M$ be $\Sigma$-weakly supplemented and $\Sigma$-copolyform. Let $f \in \nabla(M, E)$ for any injective $E \in \sigma[M]$. As $E \simeq M^{(\Lambda)}/K$ for some set $\Lambda$ and some submodule $K \subseteq M^{(\Lambda)}$, for any $\lambda \in \Lambda$ and projection $\pi_\lambda : M^{(\Lambda)} \to M$, $\pi_\lambda f \in \nabla(M^{(\Lambda)}, E)$. By 17.19 (2), $\nabla(M^{(\Lambda)}, E) = 0$ as $M^{(\Lambda)}$ is copolyform and weakly supplemented. Hence $f = 0$ and $M$ is fully non-$M$-small. $\qquad\square$

It is easy to verify that a direct summand of a copolyform module is also copolyform. However, the direct sum of two copolyform modules need not be copolyform, as seen above in 17.18. For another example, take $R = \mathbb{Z}$, $A = \mathbb{Z}/2\mathbb{Z}$ and $B = E(A)$, the injective envelope of $A$. Then $A$ and $B$ are copolyform $\mathbb{Z}$-modules but $A \oplus B$ is not a copolyform $\mathbb{Z}$-module. We prove a necessary and sufficient condition for a weakly supplemented direct sum of copolyform modules to be a copolyform module.

If $M$ is weakly supplemented, then for every $K \subset M$ there exists a weak supplement $L$ of $K$ in $M$ such that $M/K$ is a direct summand of $M/(K \cap L)$ and $K \cap L \ll M$. Hence a weakly supplemented module is copolyform if and only if $\nabla(M, M/K) = 0$ for all proper submodules $K \subset M$, or — equivalently — if $M$ is im-small projective with $\nabla(M) = \nabla(M, M) = 0$ (see 17.19).

**17.21. Weakly supplemented direct sum of copolyform modules.** *Let the module* $M = M_1 \oplus \cdots \oplus M_n$ *be weakly supplemented. Then the following are equivalent:*

(a) *$M$ is copolyform;*

(b) *the $M_i$ are copolyform for all $i \leq n$, they are relatively im-small projective and $\nabla(M_i, M_j) = 0$ for any $i \neq j$.*

(c) *each $M_i$, $i \in \{1, \ldots, n\}$, is copolyform and strongly coclosed in $M$.*

*Proof.* By Lemma 2.5, $\nabla(M) = 0$ if and only if $\nabla(M_i, M_j) = 0$ for all $i, j \leq n$.

(a)$\Leftrightarrow$(b). In view of the remarks preceding the theorem this equivalence follows from 9.19.

(a)$\Rightarrow$(c). Suppose $M$ is a copolyform module. As seen in 17.19 (2), $M$ is im-small $M$-projective and $\nabla(M) = 0$. By 4.34, for each $i$, $M_i$ is im-small $\bigoplus_{i \neq j} M_j$-projective. Moreover, $\nabla(M) = 0$ implies $\nabla(M_i) = 0$ and $\nabla(M_i, \bigoplus_{j \neq i} M_j) = 0$ for all $i$. Hence, by 17.19, $M_i$ is copolyform and strongly coclosed in $M$.

(c)$\Rightarrow$(b). Since $M_i$ is im-small $\bigoplus_{i \neq j} M_j$-projective it is also im-small $M_j$-projective for all $i \neq j$. Thus the $M_i$ are relatively im-small projective. Since, by 17.17, $\nabla(M_i, \bigoplus_{j \neq i} M_j) = 0$, we also have $\nabla(M_i, M_j) = 0$ for all $i, j \leq n$. $\qquad\square$

**17.22. Exercises.**

(1) Prove that for an $R$-module $M$ the following are equivalent (cf. [363, 21.15]):

    (a) $M$ is semilocal;

    (b) every $M$-generated module is semilocal;

    (c) for every $R$-module $N$ (in $\sigma[M]$) with $\operatorname{Rad} N = 0$, $\operatorname{Tr}(M, N)$ is a semisimple module;

    (d) for any family $\{S_i\}_I$ of (semi)simple $R$-modules (in $\sigma[M]$), $\operatorname{Tr}(M, \prod_I S_i)$ is a semisimple module.

(2) A module $M$ is called *finitely weakly supplemented*, or *fw-supplemented* for short, if every finitely generated submodule has a weak supplement in $M$. (See [6].)

    (i) Prove that if $M$ is fw-supplemented and $K \subset M$ is a finitely generated submodule, then $M/K$ is fw-supplemented.

(ii) Prove that if $K \ll M$, then $M$ is fw-supplemented if and only if $M/K$ is fw-supplemented.

(iii) If $\operatorname{Rad} M \ll M$, show that the following statements are equivalent:

(a) $M$ is fw-supplemented;

(b) every cyclic submodule has a weak supplement in $M$;

(c) every cyclic submodule of $M/\operatorname{Rad} M$ is a direct summand.

(d) $M/\operatorname{Rad} M$ is fw-supplemented.

(3) A module $M$ is called *closed weakly supplemented* if every (essentially) closed submodule has a weak supplement in $M$ ([381]).

(i) Let $M$ be closed weakly supplemented. Prove:

($\alpha$) Every direct summand of $M$ is closed weakly supplemented.

($\beta$) Any nonsingular image of $M$ is closed weakly supplemented.

($\gamma$) If $\operatorname{Rad} M = 0$, then $M$ is extending.

(ii) Let $M = M_1 \oplus M_2$ be a module having a distributive lattice of submodules. Prove that $M$ is closed weakly supplemented if and only if $M_1$ and $M_2$ are closed weakly supplemented.

(4) Recall that for $R$-modules $M$ and $N$, $M \rightsquigarrow N$ means $\nabla(M, N') = 0$ for all factor modules $N'$ of $N$ (see 4.45).

(i) Assume that $N$ is weakly supplemented. Show that $M \rightsquigarrow N$ if and only if $M$ is im-small $N$-projective and $\nabla(M, N) = 0$.

(ii) Show that any module $M$ with $M \rightsquigarrow M$ is strongly copolyform and that the converse holds provided $M$ is weakly supplemented.

**17.23. Comments.** The existence of complements in a module is due to the upper continuity of the lattice of submodules of a module. Since in general the lattice of submodules of a module is not lower continuous, the existence of supplements is not guaranteed. Zöschinger introduced the concept of a weak supplement in [394] and the structure of weakly supplemented modules is determined in some special cases. A complete characterisation of weakly supplemented modules over commutative local noetherian rings was obtained by Rudlof in [304]. Semilocal modules are investigated in [227] where also closure properties of weakly supplemented modules are studied. Corisello and Facchini study homogeneous semilocal rings in [75] showing that they share many properties with local rings.

**References.** Alizade and Büyükasik [6]; Corisello and Facchini [75]; Lomp [226, 227]; Rudlof [304, 305]; Talebi and Vanaja [330, 331]; Zeng and Shi [381]; Zöschinger [394].

# 18   Weak supplements and hollow dimension

Since a module has finite hollow dimension if and only if it is a small cover of a finite direct sum of hollow modules (see 5.2 (f)), it also follows from 17.13 that modules with finite hollow dimension are weakly supplemented.

**18.1. Corollary.** *Modules with finite hollow dimension are weakly supplemented.*

The existence of weak supplements allows us to prove

**18.2. Dimension formula.** *For any submodule $N \subset M$,*

$$\mathrm{h.dim}(M/N) \leq \mathrm{h.dim}(M) \leq \mathrm{h.dim}(M/N) + \mathrm{h.dim}(N).$$

*Proof.* As detailed in the paragraph preceding 5.4, for every submodule $N$ of $M$ we have $\mathrm{h.dim}(M/N) \leq \mathrm{h.dim}(M)$. Thus if $\mathrm{h.dim}(N) = \infty$ or $\mathrm{h.dim}(M/N) = \infty$, then $\mathrm{h.dim}(M) \leq \mathrm{h.dim}(M/N) + \mathrm{h.dim}(N)$ trivially.

We may now assume that $\mathrm{h.dim}(M/N)$ and $\mathrm{h.dim}(N)$ are both finite. Suppose first, to the contrary, that $\mathrm{h.dim}(M) = \infty$ and let $\{K_i\}_\mathbb{N}$ be an infinite coindependent family of submodules of $M$. Then let

$$L_1 = K_1, L_2 = K_2 \cap K_3, \ldots, L_n = K_{t+1} \cap \cdots \cap K_{t+n}, \ldots$$

where $t = (n-1)n/2$. For every $n \in \mathbb{N}$, we have $M/L_n \simeq \bigoplus_{i=1}^n M/K_{t+i}$ as $\{K_{t+1}/L_n, \ldots, K_{t+n}/L_n\}$ is coindependent (see 1.29). Hence $n \leq \mathrm{h.dim}(M/L_n)$. By definition, $\{L_i\}_\mathbb{N}$ is again an infinite coindependent family of submodules of $M$. Since $\mathrm{h.dim}(M/N)$ is finite, $\{(N + L_i)/N\}_\mathbb{N}$ is not coindependent and so $N + L_n = M$ for almost all $n$. Choose $n$ such that $n > \mathrm{h.dim}(N)$ and $N + L_n = M$. Then $M/L_n \simeq N/(N \cap L_n)$ is a factor module of $N$. Thus $n \leq \mathrm{h.dim}(M/L_n) \leq \mathrm{h.dim}(N) < n$, yielding a contradiction. Hence $M$ has finite hollow dimension.

By 18.1 we may choose a weak supplement $K$ of $N$. Then, by 17.9,

$$\mathrm{h.dim}(M) = \mathrm{h.dim}(M/N) + \mathrm{h.dim}(M/K) \leq \mathrm{h.dim}(M/N) + \mathrm{h.dim}(N),$$

since $M/K \simeq N/(N \cap K)$.                                                        $\square$

For later use we insert a technical lemma relating weak supplements and coindependent submodules.

**18.3. Lemma.** *Let $\{N_\lambda\}_\Lambda$ be a coindependent family of submodules of $M$ and assume that $N_\mu$ has a weak supplement $L$ in $M$ for some $\mu \in \Lambda$. Then*

$$\{[L + (N_\lambda \cap N_\mu)]/L\}_{\Lambda \setminus \{\mu\}}$$

*is a coindependent family of submodules of $M/L$.*

*Proof.* Let $\Lambda' := \Lambda \setminus \{\mu\}$ and $\lambda \in \Lambda'$. If $M = L + (N_\lambda \cap N_\mu)$ this gives $N_\mu = (N_\mu \cap L) + (N_\lambda \cap N_\mu)$. Then, since $N_\mu \cap L \ll M$, we have

$$M = N_\lambda + N_\mu = N_\lambda + (N_\mu \cap L) + (N_\lambda \cap N_\mu) = N_\lambda + (N_\lambda \cap N_\mu) = N_\lambda,$$

a contradiction since $N_\lambda$ is a proper submodule. Thus $[L + (N_\lambda \cap N_\mu)]/L$ is a proper submodule of $M/L$ for each $\lambda \neq \mu$. Moreover, for each $\lambda \in \Lambda'$ and each finite subset $F \subseteq \Lambda' \setminus \{\lambda\}$, we have

$$[L + (N_\lambda \cap N_\mu)] + \bigcap_{i \in F}[L + (N_i \cap N_\mu)] \supseteq L + (N_\lambda \cap N_\mu) + (\bigcap_{i \in F} N_i \cap N_\mu)$$
$$= L + \left(N_\mu \cap (N_\lambda + (N_\mu \cap \bigcap_{i \in F} N_i))\right)$$
$$= L + (N_\mu \cap M) = L + N_\mu = M$$

and from this $\{(L + (N_\lambda \cap N_\mu))/L\}_{\Lambda \setminus \{\mu\}}$ is a coindependent family of submodules of $M/L$. $\qquad\square$

**18.4. Definitions.** A submodule $N$ of a module $M$ is called *cofinite* if $M/N$ is finitely generated. Let us call a module $M$ *cofinitely weakly supplemented*, or *cfw-supplemented* for short, if every cofinite submodule of $M$ has a weak supplement. Obviously any finitely generated module is weakly supplemented if and only if it is cofinitely weakly supplemented.

**18.5. Cofinitely weakly supplemented modules.**

(1) *The class of cfw-supplemented modules is closed under homomorphic images, direct sums, and small covers.*

(2) *If $M$ is cfw-supplemented, then every cofinite submodule of $M/\operatorname{Rad} M$ is a direct summand.*

(3) *The following are equivalent for a module $M$:*

    (a) *$M$ is cofinitely weakly supplemented;*

    (b) *every maximal submodule of $M$ has a weak supplement in $M$;*

    (c) *$M/C$ has no maximal submodule, where $C$ is the sum of all weak supplements of maximal submodules of $M$.*

*Proof.* (1) This can be shown with standard arguments.

(2) By (1), $M/\operatorname{Rad} M$ is cfw-supplemented. If $T/\operatorname{Rad} M$ is a cofinite submodule of $M/\operatorname{Rad} M$, then $T$ is a cofinite submodule of $M$ and hence has a weak supplement $L$ in $M$. Then, since $T \cap L \ll M$, we have $T \cap L \subseteq \operatorname{Rad} M$ and so $M/\operatorname{Rad} M = T/\operatorname{Rad} M \oplus (L + \operatorname{Rad} M)/\operatorname{Rad} M$.

(3) (a)$\Rightarrow$(b)$\Rightarrow$(c) are obvious.

(c)$\Rightarrow$(a). Let $C$ be the sum of all weak supplements of maximal submodules of $M$ and let $N$ be a cofinite submodule of $M$. Since $M/(N+C)$ is finitely generated and since $M/C$ and so $M/(N+C)$ have no maximal submodules, we get $N + C =$

$M$. As $M/N$ is finitely generated we may choose finitely many weak supplements $K_1, \ldots, K_n$ of maximal submodules of $M$ such that $M = N + K_1 + \cdots + K_n$. Since for each $1 \leq i \leq n$ there exists a submodule $L_i \subseteq K_i$ such that $K_i/L_i$ is simple and hence weakly supplemented and $L_i \ll M$, we may apply Lemma 17.11 repeatedly to conclude that $N$ has a weak supplement in $M$.                    $\square$

If $M$ is a module with a composition series, we denote its (composition) length by $\ell(M)$. The finite hollow dimension of $M$ is related to the length of $M/\mathrm{Rad}\, M$.

**18.6. Finitely generated semilocal modules.** *Assume* $\mathrm{Rad}\, M \ll M$. *Then the following statements are equivalent:*

(a) *$M$ has finite hollow dimension;*

(b) *$M$ is weakly supplemented and finitely generated;*

(c) *$M$ is semilocal and finitely generated;*

(d) *$M/\mathrm{Rad}\, M$ is finitely cogenerated.*

*In this case,* $\mathrm{h.dim}(M) = \ell(M/\mathrm{Rad}\, M)$.

*Proof.* (a)$\Rightarrow$(b). As $M$ has finite hollow dimension, $M$ is weakly supplemented by 18.1. Since $\mathrm{Rad}\, M \ll M$, 17.9 (4) and 5.4 (2) show that $M/\mathrm{Rad}\, M$ is semisimple and has finite hollow dimension. Hence $M/\mathrm{Rad}\, M$ is finitely generated. It now follows that $M$ is finitely generated by 2.8 (6).

(b)$\Rightarrow$(c) follows from 17.9 (4).

(c)$\Rightarrow$(a). If (c) holds, $M/\mathrm{Rad}\, M$ is semisimple artinian. By 5.4 (2), since $\mathrm{Rad}\, M \ll M$, $\mathrm{h.dim}(M) = \mathrm{h.dim}(M/\mathrm{Rad}\, M) = \ell(M/\mathrm{Rad}\, M)$ is finite.

(c)$\Leftrightarrow$(d). This follows since $M/\mathrm{Rad}\, M$ is finitely cogenerated if and only if it is finitely generated and semisimple (see 2.8 (3)).                    $\square$

Theorem 18.6 applies in particular to $M = R$:

**18.7. Corollary.** *For a ring $R$ the following statements are equivalent:*

(a) *$R$ is semilocal;*

(b) *$R$ is weakly supplemented as a left (or right) $R$-module;*

(c) *$R$ has finite hollow dimension as a left (or right) $R$-module.*

The number of isomorphism classes of direct summands of a self-projective module is bounded if the module has finite hollow dimension. An upper bound depending on the hollow dimension can be given.

**18.8. Projective direct summands of f.g. semilocal modules.** *Let $M$ be a finitely generated semilocal module.*

(1) *The number of non-isomorphic $M$-projective direct summands of $M$ is at most $2^k$, where $k = \ell(M/\mathrm{Rad}\, M)$.*

(2) *There are only finitely many isomorphism classes of finitely $M$-generated indecomposable projective modules in $\sigma[M]$.*

*Proof.* Let $M/\mathrm{Rad}\, M \simeq E_1 \oplus \cdots \oplus E_k$ with $E_i$ simple for all $1 \leq i \leq k$. By 18.6, $M$ has finite hollow dimension $k$.

(1) Let $P$ and $Q$ be two $M$-projective direct summands of $M$. We note that $P$ and $Q$ are finitely generated and hence projective in $\sigma[M]$. Since $M$ is finitely generated semilocal, $P$ and $Q$ are also. Moreover

$$P/\mathrm{Rad}\, P \simeq E_1^{(x_1)} \oplus \cdots \oplus E_k^{(x_k)} \text{ and } Q/\mathrm{Rad}\, Q \simeq E_1^{(y_1)} \oplus \cdots \oplus E_k^{(y_k)}$$

where $x_i, y_i \in \{0, 1\}$ for all $i \in \{1, \ldots, k\}$. Suppose that $x_i = y_i$ for all $i$. As $P$ and $Q$ are projective covers of $P/\mathrm{Rad}\, P$ and $Q/\mathrm{Rad}\, Q$ respectively, we get $P \simeq Q$.

This shows that if $P \not\simeq Q$, then $x_i \neq y_i$ for some $i \in \{1, \ldots, k\}$. Since there are $2^k$ distinct $n$-tuples $(x_1, \ldots, x_k)$ with $x_i \in \{0, 1\}$, the result follows.

(2) Let $P$ and $Q$ be nonzero finitely $M$-generated indecomposable projective modules in $\sigma[M]$. Then, for some $m, l \in \mathbb{N}$, $P$ is a direct summand of $M^m$ and $Q$ is a direct summand of $M^l$ and so

$$P/\mathrm{Rad}\, P \simeq E_1^{(x_1)} \oplus \cdots \oplus E_k^{(x_k)} \text{ and } Q/\mathrm{Rad}\, Q \simeq E_1^{(y_1)} \oplus \cdots \oplus E_k^{(y_k)}$$

where $x_i$ and $y_i$ are non-negative integers. Suppose that $x_i \geq y_i$ for all $i \in \{1, \ldots, k\}$. Then $P/\mathrm{Rad}\, P$, and consequently $P$, maps onto $Q/\mathrm{Rad}\, Q$. Thus, since $P$ is $Q$-projective and the canonical projection $Q \to Q/\mathrm{Rad}\, Q$ is a small epimorphism, $P$ maps onto $Q$. Then, since $Q$ is $P$-projective, it must be isomorphic to a direct summand of $P$. Hence $P \simeq Q$ as $P$ is indecomposable. Thus we have shown:

$(*)$ $P \simeq Q \Leftrightarrow x_i = y_i$ for all $i \in \{1, \ldots, k\} \Leftrightarrow x_i \geq y_i$ for all $i \in \{1, \ldots, k\}$.

The remaining part of the argument is of a combinatorial nature.

Let $X$ denote the set of all $n$-tuples $(x_1, \ldots, x_k)$ of non-negative integers that correspond, as above, to the isomorphism classes of finitely generated indecomposable projective modules in $\sigma[M]$. Assume, by way of contradiction, that $X$ is infinite. Then $X$ must be unbounded in at least one component. Renumbering $E_1, \ldots, E_k$ if necessary, we may assume $X$ is unbounded in the first component to obtain an infinite sequence $S_1$ in $X$,

$$\{(x_{1_i}, x_{2_i}, \ldots, x_{k_i})\}_{i \in \mathbb{N}}, \text{ where } x_{1_1} < x_{1_2} < x_{1_3} < \cdots .$$

Consider the set $Y$ of $n-1$-tuples $\{(x_{2_i}, \ldots, x_{k_i})\}_{i \in \mathbb{N}}$ got from $S_1$ by deleting the first component of its elements. If the set $Y$ is finite, then there exists $i < j$, $i, j \in \mathbb{N}$ such that $(x_{2_i}, \ldots, x_{k_i}) = (x_{2_j}, \ldots, x_{k_j})$. Since $x_{1_i} < x_{1_j}$, we get a contradiction to the property $(*)$ satisfied by $X$. Hence $Y$ is an infinite set. Therefore the elements of $Y$ are unbounded in at least one component. Renumbering $E_1, \ldots, E_k$

if necessary, we may assume $Y$ is unbounded in the first component to obtain an infinite subsequence $S_2$ of $S_1$ in $X$,

$$\{(x_{1_{i_j}}, x_{2_{i_j}}, \ldots, x_{k_{i_j}})\}_{j \in \mathbb{N}}, \quad \text{with}$$

$$x_{1_{i_1}} < x_{1_{i_2}} < x_{1_{i_3}} < \cdots \text{ and } x_{2_{i_1}} < x_{2_{i_2}} < x_{2_{i_3}} < \cdots.$$

Continuing this process $k$ times we get a sequence $S_k$ in $X$,

$$\{(x_{1_{\alpha_j}}, x_{2_{\alpha_j}}, \ldots, x_{k_{\alpha_j}})\}_{j \in \mathbb{N}}, \quad \text{with}$$

$$x_{\ell_{\alpha_n}} < x_{\ell_{\alpha_{n+1}}} \quad \text{for } 1 \le \ell \le k \text{ and } n \in \mathbb{N}.$$

This contradicts the property $(*)$ and hence $X$ is a finite set. $\qquad\square$

As a corollary we get the following result of Fuller and Shutters (see [113]).

**18.9. Corollary.** *A semilocal ring has only finitely many isomorphism classes of finitely generated indecomposable projective modules.*

Besides module theoretic characterisations there are also some ring theoretic properties which set semilocal rings apart.

**18.10. Characterisation of semilocal rings.** *For $R$ the following are equivalent:*

(a) *$R$ is left semilocal;*

(b) *every finitely generated left $R$-module has finite hollow dimension;*

(c) *every finitely generated left $R$-module is weakly supplemented;*

(d) *any injective cogenerator $_RQ$ of $R$-Mod has finite uniform dimension as a right module over $T = \mathrm{End}(_RQ)$;*

(e) *any product of simple left $R$-modules is semisimple;*

(f) *for every left $R$-module $M$, $\mathrm{Soc}\, M = \{m \in M \mid (\mathrm{Jac}\, R)m = 0\}$;*

(g) *there exists a ring $S$ and an $R$-$S$-bimodule $M$ such that $\mathrm{u.dim}(M_S)$ is finite and $\mathrm{Ann}_M(r) \neq 0$ for all non-units $r \in R$;*

(h) *there exists a non-negative integer $n$ and a function $d : R \to \{0, \ldots, n\}$ such that, for all $a, b \in R$,*

   (1) *$d(a - aba) = d(a) + d(1 - ba)$ and*

   (2) *if $d(a) = 0$, then $a$ is a unit in $R$;*

(i) *there exists a partial order $\le$ on $R$ satisfying the minimum condition such that, for all $a, b \in R$, if $1 - ba$ is a non-unit in $R$, then $a - aba < a$;*

(j) *$R$ is right semilocal (and the right duals of the statements above hold).*

*In this case*

$$\mathrm{h.dim}(R) = \ell(R/\mathrm{Jac}\, R) = \mathrm{u.dim}(Q_T)) \le \min(n, \mathrm{u.dim}(M_S)),$$

*where $n$ is the integer in* (h) *and $M_S$ is the module in* (g)*.*

*Proof.* (a)⇔(b)⇔(c). By 18.6, for a finitely generated left $R$-module, the notions weakly supplemented, semilocal and finite hollow dimension are equivalent. Since, by 17.13, every finitely generated left $R$-module is weakly supplemented provided $R$ is, the equivalence of (a), (b) and (c) follows.

(a)⇔(d). This follows from $\text{h.dim}(R) = \text{u.dim}(Q_T)$ (see 5.7).

(a)⇒(f). For each left $R$-module $M$, denote $\{m \in M \mid (\text{Jac } R)m = 0\}$ by $\text{Ann}_M(\text{Jac } R)$. Then, since $(\text{Jac } R)\text{Ann}_M(\text{Jac } R) = 0$, $\text{Ann}_M(\text{Jac } R)$ is an $R/\text{Jac } R$-module hence semisimple and contained in $\text{Soc } M$. On the other hand, it is well-known that $(\text{Jac } R)\text{Soc } M = 0$ holds for all $M$ (see [363, 21.12]). Thus $\text{Soc } M = \text{Ann}_M(\text{Jac } R)$.

(f)⇒(e). If $M$ is a product of simple $R$-modules, then $(\text{Jac } R)m = 0$ for all elements $m \in M$ and by (f), we get $\text{Soc } M = M$, that is, $M$ is semisimple.

(e)⇒(a). Since $R/\text{Jac } R$ has zero radical, it is a submodule of a product of simple $R$-modules and hence by hypothesis also semisimple, that is, $R$ is semilocal.

(a)⇒(g). Take $S := R/\text{Jac } R$ and $M$ to be the $R$-$S$-bimodule $S$. Then $M_S$ is semisimple and so $\text{u.dim}(M_S)$ is finite. Let $r \in R$ be a non-unit. Then $r + \text{Jac } R$ is a non-unit in $S$ and so a zero-divisor in $S$. It follows that $\text{Ann}_M(r) \neq 0$.

(g)⇒(h). Let $\text{u.dim}(M_S) = n$. Note that, for all $a \in R$, $\text{Ann}_M(a)$ is a right $S$-submodule of $M_S$ and so we may define the function $d : R \to \{0, \ldots, n\}$ by $d(a) := \text{u.dim}(\text{Ann}_M(a)_S)$ for all $a \in R$. If $d(a) = 0$, then $\text{Ann}_M(a) = 0$ and so $a$ is a unit in $R$, verifying condition (2). Moreover (1) holds since for all $a, b \in R$, we have $\text{Ann}_M(a) \oplus \text{Ann}_M(1 - ba) = \text{Ann}_M(a - aba)$.

(h)⇒(i). For every $a, b \in R$, we define

$$a \leq b \quad \text{if either} \quad a = b \quad \text{or} \quad d(b) < d(a).$$

Then it is easily checked that $\leq$ is a partial order on $R$ satisfying the minimum condition. Moreover, if $1 - ba$ is a non-unit, then, by condition (1), $d(1 - ba) \neq 0$ and hence, by condition (2), $d(a - aba) > d(a)$. Thus $a - aba < a$.

(i)⇒(a). By way of contradiction, assume that there is a left ideal $I$ of $R$ which has no weak supplement. We will construct a strictly descending chain

$$1 > b_1 > b_2 > \cdots > b_n > \cdots$$

in the poset $(R, \leq)$ such that $I + Rb_n = R$ for all $n \in \mathbb{N}$. Since $(R, \leq)$ is artinian, this is a contradiction and so $I$ must have a weak supplement in $R$. Thus $R$ is semilocal by Corollary 18.6.

Now we construct the chain as follows. Note first that $I \not\ll R$, since otherwise $R$ is a weak supplement of $I$. Thus $I \not\subseteq \text{Jac } R$ and so there is a $b \in I$ such that $1 - b$ is a non-unit in $R$. We set $b_1 = 1 - b$. Then $1 > b_1$ and $I + Rb_1 = R$. Now assume that for some $n \geq 1$ we have constructed a chain $1 > b_1 > b_2 > \cdots > b_n$ in which $I + Rb_n = R$. By assumption, $I \cap Rb_n \not\ll R$ and so there is an $r \in R$ such that $rb_n \in I$ and $x := 1 - rb_n$ is not a unit in $R$. Hence, setting $b_{n+1} = b_n x$, we have

$$b_n > b_n(1 - rb_n) = b_{n+1}.$$

An easy calculation shows that $Rb_{n+1} = Rb_n \cap Rx$. Moreover,

$$b_n = b_n - b_n x + b_n x = b_n(1 - x) + b_{n+1} = b_n r b_n + b_{n+1} \in I + Rb_{n+1}$$

and so $Rb_n \subseteq I + Rb_{n+1}$. However, since $Rb_{n+1} \subseteq Rb_n$, the modularity law then gives

$$(I \cap Rb_n) + Rb_{n+1} = (I + R_{n+1}) \cap Rb_n = Rb_n.$$

Thus $R = I + Rb_n = I + Rb_{n+1}$ and this permits us to construct the next term in the contradictory sequence.                                                        □

**18.11. Corollary.** *Let $R$ and $S$ be rings and $f : R \to S$ be a ring homomorphism such that non-units $r \in R$ are carried to non-units $f(r) \in S$. If $S$ is semilocal, then $R$ is semilocal and* h.dim$(R) \leq$ h.dim$(S)$.

*Proof.* Let $S$ be semilocal, so that $S/\mathrm{Jac}\, S$ is left semisimple. The canonical projection $\pi : S \to S/\mathrm{Jac}\, S$ is a ring homomorphism which carries non-units of $S$ to non-units of $S/\mathrm{Jac}\, S$ and so by hypothesis, $\pi f : R \to S/\mathrm{Jac}\, S$ does likewise. Thus we may assume, without loss of generality, that $S$ is left semisimple.

We will use condition 18.10 (g) to show that $R$ is semilocal. Note first that $S$ is a left $R$-module with multiplication $rs := f(r)s$ for all $r \in R, s \in S$. Let $M := S$. Then, since $M_S$ is semisimple artinian, u.dim$(M_S) = \ell(M_S)$ is finite. Next, let $r$ be a non-unit in $R$. Then it remains to show that $\mathrm{Ann}_M(r) \neq 0$. By assumption, $f(r)$ is a non-unit in $S$. Consider the descending sequence $f(r)S \supseteq f(r)^2 S \supseteq f(r)^3 S \supseteq \cdots$. Since $S$ is artinian, for some $n \in \mathbb{N}$ and $s \in S$ we have $f(r)^n = f(r)^{n+1}s$ and so $f(r)^n(1 - f(r)s) = 0$. Since $f(r)$ is not invertible, $1 - f(r)s \neq 0$. It now follows that, for some $0 \leq k < n$, we have $f(r)f(r)^k(1 - f(r)s) = 0$ but $f(r)^k(1 - f(r)s) \neq 0$. Thus $f(r)^k(1 - f(r)s) \in \mathrm{Ann}_M(r)$. Hence, by 18.10 (g), $R$ is semilocal and h.dim$(R) \leq$ u.dim$(M_S) =$ h.dim$(S)$.                     □

By a result of Camillo (see [50], [85, 5.11]) all factor modules of a module $M$ have finite uniform dimension if and only if every factor module of $M$ has a finitely generated socle. To investigate the dual condition for modules we first observe:

**18.12. Lemma.** Soc $M$ *is finitely generated if and only if there is a submodule $K$ of $M$ such that $M/K$ is finitely cogenerated and* Soc $K = 0$.

*Proof.* First assume that Soc $M$ is finitely generated and let $K$ be a complement of Soc $M$ in $M$. Then obviously Soc $K = 0$ and $K \oplus$ Soc $M \trianglelefteq M$. It follows that (Soc $M \oplus K)/K$ is a finitely generated semisimple essential submodule of $M/K$. Hence $M/K$ is finitely cogenerated.

Conversely suppose there is a submodule $K$ with $M/K$ finitely cogenerated and $\mathrm{Soc}\,K = 0$. Then $K \cap \mathrm{Soc}\,M = 0$ and so $\mathrm{Soc}\,M \simeq (\mathrm{Soc}\,M \oplus K)/K \subseteq \mathrm{Soc}\,(M/K)$. Hence $\mathrm{Soc}\,M$ is finitely generated. □

**18.13. All submodules with finite hollow dimension.** *For an $R$-module $M$, consider the following statements.*

(i) *Every submodule of $M$ has finite hollow dimension.*

(ii) *Every submodule $N$ of $M$ is semilocal and $N/\mathrm{Rad}\,N$ is finitely generated.*

(iii) *Every factor module of $M$ has finite uniform dimension.*

*Then the implications* (i)$\Rightarrow$(ii)$\Rightarrow$(iii) *always hold, and*

(1) *If $N/\mathrm{Rad}\,N$ has essential socle for every $N \subseteq M$, then* (iii)$\Rightarrow$(ii).

(2) *If $\mathrm{Rad}\,N \ll N$ for every $N \subseteq M$, then* (ii)$\Rightarrow$(i).

*Proof.* (i)$\Rightarrow$(ii). For any submodule $N \subset M$, by assumption h.dim$(N)$ is finite and hence so too is h.dim$(N/\mathrm{Rad}\,N)$. Then $N/\mathrm{Rad}\,N$ is finitely generated semisimple by 18.6.

(ii)$\Rightarrow$(iii). For any proper submodule $N \subset M$, $\mathrm{Soc}\,(M/N) = H/N$ for some $N \subset H \subseteq M$. Since $H/N$ is semisimple it follows that $\mathrm{Rad}\,H \subseteq N$ and hence $H/N$ is finitely generated (as a factor module of the finitely generated semisimple module $H/\mathrm{Rad}\,H$). Now by Camillo's theorem (see immediately above 18.12) every factor module of $M$ has finite uniform dimension.

(1) Now assume that $N/\mathrm{Rad}\,N$ has essential socle for every $N \subseteq M$ and that (iii) holds. Then for every $N \subseteq M$, by Camillo's theorem and 18.12, $M/\mathrm{Rad}\,N$ has a finitely cogenerated factor module $M/K$ such that $K/\mathrm{Rad}\,N$ has no simple submodule. Then $(N \cap K)/\mathrm{Rad}\,N$ is a submodule of $N/\mathrm{Rad}\,N$ having zero socle. Thus, since $N/\mathrm{Rad}\,N$ has essential socle, it follows that $N \cap K = \mathrm{Rad}\,N$ and so $N/\mathrm{Rad}\,N = N/(N \cap K) \simeq (N + K)/K$. Consequently, since $M/K$ is finitely cogenerated, $N/\mathrm{Rad}\,N$ is also finitely cogenerated, that is, finitely generated and semisimple. Thus, in this case, (iii)$\Rightarrow$(ii).

(2) If $\mathrm{Rad}\,N \ll N$ for every $N \subseteq M$, then (ii)$\Rightarrow$(i) follows by 18.6. □

Notice that in general the converse of (iii)$\Rightarrow$(ii) in the theorem above is not true as shown by the $\mathbb{Z}$-module $\mathbb{Z}$.

If a module $M$ has property (i) of the theorem above, then every subfactor of $M$ has finite uniform and finite hollow dimension and hence has a semilocal endomorphism ring (see 19.6).

**18.14. Exercises.**

(1) Let the ring $R$ be given by $R = \begin{pmatrix} K & V \\ 0 & K \end{pmatrix}$

where $K$ is a field and $_K V$ a vector space over $K$. Prove:

  (i) $R$ is semilocal and hence h.dim$(_R R) < \infty$;

(ii)  u.dim($_R R$) is finite if and only if dim$_K(V)$ is finite.

Thus modules with finite hollow dimension need not have finite uniform dimension.

(2) Prove that finite matrix rings over (homogeneous) semilocal rings are again (homogeneous) semilocal.

(3) An $R$-module $M$ is said to be *2-generated* if it contains a generating set of two elements. Prove that the following are equivalent for $R$ ([165]):

  (a)  $R$ is semilocal;

  (b)  every 2-generated $R$-module $M$ with Rad $M = 0$ is self-injective;

  (c)  every 2-generated $R$-module $M$ with Rad $M = 0$ is self-projective.

(4) Let $M$ be a finitely generated, self-projective, homogeneous semilocal module and $P$ be the unique indecomposable $M$-projective $M$-cyclic module (see 17.5). Prove that a finitely $M$-generated module $N$ has a projective cover in $\sigma[M]$ if and only if h.dim($P$) divides h.dim($N$) (compare [75, Theorem 3.7]).

(5) Let $A, B$ be algebras over the field $K$. Prove ([213]):

  (i)  if $A \otimes_K B$ is semilocal, then $A$ and $B$ are semilocal and (at least) one of them is algebraic over $K$;

  (ii)  $A \otimes_K B$ is semilocal if and only if $A/\operatorname{Jac} A \otimes_K B/\operatorname{Jac} B$ is semilocal and $(\operatorname{Jac} A \otimes_K B) + (A \otimes_K \operatorname{Jac} B) \subseteq \operatorname{Jac}(A \otimes_K B)$.

**18.15. Comments.** Reiter [295] observed that modules with finite hollow dimension are weakly supplemented. The fact that a ring is semilocal if and only if it is weakly supplemented as a module over itself is due to Varadarajan [344]. Cofinitely weakly supplemented modules were characterised by Alizade and Büyükasik in [6]. The proof of 18.8 is due to Fuller and Shutters (see [110]).

Properties (g), (h) and (i) of 18.10 were shown by Camps and Dicks in [56] thereby answering a question of Menal as to whether the endomorphism ring of an artinian module is semilocal.

**References.** Alizade and Büyükasik [6]; Corisello and Facchini [75]; Fuller and Shutters [110]; Hausen [165]; Krempa and Okniński [213]; Lomp [226]; Zöschinger [388].

# 19 Semilocal endomorphism rings

Modules whose endomorphism rings are semilocal have special properties. Under some projectivity conditions we show that a module has a semilocal endomorphism ring if and only if it has finite hollow dimension.

**19.1. $\pi$-projective modules and finite hollow dimension.** *A $\pi$-projective module $M$ with semilocal endomorphism ring has finite hollow dimension and*

$$\mathrm{h.dim}(M) \leq \mathrm{h.dim}(\mathrm{End}(M)).$$

*Proof.* By 4.15 any coindependent family $\{N_i\}_I$ of submodules of $M$ induces a coindependent family $\{\mathrm{Hom}(M, N_i)\}_I$ of left ideals of $\mathrm{End}(M)$. Thus $\mathrm{h.dim}(M) \leq \mathrm{h.dim}(\mathrm{End}(M))$. $\square$

Using the characterisation of semilocal rings from 18.10, we now show that a module with finite hollow dimension has a semilocal endomorphism ring provided every epimorphism is a retraction.

**19.2. Semi-Hopfian modules with finite hollow dimension.** *A semi-Hopfian module $M$ with finite hollow dimension has a semilocal endomorphism ring and in this case*

$$\mathrm{h.dim}(\mathrm{End}(M)) \leq \mathrm{h.dim}(M).$$

*Proof.* Let $\mathrm{h.dim}(M) = n$, $S = \mathrm{End}(M)$ and define

$$d : S \to \{0, \ldots, n\}, \quad f \mapsto \mathrm{h.dim}(\mathrm{Coke}\, f).$$

For any $f, g \in S$, $\mathrm{Coke}\,(f - fgf) = \mathrm{Coke}\, f \oplus \mathrm{Coke}\,(1 - gf)$ and so $d(f - fgf) = d(f) + d(1 - gf)$. Also if $d(f) = 0$ for some $f \in S$, then $\mathrm{Coke}\, f = 0$, that is, $f$ is an epimorphism. Since $M$ has finite hollow dimension, $f$ has a small kernel. By hypothesis, $f$ splits and so must be an isomorphism. Thus the properties in 18.10 (h) are fulfilled and $S$ is semilocal with $\mathrm{h.dim}(S) \leq n = \mathrm{h.dim}(M)$. $\square$

Combining the last two results we get:

**19.3. Corollary.** *Any $\pi$-projective, semi-Hopfian module $M$ has finite hollow dimension if and only if $\mathrm{End}(M)$ is semilocal. In this case*

$$\mathrm{h.dim}(M) = \mathrm{h.dim}(\mathrm{End}(M)).$$

Notice that the corollary applies in particular to self-projective modules. Under this stronger condition we have:

**19.4. Semilocal self-projective modules.** *Let $M$ be finitely generated and self-projective and $S = \mathrm{End}(M)$.*

(1) *$M$ is semilocal if and only if $S$ is semilocal.*

(2) *$M$ is homogeneous semilocal if and only if $S$ is homogeneous semilocal.*

*Proof.* Under the given conditions $\mathrm{End}(M/\mathrm{Rad}\,M) \simeq S/\mathrm{Jac}\,S$ (e.g., [363, 22.2]). If $M/\mathrm{Rad}\,M$ is semisimple, then $S/\mathrm{Jac}\,S$ is left semisimple. Moreover, if $M/\mathrm{Rad}\,M$ is homogeneous semisimple, then $S/\mathrm{Jac}\,S$ is a simple left artinian ring. This shows '$\Rightarrow$' in (1) and (2).

The converse assertions follow by 19.3 and 18.6.                                  $\square$

We note that self-projectivity cannot be dropped for the assertion in 19.4(2) (see Exercise 19.13).

Dual to 19.2, applying the map $d : S \rightarrow \{0,\dots,n\}$, $f \mapsto \mathrm{u.dim}(\mathrm{Ke}\,f)$, one obtains for semi-co-Hopfian modules $M$ (every monomorphism is a coretraction):

**19.5. Proposition.** *Any semi-co-Hopfian module $M$ with finite uniform dimension has a semilocal endomorphism ring and*

$$\mathrm{h.dim}(\mathrm{End}(M)) \leq \mathrm{u.dim}(M).$$

The next result provides a further guarantee that certain endomorphism rings are semilocal and bounds their hollow dimension.

**19.6. Proposition.** *Any module $M$ with finite uniform and finite hollow dimension has a semilocal endomorphism ring and*

$$\mathrm{h.dim}(\mathrm{End}(M)) \leq \mathrm{h.dim}(M) + \mathrm{u.dim}(M).$$

*Proof.* Let $\mathrm{h.dim}(M) = n$ and $\mathrm{u.dim}(M) = k$. For $S = \mathrm{End}(M)$ define

$$d : S \rightarrow \{0,\dots,n+k\}, \quad f \mapsto \mathrm{u.dim}(\mathrm{Ke}\,(f)) + \mathrm{h.dim}(\mathrm{Coke}\,f).$$

Then one readily verifies the conditions in 18.10(h). Thus $S$ is semilocal and $\mathrm{h.dim}(S) \leq n + k = \mathrm{u.dim}(M) + \mathrm{h.dim}(M)$.                          $\square$

The last result shows that linearly compact, e.g., artinian modules have a semilocal endomorphism ring. It also shows that a biuniform module (see 16.13) has either a local endomorphism ring or a semilocal endomorphism ring $S$ whose factor $S/\mathrm{Jac}\,S$ has length 2.

**19.7. Finite decompositions of a module and its endomorphism ring.** *Let $M$ be an $R$-module and $S = \mathrm{End}(M)$. Then there is a bijection $\alpha$ between the set of all finite direct sum decompositions of $_RM$ and finite direct sum decompositions of $_SS$ given by*

$$\alpha : \{M_i\}_I \mapsto \{Se_i\}_I,$$

where $M = \bigoplus_I M_i$, $I$ is a finite set and, for each $i \in I$, $e_i = \pi_i \varepsilon_i$ is an idempotent (where $\pi_i : M \to M_i$ and $\varepsilon_i : M_i \to M$ denote the canonical projection and inclusion respectively). The inverse mapping $\alpha^{-1}$ is given by

$$\{S_i\}_I \mapsto \{(M)S_i\}_I,$$

where $S = \bigoplus_I S_i$ and $I$ is a finite set.

Moreover, for any decomposition $M = \bigoplus_I M_i$ and $i, j \in I$,

(i) $M_i$ is indecomposable if and only if $S_i$ is indecomposable;

(ii) $M_i \simeq M_j$ as R-modules if and only if $S_i \simeq S_j$ as S-modules.

**19.8. $n$-th root uniqueness property.** Given $n \in \mathbb{N}$, we say that a module $M$ has the *n-th root uniqueness property* if $M^n \simeq N^n$ for some module $N$ implies $M \simeq N$.

As observed in [96, Proposition 2.1] this happens when $M$ has a semilocal endomorphism ring.

**19.9. Modules with semilocal endomorphism rings.** *Any module $M$ with semilocal endomorphism ring has the n-th root uniqueness property for all $n \in \mathbb{N}$.*

*Proof.* Let $L := \bigoplus_{i=1}^n M_i = \bigoplus_{i=1}^n N_i$ with $M_i \simeq M$ and $N_i \simeq N$ for all $i \in \{1, \ldots, n\}$, and assume $\mathrm{End}(M)$ is semilocal. Then $S = \mathrm{End}(L)$ is also semilocal. From above we derive two decompositions of $S$, say

$$S = \bigoplus_{i=1}^n Se_i = \bigoplus_{i=1}^n Sf_i,$$

where $e_1, \ldots, e_n$ and $f_1, \ldots, f_n$ are orthogonal idempotents such that $\sum_{i=1}^n e_i = \sum_{i=1}^n f_i = 1_S$, $\mathrm{End}(M) \simeq Se_i$ and $\mathrm{End}(N) \simeq Sf_i$ for all $1 \leq i \leq n$.

Let $\bar{S} := S/\mathrm{Jac}\, S$ and $\bar{s} = s + \mathrm{Jac}\, S$ for any $s \in S$. Then, if $e$ and $f$ are idempotents in $S$, we have $Se \simeq Sf$ as $S$-modules if and only if $\bar{S}\bar{e} \simeq \bar{S}\bar{f}$ as $\bar{S}$-modules (see [363, 21.17 (3)]). Thus we get two decompositions $\bar{S} = \bigoplus_{i=1}^n \bar{S}\bar{e}_i = \bigoplus_{i=1}^n \bar{S}\bar{f}_i$ in which $\bar{S}\bar{e}_i \simeq \bar{S}\bar{e}_j$ and $\bar{S}\bar{f}_i \simeq \bar{S}\bar{f}_j$ for every $i$ and $j$. However, since $S$ is semilocal, $\bar{S}$ is semisimple artinian and so $\bar{S}\bar{e}_1 \simeq \bar{S}\bar{f}_1$. Thus $\mathrm{End}(M) \simeq Se_1 \simeq Sf_1 \simeq \mathrm{End}(N)$ and so $M \simeq N$. $\square$

In 18.8 we saw that the number of $M$-projective direct summands of a finitely generated semilocal module $M$ is finite and bounded by $2^k$, where $k = \ell(M/\mathrm{Rad}\, M)$. The number of direct summands of the endomorphism ring of $M$ yields a bound for the direct summands of $M$ (compare [96, Proposition 2.1]).

**19.10. Finite number of direct summands.** *Any module $M$ with semilocal endomorphism ring $S$ has a finite number of isomorphism classes of direct summands, bounded by $2^k$, where $k = \mathrm{h.dim}(S)$. If $M$ is artinian, then $k \leq \mathrm{u.dim}(M)$.*

*Proof.* The direct summands of $M$ correspond to summands of $S$. By 18.8, the number of the latter is at most $2^k$, where $k = \ell(S/\mathrm{Jac}\,S)$. If $M$ is artinian, $k = \mathrm{h.dim}(S) \leq \mathrm{u.dim}(M)$ by 19.5.                                    □

**19.11. Modules with homogeneous semilocal endomorphism rings.** *For any module $M$ with homogeneous semilocal endomorphism ring $S$, there exists an indecomposable direct summand $N$ of $M$ with $\mathrm{End}(N)$ homogeneous semilocal and $M \simeq N^k$, for some $k \in \mathbb{N}$.*

*Proof.* The functor $\mathrm{Hom}(M, -)$ gives an equivalence between the category $\mathrm{add}\,(M)$ of direct summands of finite direct sums of copies of $M$ and the category of finitely generated projective left $S$-modules. In view of this the assertion follows from 17.7.                                    □

Not every module with semilocal endomorphism ring has finite hollow dimension as can be seen from the $\mathbb{Z}$-module $\mathbb{Q}$. However, there are also cyclic modules with semilocal endomorphism ring which have infinite hollow dimension.

**19.12. Examples.**

(1) *Let $R$ be a ring that can be embedded in a local ring $S$. Then $R$ can be realised as the endomorphism ring of a local module.*

(2) *There exists a cyclic module with infinite hollow dimension whose endomorphism ring is semilocal ([168]).*

*Proof.* We start with a construction that serves both examples. Let $R \hookrightarrow S$ be a ring embedding. Consider the $(S, R)$-bimodule $M = {}_S\mathrm{Hom}_R({}_RS, {}_RS/R)_R$, where $S$ and $R$ act on $M$ by

$$sf : x \mapsto (xs)f, \qquad fr : x \mapsto (x)f \cdot r,$$

for $s \in S$, $f \in M$, $r \in R$ and $x \in S$. Note that $N := \{f \in M \,|\, (R)f = 0\}$ is an $(S, R)$-sub-bimodule of $M$. Clearly the canonical projection $\pi_R : {}_RS \to {}_RS/R$ is in $N$. Thus, if $s \in S$ is such that $sN \subseteq N$, then $(s)\pi_R = (1)s\pi_R = 0$ and so $s \in R$. Conversely, for all $s \in R$, we have $Rs \subseteq R$ and so $sf \in N$ for all $f \in N$. Thus $sN \subseteq N$ if and only if $s \in R$. Clearly we also have $f \in N$ if and only if $fR \subseteq N$. Now put

$$T = \begin{pmatrix} S & M \\ 0 & R \end{pmatrix} \quad \text{and} \quad I = \begin{pmatrix} 0 & N \\ 0 & R \end{pmatrix}.$$

Then $I$ is a right ideal of $T$. By definition, the idealiser $I'$ of $I$ in $T$ is given by $\{t \in T \,|\, tI \subseteq I\}$ and so, since $t = \begin{pmatrix} s & f \\ 0 & r \end{pmatrix}$ is in $I'$ if and only if $sN \subseteq N$ and $fR \subseteq N$, we have by the above that

$$I' = \begin{pmatrix} R & N \\ 0 & R \end{pmatrix}.$$

In this setting we have $\mathrm{End}_T(T/I) = I'/I = R$ (see, e.g., [300, Proposition 1]).

Every proper right ideal of $T$ containing $I$ is of the form $\begin{pmatrix} J & K \\ 0 & R \end{pmatrix}$, where $J$ is a proper right ideal of $S$ and $K$ is a submodule of $M$ containing $N$ such that $JM \subseteq K$.

(1) Assume $S$ to be a local ring. Then $J \subseteq \mathrm{Jac}\, S$ for $J$ as above. Hence every proper right $T$-submodule of $T/I$ is contained in $U/I$ where $U$ is the right ideal $\begin{pmatrix} \mathrm{Jac}\, S & M \\ 0 & R \end{pmatrix}$. Thus $T/I$ is a local right $T$-module with endomorphism ring $R$.

(2) Now take $R$ to be a semilocal ring and $S$ not to be semilocal; for example, let $K$ be a field and $S = R[X]$. Then $S$ has an infinite coindependent family $\{A_i\}_{\mathbb{N}}$ of right ideals. These induce the family of right ideals $\{U_i\}_{\mathbb{N}}$ of $T$, where $U_i = \begin{pmatrix} A_i & M \\ 0 & R \end{pmatrix}$ for each $i \in \mathbb{N}$ and it is easy to see that $\{U_i/I : n \in \mathbb{N}\}$ is an infinite family of coindependent submodules of $T/I$. Thus the right $T$-module $T/I$ has infinite hollow dimension, but its endomorphism ring is the semilocal ring $R$. $\qquad\square$

### 19.13. Exercise.

Let $p$ be a prime number, $m, n \in \mathbb{N}$ with $m \neq n$, and consider the $\mathbb{Z}$-module $M = \mathbb{Z}/p^n\mathbb{Z} \oplus \mathbb{Z}/p^m\mathbb{Z}$. Prove ([75]):

(i) $M$ is homogeneous semilocal and

(ii) $\mathrm{End}_{\mathbb{Z}}(M)$ is semilocal but not homogeneous semilocal.

**19.14. Comments.** Most of this short section is based on results by Herbera and Shamsuddin [168] and Facchini, Herbera, Levy, and Vamos [96]. Further information on modules with a semilocal endomorphism ring can be found in Facchini's articles [94] and [93].

Given any cardinal $\aleph$ and indecomposable modules $A$ and $B$, both with semilocal endomorphism rings, we may ask if $A^{(\aleph)} \simeq B^{(\aleph)}$ guarantees $A \simeq B$. If $\aleph$ is finite, the answer is affirmative by 19.8. However, if $\aleph = \aleph_0$ the answer may be negative. Indeed, using Eilenberg's trick, Facchini and Levy present an example in [98] of a noncommutative integral domain $R$ which is a module-finite algebra over a discrete valuation ring for which there is an indecomposable progenerator $P$ such that the left $R$-modules $_RR$ and $P$ are both indecomposable with semilocal endomorphism rings, $_RR^{(\aleph_0)} \simeq P^{(\aleph_0)}$ but $_RR \not\simeq P$. On the other hand, Příhoda [285] has recently shown that if $A$ and $B$ are uniserial modules for which $A^{(\aleph)} \simeq B^{(\aleph)}$ for some cardinal $\aleph$, then $A \simeq B$.

It follows from 15.33 that, as with semilocal rings, a unit-regular ring $R$ has stable range 1 and so finitely generated projective $R$-modules are cancellable. However, in the concluding remarks of [132], Goodearl notes that there is a simple unit-regular ring $R$ with non-isomorphic finitely generated projective $R$-modules $P_1$ and $P_2$ for which $P_1^n \simeq P_2^n$ for some $n > 1$. On the other hand, it follows from results of Baccella and Ciampella

[32] that if $R$ is a semi-artinian unit-regular ring, then every finitely generated projective $R$-module does satisfy the $n$-th root uniqueness property.

**References.** Corisello and Facchini [75]; Facchini, Herbera, Levy, and Vamos [96]; Golan [121]; Herbera and Shamsuddin [168]; Zöschinger [389, 394].

# 20   Supplemented modules

Again let $M$ denote a nonzero left $R$-module.

**20.1. Supplements.** *For submodules $N, L \subset M$, the following are equivalent:*

(a) $N$ *is minimal in the set of submodules $\{K \subset M \mid L + K = M\}$;*

(b) $L + N = M$ *and* $L \cap N \ll N$.

If these hold, then $N$ is called a *supplement of $L$ in $M$.*

*Proof.* (a)$\Rightarrow$(b). Consider any $K \subseteq N$ with $(L \cap N) + K = N$. Then $M = L + N = L + (L \cap N) + K = L + K$, hence $K = N$ by the minimality of $N$. Thus $L \cap N \ll L$.

(b)$\Rightarrow$(a). Let $L + N = M$ and $L \cap N \ll N$. For any submodule $K \subseteq N$ with $L + K = M$, we have, by modularity, $N = M \cap N = (L \cap N) + K$, implying $K = N$. Hence $N$ is minimal in the desired sense. $\qquad\qquad\square$

We say that a submodule $N \subset M$ is a *supplement* if it is a supplement for some submodule $L \subset M$. Obviously every supplement submodule is a weak supplement (for the same submodule) and the two notions are related in the following way.

**20.2. Supplement submodules.** *For a submodule $N \subset M$, the following statements are equivalent:*

(a) $N$ *is a supplement in $M$;*

(b) $N$ *is a weak supplement in $M$ that is coclosed in $M$;*

(c) $N$ *is a weak supplement in $M$ and, whenever $K \subset N$ and $K \ll M$, then $K \ll N$.*

*Proof.* (a)$\Rightarrow$(b). Assume that $N$ is a supplement of $L \subseteq M$. For all submodules $K \subseteq N$ such that $K \subset N$ is cosmall in $M$, we have that $N + L = M$ implies $K + L = M$ (see 3.2). By the minimality of $N$ with respect to this property we get $K = N$. Hence $N$ is coclosed. Obviously $N$ is also a weak supplement.

(b)$\Rightarrow$(c). Let $K \subseteq N$ and $K \ll M$. Assume $N = K + X$ for $X \subseteq N$. Then for every $Y \subseteq M$ with $N + Y = M$ we get $M = X + Y$ since $K \ll M$. Then, by 3.2 (1), $X \subseteq N$ is cosmall in $M$. Since $N$ is coclosed, we get $X = N$ and thus $K \ll N$.

(c)$\Rightarrow$(a). Let $N$ be a weak supplement of some submodule $L \subseteq M$. Then $N \cap L \ll M$ and so $N \cap L \ll N$. Thus $N$ is a supplement of $L$ in $M$. $\qquad\square$

**20.3. Supplements in weakly supplemented modules.** *For a submodule $K \subset M$ of a weakly supplemented module $M$, the following are equivalent:*

(a) $K$ *is a supplement submodule;*

(b) $K$ *is a coclosed submodule.*

*Proof.* (a)⇒(b) is shown in 20.2.

(b)⇒(a). Let $K$ be a coclosed submodule and $L$ be a weak supplement of $K$. Then $K \cap L \ll M$ and hence $K \cap L \ll K$ (see 3.7 (3)). Thus $K$ is a supplement of $L$ in $M$ and hence $K$ is a supplement submodule.                              □

In general coclosed submodules need not be supplements. To see this, take a cosemisimple module $M$ which is not semisimple. Then there exists a submodule $N \subset M$ which is not a direct summand. By 3.8, $N$ is coclosed in $M$. However, since $\mathbb{S}[M] = 0$ (see 8.2 (6)), all supplement submodules of $M$ are in fact direct summands.

**20.4. Properties of supplements.** *Let $U$ and $V$ be submodules of $M$ where $V$ is a supplement of $U$ in $M$.*

(1) *If $W + V = M$ for some $W \subset U$, then $V$ is a supplement of $W$.*

(2) *If $M$ is finitely generated (resp. a max module, 2.19), then $V$ is also finitely generated (resp. a max module).*

(3) *If $U$ is a maximal submodule of $M$, then $V$ is local and $U \cap V = \operatorname{Rad} V$ is the unique maximal submodule of $V$.*

(4) *If $K \ll M$, then $V$ is a supplement of $U + K$.*

(5) *If $\operatorname{Rad} M \ll M$, then $U$ is contained in a maximal submodule of $M$.*

(6) *If $\operatorname{Rad} M \ll M$ or $\operatorname{Rad} M \subset U$, and $p : M \to M/\operatorname{Rad} M$ is the canonical projection, then $M/\operatorname{Rad} M = Up \oplus Vp$.*

(7) *$\operatorname{Rad} V = V \cap \operatorname{Rad} M$.*

(8) *Let $W \subseteq V$. Then $W$ is coclosed in $V$ if and only if $W$ is coclosed in $M$.*

(9) *If $U$ is a supplement in $M$, then $U$ is a supplement of $V$ in $M$.*

*Proof.* The proofs for (1)–(7) can be found in [363, 41.1].

(8) First assume that $W$ is coclosed in $M$. If $Z \subset W$ is cosmall in $V$, then $Z \overset{cs}{\underset{M}{\to}} W$. Hence $Z = W$ since $W$ is coclosed in $M$.

Conversely suppose $W$ is coclosed in $V$ and that $Z \overset{cs}{\underset{M}{\to}} W$. Then, since $W$ is coclosed in $V$, $W/Z$ is not small in $V/Z$ and so, for some $Z \subset X \subset V$, we have $W + X = V$. This gives $M = U + V = U + X + W$ and so, since $Z \subset W$ is cosmall in $M$, we get $M = U + X + Z = U + X$. Since $V$ is a supplement of $U$ in $M$, we get $X = V$, a contradiction. Hence $W$ is coclosed in $M$.

(9) This follows from 20.2 (3).                              □

**20.5. Supplements in factor modules.** *Let $K \subseteq L \subset M$ be submodules.*

(1) *If $L$ is a supplement in $M$, then $L/K$ is a supplement in $M/K$.*

(2) *If $K$ is a supplement in $M$ and $L/K$ is a supplement in $M/K$, then $L$ is a supplement in $M$.*

(3) *Let $M$ be a weakly supplemented module and assume that $L/K$ is coclosed in $M/K$ and $K$ is coclosed in $M$. Then $L$ is coclosed in $M$.*

*Proof.* (1) Let $N \subset M$ be such that $L + N = M$ and $L \cap N \ll L$. Then $L/K + (N + K)/K = M/K$ and $(L \cap (N + K))/K = (L \cap (N + K))/K \ll L/K$, since the image of the small submodule $L \cap N \ll L$ is small in $L/K$.

(2) Let $L/K$ be a supplement of $L'/K$ in $M/K$ and let $K$ be a supplement of $K'$ in $M$. Then $M/K = L/K + L'/K$ and $(L/K) \cap (L'/K) \ll L/K$. Moreover $M = K + K'$ and $K \cap K' \ll K \subseteq L$. Hence $M = (L \cap L') + K'$ and $M = L + L'$ and so $M = L + (K' \cap L')$ (by 1.24).

Now $L = L \cap (K + K') = K + (L \cap K')$ and $(L \cap L')/K \ll L/K$ and so, by 2.3 (1),

$$(L \cap L' \cap K')/(K \cap K') \ll L/(K \cap K')$$

and so $L \cap L' \cap K' \ll L$. Thus $L$ is a supplement of $K' \cap L'$ in $M$.

(3) This is clear by properties of coclosed submodules and (2).                      □

**20.6. Supplements in submodules.** *Let $K \subset V \subset M$ be submodules.*

(1) *If $K$ is a supplement in $M$, then $K$ is a supplement in $V$.*

(2) *If $V \subset M$ is a supplement, the following are equivalent:*

(a) *$K$ is a supplement in $V$;*

(b) *$K$ is a supplement in $M$.*

*Proof.* (1) If $K$ is a supplement in $M$, there is some submodule $P \subset M$ with $K + P = M$ and $K' + P \neq M$ for a proper $K' \subset K$. By modularity, $K + (P \cap V) = V$. Assume $L + P \cap V = V$ for some submodule $L \subset K$. Then the equality $V + P = K + P = M$ yields $M = L + (P \cap V) + P = L + P$ and so, by the minimality of $K$, we get $L = K$. Thus $K$ is a supplement of $P \cap V$ in $V$.

(2) By assumption there is a submodule $N \subset M$ with $V + N = M$ and for any proper submodule $V' \subset V$, $V' + M \neq M$.

(a)$\Rightarrow$(b). There exists $L \subset V$ with $K + L = V$ and $K' + L \neq V$ for proper submodules $K' \subset K$. Thus $K + L + N = V + N = M$. Assume that for some $K' \subset K$, $K' + L + N = M$. However, $K' + L \subset V$ and hence, by assumption, the equality implies $K' + L = V$. By condition (a), this implies $K' = K$, showing that $K$ is a supplement of $L + N$ in $M$.

(b)$\Rightarrow$(a) follows from (1).                                               □

**20.7. The proper class of supplements.** *Let $\mathbb{E}_{supp}$ denote the class of short exact sequences $0 \to K \xrightarrow{f} L \to N \to 0$ in $\sigma[M]$ where $\mathrm{Im}\, f \subset L$ is a supplement submodule. Then*

(1) *$\mathbb{E}_{supp}$ is a proper class of short exact sequences in $\sigma[M]$;*

(2) *Every module $Q$ with* $\operatorname{Rad} Q = 0$ *is* $\mathbb{E}_{supp}$-*injective, that is, supplement submodules are co-neat submodules* (10.13);

(3) *if* $\operatorname{Rad} K \ll K$, *then $K$ is a co-neat submodule if and only if it is a supplement submodule.*

*Proof.* (1) We have to show that $\mathbb{E}_{supp}$ satisfies the properties required in 10.1. The conditions P.1 and P.2 are easily established while P.3 is shown in 20.6 and P.4 follows from 20.5.

(2) Let $K \subset L$ be a supplement submodule and $X \subset L$ such that $K + X = L$ and $K \cap X \ll K$, and so $L/(K \cap X) \simeq K/(K \cap X) \oplus X/(K \cap X)$.

Consider any $f : K \to Q$ where $\operatorname{Rad} Q = 0$. Then $K \cap X \subseteq \operatorname{Rad} K \subset \operatorname{Ke} f$ and hence $f$ factorises via $\bar{f} : K/(K \cap X) \to Q$. Since $K/(K \cap X)$ is a direct summand in $L/(K \cap X)$ there is a morphism $L/(K \cap X) \to Q$ leading to an extension of $f : K \to Q$ to $L \to Q$, as required.

(3) This is straightforward.                                                              □

**20.8. Corollary.** *If $M$ is a Bass module, then a submodule $K \subset L$ of any $L \in \sigma[M]$ is a supplement if and only if it is co-neat.*

*Proof.* If $M$ is a Bass module, every module in $\sigma[M]$ has a small radical (see 2.19).                                                                                         □

**20.9. Supplements in $\pi$-projective modules.** *If $N$ and $L$ are mutual supplements of a $\pi$-projective module $M$, then $N \cap L = 0$ and hence $M = N \oplus L$.*

*Proof.* If $N, L$ are mutual supplements, then $N \cap L \ll N$ and $N \cap L \ll L$. Then the kernel of the map $\phi : N \oplus L \to M$, $(n, l) \mapsto n + l$,

$$\operatorname{Ke} \phi = \{(n, -n) \mid n \in N \cap L\} \subseteq (N \cap L, 0) \oplus (0, N \cap L) \ll N \oplus L,$$

is small and splits by assumption. Thus $\operatorname{Ke} \phi = 0$ and so $N \cap L = 0$.            □

**20.10. Supplements and hollow dimension.** *Let $N$ be a submodule of $M$.*

(1) *If there exists a submodule $L$ of $N$ such that $L \overset{cs}{\underset{M}{\hookrightarrow}} N$ and $L$ is a supplement in $M$, then*

$$\operatorname{h.dim}(M) = \operatorname{h.dim}(M/N) + \operatorname{h.dim}(N) - \operatorname{h.dim}(N/L).$$

(2) *If $N$ is a supplement in $M$, then*

$$\operatorname{h.dim}(M) = \operatorname{h.dim}(M/N) + \operatorname{h.dim}(N).$$

(3) *If $\operatorname{h.dim}(M)$ is finite, then $N$ is a supplement in $M$ if and only if*

$$\operatorname{h.dim}(M) = \operatorname{h.dim}(M/N) + \operatorname{h.dim}(N).$$

*Proof.* (1) Let $L$ be a supplement of the submodule $K$ of $M$. Then, by 3.2 (3), we have $(L \cap K) \overset{cs}{\underset{M}{\to}} (N \cap K)$ and, since $L \cap K \ll M$, we get $N \cap K \ll M$ by 3.2(1). Hence $N$ is a weak supplement of $K$ in $M$. Then, by 17.9 (2),

$$\text{h.dim}(M) = \text{h.dim}(M/N) + \text{h.dim}(M/K).$$

Furthermore, $N \cap K$ is a weak supplement of $L$ in $N$ since, by modularity,

$$N = N \cap (L + K) = L + (N \cap K)$$

and also $(N \cap K) \cap L = L \cap K \ll L \subseteq N$. Applying 17.9 (2) again, we get

$$\text{h.dim}(N) = \text{h.dim}(N/(N \cap K)) + \text{h.dim}(N/L) = \text{h.dim}(M/K) + \text{h.dim}(N/L).$$

This together with our earlier dimension formula gives the required result.

(2) follows from (1), taking $L = N$.

(3) If $N$ is a supplement, then the formula holds by (2).

Conversely, assume that the formula holds. Since $M$ has finite hollow dimension, by 18.1, $N$ has a weak supplement $K$ and so, by 17.9,

$$\text{h.dim}(M) = \text{h.dim}(M/N) + \text{h.dim}(M/K).$$

Thus, from our assumption, $\text{h.dim}(N) = \text{h.dim}(M/K) = \text{h.dim}(N/(N \cap K))$ and, in particular, $\text{h.dim}(N)$ is finite. Applying 5.4 (2) we get $N \cap K \ll N$ and so $N$ is a supplement of $K$ in $M$. $\qquad\square$

We say a submodule $N$ of $M$ has *ample (or enough) supplements in $M$* if, for every $L \subset M$ with $N + L = M$, there is a supplement $L'$ of $N$ with $L' \subset L$.

As we will see shortly, in general supplements need not exist in $M$.

**20.11. Definitions.** The module $M$ is called

| | |
|---|---|
| *supplemented* | if any submodule has a supplement, |
| *finitely supplemented* | if finitely generated submodules have supplements, |
| *cofinitely supplemented* | if cofinite submodules have supplements, |
| *amply supplemented* | if all submodules have ample supplements in $M$. |

We will usually abbreviate *finitely supplemented* and *cofinitely supplemented* to *f-supplemented* and *cf-supplemented*, respectively.

Clearly we have the implications

amply supplemented $\Rightarrow$ supplemented

$\Rightarrow$ weakly supplemented

$\Rightarrow$ semilocal.

Note that in general all these implications are proper. In particular there are weakly supplemented modules that are not supplemented and thus there exist weak supplements that are not supplements.

**20.12. Example.** *The $\mathbb{Z}$-module $\mathbb{Q}$ is weakly supplemented but not supplemented.*

*Proof.* Let $p$ be a prime number and $\mathbb{Z}_{(p)} = \{a/b \in \mathbb{Q} \,|\, p$ does not divide $b\}$. Assume that $\mathbb{Z}_{(p)}$ has a supplement $K$ in $\mathbb{Q}$. Then $(K + \mathbb{Z})/\mathbb{Z}$ is a supplement of $\mathbb{Z}_{(p)}/\mathbb{Z}$ in $\mathbb{Q}/\mathbb{Z}$. Note that $\mathbb{Z}_{(p)}/\mathbb{Z}$ is the sum of all the $q$-components of $\mathbb{Q}/\mathbb{Z}$ where $q$ runs through all primes different from $p$. Since $(K + \mathbb{Z})/\mathbb{Z} + \mathbb{Z}_{(p)}/\mathbb{Z} = \mathbb{Q}/\mathbb{Z}$ and since the $p$-component of $\mathbb{Z}_{(p)}/\mathbb{Z}$ is zero, the $p$-component of $\mathbb{Q}/\mathbb{Z}$ must equal the $p$-component of $(K + \mathbb{Z})/\mathbb{Z}$. On the other hand, if $q$ is a prime different from $p$, then the $q$-component of $(K + \mathbb{Z})/\mathbb{Z}$ is a submodule of the $q$-component of $\mathbb{Z}_{(p)}/\mathbb{Z}$, that is,

$$[(K + \mathbb{Z})/\mathbb{Z}]_q \subseteq ((K + \mathbb{Z})/\mathbb{Z}) \cap (\mathbb{Z}_{(p)}/\mathbb{Z}) \ll (K + \mathbb{Z})/\mathbb{Z}.$$

However, the $q$-component of a torsion $\mathbb{Z}$-module is a direct summand. Hence the $q$-components of $(K + \mathbb{Z})/\mathbb{Z}$ must be all zero, that is, $(K + \mathbb{Z})/\mathbb{Z}$ is equal to its $p$-component, namely the $p$-component of $\mathbb{Q}/\mathbb{Z}$. Since $(\mathbb{Q}/\mathbb{Z})_p = \mathbb{Z}[1/p]/\mathbb{Z}$, where $\mathbb{Z}[1/p] := \{a/p^k \in \mathbb{Q} \,|\, a \in Z, k \geq 0\}$, we have that $(K + \mathbb{Z})/\mathbb{Z} = \mathbb{Z}[1/p]/\mathbb{Z}$ and so $K + \mathbb{Z} = \mathbb{Z}[1/p]$. Now $K \cap \mathbb{Z} \subseteq K \cap \mathbb{Z}_{(p)} \ll K$ and so $n\mathbb{Z} = K \cap \mathbb{Z} \ll \mathbb{Z}[1/p]$ for some nonzero number $n$. On the other hand, if $q$ is a prime that does not divide $n$ nor $p$, then $\mathbb{Z}[1/p] = n\mathbb{Z} + q\mathbb{Z}[1/p]$, because by the Euclidean algorithm, for any $k \geq 0$, there are integers $r$ and $s$ such that $1 = rq + snp^k$. Thus $1/p^k = q(r/p^k) + sn \in q\mathbb{Z}[1/p] + n\mathbb{Z}$. Since $q \neq p$, $q\mathbb{Z}[1/p] \neq \mathbb{Z}[1/p]$, that is, $n\mathbb{Z}$ is not small in $\mathbb{Z}[1/p]$ — a contradiction. Thus $\mathbb{Z}_{(p)}$ cannot have a supplement in $\mathbb{Q}$, and so, in particular, $\mathbb{Q}$ is not supplemented.                                                        □

The reader should be warned that different names have been used for the concepts listed above and that, conversely, the terminology we use here may be applied with different meaning elsewhere. For example, *amply supplemented modules* are called *perfect modules* in Miyashita [235] and Chamard [60] and are called *modules with property* $(P_2)$ in Varadarajan [345].

**20.13. Supplemented submodules and supplements.**

(1) *Let $N$, $K$ be submodules of $M$ with $N$ supplemented. If there is a supplement for $N + K$ in $M$, then $K$ also has a supplement in $M$.*

(2) *Suppose $M = N + K$. If $K$ is supplemented, then $K$ contains a supplement of $N$ in $M$.*

*Proof.* (1) This is shown in [363, 41.2 (1)].

(2) Let $L$ be a supplement of $N \cap K$ in $K$. Then $K = (N \cap K) + L$ and $N \cap L = (N \cap K) \cap L \ll L$. Also $M = N + K = N + (N \cap K) + L = N + L$.   □

**20.14. Supplements in a sum of supplemented modules.** *Suppose $M = K + N$, where $K$ and $N$ are supplemented modules. Then $M$ is supplemented and in fact any submodule $X$ of $M$ has a supplement of the form $K' + N'$ with $K' \subseteq K$ and $N' \subseteq N$.*

*If we further assume that $X$ is a supplement submodule of $M$, then $K'$ and $X + N'$ are mutual supplements in $M$ and the same is true for $N'$ and $X + K'$.*

*Proof.* Since $X + K + N = M$ trivially has a supplement, it follows from 20.13 (1), that $K + X$ has a supplement $N' \subseteq N$ and, similarly, $X + N'$ has a supplement $K' \subseteq K$. We have

$$M = K + X + N', \quad N' \cap (X + K) \ll N', \text{ and}$$
$$M = K' + X + N' \quad K' \cap (X + N') \ll K'.$$

Now $X \cap (K' + N') \subseteq [K' \cap (X + N')] + [N' \cap (K' + X)]$. Hence it follows that $X \cap (K' + N') \ll K' + N'$. Since $X + K' + N' = M$, this establishes the first claim.

Now suppose $X$ is a supplement submodule of $M$. Then, by 20.2 (c), since $X \cap (K' + N') \ll M$ we have $X \cap (K' + N') \ll X$. This gives

$$K' \cap (X + N') \subseteq [X \cap (K' + N')] + [N' \cap (X + K')] \ll X + N'.$$

Similarly $N' \cap (X + K') \ll X + K'$. $\qquad\qquad\qquad\qquad\qquad\qquad\qquad$ $\square$

**20.15. Corollary.** *If $M$ is supplemented, then every finitely $M$-generated module is supplemented and $M/\mathrm{Rad}\,M$ is a semisimple module.*

We note that this is also shown in [363, 41.2 (3)].

**20.16. Remarks.** If $K$ and $N$ are amply supplemented modules, then $K \oplus N$ need not be an amply supplemented module. For an example, see 23.7.

Any small cover of a weakly supplemented module is weakly supplemented (see 17.13) but this is not the case for (amply) supplemented modules. For example, consider a semilocal ring which is not semiperfect; then $R/\mathrm{Jac}\,R$ is semisimple and hence amply supplemented but $R$ is not supplemented.

A coclosed submodule of a weakly supplemented module is weakly supplemented (see 17.13). We do not know if a supplement submodule of a supplemented module is supplemented.

**20.17. Properties of f-supplemented modules.**

(1) *Let $M = M_1 + M_2$, with $M_1$, $M_2$ both finitely generated and finitely supplemented. Assume either*

    (i) *$M$ is coherent in $\sigma[M]$ or*

    (ii) *$M$ is self-projective and $M_1 \cap M_2 = 0$.*

    *Then $M$ is f-supplemented.*

(2) *Let $M$ be an f-supplemented $R$-module.*

    (i) *If $L \subset M$ is a finitely generated or a small submodule, then $M/L$ is also f-supplemented.*

(ii) *If* Rad $M \ll M$, *then finitely generated submodules of* $M/\mathrm{Rad}\,M$ *are direct summands.*

(iii) *If* $M$ *is either finitely generated and* $M$-*projective or coherent in* $\sigma[M]$, *then for any finitely generated or small submodule* $K \subset M^n$, *with* $n \in \mathbb{N}$, *the factor module* $M^n/K$ *is f-supplemented.*

*Proof.* This is shown in [363, 41.3]. □

Cofinitely supplemented modules $M$ are obviously cofinitely weakly supplemented (in the sense that all cofinite submodules have weak supplements) and hence any cofinite submodule of $M/\mathrm{Rad}\,M$ is a direct summand (see 18.5). There are some more interesting properties of this class of modules.

**20.18. Properties of cf-supplemented modules.** *Let* $M$ *be a module.*

(1) *Let* $N, L \subset M$ *be submodules such that* $N$ *is cofinite,* $L$ *is cf-supplemented, and* $N + L$ *has a supplement in* $M$. *Then* $N$ *has a supplement in* $M$.

(2) *All direct sums and factor modules of cf-supplemented modules are again cf-supplemented.*

(3) *The following are equivalent for* $M$:

    (a) $M$ *is cf-supplemented;*

    (b) *every maximal submodule has a supplement in* $M$;

    (c) $M/\mathrm{Loc}(M)$ *has no maximal submodule, where* $\mathrm{Loc}(M)$ *denotes the sum of all the local submodules of* $M$;

    (d) $M/\mathrm{Cfs}(M)$ *has no maximal submodule, where* $\mathrm{Cfs}(M)$ *denotes the sum of all the cf-supplemented submodules of* $M$.

*Proof.* (1) Let $K$ be a supplement of $N+L$ in $M$. Then $L \cap (N+K) \subset L$ is a cofinite submodule and so, by assumption, there exists a supplement $H$ of $L \cap (N+K)$ in $L$. Then $M = N + L + K = N + H + K$ and

$$N \cap (H + K) \ \subseteq\ [H \cap (N + K)] + [K \cap (N + H)]$$
$$\subseteq\ [H \cap (N + K)] + [K \cap (N + L)] \ \ll\ H + K,$$

and so $H + K$ is a supplement of $N$ in $M$ (see 20.1).

(2) Let $M$ be cf-supplemented and $N \subset M$ be any submodule. Any cofinite submodule of $M/N$ is of the form $L/N$ for some cofinite submodule $L \subset M$. Thus there is a submodule $K \subset M$ such that $M = L + K$ and $L \cap K \ll K$. Then $M/N = L/N + (K + N)/N$ and

$$L/N \cap (K + N)/N = (L \cap (K + N))/N \ll (K + N)/N,$$

showing that $(K + N)/N$ is a supplement of $L/N$ in $M/N$.

We show that for any family $\{N_\lambda\}_\Lambda$ of cf-supplemented submodules $N_\lambda \subset M$, the submodule $N' = \sum_\Lambda N_\lambda \subseteq M$ is cf-supplemented. For this let $L \subset N'$ be a cofinite submodule. Then $N' = L + N_{\lambda_1} + \cdots + N_{\lambda_k}$, for some $k \in \mathbb{N}$ and $\lambda_i \in \Lambda$. Applying (1) we see by induction that $L$ has a supplement in $N'$.

(3) (a)$\Rightarrow$(b) is obvious.

(b)$\Rightarrow$(c). Let $K$ be a maximal submodule of $M$ and $L \subset M$ such that $M = K + L$ and $K \cap L \ll L$. Since $L/(K \cap L) \simeq M/K$ we see that $L$ is a local submodule of $M$. Therefore $\mathrm{Loc}(M)$ is not contained in $K$ and so $M/\mathrm{Loc}(M)$ has no maximal submodules.

(c)$\Rightarrow$(d). This follows from the inclusion $\mathrm{Loc}(M) \subseteq \mathrm{Cfs}(M)$.

(d)$\Rightarrow$(a). For any cofinite submodule $N \subset M$, $N + \mathrm{Cfs}(M)$ is also cofinite in $M$ and (d) implies $N + \mathrm{Cfs}(M) = M$, and so $M = N + K_1 + \cdots + K_n$ for some cf-supplemented submodules $K_i \subset M$. Since $\mathrm{Cfs}(M)$ is cf-supplemented (by (2)) it follows by (1) that $N$ has a supplement in $M$. $\qquad\square$

Representing modules as a sum of submodules, the following property (known for internal direct sums) turns out to be of interest.

If $M = \sum_{\Lambda} M_\lambda$, then this sum is called *irredundant* if, for every $\lambda_0 \in \Lambda$, $\sum_{\lambda \neq \lambda_0} M_\lambda \neq M$.

**20.19. Sums of hollow modules.** *For $M$ the following are equivalent:*

(a) *$M$ is a sum of hollow submodules and $\mathrm{Rad}\,M \ll M$;*

(b) *every proper submodule of $M$ is contained in a maximal one, and every maximal submodule has a supplement in $M$;*

(c) *$M$ is an irredundant sum of local modules and $\mathrm{Rad}\,M \ll M$.*

*Proof.* This is shown in [363, 41.5]. $\qquad\square$

As a consequence we note that a supplemented module $M$ with $\mathrm{Rad}\,M \ll M$ is an irredundant sum of local modules. Moreover we have the corollary:

**20.20. Finitely generated supplemented modules.** *If $M$ is finitely generated, then the following are equivalent:*

(a) *$M$ is supplemented;*

(b) *every maximal submodule of $M$ has a supplement in $M$;*

(c) *$M$ is a sum of hollow submodules;*

(d) *$M$ is an irredundant (finite) sum of local submodules.*

The next observation shows that irredundant sums of submodules are closely related to the hollow dimension of the module.

**20.21. Supplemented modules with finite hollow dimension.**

(1) *The hollow dimension of an irredundant sum of $n$ hollow modules is $n$.*

(2) *Any supplemented module of hollow dimension $n$ can be written as an irredundant sum of $n$ hollow submodules.*

*Proof.* (1) Assume $M = H_1 + \cdots + H_n$, where each $H_i$ is hollow and the sum is irredundant. Consider the epimorphism

$$f : H_1 \oplus \cdots \oplus H_n \to M, \quad (h_1, \ldots, h_n) \mapsto h_1 + \cdots + h_n.$$

Then $\operatorname{Ke} f = K_1 \oplus \cdots \oplus K_n$ with $K_i := H_i \cap (H_1 + \cdots + H_{i-1} + H_{i+1} + \cdots + H_n)$. As the sum is irredundant and $H_i$ is hollow, $K_i \ll H_i$, for $1 \leq i \leq n$. Hence we get $\operatorname{Ke} f \ll H_1 \oplus \cdots \oplus H_n$. Thus, by 5.4 (2), $\operatorname{h.dim}(M) = \operatorname{h.dim}(H_1 \oplus \cdots \oplus H_n) = n$.

(2) We will prove this by induction on the hollow dimension $n$. For $n = 1$, $M$ is hollow. Let $n > 1$ and assume that all supplemented modules with hollow dimension $n - 1$ can be written as an irredundant sum of $n - 1$ hollow modules. As $M$ has finite hollow dimension, there exists a nonzero hollow factor module $M/N$ by 5.2. Since $M$ is supplemented, $N$ has a supplement $H_1$ in $M$. By 5.4 (2),

$$\operatorname{h.dim}(H_1) = \operatorname{h.dim}(H_1/(H_1 \cap N)) = \operatorname{h.dim}(M/N) = 1$$

and so $H_1$ is hollow. Let $H'$ be a supplement of $H_1$ in $M$. Since $H_1$ is hollow, $H_1$ is also a supplement of $H'$. By 20.10, we have

$$
\begin{aligned}
\operatorname{h.dim}(M) &= \operatorname{h.dim}(H') + \operatorname{h.dim}(M/H') = \operatorname{h.dim}(H') + \operatorname{h.dim}(H_1/(H_1 \cap H')) \\
&= \operatorname{h.dim}(H') + 1.
\end{aligned}
$$

Thus $\operatorname{h.dim}(H') = n - 1$ and so, by assumption, we may write $H'$ as an irredundant sum of hollow modules $H' = \sum_{i=2}^{n} H_i$. Then, since $H'$ and $H_1$ are mutual supplements, $M = \sum_{i=1}^{n} H_i$ is irredundant as required.                                                    $\square$

The following three propositions are taken from [363, 41.7–41.9].

**20.22. Amply supplemented modules.** *Let $M$ be amply supplemented.*

(1) *Every supplement of a submodule of $M$ is an amply supplemented module.*

(2) *Direct summands and factor modules of $M$ are amply supplemented.*

(3) *$M = (\sum_\Lambda L_\lambda) + K$ where $\sum_\Lambda L_\lambda$ is an irredundant sum of local modules $L_\lambda$ and $K = \operatorname{Rad} K$. Moreover, if $M/\operatorname{Rad} M$ is finitely generated, then the sum is finite.*

**20.23. Supplements of intersections.** *Let $K, L \subset M$ be submodules with $K + L = M$. If $K$ and $L$ have ample supplements in $M$, then $K \cap L$ has also ample supplements in $M$.*

**20.24. Characterisation of amply supplemented modules.** *The following properties are equivalent for a module $M$:*

(a) *$M$ is amply supplemented;*

(b) *every submodule $U \subset M$ is of the form $U = X + Y$, with $X$ a supplement and $Y \ll M$;*

(c) *for every submodule $U \subset M$, there is a supplement submodule $X \subseteq U$ with $X \subseteq U$ cosmall in $M$.*

If $M$ *is finitely generated, then* (a)–(c) *are also equivalent to:*

(d) *Every maximal submodule has ample supplements in $M$.*

The existence of coclosures in $M$ has an influence on the supplemented structure of $M$.

**20.25. Coclosures in weakly supplemented modules.** *The $R$-module $M$ is amply supplemented if and only if $M$ is weakly supplemented and every submodule of $M$ has a coclosure in $M$.*

*Proof.* $\Rightarrow$. Let $K \subseteq M$. Since $M$ is amply supplemented, $M$ is weakly supplemented as well as supplemented. Let $L$ be a supplement of $K$ in $M$. Then $M = K + L$ with $K \cap L \ll L$ and so $K \cap L \ll M$. Since $M$ is amply supplemented, there exists a supplement $N$ of $L$ such that $N \subseteq K$. Then $M = N + L$ and $N \cap L \ll N$. We claim that $N$ is a coclosure of $K$ in $M$. As a supplement submodule, $N$ is coclosed in $M$ (see 20.2). Since $K = N + (L \cap K)$ and $K \cap L \ll M$, we get $N \overset{cs}{\underset{M}{\longrightarrow}} K$ (see 3.2 (5)), as claimed.

$\Leftarrow$. Suppose that $M$ is weakly supplemented and every submodule of $M$ has a coclosure in $M$. Let $K, L \subseteq M$ with $M = K + L$. Since $M$ is weakly supplemented there exists a weak supplement $T$ of $K$ such that $T \subseteq L$ (see 17.9 (6)). Then $M = K + T$, $T \subseteq L$ and $K \cap T \ll M$. Suppose $N$ is a coclosure of $T$ in $M$. Then $M - K + N$ (see 3.2 (1)) and $N \cap K \ll N$ since $N$ is coclosed in $M$ and $N \cap K \ll M$ (see 3.7 (3)). Thus $N$ is a supplement of $K$ in $M$ such that $N \subseteq L$.  $\square$

**20.26. Coclosed submodules as direct summands.** *Let $M_1$ and $M_2$ be supplemented modules and $M = M_1 \oplus M_2$. Then the following are equivalent:*

(a) *every coclosed submodule of $M$ is a direct summand;*

(b) *for every coclosed submodule $N \subseteq M$, both $M = N + M_1$ and $M = N + M_2$ imply that $N$ is a direct summand of $M$.*

If $M_1$ *and* $M_2$ *are amply supplemented, then* (a), (b) *are equivalent to:*

(c) *for every coclosed submodule $K \subseteq M$, either*

    (i) $(K + M_1)/K \ll M/K$, *or*

    (ii) $(K + M_2)/K \ll M/K$, *or*

    (iii) $M = K + M_1 = K + M_2$

*imply that $K$ is a direct summand of $M$.*

*Proof.* (a)$\Rightarrow$(b) and (b)$\Rightarrow$(c) are obvious.

(b)$\Rightarrow$(a). Let $X$ be a coclosed submodule of $M$. By 20.14, $X$ has a supplement of the form $M_1' \oplus M_2'$, where $M_1'$ and $M_2'$ are direct summands of $M_1$ and $M_2$, respectively. Also $X + M_1'$ is a supplement of $M_2'$. As $M = (X + M_1') + M_2$, (b) implies $X + M_1'$ is a direct summand of $M$. Suppose $M = (X + M_1') \oplus K$. Then $M/X = (X + M_1')/X \oplus (K + X)/X$. Thus $(K + X)/X$ is a supplement submodule of $M/X$ and so, $X$ is a supplement submodule of $M$, we get that $K \oplus X$ is a supplement submodule of $M$ (see 20.5(2)). We have $M = (K \oplus X) + M_1' = (K \oplus X) + M_1$. By (b), $K \oplus X$ is a direct summand of $M$. Hence $X$ is a direct summand of $M$.

(c)$\Rightarrow$(b). Let $K \subseteq M$ be a coclosed submodule such that $M = K + M_2$. Since $M/K \simeq M_2/(K \cap M_2)$ and $M_2$ is amply supplemented we get that $M/K$ is amply supplemented (see 20.22 (2)). Hence there exists a coclosed submodule $N/K$ of $M/K$ such that the inclusion $N/K \subseteq (K + M_1)/K$ is cosmall in $M/K$. Since $N/K$ is coclosed in $M/K$ and $K$ is coclosed in $M$, $N$ is coclosed in $M$ (see 20.5 (3)). Also $(K + M_1)/N \ll M/N$ and $N + M_1 \subseteq K + M_1$. Hence $(N + M_1)/N \ll M/N$.

By hypothesis, there exists a submodule $N'$ of $M$ such that $M = N \oplus N'$ and so

$$M = K + M_2 = (K + N') + M_2 = (K + N') + M_1.$$

Since $M/K = N/K \oplus (K + N')/K$, $(K + N')/K$ is a coclosed submodule of $M/K$. Again by 20.5 (3), $K + N'$ is a coclosed submodule of $M$. Hence by hypothesis, $K + N'$ is a direct summand of $M$. Since $K \cap N' = 0$ we conclude that $K$ is a direct summand of $M$.

Similarly a coclosed submodule $K$ of $M$ with $M = K = M_1$ is a direct summand of $M$. $\qquad\square$

Amply supplemented modules guarantee nice hollow dimension properties.

**20.27. Proposition.** *Suppose $M$ is an amply supplemented module. For any proper submodules $K$ and $L$ of $M$ we have*

$$\mathrm{h.dim}(M/K) + \mathrm{h.dim}(M/L) \leq \mathrm{h.dim}(M/(K + L)) + \mathrm{h.dim}(M/(K \cap L)).$$

*If $M$ has finite hollow dimension, then*

$$\mathrm{h.dim}(M/K) + \mathrm{h.dim}(M/L) = \mathrm{h.dim}(M/(K + L)) + \mathrm{h.dim}(M/(K \cap L))$$

*holds if and only if $M/(K \cap L)$ is a supplement in $M/K \oplus M/L$.*

*Proof.* Consider the epimorphism

$$g : M/K \oplus M/L \to M/(K + L), \quad (x + K, y + L) \mapsto x - y + K + L.$$

Regard $T = M/(K \cap L)$ as a submodule of $M/K \oplus M/L$ under the canonical monomorphism. Then clearly $T \subseteq \mathrm{Ke}\,g$. Let $(x + K, y + L) \in \mathrm{Ke}\,g$. Then $x - y \in K + L$ and so $x = y + \ell + k$ for some $\ell \in L$ and $k \in K$. Hence we get $(x + K, y + L) = (z + K, z + L)$ for $z = y + \ell$. Hence $\mathrm{Ke}\,g \subseteq T$. Thus $\mathrm{Ke}\,g = T = M/(K \cap L)$ and we have the exact sequence

$$0 \longrightarrow M/(K \cap L) \longrightarrow M/K \oplus M/L \overset{g}{\longrightarrow} M/(K + L) \longrightarrow 0.$$

By 20.22 (2) and 20.24 (c), we may choose a supplement submodule $T'$ which is cosmall in $T$. Then, by 20.10,

$$\begin{aligned}
\mathrm{h.dim}(M/K) + \mathrm{h.dim}(M/L) &= \mathrm{h.dim}(M/K \oplus M/L) \\
&= \mathrm{h.dim}((M/K \oplus M/L)/T) + \mathrm{h.dim}(T) - \mathrm{h.dim}(T/T') \\
&= \mathrm{h.dim}(M/(K + L)) + \mathrm{h.dim}(M/(K \cap L)) - \mathrm{h.dim}(T/T') \\
&\leq \mathrm{h.dim}(M/(K + L)) + \mathrm{h.dim}(M/(K \cap L)).
\end{aligned}$$

If $\mathrm{h.dim}(M)$ is finite, then equality holds in the inequality above if and only if $\mathrm{h.dim}(T/T') = 0$, that is, $M/(K \cap L)$ is a supplement in $M/K \oplus M/L$ (see 20.10 (3)). $\qquad\square$

**20.28. Definition.** An amply supplemented module $M$ is called a *hollow dimension module* if for any two submodules $K$, $L$ of $M$,

$$\mathrm{h.dim}(M/(K + L)) + \mathrm{h.dim}(M/(K \cap L)) = \mathrm{h.dim}(M/K) + \mathrm{h.dim}(M/L).$$

It is straightforward to show that any factor module of a hollow dimension module is also a hollow dimension module.

As we now see, to show that an amply supplemented module $M$ is a hollow dimension module, it is enough to check the hollow dimension formula for coclosed submodules $K$ and $L$ of $M$ with $\mathrm{h.dim}(M/K)$ and $\mathrm{h.dim}(M/L)$ finite (this is dual to Del Valle's result [79, Proposition 1] for the uniform dimension).

**20.29. Characterisation of hollow dimension modules.** *For an amply supplemented module $M$, the following are equivalent:*

(a) *$M$ is a hollow dimension module;*

(b) *$\mathrm{h.dim}(M/(K + L)) + \mathrm{h.dim}(M/(K \cap L)) = \mathrm{h.dim}(M/K) + \mathrm{h.dim}(M/L)$ holds for any two coclosed submodules $K$ and $L$ of $M$ with $\mathrm{h.dim}(M/K)$ and $\mathrm{h.dim}(M/L)$ finite;*

(c) *if $A \overset{cs}{\underset{M}{\hookrightarrow}} K$, $B \overset{cs}{\underset{M}{\hookrightarrow}} L$, and $\mathrm{h.dim}(M/K)$ and $\mathrm{h.dim}(M/L)$ are both finite, then $(A \cap B) \overset{cs}{\underset{M}{\hookrightarrow}} (K \cap L)$;*

(d) *if $\mathrm{h.dim}(M/K)$ is finite, $A \overset{cs}{\underset{M}{\hookrightarrow}} K$ and $B \overset{cs}{\underset{M}{\hookrightarrow}} K$, then $(A \cap B) \overset{cs}{\underset{M}{\hookrightarrow}} K$;*

(e) *if* h.dim$(M/K)$ *is finite, then $K$ has a unique coclosure in $M$;*

(f) *if* $K \overset{cc}{\hookrightarrow} M$, $L \overset{cc}{\hookrightarrow} M$, *and* h.dim$(M/K)$ *and* h.dim$(M/L)$ *are both finite, then* $(K+L) \overset{cc}{\hookrightarrow} M$.

*Proof.* (a)$\Rightarrow$(b) is trivial and the proofs of (c)$\Rightarrow$(d) and (d)$\Rightarrow$(e) are straightforward.

(b)$\Rightarrow$(a). Suppose h.dim$(M/K)$ or h.dim$(M/L)$ is infinite. Then by 18.2, h.dim$(M/(K \cap L))$ is also infinite. So to show that a module is a hollow dimension module it is enough to prove that the hollow dimension formula holds for submodules $K$ and $L$ of $M$ with both h.dim$(M/K)$ and h.dim$(M/L)$ finite. Suppose $K$ and $L$ are such submodules of $M$. Let $\widetilde{K}$ and $\widetilde{L}$ be coclosures of $K$ and $L$ in $M$ respectively. We have $(\widetilde{K} + \widetilde{L}) \overset{cs}{\underset{M}{\hookrightarrow}} (K+L)$ and, by 5.4 (2),

$$
\begin{aligned}
\text{h.dim}(M/K) &= \text{h.dim}(M/\widetilde{K}), \\
\text{h.dim}(M/L) &= \text{h.dim}(M/\widetilde{L}), \text{ and} \\
\text{h.dim}(M/(K+L)) &= \text{h.dim}(M/(\widetilde{K}+\widetilde{L})).
\end{aligned}
$$

This gives, on referring to our assumption,

$$
\begin{aligned}
\text{h.dim}(M/(K \cap L)) &\leq \text{h.dim}(M/(\widetilde{K} \cap \widetilde{L})) \\
&= \text{h.dim}(M/\widetilde{K}) + \text{h.dim}(M/\widetilde{L}) - \text{h.dim}(M/(\widetilde{K}+\widetilde{L})) \\
&= \text{h.dim}(M/K) + \text{h.dim}(M/L) - \text{h.dim}(M/(K+L)) \\
&\leq \text{h.dim}(M/(K \cap L)) \text{ (by 20.27)}.
\end{aligned}
$$

Thus $M$ is a hollow dimension module.

(a)$\Rightarrow$(c). Suppose $K$ and $L$ are submodules of $M$ with $A \overset{cs}{\underset{M}{\hookrightarrow}} K$, $B \overset{cs}{\underset{M}{\hookrightarrow}} L$ and both h.dim$(M/K)$ and h.dim$(M/L)$ are finite. By 5.4 (2),

$$\text{h.dim}(M/A) = \text{h.dim}(M/K) \quad \text{and} \quad \text{h.dim}(M/B) = \text{h.dim}(M/L).$$

Since $M$ is a hollow dimension module, if h.dim$(M/A)$ and h.dim$(M/B)$ are both finite, then so too are h.dim$(M/(A+B))$ and h.dim$(M/(A \cap B))$. Again by 5.4 (2), since $(A+B) \overset{cs}{\underset{M}{\hookrightarrow}} (K+L)$ we have h.dim$(M/(A+B)) = $ h.dim$(M/(K+L))$. Thus

$$
\begin{aligned}
\text{h.dim}(M/(A \cap B)) &= -\text{h.dim}(M/(A+B)) + \text{h.dim}(M/A) + \text{h.dim}(M/B) \\
&= -\text{h.dim}(M/(K+L)) + \text{h.dim}(M/K) + \text{h.dim}(M/L) \\
&= \text{h.dim}(M/(K \cap L)).
\end{aligned}
$$

Since h.dim$(M/(A \cap B))$ is finite, we have $(K \cap L)/(A \cap B) \ll M/(A \cap B)$ (by 5.4 (2) once more). Thus $(A \cap B) \overset{cs}{\underset{M}{\hookrightarrow}} (K \cap L)$.

(e)$\Rightarrow$(f). Suppose h.dim$(M/K)$ and h.dim$(M/L)$ are finite and $L \overset{cc}{\hookrightarrow} M$ and $K \overset{cc}{\hookrightarrow} M$. It is obvious that h.dim$(M/(K+L))$ is finite. Suppose $A$ is the unique coclosure of $K + L$ in $M$. Since, by 20.22 (2), the module $M/K$ is amply supplemented, by 20.25 $(K+L)/K$ has a coclosure, say $B/K$, in $M/K$. Then $B \overset{cs}{\underset{M}{\hookrightarrow}} (K+L)$ (see 3.2). Then, since $B/K \overset{cc}{\hookrightarrow} M/K$ and $K \overset{cc}{\hookrightarrow} M$, we have $B \overset{cc}{\hookrightarrow} M$ (see 20.5). By the uniqueness of the coclosure of $K + L$ in $M$, we get $A = B$ and so $K \subseteq A$. Similarly $L \subseteq A$ implying $K + L = A$ and therefore $(K + L) \overset{cc}{\hookrightarrow} M$.

(f)$\Rightarrow$(b). Suppose that h.dim$(M/K)$ and h.dim$(M/L)$ are finite and $L \overset{cc}{\hookrightarrow} M$ and $K \overset{cc}{\hookrightarrow} M$. Then $(K + L) \overset{cc}{\hookrightarrow} M$ and h.dim$(M/(K + L))$ is finite. Moreover we have $(K + L)/L \overset{cc}{\hookrightarrow} M/L$ and $K/(K \cap L) \overset{cc}{\hookrightarrow} M/(K \cap L)$. Using 5.4 (2) we get

$$
\begin{aligned}
\text{h.dim}(M/(K \cap L)) \;&=\; \text{h.dim}(K/(K \cap L)) + \text{h.dim}(M/K) \\
&=\; \text{h.dim}((K + L)/L) + \text{h.dim}(M/K) \\
&=\; -\text{h.dim}(M/(K + L)) + \text{h.dim}(M/L) + \text{h.dim}(M/K),
\end{aligned}
$$

and now (b) follows. $\qquad\square$

We note that the proof above of (a)$\Rightarrow$(c) did not use the assumption that $M$ is amply supplemented. We use this in the proof of (a)$\Rightarrow$(b) in the next result.

**20.30. Homomorphic images in factor modules.** *For $M$ consider the conditions:*

(a) *$M$ is a hollow dimension module;*

(b) *for any $X \subseteq M$, $M/X \not\simeq A \oplus B$, where h.dim$(A)$ is finite and $B \simeq S \ll A$;*

(c) *if $f : M \to M/X$ with $X + \text{Ke}\, f = M$, $\text{Im}\, f \ll M/X$ and h.dim$(M/X)$ finite, then $f - 0$;*

(d) *if $f : M \to M/N$ is a homomorphism such that $\text{Ke}\, f + N = M$ and h.dim$(M/X)$ finite, then $\text{Im}\, f \overset{cc}{\hookrightarrow} M/N$.*

*Then (a)$\Rightarrow$(b)$\Rightarrow$(c)$\Rightarrow$(d). Moreover, if $M$ is amply supplemented, then all the above conditions are equivalent.*

*Proof.* (a)$\Rightarrow$(b). Let $M$ be a hollow dimension module with $X \subseteq M$. Then any factor module of $M$, in particular $M/X$, is also a hollow dimension module. Suppose to the contrary that $M/X \simeq A \oplus B$, where h.dim$(A)$ is finite, $S \ll A$ and $\phi : B \to S$ is an isomorphism. Set $L = A \oplus B$ and $N = S \oplus B$. Then, since h.dim$(L/B)$ is finite, h.dim$(L/N)$ is also finite. However both $B$ and $C := \{((b)\phi, b) \mid b \in B\}$ are cosmall submodules of $N$ in $L$. Using (c) of 20.29 and the remark above, this gives a contradiction since $L$ is a hollow dimension module. Thus the factor module $M/X$ is as claimed in (b).

(b)$\Rightarrow$(c). Let $f : M \to M/X$ be a homomorphism with $\text{Im}\, f = T/X \ll M/X$, $X + \text{Ke}\, f = M$, and h.dim$(M/X)$ finite. Define $N := A \oplus (M/X)$, where $A$ is such that there exists an isomorphism $g : T/X \to A$. Let $\eta : M \to M/X$ be the natural map. Then the map $h : M \to N$ given by $(m)h = ((m)fg, (m)\eta)$ is onto. Now (b) gives $A = 0$ and hence $f = 0$.

(c)$\Rightarrow$(d). Let $f : M \to M/N$ be a homomorphism with $\mathrm{Ke}\, f + N = M$, $\mathrm{Im}\, f = T/N$ and h.dim$(M/X)$ is finite. If possible let $X/N \subset T/N$ be cosmall in $M/N$, where $X \neq T$. Then $T/X \ll M/X$. Consider the natural map $g : M/N \to M/X$. Then $\mathrm{Im}\,(fg) \ll M/X$ and so, by (c), $fg = 0$. Hence $T = X$ and so $\mathrm{Im}\, f = T/N \overset{cc}{\hookrightarrow} M/N$.

(d)$\Rightarrow$(a). Now we assume that $M$ is amply supplemented. Let $A \subseteq M$ be such that h.dim$(M/A)$ is finite, let $X$ and $Y$ be two coclosures of $A$ in $M$, and let $T$ be a supplement of $A$ in $M$. Then $A + T = X + T = Y + T = M$ and, since $X$ and $Y$ are supplements of $T$ in $M$, we have $X \cap T \ll X$ and $Y \cap T \ll Y$. Now consider

$$M \overset{f}{\to} M/T \simeq (X+T)/T \simeq X/(X \cap T) \overset{g}{\to} (X+Y)/((X \cap T) + Y),$$

where $f$ and $g$ are the natural maps. Put $N := (X \cap T) + Y$. Then we have a map $h : M \to M/N$ with $\mathrm{Im}\, h = (X+Y)/N$ and $\mathrm{Ke}\, h + N = M$ (for $\mathrm{Ke}\, h \supseteq T$). Thus, by (d), $\mathrm{Im}\, h \overset{cc}{\hookrightarrow} M/N$. Since $Y \subseteq N \subseteq X + Y \subseteq A$ and $Y \subset A$ is cosmall in $M$, it follows that $N \subset X + Y$ is cosmall in $M$. Thus $\mathrm{Im}\, h \ll M/N$ and so $\mathrm{Im}\, h = 0$. As $Y \subset A$ is cosmall in $M$, h.dim$(M/A) = $ h.dim$(M/Y)$ and $Y \subseteq N$ implies that h.dim$(M/N)$ is finite. Now since $X + Y = N = (X \cap T) + Y$ and $(X \cap T) \ll X$ (hence $\ll N$), we get $Y = N$. Thus $X \subseteq Y$. Similarly $Y \subseteq X$, proving $X = Y$. It now follows from 20.29 (e) that $M$ is a hollow dimension module.   $\square$

**20.31. Every $R$-module is hollow dimension.** *The following are equivalent for $R$:*

(a) *every $R$-module is a hollow dimension module;*

(b) *every $R$-module of finite hollow dimension has no proper small cover;*

(c) *every $R$-module of finite hollow dimension is semisimple.*

*Proof.* (a)$\Rightarrow$(b). Let $M$ be an $R$-module of finite hollow dimension with a proper small cover $f : N \twoheadrightarrow M$. By 5.4 (2) applied to $N$, $N$ also has finite hollow dimension. Then, by 20.30, the module $L = N \oplus T$, where $T \simeq \mathrm{Ke}\, f$, is not a hollow dimension module, a contradiction to (a). Hence (b) follows.

(b)$\Rightarrow$(c). Let $M$ be an $R$-module of finite hollow dimension. Then, by 5.2 (e), there is a small epimorphism $f : M \to N$ where $N$ is a finite direct sum of hollow modules. Hence (b) implies that every module of finite hollow dimension is a direct sum of finitely many hollow modules. Thus it suffices to show that each hollow module is simple. Suppose to the contrary that $H$ is a hollow module which is not simple. Then there exists $0 \neq K \subset H$ such that $K \ll H$. Then, by 5.4 (2), $H/K$ is a module with finite hollow dimension and a proper small cover. This contradicts (b). Hence every hollow module is simple and (c) follows.

(c)$\Rightarrow$(a). Let $M$ be any $R$-module. If $K$ and $L$ are submodules of $M$ with either h.dim$(M/K)$ or h.dim$(M/L)$ infinite, then h.dim$(M/(K \cap L))$ is also infinite. Thus to show that $M$ is a hollow dimension module it is enough to check the hollow dimension formula of 20.28 for submodules $K$ and $L$ where both h.dim$(M/K)$ and h.dim$(M/L)$ are finite. In this case, by (c), $M/K$, $M/L$ and $M/(K+L)$ are finitely generated semisimple. Now, regarding $M/(K \cap L)$ as a submodule of $M/K \oplus M/L$ under the canonical monomorphism, we get an exact sequence

$$0 \to M/(K \cap L) \to M/K \oplus M/L \to M/(K+L) \to 0$$

with the obvious epimorphism. From this it is straightforward to see that the hollow dimension formula holds and hence $M$ is a hollow dimension module. $\square$

**20.32. im-small projectives and supplemented modules.** *Let* $M = M_1 \oplus M_2$ *where* $M_2$ *is supplemented. Then the following are equivalent:*

(a) $M_1$ *is im-small* $M_2$*-projective;*

(b) *for every submodule* $N$ *of* $M$ *such that* $(N + M_1)/N \ll M/N$, *there exists a submodule* $N'$ *of* $N$ *such that* $M = N' \oplus M_2$;

(c) *for every submodule* $N \subseteq M$ *such that* $M_2$ *is a supplement of* $N$ *in* $M$, *there exists* $N' \subseteq N$ *such that* $M = N' \oplus M_2$.

*If further* $M_2$ *has ample supplements in* $M$, *then* (a) *is equivalent to:*

(d) *if* $N$ *is a coclosed submodule of* $M$ *such that* $(N + M_1)/M_1 \ll M/M_1$ *and* $M = N + M_2$, *then* $M = N \oplus M_2$.

*Proof.* (a)$\Leftrightarrow$(b) is shown in 4.35, and (b)$\Rightarrow$(c) follows from 2.3 (2).

(c)$\Rightarrow$(b). Let $N \subseteq M$ be such that $(N + M_1)/N \ll M/N$. Since $M/M_1$ is supplemented, there exists $X \subseteq M$ with $M_1 \subseteq X$, $M/M_1 = X/M_1 + (N+M_1)/M_1$ and $(X \cap (N + M_1))/M_1 \ll X/M_1$. Then

$$M = N + X = (N + M_1) + (X \cap M_2).$$

Since $(N + M_1)/N \ll M/N$ we get $M = (X \cap M_2) + N$. As $M = X + M_2$, by 1.24, this gives $M = M_2 + (X \cap N)$. Moreover $(N \cap X) + M_1 = X \cap (N + M_1)$ and so $((N \cap X) + M_1)/M_1 \ll X/M_1$ whence $((N \cap X) + M_1)/M_1 \ll M/M_1$. Then, by hypothesis there is a submodule $N' \subseteq N \cap X$ with $M = N' \oplus M_2$.

(c)$\Rightarrow$(d). Let $N \subseteq M$ be a coclosed submodule such that $(N + M_1)/M_1 \ll M/M_1$ and $M = N + M_2$. By hypothesis, there exists a submodule $N'$ of $N$ such that $M = N' \oplus M_2$. This gives $N = N' \oplus (N \cap M_2)$. Hence $((N \cap M_2) + M_1)/M_1 \ll M/M_1$ and so $N \cap M_2 \ll M_2$. This implies $N \cap M_2 \ll M$. Now the projection $M \to M/N'$ yields $((N \cap M_2) + N')/N' = N/N' \ll M/N'$. Since $N$ is coclosed this gives $N' = N$.

(d)$\Rightarrow$(c). Let $N \subset M$ be a submodule with $(N + M_1)/M_1 \ll M/M_1$ and $M = N + M_2$. Since $M_2$ has ample supplements in $M$, $M_2$ has a supplement $N'$ such that $N' \subseteq N$. Clearly $(N' + M_1)/M_1 \subseteq (N + M_1)/M_1$ and therefore $(N' + M_1)/M_1 \ll M/M_1$. By hypothesis, $M = N' \oplus M_2$. $\square$

**20.33. Generalised projectivity and supplemented modules.** *Let* $M = K \oplus L$ *where* $K$ *is generalised* $L$-*projective.*

(1) *If* $M$ *is supplemented, then* $L$ *has ample supplements in* $M$.

(2) *If* $M$ *is amply supplemented, then for any direct summands* $K^* \subset K$ *and* $L^* \subset L$, $K^*$ *is generalised* $L^*$-*projective.*

*Proof.* (1) Let $M$ be supplemented and $M = N + L$. Since $K$ is generalised $L$-projective, by 4.42 we have

$$M = N' \oplus K' \oplus L' = N' + L \text{ with } N' \subseteq N, K' \subseteq K \text{ and } L' \subseteq L.$$

As $N'$ is supplemented, it contains a supplement $N''$ of $L$ in $M$ by 20.13 (2).

(2) Suppose $K = K^* \oplus K^{**}$ and $L = L^* \oplus L^{**}$. Put $N = K^* \oplus L$. We first show, using 4.42, that $K^*$ is generalised $L$-projective.

To this end, suppose that $N = X + L$. Since $N$ is also amply supplemented, $L$ has a supplement $Y \subseteq X$ in $N$. We have $M = (K^{**} \oplus Y) + L$. It is clear that $(K^{**}+Y) \cap L = Y \cap L$ and hence $K^{**} \oplus Y$ is a supplement of $L$. As $K$ is generalised $L$-projective and $K^{**} \oplus Y$ is a supplement of $L$, we have by 4.42,

$$M = (K^{**} \oplus Y) \oplus K' \oplus L' = (K^{**} \oplus Y) + L \text{ with } K' \subseteq K, L' \subseteq L.$$

With the projection $\pi_{K^*} : K \to K^*$ along $K^{**}$, $K^{**} \oplus K' = K^{**} \oplus (K')\pi_{K^*}$. Thus $M = K^{**} \oplus (Y \oplus (K')\pi_{K^*} \oplus L')$ and hence $N = Y \oplus (K')\pi_{K^*} \oplus L'$ with $Y \subseteq X$, $(K')\pi_{K^*} \subseteq K^*$ and $L' \subseteq L$. Thus $K^*$ is generalised $L$-projective. It now follows from 4.43 that $K^*$ is generalised $L^*$-projective.                           $\square$

For amply supplemented modules, finite hollow dimension can be characterised by chain conditions on supplements.

**20.34. Hollow dimension and chain conditions on supplements.** *Let* $M$ *be an amply supplemented module. Then the following are equivalent:*

(a) $M$ *has finite hollow dimension;*

(b) $M$ *has DCC on supplements;*

(c) $M$ *has ACC on supplements.*

*Proof.* (a)$\Rightarrow$(b), (c) are immediate from 5.3 and 20.2.

(b)$\Rightarrow$(c). Let $N_1 \subseteq N_2 \subseteq \cdots$ be an ascending chain of supplements in $M$. Then, by hypothesis, for every integer $i$ there are supplements $L_i$ of $N_i$ such that $L_1 \supseteq L_2 \supseteq \cdots$. By (b), there is an integer $n$ such that $L_n = L_k$, for every $k \geq n$. Then, by 17.9 (1), $N_n \subseteq N_k$ is cosmall in $M$ for all $k \geq n$ and thus $N_n = N_k$, since $N_k$ is coclosed.

(c)$\Rightarrow$(a). Assume that $M$ contains an infinite coindependent family $\{N_\lambda\}_\Lambda$ of submodules. We show by induction that there is a strictly ascending chain of supplements

$$L_1 \subset L_2 \subset L_3 \subset \cdots$$

in $M$ such that $M/L_k$ contains an infinite coindependent family of proper submodules for all $k \in \mathbb{N}$. Let $\mu \in \Lambda$, $\Lambda' := \Lambda \setminus \{\mu\}$, and $L_1$ be a supplement of $N_\mu$ in $M$. Then, by Lemma 18.3, $\{(L_1 + (N_\lambda \cap N_\mu))/L_1\}_{\Lambda'}$ is an infinite coindependent family of submodules of $M/L_1$. Now assume $k \geq 1$ and there exists an ascending chain $L_1 \subset L_2 \subset \cdots \subset L_k$ such that each $L_i$ is a supplement in $M$ and $M/L_i$ contains an infinite coindependent family for all $1 \leq i \leq k$.

Let $\{N_\gamma/L_k\}_\Gamma$ be an infinite coindependent family of submodules of $M/L_k$. Let $\mu \in \Gamma$ and choose a supplement $L'$ of $N_\mu$ in $M$. Let $L_{k+1} := L_k + L'$. Then, by 20.4 (6), $L_{k+1}/L_k$ is a supplement of $N_\mu/L_k$ in $M/L_k$. Hence, applying Lemma 18.3 again, $M/L_{k+1}$ contains an infinite coindependent family. Proceeding in this way we see that if $M$ contains an infinite coindependent family of submodules, then it contains a strictly ascending chain of supplements.

Notice that, for (c)$\Rightarrow$(a), $M$ was only required to be supplemented. $\qquad\square$

In amply supplemented modules some more properties of $\mathrm{Re}_\mathbb{S}{}^2$ can be observed.

**20.35. Homomorphic images of $\mathrm{Re}_\mathbb{S}{}^2(N)$.** *Let $N$ be an amply supplemented module in $\sigma[M]$ and $L$ be a subfactor of $N$.*

(1) *$\mathrm{Re}_\mathbb{S}{}^2(L)$ is the largest fully non-$M$-small submodule of $L$.*

(2) *If $f : L \to T$ an epimorphism, then $(\mathrm{Re}_\mathbb{S}{}^2(L))f = \mathrm{Re}_\mathbb{S}{}^2(T)$.*

*Proof.* Since every factor of $N$ is amply supplemented (see 20.22) and $L$ is a submodule of a factor of $N$, we can assume $L$ is a submodule of $N$.

(1) Suppose $\widetilde{L}$ is a coclosure of $L$ in $N$, which exists by 20.25. Since $N$ is amply supplemented, $\widetilde{L}$ is a supplement in $N$ and so is amply supplemented (see 20.22). Then, by 17.9, $L = \widetilde{L} + S$, where $S$ is $M$-small and so, by 8.10, $\mathrm{Re}_\mathbb{S}(\widetilde{L}) = \mathrm{Re}_\mathbb{S}(L)$ and, by 8.19, $\mathrm{Re}_\mathbb{S}{}^2(\widetilde{L}) = \mathrm{Re}_\mathbb{S}{}^3(\widetilde{L})$. Hence $\mathrm{Re}_\mathbb{S}{}^2(L)$ is the largest fully non-$M$-small submodule of $L$.

(2) As $N$ is amply supplemented, $L = \widetilde{L} + S$ where $\widetilde{L}$ is a coclosure of $L$ in $N$ and $S \ll N$. By 8.10, $\mathrm{Re}_\mathbb{S}{}^2(L) = \mathrm{Re}_\mathbb{S}{}^2(\widetilde{L})$ and $\mathrm{Re}_\mathbb{S}{}^2((L)f) = \mathrm{Re}_\mathbb{S}{}^2((\widetilde{L})f)$. Hence we can assume $L$ is amply supplemented.

Put $D = \mathrm{Re}_\mathbb{S}{}^2(T)$ and $C = \mathrm{Re}_\mathbb{S}{}^2(L)$. Applying (1) to $T$ we get that $D$ is a fully non-$M$-small module. By 8.10, $(C)f \subseteq D$. Suppose $\widetilde{A}$ is a coclosure of $A$ in $L$. By 17.9, $A = \widetilde{A} + S'$ where $S' \ll L$. Now $(S')f$ is an $M$-small module contained in the fully non-$M$-small module $D$ and so $(S')f \ll \mathrm{Re}_\mathbb{S}{}^2(T)$ (see 8.12). Hence $(\widetilde{A})f = (A)f = D$. By 20.22, $\widetilde{A}$ is amply supplemented and so $\widetilde{A} \cap \mathrm{Ke}\, f$ has a supplement $B$ in $\widetilde{A}$. Then $B + (\widetilde{A} \cap \mathrm{Ke}\, f) = \widetilde{A}$ and $(\widetilde{A} \cap \mathrm{Ke}\, f) \cap B = \mathrm{Ke}\, f \cap B \ll B$. Since $D = (\widetilde{A})f = (B)f$, we have $B/(\mathrm{Ke}\, f \cap B) \simeq D$, a fully non-$M$-small module. As $\mathrm{Ke}\, f \cap B \ll B$, $B$ is fully non-$M$-small (see 8.12). By (1), $C$ is the largest fully non-$M$-small submodule of $L$ and so $B \subseteq C$. Hence $D = (B)f \subseteq (C)f$. Thus $(C)f = D$. $\qquad\square$

Since injective modules in $\sigma[M]$ are $M$- and also $\widehat{M}$-generated, and factor modules of amply supplemented modules are amply supplemented (see 20.22), we have the following.

**20.36. Corollary.** *Suppose that either $M$ or its self-injective hull $\widehat{M}$ is $\Sigma$-amply-supplemented. Then:*

(1) *every injective module in $\sigma[M]$ is amply supplemented.*

(2) *for any module $N \in \sigma[M]$, $\mathrm{Res}^2(N)$ is the largest fully non-$M$-small submodule of $N$.*

(3) *for any epimorphism $f : N \to T$ in $\sigma[M]$, $(\mathrm{Res}^2(N))f = \mathrm{Res}^2(T)$.*

Recall that $\varrho_{\mathbb{S}}$ is defined as the idempotent radical induced by the class $\mathbb{S}$ of small modules in $\sigma[M]$ (8.16).

**20.37. Cohereditary $\varrho_{\mathbb{S}}$.** *Let $M$ be such that every injective module in $\sigma[M]$ is amply supplemented. Then:*

(1) $\varrho_{\mathbb{S}}(N) = \mathrm{Res}^2(N)$, *for all $N \in \sigma[M]$;*

(2) $\mathbb{S}^{\circ} = \{N \in \sigma[M] \mid \mathrm{Res}^2(N) = N\}$;

(3) $\mathbb{S}^{\circ\bullet} = \{N \in \sigma[M] \mid \mathrm{Res}^2(N) = 0\}$;

(4) $\varrho_{\mathbb{S}}$ *is cohereditary.*

*Proof.* Each $N \in \sigma[M]$ is a submodule of an injective module in $\sigma[M]$. By 20.35, $\varrho_{\mathrm{s}}(N) = \mathrm{Res}^2(N)$ and is the largest fully non-$M$-small submodule of $N$ and $\varrho_{\mathrm{s}}$ is cohereditary. Now (2) and (3) follow easily.                                             $\square$

**20.38. $M$-small cogenerated modules projective.** *Let $M$ be such that every injective module in $\sigma[M]$ is amply supplemented. Then the following are equivalent:*

(a) *every $M$-small-cogenerated module in $\sigma[M]$ is projective in $\sigma[M]$;*

(b) *every module in $\sigma[M]$ is a direct sum of a fully non-$M$-small module and a semisimple module.*

*Proof.* (a)$\Rightarrow$(b). By 8.21(ii), each $N$ in $\sigma[M]$ is a direct sum of a fully non-$M$-small and an $M$-small-cogenerated module. From the proof of 8.21(ii), $\mathrm{Res}_{\mathbb{S}}(N) = \mathrm{Res}^2(N)$ and so, by 20.35, any homomorphic image of an $M$-small-cogenerated module $L$ is $M$-small-cogenerated. The projectivity condition then shows that $L$ is semisimple and so (b) follows.

(b)$\Rightarrow$(a). Every $M$-small-cogenerated module is semisimple. Let $S$ be a simple $M$-small-cogenerated module and $f : N \to S$ be an epimorphism, where $N \in \sigma[M]$. Then $N = A \oplus B$, where $A$ is fully non-$M$-small and $B$ is semisimple. By 20.35, $(A)f = 0$ and so the map $f$ splits. Therefore $S$, and hence any $M$-small-cogenerated module in $\sigma[M]$, is projective in $\sigma[M]$.                                             $\square$

**20.39. Exercises.**

(1) For an $R$-module $M$, let $V$ be a supplement of the submodule $U$ in $M$, and $K$ and $T$ be submodules of $V$. Prove that $T$ is a supplement of $K$ in $V$ if and only if $T$ is a supplement of $U + K$ in $M$.

(2) Let $M$ be an amply supplemented module and $A$, $B$ and $C$ be submodules of $M$ such that $A \subseteq B$. Prove ([327]):

    (i) If $A + C = M$, then, for any supplement $B'$ of $B$ in $M$ with $B' \subseteq C$, there exists a supplement $A'$ of $A$ in $M$ with $B' \subseteq A' \subseteq C$.

    (ii) For any supplement $A'$ of $A$ (resp. $B'$ of $B$) in $M$, there exists a supplement $B'$ of $B$ (resp. $A'$ of $A$) in $M$ such that $B' \subseteq A'$.

(3) Let $A \subset M$ be a submodule. Define a *double supplement* $A''$ of $A$ in $M$ as a supplement of some supplement of $A$ in $M$ such that $A'' \subseteq A$.

    (i) Let $A \subset M$ be a submodule that has a double supplement. Prove that the following are equivalent ([327]):

        (a) $A$ is coclosed in $M$;

        (b) $A$ is a supplement in $M$;

        (c) $A = A''$ for some double supplement $A''$ of $A$ in $M$;

        (d) $A = A''$ for every double supplement $A''$ of $A$ in $M$.

    (ii) Let $M$ be an amply supplemented module and let $A \subseteq B$ be submodules of $M$. Prove that for any double supplement $A''$ of $A$ in $M$, there exists a double supplement $B''$ of $B$ in $M$ such that $A'' \subseteq B''$.

(4) Let $M$ be an amply supplemented module and $C$ a supplement in $M$. Prove that there exists a one-to-one inclusion preserving correspondence between the set of all supplements of $M$ that include $C$ and the set of all supplements of $M/C$. (See [327].)

(5) Let $M$ be a projective $R$-module and $S = \mathrm{End}(M)$. Prove that the following are equivalent ([248, Theorem 2.8]):

    (a) every supplement in $M$ is a direct summand;

    (b) every supplement in $_S S$ is a direct summand.

(6) Let $M$ be a supplemented module. A chain of submodules

$$0 = L_0 \subset L_1 \subset L_2 \subset \cdots \subset L_n = M,$$

is called a *supplement composition series* provided that, for each $1 \leq i < n$,

        (i) $L_i$ is a supplement in $M$ and    (ii) $L_{i+1}/L_i$ is hollow.

Prove ([327, 4.13], [344, 2.28]):

$$\mathrm{h.dim}(M) = \sup\{k : M \text{ has a supplement composition series of length } k\}.$$

(7) Let $R$ be a semilocal ring. Prove that for any left $R$-module $L$, a submodule $K \subset L$ is co-neat if and only every simple module is injective with respect to the inclusion $K \to L$ ([233, Theorem 3.8.7]).

(8) Call two submodules $K, L$ of a module $M$ *equivalent* if they have the same supplements in $M$. Assume $M$ to be amply supplemented. Prove ([60]):

    (i) For submodules $K \subset L$ of $M$ the following are equivalent:

        (a) $K$ and $L$ are equivalent;

        (b) $K \subset L$ is a cosmall inclusion in $M$.

    (ii) Two submodules $K, L$ of $M$ are equivalent if and only if $K \subset K + L$ and $L \subset K + L$ are cosmall inclusions in $M$.

(9) A module $M$ is said to be *totally supplemented* if every submodule of $M$ is supplemented. Prove ([317]):

    (i) Every totally supplemented module is amply supplemented.

    (ii) Let $K$ be a linearly compact submodule (e.g., [363, 29.7]) of $M$. Then $M$ is (totally) supplemented if and only if $M/K$ is (totally) supplemented.

    (iii) Consider $M = M_1 \oplus M_2$ where $M_2$ is semisimple. Then $M$ is totally supplemented if and only if $M_1$ is totally supplemented.

(10) Suppose every supplement in the module $M$ is a direct summand and $A$ is a supplement submodule of $M$. Prove that every supplement submodule of $M/A$ is a direct summand in $M/A$.

(11) An $R$-module $M$ is called $\oplus$-*supplemented* if every submodule of $M$ has a (weak) supplement that is a direct summand. Prove ([164], [174]):

    (i) Any finite direct sum of $\oplus$-supplemented modules is $\oplus$-supplemented.

    (ii) If $M$ is $\oplus$-supplemented, then $M = M_1 \oplus M_2$, where $\operatorname{Rad} M_1 \ll M_1$ and $\operatorname{Rad} M_2 = M_2$.

    (iii) If $M$ is $\oplus$-supplemented, then $M = M_1 \oplus M_2$, where $\operatorname{Tr}_\mathbb{S}(M_1) \ll M_1$ and $\operatorname{Tr}_\mathbb{S}(M_2) = M_2$.

    (iv) If $M$ is finitely generated and each direct summand of $M$ is $\oplus$-supplemented, then $M$ is a direct sum of cyclic modules.

    (v) If $R$ is a Dedekind domain, then $M$ is $\oplus$-supplemented if and only if it is supplemented.

    (vi) For a commutative local ring $R$, the following are equivalent:

        (a) every finitely presented $R$-module is $\oplus$-supplemented;

        (b) $R$ is a valuation ring.

(12) A supplemented module $M$ is called *strongly* $\oplus$-*supplemented* (also *weak lifting*) if every supplement submodule of $M$ is a direct summand. (See [251], [205].) Prove:

    (i) Every direct summand of a strongly $\oplus$-supplemented module is a strongly $\oplus$-supplemented module.

    (ii) For a supplemented module the following are equivalent:

        (a) $M$ is strongly $\oplus$-supplemented;

        (b) every supplement submodule lies above a direct summand (see 22.1);

        (c) ($\alpha$) every nonzero supplement submodule of $M$ contains a nonzero direct summand of $M$ and

($\beta$) every supplement submodule of $M$ contains a maximal direct summand of $M$.

(iii) Let $M$ be a finitely generated strongly $\oplus$-supplemented $R$-module. Then $M$ is a direct sum of cyclic submodules.

(iv) Let $M = \bigoplus_{i=1}^{n} M_i$ be a self-projective module. Then $M$ is a strongly $\oplus$-supplemented module if and only if every $M_i$ is strongly $\oplus$-supplemented.

(v) Let $R$ be a Prüfer ring. Then every finitely generated torsion free supplemented $R$-module is strongly $\oplus$-supplemented.

(13) Assume that $\varrho_\mathbb{S}$ (see 8.16) is a cohereditary radical for $R$-Mod and put $I = \varrho_\mathbb{S}(R)$. Prove that the following are equivalent ([119]):

(a) For any left $R$-module $M$, $\varrho_\mathbb{S}(M)$ is a supplement in $M$;

(b) $R = I + \mathrm{An}_R(I_R)$.

(14) Call a module $M$ *supplementing* if it has a supplement in any module in which it is contained as a submodule. These modules are called *Moduln mit Ergänzungseigenschaft* in [23]. Prove:

(i) If $M$ is supplementing, then every module isomorphic to $M$ is supplementing.

(ii) If $M$ is supplementing, then for any flat module $F$, $\mathrm{Ext}_R^1(F, M) = 0$ (i.e., $M$ is a *cotorsion* module).

(iii) A ring $R$ is left perfect if and only if every left $R$-module is supplementing.

(iv) If $M$ is linearly compact, then $M$ is supplementing (e.g. [363, 41.10]).

(v) Let $R$ be a noetherian ring. Then $R$ is left artinian if and only if $R^{(\mathbb{N})}$ is supplementing.

(vi) Let $R$ be a commutative noetherian ring. Then a finitely generated $R$-module $M$ is supplementing if and only if $M$ is linearly compact.

**20.40. Comments.** Given an element $a$ in a complete lattice $(\mathcal{L}, \wedge, \vee, 0, 1)$, a pseudo-complement $b \in \mathcal{L}$ is an element which is maximal with respect to $a \wedge b = 0$. In the case where $\mathcal{L}$ is the lattice of submodules of a module $M$, a submodule $N$ is called a complement submodule if it is a pseudo-complement in $\mathcal{L}$. Dually $N$ is a supplement submodule in $M$ if it is a pseudo-complement in the dual lattice $\mathcal{L}^0$. While the existence of complement submodules is secured by the fact that the lattice of submodules of a module is lower continuous, the existence of supplement submodules is not guaranteed, since the dual lattice need not be lower continuous.

Historically, the notion of a supplement submodule was introduced by Kasch and Mares in [193] in order to characterise semiperfect modules, that is, projective modules whose homomorphic images have projective covers. This lead to a lattice theoretical characterisation of perfect and semi-perfect rings, which represent rings with special homological properties (see [189]). Independently, Miyashita had shown a similar characterisation of perfect rings in [235]. That the intersection of mutual supplements of a projective module is zero (20.9) goes essentially back to Kasch and Mares.

Generalisations of these concepts demand the existence of (weak) supplements for just certain submodules such as maximal submodules (Alizade, Bilhan and Smith [5]), cofinitely generated submodules (Alizade and Büyükasik, [6]) or closed submodules (Zeng and Shi, [381]. (See 17.22.)

Returning to the lattice-theoretical interpretation of complement and supplement submodules, it can easily be shown that in a modular lower continuous lattice $\mathcal{L}$ the pseudo-complement elements $a$ are precisely those that are essentially closed in $\mathcal{L}$, i.e. they do not admit any element $b \in \mathcal{L}$ such that $a$ is an essential element in $[0, b]$. Those submodules of a module $M$ which are essentially closed as elements in the lattice of submodules of $M$ are termed *closed submodules*.

In [121], Golan calls a submodule $N$ of a module $M$ a coclosed submodule in $M$ if it is a closed element in the dual lattice of submodules of $M$. Again the dual terms do not behave as nicely as the original ones and in general coclosed submodules do not have to be supplement submodules. On the other hand, if weak supplements exist, then a submodule is a supplement submodule if and only if it is coclosed (Theorem 20.2). This was shown by Takeuchi in [327] and by Inoue in [175]. It was proved in Keskin [199, Lemma 1.7] that a weakly supplemented module is amply supplemented if and only if every submodule has a coclosure (see 20.25).

Zöschinger examines coclosed submodules over commutative noetherian rings $R$ in [392]. He proves that the coclosed submodules of a flat $R$-modules are precisely its pure submodules. Moreover, if $R$ is a Dedekind domain, he shows that a module $M$ is supplemented if and only if every submodule of $M$ has a coclosure in $M$.

**References.** Alizade, Bilhan and Smith [5]; Averdunk [23]; Chamard [60]; Generalov [117, 118, 119]; Golan [121]; Harmanci, Keskin and Smith [164]; Idelhadj and Tribak [174]; Keskin [198, 199]; Keskin and Tribak [205]; Lomp [226]; Miyashita [235]; Nebiyev and Pancar [251]; Smith [317]; Takeuchi [327]; Varadarajan [344]; Wisbauer [363]; Zöschinger [388, 389].

# 21 Submodules with unique coclosure

Essential closures of submodules in a module $M$ are not unique unless $M$ has some special properties, for example, if it is non-$M$-singular. A similar behaviour is observed for coclosures of submodules of $M$.

**21.1. UCC modules.** A module $M$ is called a *unique coclosure module*, denoted by UCC, if every submodule of $M$ has a unique coclosure in $M$.

**21.2. Examples.** Consider the ring $R$ given in 9.14. This is not a UCC module since for each $x \in K$, $D_x$ is a coclosure of $\operatorname{Soc} R = Re_{11} \oplus Re_{12}$.

A module $M$ is amply supplemented if and only if it is weakly supplemented and every submodule of $M$ has a coclosure (see 20.25). Hence a weakly supplemented UCC module is amply supplemented.

An amply supplemented UCC module need not be copolyform; for example, consider the $\mathbb{Z}$-module $\mathbb{Z}/8\mathbb{Z}$.

A supplemented copolyform module need not be a UCC module; for example, let $K$ be the quotient field of a discrete valuation domain $R$ which is not complete. Then the $R$-module $M = K \oplus K$ is a supplemented but not amply supplemented module (see 23.7) and hence is not a UCC module. As $R$ is a hereditary ring and $M$ is an injective $R$-module, $\operatorname{Re}_{\mathbb{S}[R]}(M) = M$ (see 8.4) and so $M$ is a (strongly) copolyform $R$-module.

Suppose $M$ is an amply supplemented module. We show that if $M$ is copolyform, then $M$ is a UCC module and if $M \oplus M$ is UCC, then $M$ is copolyform. We need some properties of UCC modules.

**21.3. Characterisation of UCC modules.** *For $M$ the following are equivalent.*

(a) *$M$ is a UCC module;*

(b) *given $N \subseteq M$, there exists a coclosure $\widetilde{N}$ of $N$ such that $\widetilde{N} \subseteq L$ whenever $L \overset{cs}{\underset{M}{\hookrightarrow}} N$;*

(c) *every factor module of $M$ is a UCC module;*

(d) *every submodule $A$ of $M$ has a coclosure $\widetilde{A}$ in $M$ and for any $A \subseteq B \subseteq M$ we have $\widetilde{A} \subseteq \widetilde{B}$.*

*Proof.* (a)$\Rightarrow$(b). Suppose $M$ is a UCC module and $\widetilde{N}$ is the unique coclosure of $N$ in $M$. If $L \overset{cs}{\underset{M}{\hookrightarrow}} N$, then the coclosure of $L$ exists and is a coclosure of $N$ also. By the uniqueness of the coclosure of $N$ in $M$ we get $\widetilde{N} \subseteq L$.

(b)$\Rightarrow$(c). It is easy to verify that (b) implies $M$ is a UCC module. Suppose $K \subseteq M$, set $N = M/K$, and consider $A/K \subseteq N$. Now $A$ has a unique coclosure, say, $\widetilde{A}$ in $M$. We claim that $(\widetilde{A} + K)/K$ is the unique coclosure of $A/K$ in $M/K$. Since $\widetilde{A} \overset{cs}{\underset{M}{\hookrightarrow}} A$ we have $(\widetilde{A} + K)/K \overset{cs}{\underset{N}{\hookrightarrow}} A/K$. Let $X/K \overset{cs}{\underset{N}{\hookrightarrow}} (\widetilde{A} + K)/K$.

Then $X/K \overset{cs}{\underset{N}{\hookrightarrow}} A/K$ and hence $X \overset{cs}{\underset{M}{\hookrightarrow}} A$. Since $M$ is a UCC module, $\widetilde{A} \subseteq X$ and so $X/K = (\widetilde{A} + K)/K$. Thus $(\widetilde{A} + K)/K$ is a coclosure of $A/K$ in $M/K$. Next we prove that this coclosure of $A/K$ in $M/K$ is unique. Let $Y/K$ be a coclosure of $A/K$ in $M/K$. Then $Y \overset{cs}{\underset{M}{\hookrightarrow}} A$ and so $\widetilde{A} \subseteq Y$. Therefore $(\widetilde{A} + K)/K \subseteq Y/K$. This implies $(\widetilde{A} + K)/K = Y/K$, as required.

(c)$\Rightarrow$(d). Suppose $A \subseteq B \subseteq M$ with $\widetilde{A}$, $\widetilde{B}$ their respective coclosures in $M$. Consider $N = M/\widetilde{A}$. By (c) $N$ is a UCC module. Suppose $L/\widetilde{A}$ is the unique coclosure of $B/\widetilde{A}$ in $N$. Since $\widetilde{A} \overset{cc}{\hookrightarrow} M$ and $L/\widetilde{A} \overset{cc}{\hookrightarrow} N$, we have $L \overset{cc}{\hookrightarrow} M$ by 20.5 (3). Since $L/\widetilde{A} \overset{cs}{\underset{N}{\hookrightarrow}} B/\widetilde{A}$ we have $L \overset{cs}{\underset{M}{\hookrightarrow}} B$. Then $L = \widetilde{B}$, since $M$ is UCC, and so $\widetilde{A} \subseteq \widetilde{B}$.

(d)$\Rightarrow$(a). Suppose that $A_1$ and $A_2$ are two coclosures of $A \subseteq M$. Then $A_2 \subseteq A$ and, since $A_1$ and $A_2$ are respectively coclosures of $A$ and $A_2$, by (d) we have $A_2 \subseteq A_1$. Similarly $A_1 \subseteq A_2$ and hence $A_1 = A_2$. $\qquad\square$

**21.4. UCC modules and cosmall submodules.** *Suppose that every submodule of $M$ has a coclosure in $M$. Then the following are equivalent:*

(a) *$M$ is a UCC module;*

(b) *if $A_i \overset{cs}{\underset{M}{\hookrightarrow}} B_i$ for all $i \in I$, then $\bigcap_I A_i \overset{cs}{\underset{M}{\hookrightarrow}} \bigcap_I B_i$;*

(c) *if $A \overset{cs}{\underset{M}{\hookrightarrow}} (A + B)$ in $M$, then $(A \cap B) \overset{cs}{\underset{M}{\hookrightarrow}} B$.*

*Proof.* (a)$\Rightarrow$(b). Let $A_i \overset{cs}{\underset{M}{\hookrightarrow}} B_i$ for every $i \in I$. For any $X \subseteq M$, let $\widetilde{X}$ denote its unique coclosure in $M$. Since $A_i \overset{cs}{\underset{M}{\hookrightarrow}} B_i$, we have $\widetilde{A}_i = \widetilde{B}_i$. Now $\bigcap_I B_i \subseteq B_i$ for every $i \in I$ and so, by 21.3 (d), $\widetilde{\bigcap_I B_i} \subseteq \widetilde{B}_i$ for every $i \in I$. This gives

$$\widetilde{\bigcap_I B_i} \subseteq \bigcap_I \widetilde{B}_i = \bigcap_I \widetilde{A}_i \subseteq \bigcap_I A_i \subseteq \bigcap_I B_i.$$

(b)$\Rightarrow$(c). Suppose $A \overset{cs}{\underset{M}{\hookrightarrow}} (A + B)$. Then $B \overset{cs}{\underset{M}{\hookrightarrow}} B$ and (c) together imply that $(A \cap B) \overset{cs}{\underset{M}{\hookrightarrow}} B$.

(c)$\Rightarrow$(a). Suppose $A_1$ and $A_2$ are coclosures of $A \subseteq M$ in $M$. Since $A_1 \overset{cs}{\underset{M}{\hookrightarrow}} A$ we have $A_1 \overset{cs}{\underset{M}{\hookrightarrow}} (A_1 + A_2)$. By (c), $(A_1 \cap A_2) \overset{cs}{\underset{M}{\hookrightarrow}} A_1$. Hence $A_1 \cap A_2 = A_1$ and so $A_1 \subseteq A_2$. Similarly $A_2 \subseteq A_1$ and so $A_1 = A_2$. $\qquad\square$

**21.5. Examples.** (1) *The sum of two coclosed submodules need not be coclosed.* For example, in the ring $R$ defined in 9.14, $D_0$ and $D_1$ are direct summands of $R$ but $D_0 \oplus D_1 = \operatorname{Soc} R$ is not a coclosed submodule of $R$ (since it has coclosures $D_x$, $x \in K$).

(2) *The epimorphic image of a coclosed submodule need not be a coclosed submodule.* For example, suppose $A$ is a module with $S \ll A$. Consider $N = A \oplus T$ where there exists an isomorphism $g : T \to S$. Define $f : A \oplus T \to A$ by $(a, t)f = a + (t)g$. Then $T$ is a coclosed submodule of $A \oplus T$, but $(T)f = S \ll A$.

We now provide various necessary and sufficient conditions to make an amply supplemented module a UCC module.

**21.6. UCC modules and (strongly) coclosed submodules.** *The following statements are equivalent for an amply supplemented module $M$:*

(a) *$M$ is a UCC module;*

(b) *every coclosed submodule of $M$ is strongly coclosed;*

(c) *the sum of any family of coclosed submodules of $M$ is coclosed;*

(d) *the sum of two coclosed submodules of $M$ is coclosed in $M$;*

(e) *for any epimorphism $f : M \to N$, if $A \overset{cc}{\hookrightarrow} M$, then $(A)f \overset{cc}{\hookrightarrow} N$;*

(f) *if for all $i \in I$, $K_i \subseteq L_i \subseteq M$ and $L_i/K_i \overset{cc}{\hookrightarrow} M/K_i$, then*

$$(\textstyle\sum_I L_i)/(\sum_I K_i) \overset{cc}{\hookrightarrow} M/\textstyle\sum_I K_i.$$

*Proof.* (a)$\Rightarrow$(b). Suppose $K \overset{cc}{\hookrightarrow} M$ and $X \subseteq M$ is such that $K \not\subseteq X$. Then we have $(K \cap X) \underset{\neq}{\subseteq} K$, so $(K \cap X) \overset{cs}{\underset{M}{\not\hookrightarrow}} K$. By 21.4 (c), $X \overset{cs}{\underset{M}{\not\hookrightarrow}} (K + X)$ and hence $K \overset{scc}{\hookrightarrow} M$.

(b)$\Rightarrow$(c). Suppose $K_i \overset{cc}{\hookrightarrow} M$, for all $i \in I$. Then $K_i \overset{scc}{\hookrightarrow} M$, for all $i \in I$. By 3.13 we get $\sum_I K_i \overset{scc}{\hookrightarrow} M$. By 3.12, $\sum_I K_i \overset{cc}{\hookrightarrow} M$.

(c)$\Rightarrow$(d) is obvious.

(d)$\Rightarrow$(a). Let $A \subseteq M$ with two coclosures $A_1$ and $A_2$ in $M$. By (d) we have $(A_1 + A_2) \overset{cc}{\hookrightarrow} M$. But $A_1 \overset{cs}{\underset{M}{\hookrightarrow}} A$ and $A_1 \subseteq (A_1 + A_2) \subseteq A$ implies $A_1 + A_2 = A_1$. Hence $A_2 \subseteq A_1$. Similarly $A_1 \subseteq A_2$ and hence $A_1 = A_2$.

(b)$\Rightarrow$(e). The proof follows from 3.12 (c).

(e)$\Rightarrow$(d). Let $K$ and $N$ be coclosed submodules of $M$. Then (e) implies that $(K + N)/N \overset{cc}{\hookrightarrow} M/N$. Since $N \overset{cc}{\hookrightarrow} M$ and $M$ is amply supplemented, 20.5 (3) gives $(K + N) \overset{cc}{\hookrightarrow} M$.

(a)$\Rightarrow$(f). Put $K = \sum_I K_i$ and $L = \sum_I L_i$. By 21.3, for all $i \in I$, $M/K_i$ and $M/K$ are UCC modules. Let $f_i : M/K_i \to M/K$ be the natural map for each $i$. Then, since $L_i/K_i \overset{cc}{\hookrightarrow} M/K_i$, (e) gives $(L_i + K)/K \overset{cc}{\hookrightarrow} M/K$. It follows from (c) that

$$\textstyle\sum_I (L_i + K)/K = L/K \overset{cc}{\hookrightarrow} M/K.$$

(f)$\Rightarrow$(c). By taking each $K_i = 0$ in (f), (c) follows. $\qquad\square$

**21.7. UCC modules and homomorphisms to factor modules.** *For an amply supplemented module $M$, the following are equivalent:*

(a) *$M$ is a UCC module;*

(b) *for any $Y \subseteq M$, $M/Y \not\simeq A \oplus B$, where $0 \neq A$ and $A$ is isomorphic to a small submodule of $B$;*

(c) *if $f : M \to M/X$ with $X + \operatorname{Ke} f = M$ and $\operatorname{Im} f \ll M/X$, then $f = 0$;*

(d) *if $f : M \to M/N$ is a homomorphism such that $\operatorname{Ke} f + N = M$, then $\operatorname{Im} f \overset{cc}{\hookrightarrow} M/N$.*

*Proof.* (a)$\Rightarrow$(b). Suppose there exists $Y \subseteq M$ such that $M/Y = A \oplus B$, where $0 \neq A$ and there exists an isomorphism $\phi : A \to C \subseteq B$ with $C \ll B$. Put $N = A \oplus C$. It is easy to see that both $A$ and $D = \{(a, (a)\phi) \mid a \in A\}$ are coclosures of $N$ in $M/Y$, which contradicts that $M/Y$ is UCC (see 21.3).

(b)$\Rightarrow$(c). Suppose $f : M \to M/X$ is a homomorphism with $\operatorname{Im} f = T/X \ll M/X$ and $X + \operatorname{Ke} f = M$. Put $N = A \oplus (M/X)$, where $A$ is such that there exists an isomorphism $g : T/X \to A$. Let $\eta : M \to M/X$ be the natural map. Then the map $h : M \to N$, where $(m)h = ((m)fg, (m)\eta)$ is an epimorphism. It now follows from (b) that $A = 0$ and hence $f = 0$.

(c)$\Rightarrow$(d). Let $f : M \to M/N$ be a homomorphism with $\operatorname{Ke} f + N = M$ and $\operatorname{Im} f = T/N$. If possible let $X/N \overset{cs}{\underset{M/N}{\hookrightarrow}} T/N$, where $X \neq T$. We have $T/X \ll M/X$. Consider the natural map $g : M/N \to M/X$. Then $\operatorname{Im}(fg) \ll M/X$. By (c), $fg = 0$ and hence $T = X$. Thus $\operatorname{Im} f = T/N \overset{cc}{\hookrightarrow} M/N$.

(d)$\Rightarrow$(a). Let $A \subseteq M$. Suppose that $X$ and $Y$ are two coclosures of $A$ in $M$ and $T$ is a supplement of $A$ in $M$. Then $A + T = X + T = Y + T = M$. Since $X$ and $Y$ are also supplements of $T$ in $M$, we have $X \cap T \ll X$ and $Y \cap T \ll Y$. Now consider

$$M \overset{f}{\to} M/T \simeq (X + T)/T \simeq X/(X \cap T) \overset{g}{\to} (X + Y)/((X \cap T) + Y),$$

where $f$ and $g$ are the natural maps. Put $N = (X \cap T) + Y$. Thus we have a map $h : M \to M/N$ with $\operatorname{Im} h = (X + Y)/N$ and $\operatorname{Ke} h + N = M$ (for $\operatorname{Ke} h \supseteq T$). By (d), $\operatorname{Im} h \overset{cc}{\hookrightarrow} M/N$. We have $Y \subseteq N \subseteq (X + Y) \subseteq A$. $Y \overset{cs}{\underset{M}{\hookrightarrow}} A$ implies $(X + Y) \overset{cs}{\underset{M}{\hookrightarrow}} N$. Therefore $\operatorname{Im} h \ll M/N$ and hence $\operatorname{Im} h = 0$. Then $X + Y = N = (X \cap T) + Y$ and $(X \cap T) \ll X$ (hence $\ll N$), giving $Y = N$. Thus $X \subseteq Y$. Similarly $Y \subseteq X$, proving $X = Y$. □

From 21.7 and 17.19 we get the following:

**21.8. UCC and copolyform modules.** *Any copolyform amply supplemented module is UCC.*

**21.9. Finitely-UCC modules need not be countably-UCC.** There exists a module $E$ such that any finite direct sum of copies of $E$ is an amply supplemented UCC (in fact a copolyform) module but the countable direct sum of copies of $E$ is not a UCC module.

For this we come back to Example 9.13. Consider the $\mathbb{Z}$-modules $A_i = \mathbb{Z}/2^i\mathbb{Z}$, for all $i \in \mathbb{N}$, and $E = E(A_1)$. Let $N = \bigoplus_\mathbb{N} E_i$, where $E_i = E$, for all $i$. Take $B = \bigoplus_\mathbb{N} A_i$ and $B_n = \bigoplus_{k>n} A_k$. From 9.13 each $B_n \overset{cs}{\underset{N}{\hookrightarrow}} B$ and $B$ is not a small submodule of $N$, that is, $0 \subset B$ is not a cosmall inclusion in $N$. Then, since $\bigcap_\mathbb{N} B_n = 0$, we get from 21.4 that $N = E^{(\mathbb{N})}$ is not a UCC module. However, any finite direct sum of copies of $E$ is a copolyform and amply supplemented module (since it is artinian). By 21.8, $E^n$ is a UCC module.

A module $M$ is called $\Sigma$-*amply-supplemented* if every direct sum of copies of $M$ is amply supplemented.

**21.10. Finitely-amply-supplemented need not be countably-amply-supplemented.** Consider the module $M$ defined above in 21.9. As noted, $M^n$ is amply supplemented for all $n \in \mathbb{N}$. Also $M^{(\mathbb{N})}$ is copolyform but not UCC. By 21.8, $M^{(\mathbb{N})}$ is not amply supplemented.

It follows from 21.3 (c) that every direct summand of a UCC module is UCC. However, even if $M$ is a UCC module and $M \oplus M$ is amply supplemented, $M \oplus M$ need not be a UCC module. For example, let $R = \mathbb{Z}$ and $M = \mathbb{Z}/8\mathbb{Z}$. Then, since $M \oplus M$ has a factor module isomorphic to $4\mathbb{Z}/8\mathbb{Z} \oplus \mathbb{Z}/8\mathbb{Z}$, $M \oplus M$ is not a UCC module (see 21.7).

**21.11. Finite direct sums of UCC modules.** *For a weakly supplemented module $M$, the following are equivalent:*

(a) *$M \oplus M$ is a UCC module;*

(b) *$M$ is copolyform and $M \oplus M$ is amply supplemented;*

(c) *$M \oplus M$ is copolyform and amply supplemented.*

*Proof.* (a)$\Rightarrow$(b). Notice that $M \oplus M$ is weakly supplemented. Hence 20.25 implies that $M \oplus M$ is amply supplemented. By 21.6, every coclosed submodule of $M \oplus M$ is strongly coclosed. In particular the direct summand $M$ is strongly coclosed in $M \oplus M$. By 17.19 (f), $M$ is copolyform.

(b)$\Leftrightarrow$(c) by 9.18, and (c)$\Rightarrow$(a) follows from 21.8. $\qquad\qquad\square$

Next we observe some relations for the hollow-dimension of UCC and copolyform modules.

**21.12. Proposition.** *If $M$ is a weakly supplemented UCC module, then $M$ is a hollow dimension module.*

*Proof.* Since a weakly supplemented UCC module is amply supplemented, the assertion follows from (e)$\Rightarrow$(a) of 20.29.                                    $\Box$

**21.13. UCC and hollow dimension modules.** *Suppose that* h.dim$(M)$ *is finite and that coclosures exist for all submodules of $M$. Then the following are equivalent:*

(a) *$M$ is a UCC module;*

(b) *$M$ is a hollow dimension module.*

*Proof.* Since a finite hollow dimension module is weakly supplemented, $M$ is an amply supplemented module. Now (a)$\Leftrightarrow$(b) follows easily from the equivalence of (a) and (e) in 20.29.                                    $\Box$

By 21.11 and 21.13 we get the following.

**21.14. Corollary.** *Suppose $M$ is a finitely-amply-supplemented module with finite hollow dimension. Then the following are equivalent:*

(a) *$M \oplus M$ is a UCC module;*

(b) *$M$ is a copolyform module;*

(c) *$M^n$ is a copolyform module for all $n \geq 2$;*

(d) *$M^n$ is a UCC module for all $n \geq 2$;*

(e) *$M^n$ is a hollow dimension module for all $n \geq 2$.*

**21.15. Lemma.** *Suppose $M$ is an $R$-module and $N \in \sigma[M]$ is such that $N = A \oplus B$, where $0 \neq A \simeq L \ll B$. Then $N$ is neither a fully non-$M$-small nor a (strongly) copolyform nor a UCC module.*

*Proof.* By 21.7, $N$ is not a UCC module. $L \ll B$ implies $L \ll N$. Hence $0 \overset{cs}{\underset{N}{\hookrightarrow}} L$. As $A \simeq L$ we get $0 \overset{cr}{\underset{M}{\not\hookrightarrow}} L$. Thus $N$ is not a (strongly) copolyform module. By 9.23 $N$ is not a fully non-$M$-small module.                                    $\Box$

By 21.8, an amply supplemented fully non-$M$-small module is a UCC module. But a supplemented fully non-$M$-small module need not be a UCC module. For example, let $K$ be the quotient field of a discrete valuation domain $R$ which is not complete. Then $M = K \oplus K$ is a supplemented but not amply supplemented module (see 23.7) and hence is not a UCC module. As $R$ is a hereditary ring and $M$ is an injective $R$-module, $M$ is a fully non-$M$-small $R$-module.

**21.16. Every $M$-subgenerated module is copolyform.** *For an $R$-module $M$, the following are equivalent:*

(a) *every (2-generated) module in $\sigma[M]$ is a UCC module;*

(b) *every (2-generated) module in $\sigma[M]$ is a fully non-$M$-small module;*

(c) *every (2-generated) module in $\sigma[M]$ is strongly copolyform;*

(d) *every (2-generated) module in $\sigma[M]$ is copolyform;*

(e) *M is a cosemisimple module.*

*Proof.* (a), (b), (c) or (d)$\Rightarrow$(e). Suppose $S$ is a simple module in $\sigma[M]$ and $\widehat{S}$ is the injective hull of $S$. Assume $S \ll \widehat{S}$. Put $N = S' \oplus \widehat{S}$ where $S' \simeq S$. By 21.15, $N$ cannot be a UCC, or fully non-$M$-small, or (strongly) copolyform module. Thus every simple module in $\sigma[M]$ is injective in $\sigma[M]$ and so $M$ is a cosemisimple module.

(e)$\Rightarrow$(a), (b), (c) or (d). If $M$ is cosemisimple, then for every $N \in \sigma[M]$, $\mathrm{Rad}\, N = 0$. Hence $M$ has no nonzero $M$-small module and, if $N \in \sigma[M]$, then every submodule of $N$ is coclosed in $M$. It is now easy to see that every module in $\sigma[M]$ is fully non-$M$-small, (strongly) copolyform and UCC. $\qquad\square$

### 21.17. Exercises.

(1) Prove that a module $M$ is a UCC module if and only if every coclosed submodule of $M$ is UCC.

(2) Recall that a module $M$ is called a *distributive module* if
$$N \cap (K + L) = (N \cap K) + (N \cap L),$$
for all submodules $N$, $K$ and $L$ of $M$. Prove that

   (i) the lattice of submodules of a distributive module also satisfies the equivalent property
$$N + (K \cap L) = (N + K) \cap (N + L);$$

  (ii) an amply supplemented distributive module is a UCC module.

(3) Prove that for an amply supplemented module $M$, the following are equivalent:

  (a) $M$ is a UCC module;

  (b) if for every $i \in I$, $X_i \overset{cs}{\underset{M}{\hookrightarrow}} A$, then $\bigcap_{i \in I} X_i \overset{cs}{\underset{M}{\hookrightarrow}} A$;

  (c) if $X \overset{cs}{\underset{M}{\hookrightarrow}} N$ and $Y \overset{cs}{\underset{M}{\hookrightarrow}} N$, then $(X \cap Y) \overset{cs}{\underset{M}{\hookrightarrow}} N$;

  (d) if $A \overset{cc}{\hookrightarrow} M$ and $X \overset{cs}{\underset{M}{\hookrightarrow}} Y$, then $(A \cap X) \overset{cs}{\underset{M}{\hookrightarrow}} (A \cap Y)$;

  (e) for every $N \ll M$ and for any two submodules $K$ and $L$ of $M$,
$$(K \cap L) \overset{cs}{\underset{M}{\hookrightarrow}} (N + K) \cap (N + L);$$

  (f) there exists $N_0 \ll M$ such that $M/N_0$ is a UCC module and for any two submodules $K$ and $L$ of $M$,
$$(K \cap L) \overset{cs}{\underset{M}{\hookrightarrow}} (N_0 + K) \cap (N_0 + L);$$

  (g) if $X$ and $Y$ are submodules of $M$ with $X + Y = M$ and $f : M/X \to M/Y$ is a monomorphism with $\mathrm{Im}\, f \ll M/Y$, then $X = M$;

  (h) if $f : M \to M/N$ is a homomorphism with $\mathrm{Ke}\, f + N = M$ and $K \overset{cc}{\hookrightarrow} M$, then $(K)f \overset{cc}{\hookrightarrow} M/N$;

(i) if $K' \overset{cs}{\underset{M}{\to}} K$, $L' \subseteq L \subseteq M$, $K + L = M$ and $K/K' \simeq L/L'$, then $L' \overset{cs}{\underset{M}{\to}} L$;

(j) if $K/K' \overset{cc}{\hookrightarrow} M/K'$, $L' \subseteq L \subseteq M$, $K + L = M$ and $K/K' \simeq L/L'$, then $L/L' \overset{cc}{\hookrightarrow} M/L'$.

**References.** Ganesan and Vanaja [114]; Generalov [119]; Talebi and Vanaja [329].

# Chapter 5

# From lifting to perfect modules

## 22 Lifting modules

In what follows the main topic will be conditions which make supplements of a module into direct summands. Again $M$ will denote a left $R$-module.

First we consider cosmall inclusions of direct summands.

**22.1. Submodules lying above direct summands.** *For a submodule $U \subseteq M$, the following are equivalent:*

(a) *there is a direct summand $X$ of $M$ with $X \subseteq U$ and $U/X \ll M/X$;*

(b) *there is a direct summand $X \subseteq M$ and a submodule $Y$ of $M$ with $X \subseteq U$, $U = X + Y$ and $Y \ll M$;*

(c) *there is a decomposition $M = X \oplus X'$, with $X \subseteq U$ and $X' \cap U \ll X'$;*

(d) *$U$ has a supplement $V$ in $M$ such that $U \cap V$ is a direct summand in $U$;*

(e) *there is an idempotent $e \in \mathrm{End}(M)$ with*
$$Me \subseteq U \text{ and } U(1-e) \ll M(1-e).$$

If these conditions hold we say that $U$ *lies above a direct summand of $M$.*

*Proof.* (a)$\Rightarrow$(b). If $M = X \oplus X'$ and $U/X \ll M/X$, then $U = X + (U \cap X')$. Let $\phi : M/X \to X'$ be the obvious isomorphism. Then, since $U/X \ll M/X$, we have $(U/X)\phi = U \cap X' \ll X'$. Hence $U \cap X' \ll M$.

(b)$\Rightarrow$(c). If $M = X \oplus X'$, then $X'$ is a supplement of $X$ in $M$ and hence a supplement of $U = X + Y$ in $M$ (see 20.4 (4)), and so $U \cap X' \ll X'$.

(c)$\Rightarrow$(d). From (c), $U = X \oplus (U \cap X')$ and $X'$ is a supplement of $U$ in $M$.

(d)$\Rightarrow$(e). By (d), $M = U + V$, $U \cap V \ll V$ and $U = (U \cap V) \oplus X$ for some $X \subseteq U$. Then $M = U + V = X + (U \cap V) + V = X + V$ and $X \cap V = 0$, and so $M = X \oplus V$. Let $e : M \to X$ be the projection map along $V$. Then $Me \subseteq U$. As $X \subseteq U$, we have $U(1-e) = U \cap M(1-e) = U \cap V \ll V = M(1-e)$, as required.

(e)$\Rightarrow$(a). Letting $X = Me$, we have $M = X \oplus M(1-e)$ and $X \subseteq U$. Taking $\phi : M/X \to M(1-e)$ to be the obvious isomorphism gives $(U/X)\phi = U(1-e) \ll M(1-e)$. As $\phi$ is an isomorphism, it follows that $U/X \ll M/X$. $\square$

Recall that an $R$-module $M$ is *extending* provided each of its submodules is essential in a direct summand of $M$. Dually we define

**22.2. Lifting modules.** The module $M$ is called *lifting* if every submodule $N$ of $M$ lies over a direct summand, that is, $N$ contains a direct summand $X$ of $M$ such that $N/X \ll M/X$. If $M$ satisfies this condition, then

(1) *any coclosed submodule of $M$ is a direct summand;*

(2) *$M$ is amply supplemented;*

(3) *$M$ is hollow if and only if it is indecomposable;*

(4) *if $N \subseteq M$ is a fully invariant submodule of $M$, then $M/N$ is a lifting module.*

*Proof.* (1) By definition, coclosed submodules have no proper cosmall inclusion in $M$ and so, by 22.1 (a), must be direct summands.

(2) Let $M = L + K$ for submodules $K, L$. We show that $K$ contains a supplement of $L$ in $M$. By hypothesis and 22.1 (b), $K = N \oplus H$ with $H \ll M$ and $N$ a direct summand of $M$. Hence $M = L + N$. By hypothesis, $L \cap N = N_1 \oplus S$ with $S \ll M$ and $N_1$ a direct summand of $M$. Hence $N_1$ is a direct summand of $N$ and $S \ll N$.

Let $N = N_1 \oplus N_2$ for some submodule $N_2$ of $N$. Then, of course, $N_2$ is a supplement of $N_1$ in $N$. By 20.4 (4) applied to $N$, $N_2$ is a supplement of $N_1 + S$ in $N$. Thus $M = L + N = L + (L \cap N) + N_2 = L + N_2$ and $L \cap N_2 = (L \cap N) \cap N_2 \ll N_2$. Therefore $N_2$ is a supplement of $L$ in $M$.

(3) is obvious.

(4) Let $N$ be a fully invariant submodule of $M$ and $U/N \subseteq M/N$. By 22.1 (e), there is an idempotent $e \in \mathrm{End}(M)$ such that $Me \subseteq U$ and $U(1-e) \ll M(1-e)$. As $N$ is fully invariant in $M$, $e$ induces an idempotent $\bar{e} \in \mathrm{End}(M/N)$. It is easy to verify that $M\bar{e} \subseteq M/N$ and $(U/N)(1 - \bar{e}) \ll M/N$. $\qquad\square$

**22.3. Characterisation of lifting modules.** *For $M$ the following are equivalent:*

(a) *$M$ is lifting;*

(b) *for every submodule $N$ of $M$ there is a decomposition $M = M_1 \oplus M_2$ such that $M_1 \subseteq N$ and $N \cap M_2 \ll M$;*

(c) *every submodule $N$ of $M$ can be written as $N = N_1 \oplus N_2$ with $N_1$ a direct summand of $M$ and $N_2 \ll M$;*

(d) *$M$ is amply supplemented and every coclosed submodule of $M$ is a direct summand of $M$;*

(e) *$M$ is amply supplemented and supplements are direct summands of $M$.*

*Proof.* (a)$\Leftrightarrow$(b) and (a)$\Leftrightarrow$(c) follow from 22.1; (a)$\Rightarrow$(d) and (d)$\Rightarrow$(e) follow from 22.2 and 20.3 respectively; (e)$\Rightarrow$(a) is a consequence of 20.24 (b) and 22.1 (b). $\qquad\square$

The following corollary is immediate.

**22.4. Corollary.** *Every hollow module is lifting.*

In general, direct sums of lifting modules (even hollow modules) are not lifting as seen by the following (see also 23.5)

**22.5. Example.** Consider the $\mathbb{Z}$-modules $A = \mathbb{Z}/8\mathbb{Z}$ and $B = \mathbb{Z}/2\mathbb{Z}$. As $A$ and $B$ are hollow modules, they are lifting. But $M = A \oplus B$ is not lifting; for let $U$ be the submodule of $M$ generated by $(2 + 8\mathbb{Z}, 1 + 2\mathbb{Z})$. Then $U \not\ll M$ and $U$ does not contain a nonzero direct summand of $M$.

We study in detail the lifting property of direct sums of lifting modules in the next two sections. In the meantime, we have

**22.6. Lemma.** *Any coclosed submodule (and so every direct summand) of a lifting module is lifting.*

*Proof.* Let $N$ be a coclosed submodule of a lifting module $M$. Then $N$ is a direct summand of $M$ by 22.2. By 20.22(2), $N$ is amply supplemented. Let $K$ be a submodule of $N$, that is coclosed in $N$. Since $N$ is a supplement in $M$, $K$ is coclosed in $M$ by 20.4(8). Hence $K$ is a direct summand of $M$ and so of $N$. Then, by 22.3, $N$ is lifting. $\qquad\square$

The definition of lifting modules requires every submodule to contain a co-small direct summand. We now consider this condition restricted to certain classes of submodules.

**22.7. Finitely lifting modules.** The module $M$ is called *finitely lifting*, or *f-lifting* for short, if every finitely generated submodule of $M$ lies above a direct summand.

**Characterisations.** *The following are equivalent for a module $M$:*

(a) $M$ *is f-lifting;*

(b) *every cyclic submodule of $M$ lies over a direct summand;*

(c) *for every finitely generated (cyclic) submodule $N \subset M$ there is an idempotent $e \in \mathrm{End}(M)$ with $Me \subset N$ and $N(1 - e) \ll M$.*

*Proof.* This is shown in [363, 41.13]. $\qquad\square$

**22.8. Hollow-lifting modules.** An $R$-module $M$ is called *hollow-lifting*, or *h-lifting* for short, provided $M$ is amply supplemented and every hollow submodule of $M$ lies over a direct summand of $M$. By 22.1, this is obviously the case if and only if $M$ is amply supplemented and every hollow, coclosed submodule of $M$ is a direct summand of $M$ (note that coclosures exist by 20.25 since $M$ is amply supplemented). It follows from 20.22 (1) and the proof of 22.6 that any coclosed submodule (and so every direct summand) of a hollow-lifting module is hollow-lifting. More generally:

**22.9. Coclosed submodules of h-lifting modules.** *Let $M$ be a hollow-lifting module and $K \subseteq M$ be a coclosed submodule with finite hollow dimension. Then $K$ is a direct summand of $M$.*

*Proof.* The proof is by induction on the finite hollow dimension of $K$. If h.dim$(K)$ $= 1$, then $K$ is a hollow coclosed submodule of $M$ and hence a direct summand.

Now assume that coclosed submodules of $M$ with hollow dimension at most $n$ are direct summands of $M$. Let $K$ be a coclosed submodule of $M$ with h.dim$(K) = n + 1$. Since $K$ has finite hollow dimension, there is a submodule $L$ of $K$ such that $K/L$ is hollow (see 2.16). By the preceding observation, $K$ is hollow-lifting. Let $N$ be a supplement of $L$ in $K$. Then $N$ is hollow, since $N/(N \cap L) \simeq K/L$ and $N \cap L \ll N$. Furthermore $N$ is coclosed in $K$ and $K$ is a supplement in $M$, so $N$ is coclosed in $M$ (cf. 20.4 (8)) and hence it is a direct summand of $M$, say $M = N \oplus N'$. Then $K = N \oplus (K \cap N')$ and

$$h.\dim(K) = h.\dim(N) + h.\dim(K \cap N').$$

Thus $K \cap N'$ is a coclosed submodule of $M$ with h.dim$(K \cap N') = n$ and hence, by the induction hypothesis, $K \cap N'$ is a direct summand of $M$, let us say $M = (K \cap N') \oplus N''$. Then $N' = (K \cap N') \oplus (N' \cap N'')$ and so

$$M = N \oplus N' = N \oplus (K \cap N') \oplus (N \cap N'') = K \oplus (N' \cap N'').$$

$\square$

Since coclosed submodules of modules with finite hollow dimension have also finite hollow dimension (see 20.10(2)), we get the following corollary.

**22.10. Corollary.** *Let $M$ be an $R$-module with finite hollow dimension. Then $M$ is hollow-lifting if and only if $M$ is lifting.*

The following proposition shows that a lifting module with a suitable finiteness condition can be decomposed into a finite direct sum of hollow modules.

**22.11. Proposition.** *Let $M$ be a nonzero $R$-module with finite uniform dimension or finite hollow dimension.*

(1) *If $M$ is lifting, then $M = \bigoplus_{i=1}^{n} H_i$ where each $H_i$ is a hollow module and $n = $ h.dim$(M)$.*

(2) *If $M$ is extending, then $M = \bigoplus_{i=1}^{n} U_i$ where each $U_i$ is a uniform module and $n = $ u.dim$(M)$.*

*Proof.* (1) Assume $M$ to have finite uniform dimension or finite hollow dimension. Since the additive dimension formula for direct summands holds for both dimensions, the result can be proved by induction on u.dim or h.dim. In the following dim will denote either u.dim or h.dim. If $M$ is indecomposable or dim$(M) = 1$, then $M$ is hollow since an indecomposable lifting module is hollow.

Now let $n \geq 1$ and suppose that our result holds for all $R$-modules with dimension at most $n$. Assume dim$(M) = n + 1$. Then, since $M$ is a lifting module which is not hollow, by 22.1 we have a decomposition $M = M_1 \oplus M_2$ where $M_1$ and

$M_2$ are nonzero submodules of $M$. Since $\dim(M) = \dim(M_1) + \dim(M_2) = n + 1$, it follows that $\dim(M_1)$ and $\dim(M_2)$ are both at most $n$. Then, by our inductive assumption, $M_1$ and $M_2$ are both finite direct sums of hollow modules. Thus the result follows.

The proof of (2) is similar to (1), since any indecomposable extending module is uniform. □

**22.12. Corollary.** *If $M$ is a lifting $R$-module that is either finitely generated or finitely cogenerated, then $M$ is a finite direct sum of hollow submodules.*

*Proof.* By 18.6, a finitely generated, weakly supplemented module has finite hollow dimension. A finitely cogenerated module has finitely generated essential socle and hence has finite uniform dimension. Thus the result follows by applying 22.11. □

In the study of extending modules, the ascending and descending chain conditions on direct summands and essential submodules are of particular interest (see [85, 18.5–18.7]). Considering the dual notions we obtain:

**22.13. Lifting modules with radical chain condition.** *Let $M$ be a lifting module such that $\operatorname{Rad} M$ has ACC on direct summands. Then $M$ is a direct sum of a semisimple module and a finite direct sum of hollow modules.*

*Proof.* Since $M$ is lifting, it is weakly supplemented and so, by 17.9, can be decomposed as $M = M_1 \oplus M_2$ where $M_1$ is semisimple and $M_2$ has essential radical. Since $\operatorname{Rad} M = \operatorname{Rad} M_2 \trianglelefteq M_2$ has ACC on direct summands, $M_2$ has ACC on direct summands by 2.9. Since $M_2$ is lifting, it is amply supplemented and each of its supplements is a direct summand by 22.3. Then, by 20.34, $M_2$ has finite hollow dimension and so, by 22.11, $M_2$ is a finite direct sum of hollow modules. □

**22.14. Corollary.** *Let $M$ be a lifting module.*

(1) *If $M$ has ACC on small submodules, then $M = S \oplus N$, where $S$ is semisimple and $N$ is noetherian.*

(2) *If $M$ has DCC on small submodules, then $M = S \oplus A$, where $S$ is semisimple and $A$ is artinian.*

*Proof.* (1) By 2.10, $\operatorname{Rad} M$ is noetherian and hence has ACC on direct summands. By 22.13, $M = S \oplus N$, where $S$ is semisimple and $N$ is a finite direct sum of hollow modules, say $N = \bigoplus_{i=1}^{n} H_i$. Then $\operatorname{Rad} H_i$ is noetherian for all $i$ and so, since $H_i / \operatorname{Rad} H_i$ is simple (or zero), it follows that $H_i$ is noetherian. Thus $N$ is noetherian.

(2) By 2.10, $\operatorname{Rad} M$ is artinian and, by 17.9, $M = S \oplus A$ where $S$ is semisimple and $\operatorname{Rad} M \trianglelefteq A$. By 2.9, $A$ has DCC on direct summands and, by 20.34, $A$ has finite hollow dimension. Hence $\operatorname{Rad} A$ and $A / \operatorname{Rad} A$ are artinian (the latter since it is semisimple and of finite hollow dimension) and so $A$ is artinian. □

**22.15. Corollary.** *Let $M$ be a lifting module with finite hollow dimension or finite uniform dimension.*

(1) *If $M$ has ACC on small submodules, then $M$ is noetherian.*

(2) *If $M$ has DCC on small submodules, then $M$ is artinian.*

We note that the corollary above can also be deduced from Proposition 22.11.

We do not know whether every lifting module is a direct sum of indecomposable modules or whether an indecomposable decomposition of a lifting module is necessarily exchangeable. We give below some sufficient conditions for a lifting module to have an indecomposable decomposition.

From 22.11, 22.12, 22.13 and 22.14 we get the following.

**22.16. Decomposition of lifting modules I.** *Let $M$ be a lifting module satisfying any one of the following conditions:*

  (i) *$M$ has finite uniform dimension;*

 (ii) *$M$ has finite hollow dimension;*

(iii) *$M$ is finitely generated;*

(iv) *$M$ is cofinitely generated;*

 (v) *Rad $M$ has ACC on direct summands;*

(vi) *$M$ has ACC on small submodules;*

(vii) *$M$ has DCC on small submodules.*

*Then $M$ is a direct sum of indecomposable modules.*

**22.17. Decomposition of lifting modules II.** *Let $N$ be a lifting module with an indecomposable exchange decomposition and let $M$ be a lifting module which is an epimorphic image of $N$. Then $M$ is a direct sum of indecomposable modules.*

*Proof.* Let $f : N \to M$ be the given epimorphism. Then, since $N$ is lifting and any direct summand of $N$ has also an indecomposable exchange decomposition, we can assume that $f$ is a small epimorphism.

First we prove that any local direct summand $\bigoplus_K M_k$ of $M$ where each $M_k$ is hollow is a direct summand of $M$. To this end, let $\bigoplus_K M_k$ be a local direct summand of $M$ where each $M_k$ is a hollow submodule of $M$. Let $T_k = (M_k)f^{-1}$ for each $k \in K$. Since $N$ is lifting, we can write $T_k = N_k \oplus S_k$ where $N_k$ is a direct summand of $N$ and $S_k \ll N$, for all $k \in K$. Since $(S_k)f \ll N$ and $M_k$ is a direct summand of $M$ we have $(S_k)f \ll M_k$. Thus $(N_k)f = M_k$, for every $k \in K$. As $f$ is a small epimorphism, each $N_k$ is a hollow module and $\sum_K N_k$ is an irredundant sum.

We claim that $\sum_K N_k$ is a direct summand of $N$. By 13.6, any local direct summand of $N$ is a direct summand of $N$. Hence the union of any ascending chain of direct summands of $N$ is also a direct summand of $N$ (see 13.2). Therefore there exists a maximal subset $F$ of $K$, with respect to set inclusion, such that $\sum_{k \in F} N_k$

is a direct summand of $N$. Suppose, by way of contradiction, that $F \neq K$ and choose $k_0 \in K \setminus F$. Set $N_F = \sum_{k \in F} N_k$, let $N = N_F \oplus A$, and let $p : N \to A$ be the projection along $N_F$. Then $N_F + N_{k_0} = N_F \oplus (N_{k_0})p$. Now, since $(N_{k_0})p$ is a hollow submodule of $N$, it is either a small submodule of $N$ or a coclosed submodule and so a direct summand of $N$ (see 3.7 (4)). If it is a direct summand of $N$, then we get a contradiction to the choice of $F$.

Thus we suppose that $(N_{k_0})p \ll N$. As $N_F$ is a direct summand of $N$, it is coclosed in $N$ and so, since $f$ is a small epimorphism, we get that $(N_F)f = M_F :=$ $\bigoplus_F M_k$ is coclosed in $M$ (see 3.7 (5)). As $M$ is a lifting module, $M_F$ is a direct summand of $M$. Let $\pi_F : M \to M_F$ and $\pi_{k_0} : M \to M_{k_0}$ be the projection maps and set $S = (N_{k_0})pf$. Then $(S)\pi_F \ll M_F$ and hence $(S)\pi_F \ll M_F \oplus M_{k_0}$. Similarly $(S)\pi_{k_0} \ll M_F \oplus M_{k_0}$. Since $S \subseteq (S)\pi_F + (S)\pi_{k_0}$, this gives $S \ll M_F \oplus M_{k_0}$. This implies that $M_F \oplus M_{k_0} = (N_F + N_{K_0})f = (N_F)f = M_F$, a contradiction. This contradiction gives $F = K$ and so $\sum_K N_k$ is a direct summand of $N$.

Now, since $\sum_K N_k$ is a coclosed submodule of $N$ and $f$ is a small epimorphism, 3.7 (5) shows that $\bigoplus_K M_k = (\sum_K N_k)f$ is a coclosed submodule of $M$ and hence a direct summand of $M$, as required.

By Zorn's lemma we can choose a local summand $A = \bigoplus_I A_i$ of $M$ in which each $A_i$ is indecomposable and $A$ is maximal with respect to inclusion and these properties. By what we have proved, $M = A \oplus B$ for some $B \subseteq M$. Suppose $B \neq 0$. Then $(B)f^{-1}$ is a nonsmall submodule of $N$ and since $N$ is lifting, $(B)f^{-1}$ contains a nonzero direct summand of $N$. As $N$ has an indecomposable exchange decomposition, any nonzero direct summand of $N$ contains an indecomposable direct summand of $N$. Thus $(B)f^{-1}$ contains a hollow direct summand $H$ of $N$. As $f$ is a small epimorphism, $(H)f \subseteq B$ is a coclosed submodule of $M$ and hence is a direct summand of $M$. This contradicts the choice of $A$. Thus $B = 0$. $\square$

**22.18. Decomposition of lifting modules III.** *Let $M$ be a lifting module such that $\operatorname{Rad} N \ll N$ for every submodule $N \subseteq M$. Then every local direct summand of $M$ is a direct summand and hence $M$ is a direct sum of hollow modules.*

*Proof.* Let $\bigoplus_K M_k$ be a local direct summand of $M$ and set $N = \bigoplus_K M_k$. Since $M$ is lifting, we have $N = T \oplus S$ where $T$ is a direct summand of $M$ and $S \ll M$. Given any $x \in S$, we have $xR \subseteq \bigoplus_{k \in F} M_k$ where $F$ is a finite subset of $K$. Then, since $xR \ll M$ and $\bigoplus_{k \in F} M_k$ is a direct summand of $M$, we have $xR \ll \bigoplus_{k \in F} M_k$ and hence $xR \ll N$. This shows that $S \subseteq \operatorname{Rad}(N)$. However, by hypothesis, $\operatorname{Rad} N \ll N$ and hence $S = 0$. Thus $N = T$ and so $N$ is a summand of $M$, as required. It now follows from 13.3 that $M$ is a direct sum of hollow modules. $\square$

We mention some more conditions which imply decomposability of lifting modules. Recall that *continuous* modules are direct injective and $\pi$-injective (or extending, see [85, 2.12]).

**22.19. Continuous lifting modules.** *Let $M$ be a continuous lifting $R$-module. Then any local direct summand of $M$ is a direct summand of $M$ and $M$ is a direct sum of indecomposable modules.*

*Proof.* Let $M$ be a continuous lifting module and $N = \bigoplus_I M_i$ be a local direct summand of $M$. Since $M$ is lifting we have $M = A \oplus B$ where $A \subseteq N$, $N = A \oplus (B \cap N)$ and $B \cap N \ll M$. Suppose $0 \neq x \in B \cap N$. Then there exists a finite subset $K$ of $I$ such that $x \in \bigoplus_K M_k$. Put $M_K = \bigoplus_K M_k$. Then, by assumption, $M_K$ is a direct summand of $M$ and hence is extending. Thus there exists a direct summand $T$ of $M_K$ such that $Rx$ is essential in $T$. Since $Rx \cap A = 0$, we have $A \cap T = 0$. Thus $B \cap N$ contains a submodule $C$ isomorphic to $T$. Since $M$ is direct injective, $C$ is a direct summand of $M$, which is impossible since $B \cap N \ll M$. This contradiction shows that $B \cap N = 0$ and so $N$ is a direct summand of $M$, as required. By 13.3, $M$ is a direct sum of indecomposable modules.     $\Box$

Since self-injective modules are continuous, we get the following.

**22.20. Corollary.** *Let $M$ be a self-injective lifting module. Then any local direct summand of $M$ is a direct summand of $M$ and $M$ is a direct sum of indecomposable modules.*

**22.21. UCC lifting modules.** *Any UCC lifting module is a direct sum of indecomposable modules.*

*Proof.* Suppose that $M$ is a UCC lifting module. Then $M$ is amply supplemented and any coclosed submodule of $M$ is a direct summand (see 22.2). By 21.6(c), any local direct summand of $M$ is a direct summand and hence $M$ is a direct sum of indecomposable modules (see 13.3).     $\Box$

From 21.8 we get

**22.22. Corollary.** *A copolyform lifting module is a direct sum of indecomposable modules.*

Next we consider the endomorphism ring of a copolyform lifting module. Recall that a ring $R$ is a right PP-ring (principal projective ring) if every cyclic right ideal of $R$ is projective.

**22.23. Endomorphism ring of a copolyform lifting module.** *Let $M$ be a copolyform module and $S = \mathrm{End}(M)$.*

(1) *If $M$ is lifting, then $S$ is a right PP-ring.*

(2) *If $M^n$ is lifting for every $n \in \mathbb{N}$, then $S$ is a right semihereditary ring.*

(3) *If $M$ is direct projective and lifting, then $S$ is regular.*

*Proof.* Suppose $f \in S$. Since $M$ is amply supplemented and copolyform, $M$ is strongly copolyform and $\mathrm{Im}\, f \overset{cc}{\hookrightarrow} M$ (see 17.19). As $M$ is lifting, $\mathrm{Im}\, f$ is a direct summand of $M$.

(1) By [363, 39.11], $S$ is a right PP-ring.

(2) By 9.18, $M^n$ is copolyform for every $n \in \mathbb{N}$. Hence, by (1), each $\text{End}(M^n)$ $\simeq S^{n \times n}$ is a right $PP$-ring and thus $S$ is right semihereditary by [363, 39.13].

(3) Let $f \in S$. By our opening remarks, $\text{Im} f$ is a direct summand of $M$. Since $M$ is direct projective, $\text{Ke} f$ is also a direct summand of $M$. Thus $S$ is a regular ring (see 15.19 (1) or [363, 37.7]). $\qquad\square$

It is known that a module $M$ is polyform if and only if $\text{End}(\widehat{M})$ is a regular ring where $\widehat{M}$ is the $M$-injective hull of $M$. We now show, with the aid of 9.26, that the dual is true when $M$ has a projective cover which is supplemented.

**22.24. Corollary.** *Let $P$ be a projective cover of $M$ in $\sigma[M]$ which is supplemented. Then $M$ is copolyform if and only if $\text{End}(P)$ is regular.*

*Proof.* By [363, 41.15], $P$ is lifting. Moreover, it follows from 9.26 that $M$ is copolyform if and only if $P$ is copolyform. Thus, if $M$ is copolyform, then $\text{End}(P)$ is regular by 22.23 (3). Since regular rings have zero Jacobson radical, the converse follows from 9.20. $\qquad\square$

**22.25. Exercises.**

(1) The module $M$ is called *semiregular* if every cyclic submodule lies over a projective direct summand of $M$. Prove (see [255]):

  (i) For $M$ the following are equivalent:

    (a) $M$ is semiregular;

    (b) for finitely generated submodules $N \subset M$, there exists $\gamma \in \text{Hom}(M, N)$ such that $\gamma^2 = \gamma$, $M\gamma$ is projective, and $N(1-\gamma) \subseteq \text{Rad} M$;

    (c) every finitely generated submodule of $M$ lies over a projective summand of $M$.

  (ii) For a projective module $M$, the following are equivalent:

    (a) $M$ is semiregular;

    (b) $M$ is f-lifting;

    (c) for every finitely generated (cyclic) submodule $N \subset M$, $M/N$ has a projective cover.

(2) For a given submodule $U \subseteq M$, an element $x \in M$ is said to be $U$-*semiregular* if there exists a decomposition $M = A \oplus B$ such that $A$ is projective and $A \subseteq Rx$ and $Rx \cap B \subseteq U$. The module $M$ is called $U$-*semiregular* provided each element of $M$ is so. Prove (see [7]):

  (i) For a fully invariant submodule $U \subset M$, the following are equivalent:

    (a) $M$ is $U$-semiregular;

    (b) for finitely generated submodules $N \subset M$, there exists $\gamma \in \text{Hom}(M, N)$ such that $\gamma^2 = \gamma$, $M\gamma$ is projective, and $N(1-\gamma) \subseteq U$;

(c) for any finitely generated submodule $N \subset M$, there is a decomposition $M = A \oplus B$ where $A \subseteq N$ is a projective submodule and $N \cap B \subseteq U$.

(ii) Let $M$ be $U$-semiregular for a fully invariant submodule $U \subset M$. Then:

(α) $\operatorname{Rad} M \subseteq U$ and $Z(M) \subseteq U$;

(β) every finitely generated submodule of $M/U$ is a direct summand.

(iii) Assume $M$ to be projective and choose $U = \operatorname{Rad} M$.

(α) If $U \ll M$, then $M$ is semiregular if and only if it is $U$-semiregular.

(β) If $M$ is finitely generated, then $M$ is $U$-semiregular if and only if it is fw-supplemented (see 17.22(1)).

(iv) Assume $M$ to be finitely generated and projective and choose $U = Z(M)$. If $M$ is $U$-semiregular, then $Z(M) = \operatorname{Rad} M$.

(v) Let $U = \operatorname{Soc} M$ and assume $M$ to be $U$-semiregular. Then:

(α) every cyclic submodule of $M$ is a direct sum of a projective module and a singular module;

(β) if $M$ is noetherian, then it is artinian.

(3) Let $R$ be a left hereditary ring and let $M = M_1 \oplus M_2$ be an amply supplemented module where $M_1$ is injective. Then $M$ is lifting if and only if $M_1$ and $M_2$ are lifting and every coclosed submodule $N \subset M$ with $M = N + M_1$ is a direct summand ([199]).

**22.26. Comments.** Looking ahead to 27.21 and 27.24, we see that a module $M$ which is projective in $\sigma[M]$ is perfect (semiperfect) if and only if $M^{(I)}$ is lifting for any (finite) index set $I$. From this regard, the concept of a lifting module has roots in the theory of (semi)perfect rings and modules with projective covers. It almost goes without saying that the standard reference for these predecessors is the oft-cited 1960 paper [34] by Bass. (We note that Bass remarks that the term "perfect" is used earlier by Eilenberg in [87] to describe a category of modules (over a semilocal ring) each of which have a projective cover, while projective covers, under the name "minimal epimorphisms", were considered earlier still by Eilenberg and Nakayama for modules over semiprimary rings in [88].)

Following Bass, Mares [229] extended the concept of a (semi)perfect ring to (semi)-perfect projective modules. In particular she proved that if $P$ is a semiperfect projective module, then $P/\operatorname{Rad} P$ is semisimple, decompositions of $P/\operatorname{Rad} P$ can be lifted to a decomposition of $P$, and $P$ is a direct sum of cyclic hollow LE-modules.

Lifting modules were first introduced and studied by Takeuchi [327] in 1976, but under the name *codirect modules*, and he obtains some of Mares' results as corollaries. Also motivated by Mares' results, in his 1980 paper [155] Harada introduced *the lifting property of simple modules* and of *direct sums of decompositions*. He defined a module $M$ to have *the lifting (extending) property of simple modules* if, given a simple submodule $A$ of $M/\operatorname{Rad} M$ (resp. $\operatorname{Soc} M$), there exists an LE-, cyclic, hollow (resp. uniform) direct summand $N$ of $M$ such that $(N + \operatorname{Rad} M)/\operatorname{Rad} M = A$ (resp. $A$ is essential in $N$). He then pursued these notions extensively in [156] and [157]. (See also Chapter 9 of Harada's monograph [158].)

In a 1983 paper, Oshiro [269] defined the extending property and lifting property of a module $M$ for any family $\mathcal{A}$ of submodules of $M$. Specifically, he said that $M$ has *the lifting (extending) property for $\mathcal{A}$* if, for every $A \in \mathcal{A}$, there exists a direct summand $A^*$ of $M$ such that $A/A^*$ is small in $M/A^*$ ($A^*$ is essential in $A$). We note that Nicholson [255] had earlier defined another type of lifting property: he called a module $M$ *semiregular* if, for every cyclic submodule $A$ of $M$, there exists a projective direct summand $A^*$ of $M$ such that $A \subseteq A^*$, $M = A^* \oplus A^{**}$ and $A \cap A^{**}$ is small in $A^{**}$. It is easy to see that a projective module is semiregular if and only if $M$ is lifting for the family of all cyclic submodules of $M$ in Oshiro's sense. Oshiro also noted that a module $M$ has the lifting property of simple modules (as defined by Harada) if and only if $M$ is lifting for the family $\mathcal{A}$ of submodules $A$ of $M$ for which $\operatorname{Rad} M \subset A$ and $A/\operatorname{Rad} M$ is simple.

Okado and Oshiro [266] observe that the extending property of simple modules can be dualised in two ways for a module $M$ which is a direct sum of cyclic hollow modules. We can require, on the one hand, that every simple submodule of $\operatorname{Rad} M$ is induced from a direct summand of $M$ (as defined by Harada [155]) or, on the other hand, that every maximal submodule of $M$ lies above a direct summand. Okado and Oshiro provide characterisations for both alternatives.

Recall that a module $M$ is said to have *the extending property for uniform submodules* or to be *uniform-extending* if every uniform submodule of $M$ is essential in a direct summand of $M$ (see [85]). This property can also be dualised in two ways. Firstly, in [272], Oshiro defines a module $M$ to have *the lifting property of hollow modules* if every submodule $A$ of $M$ with $M/A$ hollow lies above a direct summand and he proves that, over a generalised uniserial ring $R$, every module with the lifting property of hollow modules is uniform-extending. Secondly, as in 22.8, Lomp [226] defines a module to be *hollow-lifting* if $M$ is amply supplemented and every hollow module lies above a direct summand of $M$. He proves that, if $M$ is a hollow-lifting module, then every coclosed submodule of $M$ of finite hollow dimension is a direct summand, and that a hollow-lifting module $M$ with finite hollow dimension is lifting (22.9 and its corollary), thus dualising the corresponding results for uniform-extending modules given in [85, 7.7 and 7.8].

Turning to the key question of when a lifting module is a direct sum of indecomposable modules, Lomp [226] gave various conditions on a lifting module to ensure this (see 22.11, 22.12, 22.13) and uses them to give conditions for a module with finite hollow dimension to be artinian or noetherian (see corollary to 22.14). Oshiro established 22.19 in [271, lemmas 2.4 and 2.5] while the corollary to 22.23 was proved by Lomp [226, 5.2.2].

As already detailed, extending modules or CS modules are modules whose complement submodules are direct summands, while lifting modules provide a dualisation of extending modules. Since the definition of a lifting module $M$ demands, in the first instance, that $M$ is supplemented, one might ask how one can characterise modules that are not necessarily supplemented, but all of whose supplement submodules are direct summands. In this regard, Zöschinger [391] examines those rings $R$ such that all finitely generated projective left $R$-modules satisfy the property that all supplement submodules are direct summands. He shows that in $_RR$ every supplement is a direct summand if and only if every supplement in $R_R$ is a direct summand if and only if, whenever $Rx$ is a supplement in $_RR$, then $xR$ is a supplement in $R_R$. Hence for any commutative ring $R$, supplements in $R$ coincide with direct summands. Zöschinger relates this concept with a property considered by Lazard, namely that supplements are direct summands in every finitely generated projective left $R$-module if and only if every projective left $R$-module $P$ with $P/\mathrm{Rad}\,P$ finitely generated is finitely generated. Zöschinger calls these rings *left L-rings* and shows that the notion is left-right symmetric, that is, $R$ is a left $L$-ring if and only if $R$ is a right $L$-ring. Furthermore he shows that $R$ is a left $L$-ring if and only if, for every finitely generated submodule $F$ of a projective left $R$-module $P$, $\mathrm{Rad}\,(P/F) \neq P/F$ if and only if every finitely generated flat left $R$-module $P$ is projective provided $P/PJ(R)$ is $R/J(R)$-projective.

Commutative rings and, more generally, PI-rings are $L$-rings, as are all rings $R$ such that $R/Q$ is a right Goldie ring for each prime ideal $Q$. The first example of a semilocal ring which is not an $L$-ring was given by Gerasimov and Sakhaev in [120] where they construct a semilocal ring $R$ and a non-finitely generated projective left $R$-module $P$ such that $P/\mathrm{Rad}\,P$ is finitely generated.

It was asked in [227] whether a projective module $P$ is finitely generated provided $\mathrm{End}(P)$ is semilocal (or equivalently $P$ has finite dual Goldie dimension). From above, if $R$ is an $L$-ring, the answer is "yes". Note that Gerasimov and Sakhaev's counterexample $P$ does not have finite hollow dimension. However, Facchini, Herbera and Sakhaev give a counterexample in [97] by constructing a non-finitely generated projective $R$-module $P$ having a semilocal endomorphism ring. The ring $R$ in their construction is realized as the (semilocal) endomorphism ring of a cyclic module over a "nearly simple" chain domain using results of Puninski [288].

**References.** Alkan and Öczan [7]; Keskin [198, 199]; Lomp [227, 226]; Mohamed and Müller [241]; Nicholson [255]; Özcan and Alkan [279]; Talebi and Vanaja [330]; Wisbauer [363]; Yousif and Zhou [374]; Zöschinger [391].

# 23 Finite direct sums of lifting modules

As we saw in 22.5, (finite) direct sums of lifting modules need not be lifting. In this section we look at the question of when lifting *is* preserved by finite direct sums. We will prove (see 23.12) that a module $M = K \oplus L$ is lifting if both $K$ and $L$ are lifting and $K$ and $L$ are mutually projective modules. However, this is not a necessary condition for $M$ to be lifting. For example $\mathbb{Z}/4\mathbb{Z} \oplus \mathbb{Z}/2\mathbb{Z}$ is lifting but $\mathbb{Z}/2\mathbb{Z}$ is not $\mathbb{Z}/4\mathbb{Z}$-projective. We show in 23.6 that if $M = K \oplus L$ is a lifting module and the decomposition $M = K \oplus L$ is exchangeable, then $K$ and $L$ are relatively im-small projective (4.33).

**23.1. Lemma.** *Let* $M = M_1 \oplus M_2$ *where* $M_2$ *is lifting. If* $M_1$ *is im-small* $M_2$-*projective, then every coclosed submodule* $N \subset M$ *with* $(N + M_1)/N \ll M/N$ *is a direct summand of* $M$.

*Proof.* Let $N \subset M$ be a coclosed submodule with $(N+M_1)/N \ll M/N$. By 20.32, there exists a submodule $N' \subseteq N$ such that $M = N' \oplus M_2$. Clearly (by 22.6) $M/N'$ is lifting, and since $N$ is coclosed in $M$, by 3.7 (1) $N/N'$ is coclosed and hence a direct summand of $M/N'$, that is, there is some $N' \subseteq X \subseteq M$ with $N + X = M$ and $X \cap N = N'$. Thus

$$M = N + X = N + N' + (X \cap M_2) = N + (X \cap M_2)$$

where $N \cap (X \cap M_2) = (N \cap X) \cap M_2 = 0$. $\square$

**23.2. Coclosed submodules as direct summands.** *Let* $M = M_1 \oplus M_2$ *be an amply supplemented module where* $M_1$ *and* $M_2$ *are lifting modules. Then every coclosed submodule of* $M$ *is a direct summand provided one of the following conditions holds:*

(i) $M_1$ *is im-small* $M_2$-*projective and every coclosed submodule* $N \subset M$ *with* $M = N + M_1$ *is a direct summand.*

(ii) $M_1$ *and* $M_2$ *are relatively im-small projective and every coclosed submodule* $N \subseteq M$ *with* $M = N + M_1 = N + M_2$ *is a direct summand.*

(iii) $M_2$ *is* $M_1$-*projective and* $M_1$ *is im-small* $M_2$-*projective.*

(iv) $M_1$ *is semisimple and im-small* $M_2$-*projective.*

*Proof.* (i) By 20.26, it is enough to prove that every coclosed submodule $K$ of $M$ with $M = K + M_2$ is a direct summand of $M$. Since $M/K \simeq M_2/(K \cap M_2)$ and $M_2$ is amply supplemented we get that $M/K$ is amply supplemented. Hence there exists a coclosed submodule $N/K$ of $M/K$ such that the inclusion $N/K \subseteq (K+M_1)/K$ is cosmall in $M/K$. Since $N/K$ is coclosed in $M/K$ and $K$ is coclosed in $M$, 3.7 (2) implies that $N$ is coclosed in $M$. Also $(K + M_1)/N \ll M/N$ and $N+M_1 \subseteq K+M_1$. Hence $(N+M_1)/N \ll M/N$. As $M_1$ is im-small $M_2$-projective, it follows from 23.1 that $N$ is a direct summand of $M$, say $M = N \oplus N'$. Since

$M/K = N/K \oplus (K + N')/K$, $(K + N')/K$ is a coclosed submodule of $M/K$ and, again by 3.7 (2), $K + N'$ is a coclosed submodule of $M$. Now

$$M = (M_1 + N) + N' = (K + M_1) + N' = (K + N') + M_1.$$

By hypothesis, $K + N'$ is a direct summand of $M$. As $K \cap N' = 0$, we get $K$ is a direct summand of $M$.

(ii) This follows by applying 20.26(c) and 23.1.

(iii) Let $N \subset M$ be a coclosed submodule with $M = N + M_1$. By 4.12, there exists a submodule $N'$ of $N$ such that $M = N' \oplus M_1$. Since $N/N'$ is coclosed in $M/N'$ and $M/N'$ is lifting, it is in fact a direct summand of $M/N'$. Then $M = N + X$ with $X \cap N = N'$. Now

$$M = N + X = N + N' + (X \cap M_1) = N + (X \cap M_1)$$

where $N \cap (X \cap M_1) = (N \cap X) \cap M_2 = 0$. Therefore $N$ is a direct summand of $M$. Now (iii) follows from (i).

(iv) follows from (iii).                                                                        □

We now consider some sufficient conditions for a module $M_1$ to be im-small $M_2$-projective when $M = M_1 \oplus M_2$ is lifting.

**23.3. Direct summands with the finite exchange property.** *Suppose $M = M_1 \oplus M_2$ to be a module whose coclosed submodules are direct summands, $M_1$ has the finite exchange property, and $M_2$ is supplemented. Then $M_1$ is im-small $M_2$-projective.*

*Proof.* Let $K$ be a coclosed submodule of $M$ such that $M = K + M_2$ and also $(K + M_1)/M_1 \ll M/M_1$. By 2.3 (2), $(K + M_1)/K \ll M/K$. By hypothesis, $M = K \oplus K'$ for some submodule $K'$ of $M$. Since $M_1$ has the finite exchange property, $M = M_1 \oplus A \oplus B$, for some $A \subseteq K$ and $B \subseteq K'$. Since $(K + M_1)/K \ll M/K$, we have $M = K \oplus B$. Hence $B = K'$. As $(K + M_1)/M_1 \ll M/M_1$, we get $M = K' \oplus M_1$.

As $K \simeq M_1$, $K$ has the finite exchange property (see 11.10). Therefore there exist $C \subseteq M_1$ and $D \subseteq M_2$ such that $M = K \oplus C \oplus D$. Since $(K + M_1)/K \ll M/K$, we conclude that $M = K \oplus D$. Now, since $(K + M_1)/M_1 \ll M/M_1$, we have $M = M_1 \oplus D$ and so, since $D \subseteq M_2$, we get $D = M_2$. Thus $M = K \oplus M_2$ and, by 20.32, $M_1$ is small $M_2$-projective.                                                          □

**23.4. Corollary.** *Suppose that $M = M_1 \oplus M_2$ is a lifting module and $M_1$ has the finite exchange property. Then $M_1$ is im-small $M_2$-projective.*

The corollary helps to establish another example of a direct sum of lifting modules which is not lifting (generalising 22.5).

**23.5. Example.** *Let $M$ be a uniserial module with unique composition series $0 \neq V \subset U \subset M$. Then the module $M \oplus (U/V)$ is not lifting.*

*Proof.* Since $M$ and $U/V$ are both hollow, they are lifting modules. In fact, they have local endomorphism rings and thus $M \oplus (U/V)$ has the exchange property.

Assume to the contrary that $M \oplus (U/V)$ is lifting. Consider the diagram with the obvious maps

$$
\begin{array}{c}
U/V \\
\downarrow \\
M \xrightarrow{\ p\ } M/V \longrightarrow 0 .
\end{array}
$$

By 23.4, $U/V$ is im-small $M$-projective and hence there exists a map $h : U/V \to M$ extending the diagram commutatively. However, since $U/V$ is simple the image of $h$ must be contained in $V$ and thus $h \diamond p = 0$. $\qquad\square$

Exchange decompositions, defined in 11.32, play a prominent rôle in the rest of this section.

**23.6. Relative im-small projectivity of direct summands.** *If $M$ is a lifting module with exchange decomposition $M = K \oplus L$, then $K$ and $L$ are relatively im-small projective modules.*

*Proof.* Since $M$ is lifting, $M$ is amply supplemented. Let $X$ be a coclosed submodule of $M$ such that $(X + K)/K \ll M/K$ and $M = X + L$. By 20.32 it is enough to prove that $M = X \oplus L$. As the decomposition $M = K \oplus L$ is exchangeable, we have $M = X \oplus K' \oplus L'$ for submodules $K' \subseteq K$ and $L' \subseteq L$. Since $(X + K)/K \ll M/K$ we get $L' = L$. Now $X + L = M$ implies $K' = 0$. Thus $M = X \oplus L$. $\qquad\square$

The converse of the above is not true as can be seen by the following example which also shows that strongly $\oplus$-supplemented modules need not be lifting.

**23.7. Example.** *Let $K$ be the quotient field of a discrete valuation domain $R$ which is not complete. Then $M = K \oplus K$ is a supplemented module with the exchange property and $K$ is im-small $K$-projective. Moreover $M$ is a direct projective module whose coclosed submodules are direct summands. However $K$ is not $K$-projective and hence $M$ is not lifting. Consequently $M$ is not amply supplemented.*

*Proof.* Let $R$ be a discrete valuation domain which is not complete. Then the quotient field $K$ of $R$ is not self-projective. Let $M = K \oplus K$.

As $K$ is supplemented, $M$ is a supplemented module (see 20.14). Also $K$ is trivially im-small $K$-projective and, since $M$ is injective, $M$ has the exchange property (see [357, p. 265] or [241, Theorem 1.21]).

Now suppose $M/A \simeq B$, a direct summand of $M$. As $M$ is injective and torsion free, so too is $B$. This implies that $A$ is closed in $M$ and hence a direct summand of $M$. Thus $M$ is direct projective.

Now let $A$ be a supplement of $B$ in $M$. Then, since $M$ is a divisible $R$-module, for every $0 \neq r \in R$, we have $K \oplus K = r(K \oplus K) = r(A + B) = rA + rB \subseteq rA + B$.

Thus, since $A$ is a supplement of $B$ in $M$, we get $A = rA$. In other words, $A$ is divisible, hence injective and so a direct summand of $M$.

Since $K$ is not $K$-projective, $M$ is not a quasi-discrete module (see 26.7) and hence is not a lifting module (see 27.1). Thus, by 22.3, $M$ is not amply supplemented.                                                                                 □

**23.8. Generalised projectivity and im-small projectivity.** *Suppose $L$ is a lifting module. If $K$ is a generalised $L$-projective module, then $K$ is im-small $L$-projective.*

*Proof.* Let $g : L \to X$ be an epimorphism and let $f : K \to X$ be a homomorphism with $\operatorname{Im} f \ll X$. Since $L$ is lifting, there exists a decomposition $L = L' \oplus L''$ such that $L' \subseteq \operatorname{Ke} g$ and $\operatorname{Ke} g \cap L'' \ll L''$. This gives $(L)g = (L')g + (L'')g = (L'')g$ and $\operatorname{Ke}(g|_{L''}) = \operatorname{Ke} g \cap L'' \ll L''$. Then, by 4.43, $K$ is generalised $L''$-projective. Thus we may assume that $\operatorname{Ke} g \ll L$. Since $g : L \to X$ is now a small epimorphism, given any submodule $C$ of $L$, $(C)g$ is small in $X$ if and only if $C$ is small in $L$. Hence we cannot have a map from a direct summand of $L$ to $K$ satisfying the condition for $K$ to be generalised $L$-projective. Hence $K$ is an im-small $L$-projective module.     □

**23.9. Generalised projectivity with respect to lifting modules.** *Suppose $M = K \oplus L$ where $L$ is a lifting module. Then $K$ is generalised $L$-projective if and only if given any submodule $X$ of $M$ with $M = X + L$, there exists $X' \overset{cs}{\underset{M}{\hookrightarrow}} X$, $K' \subseteq K$ and $L' \subseteq L$ such that $M = X' \oplus K' \oplus L'$.*

*Proof.* ($\Leftarrow$). This follows from 4.42.

($\Rightarrow$). Let $X$ be a submodule of $M$ with $M = X + L$. Since $L$ is lifting, there exists a decomposition $L = L_1 \oplus L_2$ such that $L_2 \subseteq L \cap X$ and $(L \cap X) \cap L_1 = L_1 \cap X \ll L_1$. Hence we have

$$M = X + L = X + L_1 \text{ and } L_1 \cap X \ll L_1.$$

As $L_2 \subseteq X$, $X = L_2 \oplus ((L_1 \oplus K) \cap X)$. Put $N = L_1 \oplus K$. Since $M = X + L_1$, $N = (N \cap X) + L_1$. By 4.43 and 4.42, there exists a decomposition

$$N = T \oplus K' \oplus L_1' = T + L_1 \text{ with } T \subseteq N \cap X, \ K' \subseteq K \text{ and } L_1' \subseteq L_1.$$

Since $T \subseteq N \cap X$ and $X = L_2 + (N \cap X)$, we get $L_2 \oplus T \subseteq X$. As $M = (L_2 + T) + L_1$ and $X \cap L_1 \ll M$, we get, by 3.2(6), that $(L_2 \oplus T) \overset{cs}{\underset{M}{\hookrightarrow}} X$. Since $M = L_2 \oplus N$, this completes the proof.                                                                              □

**23.10. Generalised projective lifting modules.** *Suppose $M = K \oplus L$, where $K$ and $L$ are lifting modules. If $K$ and $L$ are relatively generalised projective modules, then every coclosed submodule of $M$ is a direct summand of $M$ and $M = K \oplus L$ is an exchange decomposition.*

*Proof.* Let $X$ be a coclosed submodule of $M$. As $K$ and $L$ are supplemented, $M$ is a supplemented module and $X$ has a supplement $K' \oplus L'$, where $K' \subseteq K$ and $L' \subseteq L$ (see 20.14). Since $M$ is supplemented, the coclosed submodules are precisely the supplements of $M$ and $X + K'$ and $X + L'$ are supplement submodules of $M$ (see 20.14 again).

Since $K$ is generalised $L$-projective, $L$ is lifting and $X + K'$ is a supplement, it follows that there exists a decomposition $M = (X + K') \oplus K'' \oplus L''$, with $K'' \subseteq K$ and $L'' \subseteq L$ (see 23.9). Put $X_1 = K' + X$ and $N = K'' \oplus L''$. Then $M = X_1 \oplus N$ and $M/X = X_1/X \oplus (N + X)/X$. Hence $(N + X)/X$ is a supplement of $M/X$. Now $X$ is a supplement of $M$ and so, by 20.5 (1), $N + X$ is a supplement of $M$. Then, since $N + X = X \oplus K'' \oplus L''$, by 20.4 (8), $X \oplus L''$ is a supplement of $M$ and

$$(X \oplus L'') + K \supseteq X + L'' + K' + K'' = M.$$

Applying 23.9 we get

$$M = (X \oplus L'') \oplus K^* \oplus L^*, \quad \text{with} \quad K^* \subseteq K \quad \text{and} \quad L^* \subseteq L.$$

Hence $M = X \oplus K^* \oplus (L^* \oplus L'')$, where $K^* \subseteq K$ and $L^* \oplus L'' \subseteq L$. Thus $X$ is a direct summand of $M$ and since any direct summand of $M$ is a coclosed submodule we get that $M = K \oplus L$ is an exchange decomposition.  □

**23.11. Lifting modules and exchange decompositions.** *Let $M_1$ and $M_2$ be lifting modules and $M = M_1 \oplus M_2$. Then $M$ is lifting and $M = M_1 \oplus M_2$ is an exchange decomposition if and only if any direct summand of $M_1$ is generalised $M_2$-projective and vice versa.*

*Proof.* Suppose $M$ is lifting and $M = M_1 \oplus M_2$ is an exchange decomposition. By 11.33, it is enough to show that $M_1$ is generalised $M_2$-projective. Suppose $M = X + M_2$. Since $M$ is a lifting module it is amply supplemented (see 22.3) and hence has a supplement $K$ of $M_2$ in $M$ such that $K \subseteq X$. Since any supplement of a lifting module is a direct summand and $M = M_1 \oplus M_2$ is an exchange decomposition, we have $M = K + M_2 = K \oplus A \oplus B$, where $A \subseteq M_1$ and $B \subseteq M_2$. Hence $M_1$ is generalised $M_2$-projective by 4.42.

Conversely, assume that $M = M_1 \oplus M_2$, where $M_1$ and $M_2$ are lifting modules, any direct summand of $M_1$ is generalised $M_2$-projective and vice versa. By 23.10, every coclosed submodule of $M$ is a direct summand and $M = M_1 \oplus M_2$ is an exchange decomposition. Hence it is enough to prove that $M$ is amply supplemented (see 22.3) and, since $M_1 \oplus M_2$ is weakly supplemented (by 17.13), this will follow if we show that every submodule of $M$ has a coclosure in $M$ (see 20.25).

Let $X$ be a submodule of $M$. Since $M_1$ is lifting, there exists a decomposition $M_1 = M_1' \oplus M_1''$ such that $M_1'$ is a coclosure of $(X)\pi_{M_1}$ in $M_1$ ($\pi$ will denote the obvious projections). Put $X' = (M_1'' \oplus M_2) \cap X$. Since $M_2$ is lifting, there exists a decomposition $M_2 = M_2' \oplus M_2''$ such that $M_2'$ is a coclosure of $(X')\pi_{M_2}$ in $M_2$. Setting $X'' = (M_1'' \oplus M_2'') \cap X$, we see that $(X'')\pi_{M_i''} \ll M_i''$ for $i = 1, 2$,

$$
\begin{aligned}
M &= (X)\pi_{M_1} + M_1'' + M_2 = X + M_1'' + M_2 \\
  &= X + M_1'' + (X')\pi_{M_2} + M_2'' = X + M_1'' + X' + M_2'' \\
  &= X + (M_1'' \oplus M_2'')
\end{aligned}
$$

and so we see that

$$
X'' \subseteq (X'')\pi_{M_1''} \oplus (X'')\pi_{M_2''} \ll M_1'' \oplus M_2''.
$$

By assumption and 4.43, direct summands of $M_1$ and $M_2$ are relatively generalised projective. Put $N = M_1' \oplus M_2''$. Then $N = M_2'' + [(X + M_1'') \cap N]$. Define $L = (X + M_1'') \cap N$. Then $N = L + M_2''$. By 23.9 applied to $N$, we get

$$
N = U \oplus \widetilde{M_1'} \oplus \widetilde{M_2''},
$$

where $U \overset{cs}{\underset{N}{\hookrightarrow}} L$, $\widetilde{M_1'} \subseteq M_1'$ and $\widetilde{M_2''} \subseteq M_2''$. By 3.2 (1), we have $N = U + M_2''$. Let $M_1' = \overline{M_1'} \oplus \widetilde{M_1'}$ and $M_2'' = \overline{M_2''} \oplus \widetilde{M_2''}$. We see

$$
M = M_1'' \oplus M_2' \oplus U \oplus \widetilde{M_1'} \oplus \widetilde{M_2''}.
$$

Next set $T = U \oplus M_1'' \oplus M_2'$. Then $T$ is a direct summand of $M$. Now $M_2' \subseteq (X')\pi_{M_2} \subseteq X + M_1''$. Then $M = N \oplus M_1'' \oplus M_2'$ and $(X + M_1'') \supseteq M_1'' \oplus M_2'$ imply $X + M_1'' = L \oplus M_1'' \oplus M_2'$. As $U \overset{cs}{\underset{N}{\hookrightarrow}} L$ and $N$ is a direct summand of $M$, we have $T \overset{cs}{\underset{M}{\hookrightarrow}} (X + M_1'')$. Also $T = M_1'' + (X \cap T)$ and $M = T + M_2''$. Thus $M = (X \cap T) + M_1'' + M_2''$. Since

$$
X = (X \cap T) + X \cap (M_1'' \oplus M_2'') = (X \cap T) + X''
$$

and $X'' \ll M$, we see that $(X \cap T) \overset{cs}{\underset{M}{\hookrightarrow}} X$ by 3.2 (6).

Put $A = M_1'' \oplus \overline{M_1'}$ and $B = M_2' \oplus \overline{M_2''}$. Then $A$ and $B$ are relatively generalised projective. Let $f = \pi_{A \oplus B}|_T$. Then $f : T \to A \oplus B$ is an isomorphism and

$$
A \oplus B = (T)f = (X \cap T)f + M_1'' = (X \cap T)f + A.
$$

Applying 23.9 to $A \oplus B$, we easily get that $T$ contains a direct summand $T'$ such that $T' \overset{cs}{\underset{M}{\hookrightarrow}} (X \cap T)$. As $(X \cap T) \overset{cs}{\underset{M}{\hookrightarrow}} X$ in $M$, we have $T' \overset{cs}{\underset{M}{\hookrightarrow}} X$ by 3.2 (2). Hence $T'$ is a coclosure of $X$ in $M$ and this completes the proof. $\qquad \square$

**23.12. Corollary.** Let $M = K \oplus L$, where $K$ and $L$ are lifting and are relatively projective. Then $M$ is lifting and $M = K \oplus L$ is an exchange decomposition.

**23.13. Amply supplemented direct sums of lifting modules.** *Let $M_1$ and $M_2$ be lifting modules, $M = M_1 \oplus M_2$, and assume that every submodule $X$ of $M$ has a coclosure in $M$. Then $M$ is lifting and $M = M_1 \oplus M_2$ is an exchange decomposition if and only if $M_i$ is generalised $M_j$-projective $(i \neq j)$.*

*Proof.* By 23.10, every coclosed submodule of $M$ is a direct summand. By hypothesis and 20.25, $M$ is amply supplemented. Now the proof follows from 20.33 and 23.11. $\qquad\square$

**23.14. Finite direct sums of lifting modules.** *Let $M_1, \ldots, M_n$ be lifting modules and set $M = M_1 \oplus \cdots \oplus M_n$. Then the following are equivalent:*

(a) *$M$ is lifting and $M = M_1 \oplus \cdots \oplus M_n$ is an exchange decomposition;*

(b) *$K$ and $L$ are relatively generalised projective, for any direct summand $K$ of $\bigoplus_I M_i$ and any direct summand $L$ of $\bigoplus_J M_j$, for any nonempty disjoint subsets $I, J \subset \{1, 2, \ldots, n\}$;*

(c) *$M_i'$ and $T$ are relatively generalised projective for any direct summand $M_i'$ of $M_i$ and any direct summand $T$ of $\bigoplus_{j \neq i} M_j$, for any $1 \leq i \leq n$.*

*Proof.* (a)$\Rightarrow$(b) follows from 11.33 and 23.11, and (b)$\Rightarrow$(c) is trivial.

(c)$\Rightarrow$(a). We prove this by induction on $n$. For $n = 1$, (a) is obvious. Assume that for $L = L_1 \oplus \cdots \oplus L_{n-1}$ where each $L_i$ is lifting, $L_i'$ and $S$ are relatively generalised projective for any direct summand $L_i'$ of $L_i$ and any direct summand $S$ of $\bigoplus_{j \neq i} L_j$, for $1 \leq i \leq n - 1$ and $1 \leq j \leq n - 1$, implies that $L$ is lifting and the decomposition $L = L_1 \oplus \cdots \oplus L_{n-1}$ is an exchange decomposition. Now (c) implies that $N = M_1 \oplus \cdots \oplus M_{n-1}$ is lifting and an exchange decomposition.

By 23.11, $M$ is lifting and $M = M_n \oplus N$ is an exchange decomposition. It remains to show that $M = M_1 \oplus \cdots \oplus M_n$ is an exchange decomposition.

By 11.42 (7) it follows that if $N = M_1 \oplus \cdots \oplus M_{n-1}$ and $M = N \oplus M_n$ are exchange decompositions, then $M = M_1 \oplus \cdots \oplus M_n$ is an exchange decomposition. For completeness we give a proof. Let $X$ be a direct summand of $M$. Since $M = N \oplus M_n$ is an exchange decomposition, there exist a direct summand $M_n'$ of $M_n$ and a direct summand $N'$ of $N$ such that $M = X \oplus N' \oplus M_n'$. Now we see

$$N = N' \oplus [(X \oplus M_n') \cap N].$$

As $(X \oplus M_n') \cap N$ is a direct summand of $N$, by assumption, there exist direct summands $M_i'$ of $M_i$, for $1 \leq i \leq n - 1$, such that

$$N = M_1' \oplus \cdots \oplus M_{n-1}' \oplus [(X \oplus M_n') \cap N].$$

Denote the corresponding projections by

$$p_1 : N \to M_1' \oplus \cdots \oplus M_{n-1}' \quad \text{and} \quad p_2 : N \to (X \oplus M_n') \cap N.$$

Now $(N')p_1 = M'_1 \oplus \cdots \oplus M'_{n-1}$ and $p_1$ is bijective on $N'$. Then, since we have $(N')p_1 \cap [(X \oplus M'_n) \cap N] = 0$ and $(N')p_1 \subseteq N$, also $(N')p_1 \cap (X \oplus M'_n) = 0$. Now $p_1$ and $p_2$ induce an epimorphism $\phi : (N')p_1 \to (N')p_2$. We have

$$N' = \{x + (x)\phi \mid x \in (N')p_1\}.$$

Moreover $\phi$ can also be considered as a map from $(N')p_1 \to X \oplus M'_n$. Therefore $(N')p_1 \oplus (X \oplus M'_n) = N' \oplus (X \oplus M'_n)$ and so

$$
\begin{aligned}
M &= X \oplus N' \oplus M'_n = N' \oplus (X \oplus M'_n) \\
&= (N')p_1 \oplus (X \oplus M'_n) \\
&= X \oplus M'_1 \oplus \cdots \oplus M'_{n-1} \oplus M'_n.
\end{aligned}
$$

Consequently $M = M_1 \oplus \cdots \oplus M_n$ is an exchange decomposition.    $\square$

As an immediate consequence of 23.14 we obtain the following.

**23.15. Direct sums of relatively projective lifting modules.** *Let $M_1, \ldots, M_n$ be lifting modules and put $M = M_1 \oplus \cdots \oplus M_n$.*

*If $M_i$ and $M_j$ are relatively projective for each $1 \leq i, j \leq n$, $i \neq j$, then $M$ is lifting and $M = M_1 \oplus \cdots \oplus M_n$ is an exchange decomposition.*

**23.16. Direct sums of hollow modules.** *Let $M = M_1 \oplus \cdots \oplus M_n$ where each $M_i$ is a hollow module. Then the following are equivalent:*

(a) *$M$ is lifting and $M = M_1 \oplus \cdots \oplus M_n$ is an exchange decomposition;*

(b) *$M_i$ is generalised $M_j$-projective for $i \neq j$, $1 \leq i, j \leq n$;*

(c) *$M_i$ is almost $M_j$-projective for $i \neq j$, $1 \leq i, j \leq n$.*

*Proof.* (a)$\Rightarrow$(b). By 23.14, for $i \neq j$, $M_i$ is generalised $M_j$-projective.

(b)$\Leftrightarrow$(c) holds by 4.44.

(b)$\Rightarrow$(a). We prove this by induction on $n$. For $n = 2$ the result follows from 23.14. Put $I = \{1, 2, \ldots, n\}$ and $I_j = I \setminus \{j\}$, for every $j \in I$. Assume that any direct sum $\bigoplus_{k=1}^{n-1} H_k$ of $n - 1$ relatively generalised projective hollow modules is lifting and an exchange decomposition. Hence, for every $j \in I$, $\bigoplus_{k \in I_j} M_k$ is lifting and an exchange decomposition.

By 23.14, it is enough to prove that for every $j \in \{1, 2, \ldots, n\}$, $M_j$ and $T_j$ are relatively generalised projective where $T_j$ is any direct summand of $\bigoplus_{k \in I_j} M_k$. Since the latter is lifting and an exchange decomposition and the $M_k$'s are hollow, any direct summand is isomorphic to $\bigoplus_K M_k$, where $K \subseteq I_j$. Again using 23.14 and the induction hypothesis, we see that it is enough to prove that $M_j$ and $\bigoplus_{k \in I_j} M_k$ are relatively generalised projective, for every $j \in I$.

Fix $j \in I$. Put $N = \bigoplus_{k \in I_j} M_k$. We first prove that $N$ is generalised $M_j$-projective. Let $g : M_j \to M_j/X$ be an epimorphism and $f : N \to M_j/X$ be a homomorphism. Put $f|_{M_i} = f_i$, $i \in I_j$.

Suppose first that $(N)f \ll M_j/X$. Then for each $i \in I_j$, $(M_i)f \ll M_j/X$. Since $M_i$ is almost $M_j$-projective, $g$ is an epimorphism and $f$ is not an epimorphism, there exists $h_i : M_i \to M_j$ such that $h_i g = f_i$. Define $h : N \to M_j$ by $h|_{M_i} = h_i$, $i \in I_j$. Then $hg = f$, as required.

Now suppose that $(N)f$ is not small in $M_j/X$. This implies $(N)f = M_j/X$. By the induction hypothesis, $N$ is a lifting module. Hence there exists a decomposition $N = N_1 \oplus N_2$ with $\operatorname{Ke} f = N_1 \oplus (N_2 \cap \operatorname{Ke} f)$ and $(N_2 \cap \operatorname{Ke} f) \ll N_2$. Then $f|_{N_2} : N_2 \to M_j/X$ is a small epimorphism and so, since $M_j/X$ is hollow, we get that $N_2$ is hollow (by 2.2 (3)). Then, since $N = \bigoplus_{i \in I_j} M_i$ is an exchange decomposition, $N_2 \simeq M_k$, for some $k \in I_j$. As $M_k$ is almost $M_j$-projective, there is either an epimorphism $h : N_2 \to M_j$ or an epimorphism $h : M_j \to N_2$ such that $hg = f|_{N_2}$ or $hf|_{N_2} = g$ respectively. Writing $M_j = M_j \oplus 0$ and taking the zero map from $N_1 \to 0$, it follows that $N$ is generalised $M_j$-projective.

Next we show that $M_j$ is generalised $N$-projective. Let $g : N \to N/X$ be the canonical epimorphism and $f : M_j \to N/X$ be a map. Since $M_j$ is im-small $M_i$-projective for every $i \in I_j$, $M_j$ is im-small $N$-projective by 4.36 (2). Thus if $(M_j)f \ll N/X$, there exists a map $h : M_j \to N$ such that $hg = f$, as required.

Now suppose $(M_j)f$ is not small in $N/X$. Since $N$ is a lifting module, there exists a decomposition $N = N_1 \oplus N_2$ with $X = N_1 \oplus (N_2 \cap X)$ and $(N_2 \cap X) \ll N_2$. Put $g_2 = g|_{N_2}$. Then $g_2 : N_2 \to N/X$ is a small epimorphism and $Y := (M_j/X)g_2^{-1}$ is hollow and not a small submodule of $N_2$. By 3.7 (4), $Y$ is coclosed in $N_2$ and so, as $N_2$ is lifting, it is a direct summand of $N_2$, say $N_2 = Y \oplus Z$. We have $(Y)g = (M_j)f$. By our exchange decomposition of $N$, $Y \simeq M_k$, for some $k \in I_j$. As $M_j$ is almost $M_k$-projective, we get either an epimorphism $h : Y \to M_j$ or an epimorphism $h : M_j \to Y$ such that $hf = g|_Y$ or $hg = f$ respectively. Writing $M_j = M_j \oplus 0$ and taking the zero map from $0 \to N_1 \oplus Z$, it follows that $M_j$ is generalised $N$-projective. $\square$

**23.17. Corollary.** *Suppose $H_1, \ldots, H_n$ are hollow modules. If $H_i$ and $H_j$ are relatively generalised (almost) projective for every $i \neq j$, then for any nonempty disjoint subsets $K, T \subset \{1, \ldots, n\}$, $\bigoplus_K H_k$ and $\bigoplus_T H_t$ are relatively generalised projective and are also relatively almost projective.*

*Proof.* Note that $H_i$ is almost $H_j$-projective if and only if $H_i$ is generalised $H_j$-projective (see 4.44).

Suppose $H_i$ is generalised $H_j$-projective for each $i \neq j$. Then, by 23.16, $H = \bigoplus_{i=1}^n H_i$ is lifting and $H = \bigoplus_{i=1}^n H_i$ is an exchange decomposition. Let $K$ and $T$ be any two nonempty disjoint subsets of $\{1, \ldots, n\}$ and set $H_T = \bigoplus_T H_t$ and $H_K = \bigoplus_K H_k$. Then $N = H_T \oplus H_K$ is lifting and $N = H_T \oplus H_K$ is an exchange decomposition. Thus, by 23.14, $H_K$ and $H_T$ are relatively generalised projective. In particular, for every $t \in T$, $H_t$ is generalised $H_K$-projective. Therefore, for every $t \in T$, $H_t$ is almost $H_K$-projective and hence $H_T$ is almost $H_K$-projective (see 4.44). $\square$

Note that if a decomposition $M = \bigoplus_I M_i$ complements direct summands (see 12.4), then each $M_i$ is indecomposable. Hence an indecomposable decomposition $M = \bigoplus_I M_i$ complements direct summands if and only if the decomposition is an exchange decomposition. It then follows from 12.5 (3) that if $N$ is an indecomposable module and $M = N \oplus N$ is an exchange decomposition, then $N$ is an LE-module. Consequently we have the following corollary.

**23.18. Corollary.** *Suppose $H$ is a hollow module and $H$ is generalised (almost) $H$-projective. Then $H$ is an LE-module.*

**23.19. Example.** In general, if a finite direct sum of hollow modules is lifting, then the summands need not be LE-modules. This is seen by the following example from [99].

Consider the ring $R$ and its left ideal $L$ defined in 12.7, thus

$$R = \begin{bmatrix} \mathbb{Z}_{(p)} & \mathbb{Q} \\ 0 & \mathbb{Z}_{(q)} \end{bmatrix} \quad \text{and} \quad L = \begin{bmatrix} \mathbb{Z}_{(p)} & \mathbb{Z}_{(q)} \\ 0 & 0 \end{bmatrix}.$$

Now define the ring $S$ by $S = R_1 \times R_2$ where $R_1 = R_2 = R$. Let $M_k$ be an $R_k$-module isomorphic to $R/L$, for $k = 1, 2$, and $_SN = M_1 \oplus M_2$. Then $M_1$ and $M_2$ are local and relatively projective as $S$-modules (they are not isomorphic as $S$-modules) and hence, by 23.16, $N$ is lifting. However $M_1$ and $M_2$ are not LE-modules (see 12.7).

From 11.35 we get that if $M = \bigoplus_{i=1}^n M_i$ where each $M_i$ is an LE-module, then $M$ has the internal exchange property and hence the decomposition $M = \bigoplus_{i=1}^n M_i$ is exchangeable. Hence 23.16 implies the following.

**23.20. Direct sums of LE-modules.** *Let $M = M_1 \oplus \cdots \oplus M_n$ where each $M_i$ is a hollow LE-module. Then $M$ is a lifting module if and only if $M_i$ is almost (generalised) $M_j$-projective, for all $i \neq j$, $1 \leq i, j \leq n$.*

It is known (see [85, 7.11]) that a module $N \in \sigma[M]$ is extending if and only if $N = Z_M^2(N) \oplus N'$ for some submodule $N'$ of $N$, such that $N'$ and $Z_M^2(N)$ are both extending and $Z_M^2(N)$ is $N'$-injective. We now show that the dual is also true, with $\mathrm{Re}_S$ replacing $Z_M$. (See 8.16 for the definition of $\mathrm{Re}_S^\alpha$.)

**23.21. Lifting modules and $\mathrm{Re}_S^2$.** *Let $M$ be an $R$-module. A module $N \in \sigma[M]$ is lifting if and only if $N = \mathrm{Re}_S^2(N) \oplus N'$ for some submodule $N'$ of $N$, such that $N'$ and $\mathrm{Re}_S^2(N)$ are lifting, $N'$ is $\mathrm{Re}_S^2(N)$-projective, and $N$ is amply supplemented.*

*Proof.* Let $N \in \sigma[M]$ be a lifting module. By 22.2, $N$ is amply supplemented and every coclosed submodule of $N$ is a direct summand. As $N$ is amply supplemented, every submodule of $N$ has a coclosure in $N$. Then, by 8.19, $\mathrm{Re}_S^2(N)$ is coclosed in $N$ and is the largest fully non-$M$-small submodule of $N$. Consequently $N =$

$\text{Res}_{\mathcal{S}}^2(N) \oplus N'$, where $\text{Res}_{\mathcal{S}}^2(N') = 0$. It now suffices to prove that $N'$ is $\text{Res}_{\mathcal{S}}^2(N)$-projective. For this, let $f : \text{Res}_{\mathcal{S}}^2(N) \to \text{Res}_{\mathcal{S}}^2(N)/B$ be an epimorphism and let $g : N' \to \text{Res}_{\mathcal{S}}^2(N)/B$ be any homomorphism. Define

$$L := \{x + y \mid x \in N', \, y \in \text{Res}_{\mathcal{S}}^2(N) \text{ and } (x)g + (y)f = 0\}.$$

Then $N = L + \text{Res}_{\mathcal{S}}^2(N)$, and so we may assume that $L \not\ll N$. Since $N$ is lifting, there exists a direct summand $X$ of $N$ such that $X \subseteq L$, $N = X \oplus Y$ say, and $L/X \ll N/X$. Then $N = X + \text{Res}_{\mathcal{S}}^2(N)$ (by 3.2 (1)) and so $Y$ is an epimorphic image of $\text{Res}_{\mathcal{S}}^2(N)$ and hence fully non-$M$-small by 8.12 (5). By 20.35, $Y \subseteq \text{Res}_{\mathcal{S}}^2(N)$. Hence $\text{Res}_{\mathcal{S}}^2(N) = Y \oplus (X \cap \text{Res}_{\mathcal{S}}^2(N))$. Then there is a summand $Z$ of $N$ such that

$$X = Z \oplus (X \cap \text{Res}_{\mathcal{S}}^2(N)) \text{ and } N = Z \oplus \text{Res}_{\mathcal{S}}^2(N).$$

Let $\pi : N \to \text{Res}_{\mathcal{S}}^2(N)$ be the projection map along $Z$ and let $h = \pi|_{N'}$. Then $hf = g$ and so $N'$ is $\text{Res}_{\mathcal{S}}^2(N)$-projective.

Conversely, suppose $N$ is an amply supplemented module such that $N = \text{Res}_{\mathcal{S}}^2(N) \oplus N'$, where $\text{Res}_{\mathcal{S}}^2(N)$ and $N'$ are lifting and $N'$ is $\text{Res}_{\mathcal{S}}^2(N)$-projective. Then $\text{Res}_{\mathcal{S}}^2(N') = 0$ and so, by 20.35, $\text{Res}_{\mathcal{S}}^2(N'/L) = 0$ for any $L \subseteq N'$. Since $\text{Res}_{\mathcal{S}}^2(N)$ is fully non-$M$-small, each of its epimorphic images is also fully non-$M$-small (see 8.12 (5)) and so $\text{Hom}(\text{Res}_{\mathcal{S}}^2(N), N'/L) = 0$. Thus $\text{Res}_{\mathcal{S}}^2(N)$ is $N'$-projective. By 23.15, $N$ is a lifting module. $\qquad\square$

### 23.22. Exercises.

(1) Suppose $M$ is a supplemented module and every supplement submodule is a direct summand ($M$ is strongly $\oplus$-supplemented). Show that $M$ is lifting if and only if every direct summand of $M$ has ample supplements in $M$.

(2) Let $M = K \oplus L$. We say $K$ *is *generalised $L$-projective* if for any supplement $P$ of $L$ in $M$, $M$ decomposes as $M = P \oplus K' \oplus L'$, with $K' \subseteq K$ and $L' \subseteq L$. Prove:

   (i) Let $K'$ be a direct summand of $K$ and $L'$ be a direct summand of $L$. If $K$ is *generalised $L$-projective, then $K'$ is *generalised $L'$-projective.

   (ii) Let $M = K \oplus L$ where $K$ is *generalised $L$-projective and $L$ is lifting. If $X$ is a supplement in $M$ such that $X + L = M$, then $M = X \oplus K' \oplus L'$, with $K' \subseteq K$ and $L' \subseteq L$.

   (iii) Let $M = K \oplus L$ where $K$ and $L$ are lifting modules. Then $K$ and $L$ are mutually *generalised projective if and only if every coclosed submodule of $M$ is a summand and $M$ has the internal exchange property for the decomposition $M = K \oplus L$.

   (iv) Let $M = K \oplus L$. If $K$ is generalised $L$-projective, then $K$ is *generalised $L$-projective. The converse holds good if $L$ has ample supplements in $M$.

   (v) Let $M = K \oplus L$ be supplemented. Then $K$ is generalised $L$-projective if and only if $K$ is *generalised $L$-projective and $L$ has ample supplements in $M$.

**23.23. Comments.** As we have observed, direct sums of lifting modules need not necessarily be lifting. In order to express the lifting property of a direct sum of modules in terms of homomorphisms, the concept of almost $M$-projectivity was introduced by Harada and Tozaki [163] and they show that if $M = \bigoplus_I M_i$ is a lifting module where each $M_i$ is a LE hollow module, then $M_i$ is almost $M/M_i$-projective for all $i \in I$. Baba and Harada [30] proved that a module $M = \bigoplus_{i=1}^n M_i$ where each $M_i$ is a hollow LE-module is lifting if and only if $M_i$ is almost $M_j$-projective for $i \neq j$.

The concept of generalised relative injectivity (dual of the concept of generalised relative projectivity) was introduced by Hanada, Kuratomi and Oshiro in [143] to study the extending property of a module $M = \bigoplus_{i=1}^n M_i$. Mohamed and Müller prove in [244, Theorem 11] that a module $M = \bigoplus_{i=1}^n M_i$ is an extending module and the decomposition $M = \bigoplus_{i=1}^n M_i$ is an exchange decomposition if and only if each $M_i$ is extending and $M_i$ and $M/M_i$ are relatively generalised injective for $1 \leq i \leq n$ (see also [144, Theorem 2.11]). Generalised relative injectivity was renamed as *ojective* by Mohamed and Müller (in [244]) in honour of Oshiro. They also dualised the concept of ojective modules in two ways, namely to *cojective* and *\*cojective* modules (see [245]). $M$-cojectivity is precisely the same as generalised $M$-projectivity as used by Kuratomi [214] and $M$-\*cojectivity is precisely the same as *generalised $M$-projectivity (see 23.22(2)). The results in 23.22(2) are from [245]. If $A$ is generalised $B$-injective and $A', B'$ are direct summands of $A$ and $B$ respectively, then $A'$ is generalised $B'$-injective. We do not know whether the dual is true for generalised relative projective modules and hence we do not know whether the dual of [244, Theorem 11]) is true. Kuratomi [214] has proved equivalent conditions for a module with exchange decomposition $M = \bigoplus_{i=1}^n M_i$ to be lifting in terms of relative generalised projectivity of direct summands of $M$ (see 23.14). Using this we prove equivalent conditions for a finite direct sum of hollow modules to be lifting in 23.16. By example 23.19, this is a generalisation of Baba–Harada's result [30, Theorem 1]. We have neither an example of a lifting module $M = M_1 \oplus M_2$ where the decomposition $M = M_1 \oplus M_2$ is not an exchange decomposition nor an example of a lifting module which is not a direct sum of indecomposables.

**References.** Baba and Harada [30]; Hanada, Kuratomi and Oshiro in [143]; Harada [156]; Harada and Mabuchi [161]; Harada and Tozaki [163]; Keskin [199]; Kuratomi [214]; Mohamed and Müller [244, 245]; Orhan and Keskin Tütüncü [268].

# 24 The lifting property for infinite sums

In this section we study the lifting property for modules with LE-decompositions. In view of the material in § 12, it is not surprising that local semi-T-nilpotency and local direct summands play significant rôles. First we consider the lifting property of a module $M$ with an indecomposable decomposition that complements maximal direct summands. Recall the following result [363, 43.3] regarding a sequence of homomorphisms.

**24.1. Sequences of homomorphisms.** *Let* $\{f_i : N_i \to N_{i+1}\}_{i \in \mathbb{N}}$ *be a family of $R$-module homomorphisms and* $N = \bigoplus_{i \in \mathbb{N}} N_i$.

*With the canonical inclusions* $\varepsilon_i : N_i \to N$, *define*

$$g_i : N_i \to N \quad \text{by} \quad g_i = \varepsilon_i - f_i \varepsilon_{i+1} \quad \text{and}$$
$$g : N \to N \quad \text{with} \quad \varepsilon_i g = g_i, \ i \in \mathbb{N}.$$

(1) $\{\operatorname{Im} g_i\}_{i \in \mathbb{N}}$ *is an independent family of submodules and, for every $k \in \mathbb{N}$, the partial sum* $\bigoplus_{i \le k} \operatorname{Im} g_i$ *is a direct summand of $N$.*

(2) *If* $\sum_{i \in \mathbb{N}} \operatorname{Im} f_i \varepsilon_{i+1} \ll N$, *then, for every $m \in N_1$, there exists $r \in \mathbb{N}$ with* $(m)f_1 \cdots f_r = 0$.

(3) *If $\operatorname{Im} g$ is a direct summand of $N$, then, for any $m_1, \ldots, m_t \in N_1$, there exists $r \in \mathbb{N}$ and $h_{r+1,r} \in \operatorname{Hom}(N_{r+1}, N_r)$ with*

$$(m_i)f_1 \cdots f_{r-1} = (m_1)f_1 \cdots f_r h_{r+1,r} \quad \text{for } i = 1, \ldots, t.$$

*If $N_i$ is finitely generated, then* $f_1 \ldots f_{r-1} = f_1 \cdots f_r h_{r+1,r}$ *for some $r \in \mathbb{N}$.*

The following result is a variation on 12.16 and its proof uses similar techniques. Recall that for a decomposition $M = \bigoplus_I M_i$ and any nonempty subset $K \subseteq I$, we denote $M_K := \bigoplus_K M_k$.

**24.2. Sequences of homomorphisms for a lifting module.** *Let $\{M_i\}_{i \in \mathbb{N}}$ be a family of indecomposable modules and $M = \bigoplus_{i \in \mathbb{N}} M_i$ satisfy the following conditions:*

(i) *$M$ is a lifting module,*

(ii) *every nonzero direct summand of $M$ contains an indecomposable direct summand, and*

(iii) *the decomposition $M = \bigoplus_{i \in \mathbb{N}} M_i$ complements maximal direct summands.*

*Then, given a sequence of non-isomorphisms $f_i : M_i \to M_{i+1}$ and $m \in M_1$, there exists $t \in \mathbb{N}$ such that $(m)f_1 \cdots f_t = 0$.*

*Proof.* For each $i \in \mathbb{N}$, define $M_i' = \{x - (x)f_i \mid x \in M_i\}$. Then (cf. the proof of 12.16), $M_i'$ is a direct summand of $M$, with

$$M_i' \simeq M_i \quad \text{and} \quad M_i \oplus M_i' = M_i \oplus M_{i+1}.$$

Moreover $M_i' \cap M_i = 0$ if and only if $f_i$ is a monomorphism.

Now assume that none of the $f_i$ are monomorphisms. Put $X = \bigoplus_{i \in \mathbb{N}} M_i'$. As $M$ is lifting and $X$ is nonsmall in $M$ (since $M_1 + X = M$), we have $M = X_1 \oplus X_2$ with $X = X_1 \oplus (X_2 \cap X)$ and $X_2 \cap X \ll M$.

Since $M_1 + X = M = X_1 \oplus X_2$, we have $M/X \simeq M_1/(M_1 \cap X) \simeq X_2/(X_2 \cap X)$. Then, since $M_1$ is hollow and $X_2 \cap X \ll X_2$, it follows that $X_2$ is hollow and so indecomposable.

We claim that $M = X$. By way of contradiction, suppose $0 \neq x \in X_2 \cap X$. Then there exists $k \in \mathbb{N}$ such that $x \in M_1 \oplus \cdots \oplus M_k$. It follows from 12.5 (4) (taking there $t = 1$ and $N_1 = X_2$), that $X_1$ has an indecomposable decomposition, say $X_1 = \bigoplus_J D_j$. Then, by 12.5 (2) the indecomposable decomposition $M = \bigoplus_J D_j \oplus X_2$ also complements maximal direct summands and so, by 12.5 (4), we have

$$M = M_1 \oplus \cdots \oplus M_k \oplus \bigoplus_{j \in J'} D_j \oplus X_2',$$

where $J' \subseteq J$ and $X_2'$ is either 0 or $X_2$. Since $X_2 \cap (M_1 \oplus \cdots \oplus M_k) \neq 0$ by our assumption, we must have $X_2' = 0$. Thus

$$M = M_1 \oplus \cdots \oplus M_k \oplus \bigoplus_{j \in J'} D_j,$$

yielding another indecomposable decomposition, which again complements maximal direct summands by 12.5 (2). Now $M_1' \oplus \cdots \oplus M_k'$ is a direct summand of $M$ and is a finite direct sum of indecomposable modules. Note that $M_i \cap M_i' \neq 0$ for each $i \in \mathbb{N}$. Hence applying 12.5 (4) again, to our last decomposition of $M$, we get

$$M = M_1' \oplus \cdots \oplus M_k' \oplus D_{J''},$$

where $J'' \subseteq J'$. Since $D_{J'} \subseteq X$ and each $M_i' \subseteq X$, we have $M = X$, as claimed.

Now let $m \in M_1$. Then $m = m_1' + m_2' + \cdots + m_n'$, where $m_i' \in M_i'$, for $1 \leq i \leq n$. Then it quickly follows from the definition of the $M_i'$ that $(m)f_1 f_2 \cdots f_n = 0$. Thus if there exists $n \in \mathbb{N}$ such that for all $i \geq n$, $f_i$ is an epimorphism and $m \in M_1$, then there exists $t \in \mathbb{N}$ such that $(m)f_1 f_2 \cdots f_t = 0$. Assume the contrary. Hence there are infinitely many $i \in \mathbb{N}$ such that $f_i$ is not an epimorphism. Thus there exist $i_1 < i_2 < \cdots i_n < \cdots$, where $i_j \in \mathbb{N}$ is such that $f_{i_j}$ is not an epimorphism for each $j \in \mathbb{N}$. Put $g_k = f_{i_k} f_{i_k+1} \cdots f_{i_{k+1}-1} : M_{i_k} \to M_{i_{k+1}}$. As each $M_i$ is a hollow module, $g_k$ is not an epimorphism for all $k \in \mathbb{N}$.

Put $N_k = M_{i_k}$ and $N = \bigoplus_{k \in \mathbb{N}} N_k$. It is easy to see that it is enough to prove that given $n_1 \in N_1$, there exists a $t \in \mathbb{N}$ such that $(n_1)g_1 g_2 \cdots g_t = 0$. It follows easily from our hypothesis (see, for example, 12.5 (5)) that $N$ is a lifting module, the decomposition $N = \bigoplus_{k \in \mathbb{N}} N_k$ complements maximal direct summands, and any nonzero direct summand of $N$ has an indecomposable direct summand.

For each $k \in \mathbb{N}$, let $\varepsilon_k : N_k \to N$ and $\pi_k : N \to N_k$ denote the canonical inclusions and projections. By 24.1 (2) it is enough to show that $Y = \sum_{\mathbb{N}} \mathrm{Im}\, g_i \varepsilon_{i+1} \ll N$. We note that $(Y)\pi_k \ll N_k$, for all $k$. Suppose $Y$ is not small in $N$. Then, by hypothesis, there exists a nonzero indecomposable direct summand $T$ of $N$ such that $T \subseteq Y$. By 12.8, $N = T \oplus N_{\mathbb{N}\setminus\{j\}}$, for some $j \in \mathbb{N}$. Then $N = T \oplus N_{\mathbb{N}\setminus\{j\}} = N_j \oplus N_{\mathbb{N}\setminus\{j\}}$. Denote by $f$ the restriction of $\pi_j$ to $T$. Given $x \in N_j$, we can write $x = a + b$ with $a \in T$ and $b \in N_{\mathbb{N}\setminus\{j\}}$. Then $x = (x)\pi_j = (a)\pi_j + (b)\pi_j = (a)\pi_j$ and hence $f$ is an epimorphism which contradicts that $(Y)\pi_j \ll N_j$. Thus $Y \ll N$. $\quad\square$

**24.3. Example.** The converse of 24.2 is not true even when each $M_i$ is an LE-module. For example, consider the $\mathbb{Z}$-module $M = \bigoplus_{i \in \mathbb{N}} M_i$, where for all $i \in \mathbb{N}$, $M_i = \mathbb{Z}_{p^\infty}$, the injective hull of the $\mathbb{Z}$-module $\mathbb{Z}/p\mathbb{Z}$, $p$ a prime. By 25.2, $M$ is not a lifting module, even though the family $\{M_i\}_{i \in \mathbb{N}}$ is lsTn.

For ease of reference, we make the following ad hoc definitions.

**24.4. Notation for properties.** We define properties P1, P2, P3 and P4 for a module $M$ as follows:

P1. any indecomposable decomposition of $M$ is an exchange decomposition;

P2. for any indecomposable decomposition $M = \bigoplus_I M_i$, the family $\{M_i\}_I$ is locally semi-T-nilpotent;

P3. every local direct summand of $M$ is a direct summand;

P4. for any submodule $N$ of $M$, the family of all direct summands of $M$ contained in $N$ has a maximal element under set inclusion.

**24.5. Lemma.** *Suppose that $M$ is either lifting or every local direct summand of $M$ is a direct summand. Then $M$ satisfies property* P4.

*Proof.* Let $M$ be a lifting module and $N \subseteq M$. Then we can write $N = A \oplus S$, where $M = A \oplus B$ and $S \ll M$. It is easy to see that $M$ cannot contain a direct summand $C$ such that $A \subsetneq C \subseteq N$. Thus $M$ satisfies P4.

Suppose that every local direct summand of $M$ is a direct summand of $M$. Let $N \subseteq M$. By Zorn's Lemma there exists a maximal local direct summand $\bigoplus_I A_i$ such that each $A_i \subseteq N$. It is clear that $A = \bigoplus_I A_i$ is a direct summand of $M$ maximal with respect to the property $A \subseteq N$. $\quad\square$

However a module $M$ satisfying P4 need not be a lifting module. For example, consider the $\mathbb{Z}$-module $M = \bigoplus_{i \in \mathbb{N}} M_i$, where $M_i = \mathbb{Z}_{p^\infty}$ for all $i \in \mathbb{N}$, as in 24.3 above.

As noted in 12.4, an indecomposable decomposition is an exchange decomposition if and only if it complements direct summands. Thus, combining 24.2, 13.6, and 12.5 (2) we get the following.

**24.6. Corollary.** *Suppose the module $M$ has the following properties:*

(i) *$M$ is a lifting module,*

(ii) *every nonzero direct summand of $M$ contains an indecomposable direct summand, and*

(iii) *there exists an indecomposable decomposition which complements maximal direct summands.*

*Then $M$ satisfies* P1, P2, P3 *and* P4.

Now applying 12.6, 24.6 and 12.14 to an LE-decomposition we get

**24.7. Corollary.** *Let $M$ be a lifting module with LE-decomposition $M = \bigoplus_I M_i$. Then $M$ satisfies* P1, P2, P3 *and* P4.

**24.8. Corollary.** *Let $M$ be a lifting module which is a direct sum of hollow modules. Then $M$ has the exchange property if and only if any hollow direct summand of $M$ has a local endomorphism ring.*

*Proof.* Suppose $M$ has the exchange property and $H$ is a hollow direct summand of $M$. By 11.8, $H$ has the exchange property and so, by 12.2, $H$ has local endomorphism ring.

Conversely, if every hollow direct summand of $M$ is an LE-module, then we have an LE-decomposition $M = \bigoplus_I M_i$. Then by 12.6, 24.2 and 12.14, $M$ has the exchange property. $\qquad\square$

Note that we can rephrase (part of) 13.6 as follows.

**24.9. Complementing maximal summands.** *Let $M = \bigoplus_I M_i$ be an indecomposable decomposition which complements maximal direct summands and suppose that every nonzero direct summand of $M$ contains an indecomposable direct summand. Then the three properties* P1, P2, *and* P3 *are equivalent for the module $M$.*

We now look at when the direct sum of a family $\{M_i\}_I$ of hollow modules is lifting. (The finite case was considered in 23.16.) As before, we let $M_K = \bigoplus_K M_k$ for any nonempty subset $K$ of $I$.

**24.10. Lifting property of direct sums of hollow modules.** *Let $M = \bigoplus_I M_i$ where each $M_i$ is a hollow module; suppose that this decomposition complements maximal direct summands, and that every nonzero direct summand of $M$ contains an indecomposable direct summand. Then the following are equivalent:*

(a) *$M$ is a lifting module;*

(b) *$M$ is a lifting module and $M = \bigoplus_I M_i$ is an exchange decomposition;*

(c) *for any nonempty disjoint subsets $K$ and $J$ of $I$, $M_K$ is generalised (almost) $M_J$-projective and $M$ satisfies* P1, P2, P3 *and* P4;

(d) *for each $i \in I$, $M_i$ is generalised (almost) $M_{I\setminus\{i\}}$-projective and $M$ satisfies* P1, P2, P3, *and* P4;

(e) *for each $i \in I$, $M_i$ is generalised (almost) $M_{I\setminus\{i\}}$-projective and $M$ satisfies any one of the properties* P1, P2, P3 *and* P4.

*If further $M_i$ is cyclic for each $i \in I$, then the above are equivalent to:*

(f) *for all $i \neq j$ in $I$, $M_i$ is generalised (almost) $M_j$-projective and $M$ satisfies any one of the properties* P1, P2, P3 *and* P4.

*Proof.* (a)⇔(b) follows from 24.6.

(a)⇒(c). By 24.6, $M$ satisfies each of the properties P1, P2, P3 and P4. Since the decomposition $M = \bigoplus_I M_i$ is exchangeable, it follows from 24.6 that the decomposition $N = M_K \oplus M_J$ is also exchangeable. Then, since $N$ is also lifting, 23.11 shows that $M_K$ and $M_J$ are relatively generalised projective. Similarly, for any $k \in K$, $M_k$ is generalised $M_J$-projective. Since $M_k$ is a hollow module, it is then almost $M_J$-projective. From the definition of almost projectives, it is now easy to deduce that $M_K$ is almost $M_J$-projective.

(c)⇒(d) and (d)⇒(e) are obvious.

(e)⇒(a). Using 24.9 and 24.5 we see that it is enough to prove the case when $M$ satisfies property P4. As a hollow module $H$ is generalised $T$-projective if and only if $H$ is almost $T$-projective, it is enough to prove (a) assuming for all $i \in I$, $M_i$ is generalised $M_{I\setminus\{i\}}$-projective. Let $N$ be a submodule of $M$. By hypothesis, there exists a submodule $A$ of $N$ such that $M = A \oplus B$, $N = A \oplus (N \cap B)$ and $N \cap B$ does not contain any nonzero direct summand of $M$.

To show that $M$ is lifting it is enough to show that $N \cap B \ll M$. Suppose to the contrary that $N \cap B$ is not small in $M$. Then there exists $C \neq M$ such that $M = (N \cap B) + C$. Again by hypothesis, we get a direct summand $D \subseteq C$ of $M$ such that $M = D \oplus E$, $C = D \oplus (C \cap E)$ and $C \cap E$ does not contain any nonzero direct summand of $M$. Since $D \neq M$, $E \neq 0$ and so contains a nonzero indecomposable direct summand $X$ of $M$. Hence we can write $M = D \oplus Y \oplus X$. Let $p : M \to X$ be projection onto $X$ along $D \oplus Y$. By 12.8, $X \simeq M_k$ for some $k \in I$ and $D \oplus Y \simeq M_{I\setminus\{k\}}$. Hence by hypothesis, $X$ is generalised $\mathrm{Ke}\,p$-projective. As $M = (N \cap B) + C$, $X = (N \cap B)p + (C)p$ and since $X$ is hollow, either $X = (N \cap B)p$ or $X = (C)p$. Thus either $M = (N \cap B) + \mathrm{Ke}\,p$ or $M = C + \mathrm{Ke}\,p$.

Suppose first that $M = (N \cap B) + \mathrm{Ke}\,p$. Then, by 4.42, we have

$$M = \overline{N \cap B} \oplus \overline{X} \oplus \overline{\mathrm{Ke}\,p} = \overline{N \cap B} + \mathrm{Ke}\,p \text{ with}$$
$$\overline{N \cap B} \subseteq N \cap B, \ \overline{X} \subseteq X \text{ and } \overline{\mathrm{Ke}\,p} \subseteq \mathrm{Ke}\,p.$$

Since $\overline{N \cap B}$ is a nonzero direct summand of $N \cap B$, this contradicts the choice of $A$. Thus we must have $M = C + \mathrm{Ke}\,p$. As $C = D \oplus (C \cap E)$ and $D \subseteq \mathrm{Ke}\,p$, this gives $M = (C \cap E) + \mathrm{Ke}\,p$. Again by 4.42, we have

$$M = \overline{C \cap E} \oplus \overline{M_k} \oplus \overline{\mathrm{Ke}\,p} = \overline{C \cap E} + \mathrm{Ke}\,p \text{ where}$$
$$\overline{C \cap E} \subseteq (C \cap E), \ \overline{M_k} \subseteq M_k \text{ and } \overline{\mathrm{Ke}\,p} \subseteq \mathrm{Ke}\,p.$$

Then $\overline{C \cap E}$ is a nonzero direct summand of $C \cap E$, which contradicts the choice of $D$. This last contradiction shows that $N \cap B \ll M$ and so $M$ is a lifting module.

(e)$\Rightarrow$(f) is obvious.

(f)$\Rightarrow$(e). Fix $i \in I$. By 23.17, for any finite subset $K \subseteq I$ such that $i \notin K$, $M_i$ is generalised (almost) $M_K$-projective. Since $M_i$ is local, it is easy to verify that $M_i$ is generalised (almost) $M_{I \setminus \{i\}}$-projective.  $\square$

Applying 13.6 to a module $M$ with a decomposition that complements direct summands we infer that $M$ satisfies properties P1, P2, P3 and every nonzero direct summand of $M$ contains an indecomposable direct summand. Also, by 24.5, P3 implies P4. Hence we get the following corollary.

**24.11. Corollary.** *Let $\{M_i\}_I$ be a family of hollow modules such that the decomposition $M = \bigoplus_I M_i$ complements direct summands. Then the following are equivalent:*

(a) *$M$ is lifting;*

(b) *for any nonempty disjoint subsets $K$ and $J$ of $I$, $M_K$ is generalised (almost) $M_J$-projective;*

(c) *for each $i \in I$, $M_i$ is generalised (almost) $M_{I \setminus \{i\}}$-projective.*

*If further $M_i$ is cyclic for each $i \in I$, then the above are equivalent to:*

(d) *for all $i \in I$, $M_i$ is generalised (almost) $M_j$-projective for all $j \in I$ and $j \neq i$.*

If a module $M$ has an indecomposable exchange decomposition, then any two indecomposable decompositions of $M$ are equivalent. Hence we get

**24.12. Corollary.** *Suppose $M$ is a module with an indecomposable exchange decomposition. Then $M$ is lifting if and only if, for each indecomposable direct summand $H$ of $M$, $H$ is hollow and $H$ is generalised (almost) $M/H$-projective.*

By 12.6 and 24.10, we get the following result.

**24.13. Lifting property for a direct sum of hollow LE-modules.** *Let $\{M_i\}_I$ be a family of hollow LE-modules and set $M = \bigoplus_I M_i$. Then the conditions (a) through (f) of 24.10 are equivalent.*

**24.14. Direct sum of mutually projective modules.** *Suppose $M = \bigoplus_I M_i$ is a direct sum of hollow modules such that $\operatorname{Rad} M \ll M$ and $M_i$ is $M_j$-projective for $i \neq j$. Then $M$ is lifting and the decomposition $M = \bigoplus_I M_i$ complements direct summands.*

*Proof.* As $\operatorname{Rad} M \ll M$ each $M_i$ is a local module and $M/\operatorname{Rad} M$ is semisimple. Also for any nonempty disjoint subsets $K$ and $T$ of $I$, $M_K$ is $M_T$-projective. Suppose $A$ is a proper nonsmall submodule of $M$ and $p : M \to M/\operatorname{Rad} M$ is the projection map. As $M/\operatorname{Rad} M = \bigoplus_I (M_i)p$ is semisimple we get $M/\operatorname{Rad} M = (A)p \oplus \bigoplus_K (M_k)p$ for some $\emptyset \neq K \subset I$. Then $I \setminus K \neq \emptyset$. As $\operatorname{Rad} M \ll M$ we have

$M = A + M_K$ with $A \cap M_K \ll M$. Since $M_T$ is $M_K$-projective, there exists a submodule $A'$ of $A$ such that $M = A' \oplus M_K$ (see 4.12). Thus $A = A' \oplus (A \cap M_K)$. By 22.3, $M$ is a lifting module. If $A$ is a direct summand, then $A = A'$ and hence the decomposition $M = \bigoplus_I M_i$ complements direct summands. $\square$

**24.15. Comments.** Baba and Harada [30, Theorem 2] gave equivalent conditions for a direct sum of cyclic hollow LE-modules to be lifting. Suppose $M$ is a module with a decomposition complementing direct summands and such that every nonzero direct summand has an indecomposable direct summand. Using Dung's theorem (see 13.6) we give equivalent conditions for M to be a lifting module (see 24.10). By Azumaya's theorem (see 12.6), this is an improvement of [30, Theorem 2].

**References.** Baba and Harada [30]; Dung [83].

# 25   Σ-lifting modules

**25.1. Definitions.** An $R$-module $M$ is called Σ-*extending* provided any direct sum of copies of $M$ is an extending module, and $M$ is called Σ-*lifting* provided any direct sum of copies of $M$ is a lifting module.

For the properties of Σ-extending modules we refer to [85, Section 11]. In this section we study the properties of Σ-lifting modules and then apply these to Σ-lifting modules $M$ which are either copolyform modules or self-injective modules in $\sigma[M]$.

To begin with we look at the lifting property for a direct sum of indecomposable modules.

**25.2. Σ-lifting indecomposable modules.** *Let $M$ be an LE-module and let $X$ be a generating set for $M$. If $M^{(X)}$ is lifting, then $M$ is cyclic.*

*Proof.* Consider $N = \bigoplus_{x \in X} M_x$, where each $M_x = M$. Define the map $f : N \to M$ by $f|_{M_x} = 1_{M_x}$. Let $T = \bigoplus_{x \in X} T_x$, where $T_x = Rx \subseteq M_x$, for all $x \in X$. As $(T)f = M$, $T$ is a nonsmall submodule of $N$ and hence contains a nonzero direct summand of $N$. Since $N$ is a direct sum of LE-modules, $T$ contains a direct summand $L \simeq M$ (see 12.6). Now as $L$ is an LE-module, $L$ has the exchange property by 12.2. Thus $T = L \oplus \bigoplus_{x \in X} T'_x$, where each $T'_x \subseteq T_x$. Each $T_x$ is cyclic and hence $T_x/T'_x$ is cyclic and $L \simeq \bigoplus_{x \in X} T_x/T'_x$. Hence $M \simeq L$ is cyclic.   □

**25.3. Corollary.** *Let $M$ be an LE-module. If every direct sum of copies of $M$ is lifting, then $M$ is cyclic.*

We now consider the interaction between the Σ-lifting and Σ-extending properties. First we note the following.

**25.4. Indecomposable Σ-extending modules.** *An indecomposable Σ-extending module is Σ-self-injective.*

*Proof.* Suppose $U$ is an indecomposable Σ-extending module. Then $U$ is uniform. Letting $\widehat{U}$ denote the injective hull of $U$ in $\sigma[U]$, there exists an index set $I$ and an epimorphism $f : U^{(I)} \to \widehat{U}$. Since $U \subseteq \widehat{U}$ we can assume that $f$ restricted to a direct summand $U$ of $U^{(I)}$ is the inclusion map and hence $\text{Ke}\, f$ is not essential in $U^{(I)}$. Since $U^{(I)}$ is extending and $\widehat{U}$ is indecomposable, $\widehat{U}$ must be isomorphic to a direct summand of $U^{(I)}$. As $\widehat{U}$ is an LE-module [363, 19.9] it has the exchange property by 12.2 and hence $\widehat{U} \simeq U$. Thus $U$ is self-injective. By [85, 8.10], an indecomposable self-injective module that is Σ-extending is Σ-self-injective.   □

**25.5. Indecomposable Σ-lifting and Σ-extending modules.** *Let $M$ be an indecomposable module which is both Σ-lifting and Σ-extending. Then $M$ is a biuniform Σ-self-injective local LE-module.*

*Proof.* Since $M$ is indecomposable lifting and extending, it is both hollow and uniform. As $M$ is $\Sigma$-extending, by 25.4, it is $\Sigma$-self-injective. Then, since $M$ is an indecomposable quasi-injective module, $\mathrm{End}(M)$ is local. By 25.2, $M$ is a local module. $\qquad\square$

**25.6. Projective $\Sigma$-extending modules.** *Let $M$ be a $\Sigma$-extending module which is projective in $\sigma[M]$. Then $M$ is a $\Sigma$-lifting module and $\widehat{M}$ is a $\Sigma$-$M$-injective projective module in $\sigma[M]$.*

*Proof.* From [85, 11.6 and 11.7 (2)] it follows that $M$ is a direct sum of uniform modules. Given an index set $I$, we may then write $M^{(I)} = \bigoplus_J M_j$ where each $M_j$ is indecomposable and $J$ is some index set. By 25.4 and its proof, each $M_j$ is a self-injective LE-module. Now [85, 8.13] implies that the family $\{M_j\}_J$ is lsTn. Since each $M_j$ is projective in $\sigma[M]$, [363, 41.4 (3)] implies that each $M_j$ is a local module. Then, by 24.13, $M^{(I)}$ is a lifting module and so $M$ is a $\Sigma$-lifting module. From [85, 11.7 (2)] it follows that $\widehat{M}$ is a $\Sigma$-$M$-injective projective module in $\sigma[M]$. $\qquad\square$

**25.7. Self-projectivity of $\Sigma$-lifting LE-modules.** *Let $M$ be a $\Sigma$-lifting LE-module. Then $M$ is local and self-projective.*

*Proof.* By 25.2 and 22.2, $M$ is a hollow cyclic module and hence local. We claim that any epimorphism from $M$ to $M$ is an isomorphism. Suppose to the contrary that $f : M \to M$ is an onto map which is not injective. For all $n \in \mathbb{N}$, define $M_n = M$ and $f_n = f : M_n \to M_{n+1}$. Then, by 24.2, there exists a positive number $k$ such that $f^k : M_1 \to M_k$ is a zero epimorphism, a contradiction. Now, since $M \oplus M$ is lifting and $\mathrm{End}(M)$ is local, by 24.13, $M$ is almost $M$-projective. As every epimorphism $M \to M$ is bijective, the proof of 4.40 readily shows that $M$ is self-projective. $\qquad\square$

**25.8. Lemma.** *Let $M$ be an indecomposable self-projective lifting $R$-module. If $A$ and $B$ are fully invariant submodules of $M$ such that $M/A \oplus M/B$ is lifting, then either $A \subseteq B$ or $B \subseteq A$.*

*Proof.* As $A$ and $B$ are fully invariant submodules of a self-projective module $M$, $M/A$ and $M/B$ are also self-projective by 4.9. Since $M/A$ and $M/B$ are hollow and self-projective, $\mathrm{End}(M/A)$ and $\mathrm{End}(M/B)$ are local rings. Consider the natural maps $f : M/A \to M/(A+B)$ and $g : M/B \to M/(A+B)$. Since the module $M/A \oplus M/B$ is lifting, $M/A$ is almost $M/B$-projective by 23.20. Thus, since $M/B$ is an indecomposable module, we get either a map $h : M/A \to M/B$ or a map $h' : M/B \to M/A$, with $hg = f$ or $h'f = g$.

As $f$ and $g$ are small epimorphisms, the maps $h$ and $h'$ (if they exist) will be epimorphisms. Hence there is either an epimorphism $h : M/A \to M/B$ with $hg = f$ or an epimorphism $h' : M/B \to M/A$ with $h'f = g$. By 4.10 (1), either $A \subseteq B$ or $B \subseteq A$. $\qquad\square$

Recall that a submodule $B$ of a module $A$ is called a *waist* if for every submodule $C$ of $A$ we have either $B \subseteq C$ or $C \subseteq B$.

**25.9. Indecomposable summands of $\Sigma$-lifting modules I.** *Let $M = \bigoplus_I M_i$ be a $\Sigma$-lifting module, where each $M_i$ is an LE-module. Set $S = \mathrm{End}(M)$. Let $N$ be a nonzero indecomposable direct summand of $M$ and define $A = \mathrm{Hom}(M, N)$ and $B = \{f \in A \mid \mathrm{Im}\, f \ll N\}$. Then $B$ is a waist of the left $S$-module $A$ such that $A/B$ is a uniserial left $S$-module.*

*Proof.* Since each $M_i$ is indecomposable $\Sigma$-lifting, by 25.7, $M_i$ is local and self-projective. We note that any nonzero $f \in A$ such that $f \notin B$ is an epimorphism. To prove that $B$ is a waist of $A$, it is enough to prove that, for any epimorphism $f \in A$ and for any map $g \in B$, $Sg \subseteq Sf$.

Now consider such an $f$ and $g$. Since $N$ is a local module, there exists $i_0 \in I$ such that $f_{i_0} : M_{i_0} \to N$ is onto, where $f_{i_0}$ is the restriction of $f$ to $M_{i_0}$. For every $i \in I$, let $g_i : M_i \to N$ be the restriction of $g$ to $M_i$. As $M_i \oplus M_{i_0}$ is lifting, $M_i$ and $M_{i_0}$ are relatively almost projective modules (see 23.20). Also $g_i$ is not an epimorphism. Hence, for every $i \in I$, there exists $\phi_i : M_i \to M_{i_0}$ such that $\phi_i f_{i_0} = g_i$. Define $\phi : M \to M$ by $\phi|_{M_i} = \phi_i$ for every $i \in I$. Then $\phi f = g$ and this gives $Sg \subseteq Sf$. Thus $B$ is a waist of $A$.

To prove that $A/B$ is a uniserial left $S$-module, it is enough to show that whenever $f, g : M \to N$ are onto maps, either $Sf \subseteq Sg$ or $Sg \subseteq Sf$. For $i \in I$, put $[i] = \{j \in I \mid M_i \simeq M_j\}$ and $\mathcal{F} = \{[i] \mid i \in I\}$. Define an order on $\mathcal{F}$ by

$$[i] \leq [j] \text{ if and only if there exists an onto map } M_i \to M_j.$$

We claim that $(\mathcal{F}, \leq)$ is a partially ordered set. Assume that $[i] \leq [j]$ and $[j] \leq [i]$ for some $i, j \in I$. Then there exist onto maps $\theta : M_i \to M_j$ and $\psi : M_j \to M_i$. Since $M_i$ is self-projective, the map $\theta\psi : M_i \to M_i$ splits. Therefore $M_i \simeq M_j$ and hence $[i] = [j]$. Now it is easy to see that $(\mathcal{F}, \leq)$ is a partially ordered set. As each $M_i$ is local and $M$ is lifting, by applying 24.2, we get that every nonempty subset $\mathcal{G}$ of $\mathcal{F}$ has a maximal element. Put

$$I_f = \{i \in I \mid f_i = f|_{M_i} \text{ is onto }\} \text{ and } I_g = \{j \in I \mid g_j = g|_{M_j} \text{ is onto }\}.$$

Since $N$ is local, $I_f \neq \emptyset$ and $I_g \neq \emptyset$. Define

$$\overline{I}_f := \{[i] \mid i \in I_f\} \text{ and } \overline{I}_g := \{[i] \mid i \in I_g\}.$$

Suppose $[i_0]$ is a maximal element of $\overline{I}_f \cup \overline{I}_g$. Without loss of generality we can assume that $i_0 \in I_f$. We claim that $Sg \subseteq Sf$.

Let $f_{i_0} : M_{i_0} \to N$ be the restriction of $f$ to $M_{i_0}$ and $g_i : M_i \to N$ be the restriction of $g$ to $M_i$ for every $i \in I$. As before $M_i$ and $M_{i_0}$ are relatively almost projective modules for every $i \in I$. Hence if $g_i$ is not onto, then there exists $\phi_i : M_i \to M_{i_0}$ such that $\phi_i f_{i_0} = g_i$. If $g_i$ is onto, then either there exists an onto map $\phi_i : M_i \to M_{i_0}$ such that $\phi_i f_{i_0} = g_i$ or an onto map $\psi_i : M_{i_0} \to M_i$ such that $\psi_i g_i = f_{i_0}$.

By the choice of $i_0$, the existence of an onto map $\psi_i$ from $M_{i_0}$ to $M_i$ will imply that $M_i \simeq M_{i_0}$. Since $M_i$ is self-projective, the map $\psi_i$ is an isomorphism. Hence we always get a map $\phi_i : M_i \to M_{i_0}$ such that $\phi_i f_{i_0} = g_i$. Define $\phi : M \to M$ by $\phi|_{M_i} = \phi_i$ for every $i \in I$. It is obvious that $\phi f = g$. Hence $Sg \subseteq Sf$.  □

**25.10. Indecomposable summands of Σ-lifting modules II.** *Let $M$ be a Σ-lifting module such that every indecomposable direct summand of $M$ is an LE-module. Let $N$ be an indecomposable direct summand of $M$, $\mathcal{K} = \{f : N \to M \mid \operatorname{Im} f \not\ll M\}$ and $\mathcal{A} = \{\operatorname{Ke} f \mid f \in \mathcal{K}\}$.*

(1) *$\mathcal{A}$ is linearly ordered by inclusion.*

(2) *$N$ has ACC on $\mathcal{A}$.*

(3) *$N$ has DCC on $\mathcal{A}$, provided $M$ has only finitely many non-isomorphic indecomposable direct summands.*

*Proof.* By 25.7, any indecomposable direct summand of $M$ is a self-projective local module.

(1) Let $f \in \mathcal{K}$. As $\operatorname{Im} f$ is hollow, by 3.7 (4), $\operatorname{Im} f$ is coclosed in $M$. Since $M$ is lifting, $\operatorname{Im} f$ is an indecomposable direct summand of $M$ (see 22.2). Hence $\operatorname{Im} f$ is local and self-projective. Since $\operatorname{Ke} f \ll N$ and $N/\operatorname{Ke} f$ is self-projective, $\operatorname{Ke} f$ is fully invariant in $N$ (see 4.9 (2)).

Now suppose that $f, g \in \mathcal{K}$. Then, as shown, $\operatorname{Ke} f$ and $\operatorname{Ke} g$ are fully invariant in $N$ and $N/\operatorname{Ke} f \oplus N/\operatorname{Ke} g$ (as it is isomorphic to a direct summand of $M \oplus M$) is a lifting module. By 25.8, either $\operatorname{Ke} f \subseteq \operatorname{Ke} g$ or $\operatorname{Ke} g \subseteq \operatorname{Ke} f$. Hence $\mathcal{A}$ is linearly ordered by inclusion.

(2) Now suppose that there exists a strictly ascending chain

$$X_1 \subsetneqq X_2 \subsetneqq \cdots \subsetneqq X_i \subsetneqq X_{i+1} \subsetneqq \cdots$$

of elements in $\mathcal{A}$. Then there exists $f_i : N \to M$ such that $\operatorname{Ke} f_i = X_i$ and $\operatorname{Im} f_i \not\ll M$ for every $i \in \mathbb{N}$. For each $i \in \mathbb{N}$, $N/X_i$ is isomorphic to a direct summand of $M$ and hence is a Σ-lifting, self-projective, local LE-module. As $M$ is Σ-lifting we have $\bigoplus_{i \in \mathbb{N}} N/X_i$ is a lifting module with $\operatorname{End}(N/X_i)$ local for all $i \in \mathbb{N}$. With the natural maps $\eta_i : N/X_i \to N/X_{i+1}$ for all $i \in \mathbb{N}$ and the generator of $N/X_1$, we get a contradiction to 24.2. Hence $N$ satisfies ACC on $\mathcal{A}$.

(3) Suppose that

$$Y_1 \supsetneqq Y_2 \supsetneqq \cdots \supsetneqq Y_i \supsetneqq Y_{i+1} \supsetneqq \cdots$$

is a strictly descending chain of elements in $\mathcal{A}$. Each $N/Y_i$ is isomorphic to some indecomposable direct summand of $M$. As there are only finitely many non-isomorphic indecomposable direct summands of $M$, we get $N/Y_\ell \simeq N/Y_k$ for some

$k$ and $\ell$. Suppose $k < \ell$. Then $Y_\ell \subsetneq Y_k$. Since $N/Y_\ell \simeq N/Y_k$ is self-projective, the natural map $f : N/Y_\ell \to N/Y_k$ splits. Therefore $Y_k/Y_\ell$ is a nonzero proper direct summand of the hollow module $N/Y_\ell$, which is a contradiction.                                    $\square$

We apply the above results to a copolyform $\Sigma$-lifting module. First we will show that a copolyform $\Sigma$-lifting module $M$ is a fully non-$M$-small module.

Recall that for any $R$-modules $M$, $N$ and subset $X \subseteq \mathrm{Hom}(N, M)$,

$$\mathrm{Ke}\,(X) = \bigcap \{\mathrm{Ke}\,g \mid g \in X\}$$

is called an *M-annihilator submodule* of $N$. We denote the set of all these sub-modules of $N$ by $\mathcal{K}(N, M)$ (see [363, § 28]).

**25.11. Lemma.** *Let $M$ be a copolyform $\Sigma$-lifting module. Define $N := M^{(I)}$ for some set $I$. If $X$ is a direct summand of $N$ and $f : N \to X$ a homomorphism, then $(A)f$ is coclosed in $X$ whenever $A$ is coclosed in $N$.*

*Proof.* Suppose $A$ is coclosed in $N$. Since $N$ is a lifting module, $A$ is a direct summand of $N$ (see 22.2) and hence $N = A \oplus B$. Consider the homomorphism $g : N \to X$ such that $g = f$ on $A$ and $g = 0$ on $B$. Then $(A)g = (A)f = (N)g$. Moreover, since $X$ is a lifting module, there exists a direct summand $Y \subseteq (A)g$ such that $X = Y \oplus Z$ and $(A)g = Y \oplus [(A)g \cap Z]$ with $(A)g \cap Z \ll Z$. This gives a map $h := gp : N \to Z$, where $p : X \to Z$ is the projection map along $Y$. Now $\mathrm{Im}\,h = (N)gp = (A)g \cap Z \ll Z$.

Write $N = \bigoplus_I M_i$ with $M_i := M$. Suppose $\pi_i : N \to M_i$ and $\varepsilon_i : M_i \to N$ are the natural projection and inclusion maps for any $i \in I$. Let $p_i = \pi_i|_Z$. Then we get a homomorphism $h_{ij} = \varepsilon_j h p_i : M_j \to M_i$, for each $i, j \in I$. Since $M$ is copolyform and $\mathrm{Im}\,h_{ij} \ll M_i$ we get $h_{ij} = 0$, for all $i, j \in I$. Hence $h = 0$. Therefore $(A)g \cap Z = 0$. Thus $(A)f = (A)g = Y$ is a direct summand of $X$, and hence $(A)f$ is coclosed in $X$.                                    $\square$

**25.12. Copolyform and fully non-$M$-small modules.** *Suppose $M$ is a copolyform $R$-module. In the following cases $M$ is a fully non-$M$-small module.*

(i) *$M$ is a $\Sigma$-UCC and $\Sigma$-weakly supplemented module.*

(ii) *$M$ is a $\Sigma$-lifting $R$-module.*

*Proof.* (i) Suppose $M$ is a $\Sigma$-UCC and $\Sigma$-weakly supplemented module. Then $M$ is $\Sigma$-amply supplemented by 20.25. By 21.11, $M$ is $\Sigma$-copolyform. Now (i) follows from 17.20.

(ii) As $M$ is $\Sigma$-lifting, $M$ is $\Sigma$-amply supplemented. By (i) it is enough to prove that $M$ is $\Sigma$-UCC. Suppose $I$ is any indexing set and $N = \bigoplus_I M_i$, where $M_i = M$ for every $i \in I$. To show that $N$ is a UCC module it is enough to prove that if $f : N \to N/K$ is an epimorphism and $A$ is coclosed in $N$, then $(A)f$ is coclosed in $N/K$ (see 21.6).

Suppose that $f : N \to N/K$ is an epimorphism and $A$ is coclosed in $N$. As $N$ is a lifting module, $N = L \oplus L'$, where $L \subseteq K$ and $K = L \oplus (K \cap L')$ with $K \cap L' \ll L'$. We have an isomorphism $\phi : N/K \to L'/(L' \cap K)$. Also $f\phi = p\eta$, where $p : N \to L'$ is the projection along $L$, and $\eta : L' \to L'/(L' \cap K)$ the natural map. Our aim is to prove that $(A)f$ is coclosed in $N/K$. It is enough to show that $(A)f\phi = (A)p\eta$ is coclosed in $L'/(L' \cap K)$, as $\phi$ is an isomorphism. By 25.11, $(A)p$ is coclosed in $L'$. Now since $\operatorname{Ke} \eta \ll L'$, $(A)p\eta$ is coclosed in $L'/(L' \cap K)$ (see 3.7). Thus $N$ is UCC module.                                                               □

By 22.22 a copolyform lifting module is a direct sum of indecomposables. We consider copolyform Σ-lifting modules with an LE-decomposition.

**25.13. Indecomposable summands of Σ-lifting copolyform modules I.** *Suppose $M$ is a copolyform Σ-lifting module with an LE-decomposition and $N$ is an indecomposable direct summand of $M$. Then we have the following:*

(1) *$N$ is local and self-projective and $\operatorname{End}(N)$ is a division ring.*

(2) *$\{\operatorname{Ke} f \mid f : N \to M\}$ is linearly ordered by set inclusion.*

(3) *$\mathcal{K}(N, M)$ is linearly ordered by set inclusion.*

(4) *$N$ has ACC on $\mathcal{K}(N, M)$.*

(5) *$N$ has DCC on $\mathcal{K}(N, M)$, provided $M$ has only finitely many non-isomorphic indecomposable direct summand submodules.*

*Proof.* Since $M$ is copolyform and Σ-lifting, by 25.12, $M$ is a fully non-$M$-small module and hence any direct sum of copies of $M$ is also fully non-$M$-small. Thus $M$ is Σ-copolyform (see 9.23) and amply supplemented. By 21.8, $M$ is then a Σ-UCC module.

(1) Let $N$ be an indecomposable direct summand of $M$. By 24.7 any LE-decomposition of $M$ is an exchange decomposition. Hence $N$ is an LE-module. Therefore $N$ is local and self-projective (see 25.7). Suppose that $f : N \to N$ is a nonzero homomorphism. As $N$ is copolyform, $f$ is an epimorphism by 9.18. Let $L = \bigoplus_{i \in \mathbb{N}} N_i$ with $N_i = N$ for every $i \in \mathbb{N}$. Then $L$ is a UCC lifting module, and hence the sum of any family of coclosed submodules of $L$ is coclosed in $L$ (see 21.6). As $L$ is also lifting, any local direct summand of $L$ is a direct summand of $L$. Now consider $f_i = f : N_i \to N_{i+1}$ for every $i \in \mathbb{N}$. By 24.1, there exist $r \in \mathbb{N}$ and a nonzero map $h_r : N_{r+1} \to N_r$ such that $f_1 \ldots f_{r-1} = f_1 \ldots f_r h_r$. Since $f_i$ is onto for every $i \in \mathbb{N}$, $f_r h_r$ is the identity map on $N_r$. Hence $f_r$ is bijective and thus $\operatorname{End}(N)$ is a division ring. Now, by 25.7, $N$ is local and self-projective.

(2) As $M$ is fully non-$M$-small, every direct summand $N$ of $M$ is also fully non-$M$-small (see 8.12). Therefore, if $f \in \operatorname{Hom}(N, M)$ is a nonzero map, then $\operatorname{Im} f \not\ll M$. Hence if $\mathcal{A} = \{\operatorname{Ke} f \mid \operatorname{Im} f \not\ll M \}$, then $\mathcal{A} \cup N$ is linearly ordered (25.10). That is, for $f, g \in \operatorname{Hom}(N, M)$, either $\operatorname{Ke} f \subseteq \operatorname{Ke} g$ or $\operatorname{Ke} g \subseteq \operatorname{Ke} f$.

(3) Let $I_1, I_2$ be nonzero subsets of $\mathrm{Hom}(N, M)$ such that $\mathrm{Ke}\,(I_1) \not\subseteq \mathrm{Ke}\,(I_2)$. Choose a nonzero $f \in I_1$ and suppose that $\mathrm{Ke}\,g \not\subseteq \mathrm{Ke}\,f$ for all $g \in I_2$. Then, by (2), we must have $\mathrm{Ke}\,f \subseteq \mathrm{Ke}\,g$ for all such $g$ and so $\mathrm{Ke}\,f \subseteq \mathrm{Ke}\,(I_2)$. Since this implies that $\mathrm{Ke}\,(I_1) \subseteq \mathrm{Ke}\,(I_2)$, we have a contradiction. Hence, given any $f \in I_1$, there exists $g \in I_2$ such that $\mathrm{Ke}\,g \subsetneq \mathrm{Ke}\,f$ and so, by (2), $\mathrm{Ke}\,f \subseteq \mathrm{Ke}\,g$. From this we deduce that $\mathrm{Ke}\,(I_2) \subseteq \mathrm{Ke}\,(I_1)$. Thus $\mathcal{K}(N, M)$ is linearly ordered by set inclusion.

(4) Suppose

$$\mathrm{Ke}\,(I_1) \subsetneq \mathrm{Ke}\,(I_2) \subsetneq \cdots \subsetneq \mathrm{Ke}\,(I_r) \subsetneq \mathrm{Ke}\,(I_{r+1}) \subsetneq \cdots$$

is a strictly ascending chain of $M$-annihilator submodules. Consider $r \in \mathbb{N}$. As $\mathrm{Ke}\,(I_r) \subsetneq \mathrm{Ke}\,(I_{r+1})$, there exists $f_r \in I_r$ such that for every $g \in I_{r+1}$, $\mathrm{Ke}\,f_r \subsetneq \mathrm{Ke}\,g$; for if not, then by (2), for every $f \in I_r$, there exists some $g \in I_{r+1}$ such that $\mathrm{Ke}\,g \subseteq \mathrm{Ke}\,f$ and hence $\mathrm{Ke}\,(I_{r+1}) \subseteq \mathrm{Ke}\,(I_r)$, a contradiction. Hence for each $r \in \mathbb{N}$ we get $f_r \in I_r$ such that

$$\mathrm{Ke}\,f_1 \subsetneq \mathrm{Ke}\,f_2 \subsetneq \cdots \subsetneq \mathrm{Ke}\,f_r \subsetneq \mathrm{Ke}\,f_{r+1} \subsetneq \cdots,$$

which is a contradiction to 25.10 (2). So $N$ satisfies ACC on $\mathcal{K}(N, M)$.

(5) Suppose that

$$\mathrm{Ke}\,(I_1) \supsetneq \mathrm{Ke}\,(I_2) \supsetneq \cdots \supsetneq \mathrm{Ke}\,(I_r) \supsetneq \mathrm{Ke}\,(I_{r+1}) \supsetneq \cdots$$

is a strictly descending chain of $M$-annihilator submodules of $N$. Choose any $f_1 \in I_1$. For $r \geq 2$, as in the proof of (3), we can choose $f_{r+1} \in I_{r+1}$ such that for every $f_r \in I_r$, $\mathrm{Ke}\,f_{r+1} \subsetneq \mathrm{Ke}\,f_r$. This produces a strictly descending chain

$$\mathrm{Ke}\,f_1 \supsetneq \mathrm{Ke}\,f_2 \supsetneq \cdots \supsetneq \mathrm{Ke}\,f_r \supsetneq \mathrm{Ke}\,f_{r+1} \supsetneq \cdots,$$

which is a contradiction to 25.10 (3).                                                       $\square$

**25.14. Indecomposable summands of $\Sigma$-lifting copolyform modules II.** *Suppose $M$ is a copolyform $\Sigma$-lifting module with an LE-decomposition and $S = \mathrm{End}(M)$. Then, for any indecomposable direct summand $N$ of $M$, the following hold:*

(1) *$\mathrm{Hom}(N, M)$ is a uniserial, artinian right $S$-module.*

(2) *$\mathrm{Hom}(N, M)$ is a uniserial right $S$-module of finite length, provided $M$ has only finitely many non-isomorphic indecomposable direct summands.*

(3) *$\mathrm{Hom}(M, N)$ is a uniserial left $S$-module.*

(4) *If $M$ is finitely generated, then*

   (i) *$S$ is an artinian serial and right hereditary ring,*

   (ii) *$M$ has a projective cover $P$ in $\sigma[M]$ and $\mathrm{End}(P)$ is a semisimple ring.*

*Proof.* By 25.13 (1), any indecomposable direct summand of $M$ is a local, self-projective LE-module.

(1) Suppose $f, g : N \to M$ are nonzero homomorphisms. By 25.13 (2), either $\text{Ke}\, f \subseteq \text{Ke}\, g$ or $\text{Ke}\, g \subseteq \text{Ke}\, f$. Assume $\text{Ke}\, f \subseteq \text{Ke}\, g$. There exists an onto map $\psi : \text{Im}\, f \to \text{Im}\, g$ such that $f\psi = g$. As $M$ is copolyform and $\Sigma$-lifting, $M$ and hence $N$ is fully non-$M$-small. Therefore both $\text{Im}\, f$ and $\text{Im}\, g$ are coclosed in $M$ (being hollow and not small in $M$) and hence are direct summands of $M$ (for $M$ is lifting). We can then extend $\psi$ to $\phi : M \to M$ such that $f\phi = g$. Therefore $gS \subseteq fS$. Thus $\text{Hom}(N, M)$ is a uniserial right $S$-module.

Let $I_1 \supsetneqq I_2 \supsetneqq \cdots \supsetneqq I_n \supsetneqq \cdots$ be a strictly descending chain of $S$-submodules of $\text{Hom}(N, M)$. For each $n \in \mathbb{N}$, choose $f_n : N \to M$ such that $f_n \in I_n$ and $f_n \notin I_{n+1}$. From above, we have either $f_n S \subseteq f_{n+1} S$ or $f_{n+1} S \subseteq f_n S$ and $f_n \notin I_{n+1}$ implies $f_{n+1} S \subsetneqq f_n S$. This along with either $\text{Ke}\, f_{n+1} \subseteq \text{Ke}\, f_n$ or $\text{Ke}\, f_n \subseteq \text{Ke}\, f_{n+1}$ (see 25.10 (1)) gives us $\text{Ke}\, f_n \subsetneqq \text{Ke}\, f_{n+1}$. Thus we get a strictly ascending chain

$$\text{Ke}\, f_1 \subsetneqq \text{Ke}\, f_2 \subsetneqq \cdots \subsetneqq \text{Ke}\, f_n \subsetneqq \text{Ke}\, f_{n+1} \subsetneqq \cdots,$$

which contradicts 25.10 (2). Therefore $\text{Hom}(N, M)$ is an artinian right $S$-module.

(2) By (1), it is enough to prove that $\text{Hom}(N, M)$ satisfies ACC on right $S$-submodules. Suppose that

$$I_1 \subsetneqq I_2 \subsetneqq \cdots \subsetneqq I_n \subsetneqq \cdots$$

is a strictly ascending chain of $S$-submodules of $\text{Hom}(N, M)$. For each $n \in \mathbb{N}$, there exists $f_{n+1} \subset I_{n+1}$ such that $f_{n+1} \notin I_n$. As in the proof of (1), we get a strictly descending chain

$$\text{Ke}\, f_2 \supsetneqq \text{Ke}\, f_3 \supsetneqq \cdots \supsetneqq \text{Ke}\, f_n \supsetneqq \cdots,$$

which contradicts 25.10 (3). Thus $\text{Hom}(N, M)$ is a uniserial right $S$-module of finite length.

(3) As $M$ is fully non-$M$-small and $N$ is a hollow module, any nonzero map from $M$ to $N$ is an epimorphism. Now (3) follows from 25.9.

(4) As $M$ is finitely generated, $M = \bigoplus_{i=1}^{k} M_i$, where each $M_i$ is a local module with local endomorphism ring.

Since $S = \bigoplus_{i=1}^{k} \text{Hom}(M, M_i)$, it follows from (3) that $S$ is a left serial ring. Since we also have $S = \bigoplus_{i=1}^{k} \text{Hom}(M_i, M)$, it follows from (1) and (2) that $S$ is also a right serial and right artinian ring. Now every left and right serial, right artinian ring is a left artinian ring (see [363, 55.16]). Thus $S$ is a left and right artinian serial ring, as required. By 22.23, $S$ is a right semihereditary ring and so, as $S$ is right artinian, it is a right hereditary ring.

Let $I := \{1, 2, \ldots, k\}$. For every $i \in I$, put $[i] = \{j \in I \mid M_i \simeq M_j\}$ and $\mathcal{F} = \{[i] \mid i \in I\}$. Define an order on $\mathcal{F}$ by

$[i] \leq [j]$ if and only if there exists an onto map $M_i \to M_j$.

Then $(\mathcal{F}, \leq)$ is a partially ordered set (see the proof of 25.9); put

$$J = \{j \in I \mid [j] \text{ is a minimal element of } \mathcal{F}\}.$$

Let $L = \bigoplus_J M_j$. For $k, \ell \in J$, since $M_k \oplus M_\ell$ is lifting we have that $M_k$ is almost $M_\ell$-projective (see 23.16). Then, since any epimorphism from $M_\ell$ to $M_k$ is an isomorphism, $M_k$ must be $M_\ell$-projective. Consequently, since each $M_k$ is self-projective, $L$ is self-projective.

For any $i \in I$, there exists $j \in J$ such that $[j] \leq [i]$ and hence there exists an epimorphism from $M_j$ to $M_i$. Thus $\sigma[M] = \sigma[L]$. Since $L$ is finitely generated and self-projective, $L$ is projective in $\sigma[L]$ and hence in $\sigma[M]$. Given $i \in I$, we can choose $j \in J$ such that there exists an epimorphism from $M_j$ to $M_i$ and hence $M_j$ is a projective cover of $M_i$. Thus $M$ has a projective cover $P$ which is a direct summand of $L^k$ for some $k \in \mathbb{N}$. Since $P$ is finitely generated and weakly supplemented, $P/\mathrm{Rad}\, P$ is a semisimple module and hence $\mathrm{End}(P/\mathrm{Rad}\, P)$ is a semisimple ring. By 17.21, $L^k$ is copolyform and hence $P$ is copolyform. Thus $\mathrm{Rad}\,(\mathrm{End}\, P) = 0$ by 9.26. Then, as $P$ is self-projective, $\mathrm{End}(P)/\mathrm{Rad}\,(\mathrm{End}\, P) \simeq \mathrm{End}(P/\mathrm{Rad}\, P)$ (see [363, 22.2]) and hence $\mathrm{End}(P)$ is a semisimple ring. $\square$

Suppose $A$ is a $\Sigma$-lifting self-injective module. By 22.20, $A = \bigoplus_I A_i$ where each $A_i$ is indecomposable, self-injective and $\Sigma$-lifting and hence, by 25.7, is a local, uniform self-projective LE-module.

**25.15. $\Sigma$-lifting self-injective modules I.** *Let* $A = \bigoplus_I A_i$ *be a $\Sigma$-lifting, self-injective module in* $\sigma[M]$, *where each $A_i$ is indecomposable. Suppose $N$ is an indecomposable direct summand of $A$. Then we have the following:*

(1) $\{\mathrm{Ke}\, f \mid f : \mathrm{Re}_\mathbb{S}(N) \to A\}$ *is linearly ordered by set inclusion.*

(2) $\mathcal{K}(\mathrm{Re}_\mathbb{S}(N), A)$ *is linearly ordered by set inclusion.*

(3) *The family* $\{\mathrm{Re}_\mathbb{S}(A_i) \mid i \in I\}$ *is locally semi-T-nilpotent.*

*Proof.* (1) Let $N$ be an indecomposable direct summand of $A$. Then $N$ is a local uniform LE-module. Let $f : \mathrm{Re}_\mathbb{S}(N) \to A$ be a nonzero map. Since $A$ is self-injective, $f$ has an extension $\overline{f}$ to $N$. We have $\mathrm{Ke}\, f = \mathrm{Ke}\, \overline{f} \cap \mathrm{Re}_\mathbb{S}(N)$. If $\mathrm{Im}\, \overline{f} \ll A$, then $\mathrm{Im}\, \overline{f}$ is an $M$-small module and hence $\mathrm{Re}_\mathbb{S}(N) \subseteq \mathrm{Ke}\, \overline{f}$. This implies $f = 0$, a contradiction. Hence $\mathrm{Im}\, \overline{f} \not\ll A$.

Let $f$ and $g$ be nonzero maps from $\mathrm{Re}_\mathbb{S}(N)$ to $A$ and let $\overline{f}$ and $\overline{g}$ respectively denote their extensions to $N$. By 25.10 (1), either $\mathrm{Ke}\, \overline{f} \subseteq \mathrm{Ke}\, \overline{g}$ or $\mathrm{Ke}\, \overline{g} \subseteq \mathrm{Ke}\, \overline{f}$ and hence either $\mathrm{Ke}\, f \subseteq \mathrm{Ke}\, g$ or $\mathrm{Ke}\, g \subseteq \mathrm{Ke}\, f$, as required.

(2) Let $I_1, I_2$ be two nonzero subsets of $\mathrm{Hom}(\mathrm{Re}_\mathbb{S}(N), A)$ and suppose that $\mathrm{Ke}\,(I_1) \not\subseteq \mathrm{Ke}\,(I_2)$. Then, for any nonzero $f \in I_1$, there exists a nonzero $g \in I_2$ such that $\mathrm{Ke}\, g \subsetneq \mathrm{Ke}\, f$; for if not, then by (1), $\mathrm{Ke}\, f \subseteq \mathrm{Ke}\, g$ for every $g \in I_2$, giving

Ke $f \subseteq$ Ke $(I_2)$ and hence Ke $(I_1) \subseteq$ Ke $(I_2)$, a contradiction. Hence, given any $f \in I_1$, there exists $g \in I_2$ such that Ke $g \subsetneq$ Ke $f$. It follows that Ke $(I_2) \subseteq$ Ke $(I_1)$. Thus $\mathcal{K}(N, M)$ is linearly ordered by set inclusion.

(3) Let $f : \mathrm{Re}_{\mathbb{S}}(A_i) \to \mathrm{Re}_{\mathbb{S}}(A_j)$ be a nonzero non-isomorphism. As $A$ is self-injective, $A_j$ is $A_i$-injective and hence $f$ can be extended to a homomorphism $\overline{f} : A_i \to A_j$. We claim that $\overline{f}$ is also a non-isomorphism.

If $\overline{f}$ is not onto, then Im $\overline{f}$ is an $M$-small module hence $f = 0$, a contradiction. Thus $\overline{f}$ is an onto map. Suppose Ke $\overline{f} = 0$. Then $\overline{f}$ is an isomorphism. Hence $A_i \simeq A_j$ and so $\mathrm{Re}_{\mathbb{S}}(A_i) \simeq \mathrm{Re}_{\mathbb{S}}(A_j)$ and Ke $f = 0$. Now, since $\mathrm{Re}_{\mathbb{S}}(A_i)$ is fully invariant in $A_i$, it is self-injective. Thus Im $f \simeq \mathrm{Re}_{\mathbb{S}}(A_i)$ is $\mathrm{Re}_{\mathbb{S}}(A_j)$-injective and so Im $f$ is a direct summand of $\mathrm{Re}_{\mathbb{S}}(A_j)$. As $\mathrm{Re}_{\mathbb{S}}(A_j)$ is uniform, this gives Im $f = \mathrm{Re}_{\mathbb{S}}(A_j)$. Thus $f$ is an isomorphism, a contradiction. Hence Ke $\overline{f} \neq 0$.

By 24.7, the family $\{A_i\}_I$ is locally semi-T-nilpotent. As any nonzero non-isomorphism from $\mathrm{Re}_{\mathbb{S}}(A_i) \to \mathrm{Re}_{\mathbb{S}}(A_j)$ can be extended to a non-isomorphism from $A_i \to A_j$, it follows that the family $\{\mathrm{Re}_{\mathbb{S}}(A_i)\}_I$ is also locally semi-T-nilpotent. $\square$

**25.16. Σ-lifting self-injective modules II.** *Let $A = \bigoplus_I A_i$ be a Σ-lifting, self-injective module in $\sigma[M]$, where each $A_i$ is indecomposable. Let $N$ be an indecomposable direct summand of $A$ and $\overline{S} = \mathrm{End}(\mathrm{Re}_{\mathbb{S}}(A))$. Then:*

(1) $\mathrm{Hom}(\mathrm{Re}_{\mathbb{S}}(N), \mathrm{Re}_{\mathbb{S}}(A))$ *is a uniserial, artinian right $\overline{S}$-module.*

(2) $\mathrm{Hom}(\mathrm{Re}_{\mathbb{S}}(N), \mathrm{Re}_{\mathbb{S}}(A))$ *is a uniserial right $\overline{S}$-module of finite length, provided $\mathrm{Re}_{\mathbb{S}}(A)$ contains only a finite number of non-isomorphic direct summands.*

(3) $\mathrm{Hom}(\mathrm{Re}_{\mathbb{S}}(A), \mathrm{Re}_{\mathbb{S}}(N))$ *is a uniserial left $\overline{S}$-module.*

*Proof.* By 25.7, $N$ is a local self-projective LE-module. As $\mathrm{Re}_{\mathbb{S}}(A)$ is fully invariant in $A$, $\mathrm{Re}_{\mathbb{S}}(A)$ is a self-injective module.

(1) Consider nonzero $f, g \in \mathrm{Hom}(\mathrm{Re}_{\mathbb{S}}(N), \mathrm{Re}_{\mathbb{S}}(A))$. By the self-injectivity of $A$ these can be extended to $\overline{f}, \overline{g} : N \to A$. As $N$ is a hollow module (and $f$, $g$ are nonzero), Im $\overline{f}$ and Im $\overline{g}$ are not small in $A$. Then, by 25.10 (1), either Ke $\overline{f} \subseteq$ Ke $\overline{g}$ or Ke $\overline{g} \subseteq$ Ke $\overline{f}$.

Suppose Ke $\overline{f} \subseteq$ Ke $\overline{g}$. There exists an onto map $\psi :$ Im $\overline{f} \to$ Im $\overline{g}$ such that $\overline{f}\psi = \overline{g}$. Since Im $\overline{f}$ is a direct summand of $A$, we can extend $\psi$ to $f : A \to A$ such that $\overline{f}f = \overline{g}$. Therefore $\overline{g}S \subseteq \overline{f}S$, where $S = \mathrm{End}(A)$. As any homomorphism $\mathrm{Re}_{\mathbb{S}}(A) \to \mathrm{Re}_{\mathbb{S}}(A)$ can be extended to a homomorphism in $S$, and the restriction of any map in $S$ to $\mathrm{Re}_{\mathbb{S}}(A)$ can be considered as a map from $\mathrm{Re}_{\mathbb{S}}(A) \to \mathrm{Re}_{\mathbb{S}}(A)$, we get $g\overline{S} \subseteq f\overline{S}$. Hence $\mathrm{Hom}(\mathrm{Re}_{\mathbb{S}}(N), \mathrm{Re}_{\mathbb{S}}(A))$ is a uniserial right $\overline{S}$-module.

We prove that $\mathrm{Hom}(\mathrm{Re}_{\mathbb{S}}(N), \mathrm{Re}_{\mathbb{S}}(A))$ is an artinian right $\overline{S}$-module. Let

$$I_1 \supsetneq I_2 \supsetneq \cdots \supsetneq I_r \supsetneq \cdots$$

be a strictly descending chain of $\overline{S}$-submodules of $\mathrm{Hom}(\mathrm{Re}_{\mathbb{S}}(N), \mathrm{Re}_{\mathbb{S}}(A))$. For any $r \in \mathbb{N}$, there exists $f_r \in I_r$ such that $f_r \notin I_{r+1}$. As $\mathrm{Hom}(\mathrm{Re}_{\mathbb{S}}(N), \mathrm{Re}_{\mathbb{S}}(A))$ is a uniserial right $\overline{S}$-module, we have either $f_r\overline{S} \subseteq f_{r+1}\overline{S}$ or $f_{r+1}\overline{S} \subseteq f_r\overline{S}$. Since $f_r \notin I_{r+1}$ we get $f_r\overline{S} \nsubseteq f_{r+1}\overline{S}$. Hence $f_{r+1}\overline{S} \subsetneq f_r\overline{S}$. As $f_{r+1} \in f_r\overline{S}$, there exists $\overline{\alpha} \in \overline{S}$ such that $f_{r+1} = f_r\overline{\alpha}$. As $f_r\overline{S} \nsubseteq f_{r+1}\overline{S}$, $\operatorname{Ke} f_r \subsetneq \operatorname{Ke} f_{r+1}$, implying

$$\operatorname{Ke}\overline{f}_r \cap \operatorname{Res}(N) \subsetneq \operatorname{Ke}\overline{f}_r \cap \operatorname{Res}(N),$$

where $\overline{f}_r$ and $\overline{f}_{r+1}$ are extensions of $f_r$ and $f_{r+1}$ respectively from $N$ to $A$. By 25.10(1), either $\operatorname{Ke}\overline{f}_r \subseteq \operatorname{Ke}\overline{f}_{r+1}$ or $\operatorname{Ke}\overline{f}_{r+1} \subseteq \operatorname{Ke}\overline{f}_r$. As

$$\operatorname{Ke}\overline{f}_r \cap \operatorname{Res}(N) = \operatorname{Ke} f_r \subsetneq \operatorname{Ke} f_{r+1} = \operatorname{Ke}\overline{f}_{r+1} \cap \operatorname{Res}(N),$$

we get $\operatorname{Ke}\overline{f}_r \subsetneq \operatorname{Ke}\overline{f}_{r+1}$. Thus we have a strictly ascending chain

$$\operatorname{Ke}\overline{f}_1 \subsetneq \operatorname{Ke}\overline{f}_2 \subsetneq \cdots \subsetneq \operatorname{Ke}\overline{f}_r \subsetneq \cdots.$$

However, for all $r \in \mathbb{N}$, $\operatorname{Im}\overline{f}_r$ is not small in $A$ and so, by 25.10(2), the above chain becomes stationary after finitely many steps and hence this is true for our original chain

$$I_1 \supsetneq I_2 \supsetneq \cdots \supsetneq I_r \supsetneq \cdots.$$

Therefore $\operatorname{Hom}(\operatorname{Res}(N), \operatorname{Res}(A))$ is an artinian right $\overline{S}$-module.

(2) By (1), it is enough to prove that $\operatorname{Hom}(\operatorname{Res}(N), \operatorname{Res}(A))$ satisfies ACC on right $\overline{S}$-submodules. Suppose that

$$I_1 \subsetneq I_2 \subsetneq \cdots \subsetneq I_r \subsetneq \cdots$$

is a strictly ascending chain of $\overline{S}$-submodules of $\operatorname{Hom}(\operatorname{Res}(N), \operatorname{Res}(A))$. As in the proof of (1), we get homomorphisms $\overline{f}_r : N \to A$ such that $\operatorname{Im}\overline{f}_r \not\ll A$ and

$$\operatorname{Ke}\overline{f}_1 \supsetneq \operatorname{Ke}\overline{f}_2 \supsetneq \cdots \supsetneq \operatorname{Ke}\overline{f}_r \supsetneq \cdots.$$

By 25.10(3) this chain must become stationary. From this contradiction we infer that $\operatorname{Hom}(\operatorname{Res}(N), \operatorname{Res}(A))$ is a uniserial $\overline{S}$-module of finite length.

(3) As $A$ is self-injective, any two nonzero maps $f, g : \operatorname{Res}(A) \to \operatorname{Res}(N)$ can be extended to nonzero maps $\overline{f}, \overline{g} \in \operatorname{Hom}(A, N)$. As $f$ and $g$ are nonzero maps, $\operatorname{Im}\overline{f}$ and $\operatorname{Im}\overline{g}$ are not small in $N$ and hence $\overline{f}, \overline{g}$ are onto maps. By 25.9, either $S\overline{f} \subseteq S\overline{g}$ or $S\overline{g} \subseteq S\overline{f}$ where $S = \operatorname{End}(A)$.

Suppose that $S\overline{f} \subseteq S\overline{g}$. As any homomorphism $\operatorname{Res}(A) \to \operatorname{Res}(A)$ can be extended to a homomorphism in $S$, and the restriction of any map in $S$ to $\operatorname{Res}(A)$ can be considered as a map from $\operatorname{Res}(A) \to \operatorname{Res}(A)$, we have $\overline{S}f \subseteq \overline{S}g$. Hence $\operatorname{Hom}(\operatorname{Res}(A), \operatorname{Res}(N))$ is a uniserial left $S$-module.                         $\square$

**References.** Al-attas and Vanaja [2]; Baba and Harada [30]; Talebi and Vanaja [331]; Jayaraman and Vanaja [182].

# 26 Semi-discrete and quasi-discrete modules

In this and the next section we will investigate some classes of supplemented modules which satisfy additional (projectivity) conditions. As before $M$ denotes a left $R$-module.

**26.1. Definition.** A lifting module with the finite internal exchange property is called a *semi-discrete module*.

Recall that a module $M$ has the finite internal exchange property if and only if it has the 2-internal exchange property (11.38).

As our first result shows, a semi-discrete module $M$ is $\oplus$-*supplemented*, that is, every submodule of $M$ has a (weak) supplement that is a direct summand (see Exercise 20.39 (6)).

**26.2. Semi-discrete modules.** *The following are equivalent for $M$:*

(a) $M$ *is semi-discrete;*

(b) $M$ *is supplemented, every supplement is a direct summand, and, for every decomposition $M = K \oplus L$, $K$ and $L$ are relatively generalised projective;*

(c) $M$ *is $\oplus$-supplemented and, for every decomposition $M = K \oplus L$, $K$ and $L$ are relatively generalised projective;*

(d) $M$ *is lifting and, for every decomposition $M = K \oplus L$, $K$ and $L$ are relatively generalised projective.*

*Proof.* (a)$\Rightarrow$(b). As $M$ is lifting it is supplemented and every supplement is a direct summand. Let $M = X + L$. Then $X$ contains a supplement $X'$ of $L$ which is a direct summand of $M$. Then, since $M$ has the finite internal exchange property, we have $M = X' \oplus K' \oplus L'$ with $K' \subseteq K$ and $L' \subseteq L$. Thus, by 4.42, $K$ is generalised $L$-projective. Similarly $L$ is generalised $K$-projective.

(b)$\Rightarrow$(c) is trivial.

(c)$\Rightarrow$(d). It is enough to prove that $M$ is lifting. Let $N \subseteq M$. Then $N$ has a supplement $K$ which is a direct summand of $M$, let us say $M = L \oplus K$. Then, by hypothesis, $L$ is generalised $K$-projective and so, since $M = N + K$, we have, by 4.42, $M = N' \oplus L' \oplus K' = N' + K$, where $N' \subseteq N$, $K' \subseteq K$ and $L' \subseteq L$. Then $N = N' + (N \cap K)$. Thus, since $N \cap K \ll K$, it follows from 22.1(b) that $M$ is a lifting module.

(d)$\Rightarrow$(a) Suppose $M = K \oplus L$. Then, by 22.6, $K$ and $L$ are lifting modules and by hypothesis are relatively generalised projective. Thus it follows from 23.10 that $M$ has the 2-internal exchange property. $\square$

Recall that for a decomposition $M = \bigoplus_I M_i$ and any nonempty $K \subset I$, we write $M_K = \bigoplus_K M_k$.

**26.3. Semi-discreteness of finite direct sums.** *Let* $M = M_1 \oplus \cdots \oplus M_n$ *where each* $M_i$ *is semi-discrete. Then the following conditions are equivalent:*

(a) *M is semi-discrete;*

(b) *M is lifting and* $M = M_1 \oplus \cdots \oplus M_n$ *is an exchange decomposition;*

(c) *K and L are relatively generalised projective, for any direct summand K of* $\bigoplus_I M_i$ *and any direct summand L of* $\bigoplus_J M_j$, *for any nonempty disjoint subsets* $I, J \subset \{1, 2, \ldots, n\}$;

(d) $M_i'$ *and T are relatively generalised projective for any direct summand* $M_i'$ *of* $M_i$ *and any direct summand T of* $\bigoplus_{j \neq i} M_j$, *for any* $1 \leq i \leq n$.

*Proof.* (a)$\Rightarrow$(b) is clear and (b)$\Rightarrow$(c) follows from 23.14. (c)$\Rightarrow$(d) is obvious.

(d)$\Rightarrow$(a) follows from 23.14 and 11.40.                                    $\square$

As an immediate consequence of 26.3 we obtain

**26.4. Corollary.** *Let* $M = M_1 \oplus \cdots \oplus M_n$ *where each* $M_i$ *is a semi-discrete module. If* $M_i$ *and* $M_j$ *are relatively projective for each* $i \neq j$, *then M is semi-discrete.*

**26.5. Semi-discreteness of direct sums of hollow modules.** *Let M be a lifting module with an indecomposable decomposition* $M = \bigoplus_I M_i$. *Then M is a semi-discrete module if one of the following conditions is satisfied.*

(i) $M_i$ *is an LE-module, for all* $i \in I$;

(ii) *every nonzero direct summand of M contains a nonzero indecomposable direct summand and the decomposition* $M = \bigoplus_I M_i$ *complements maximal direct summands.*

*Proof.* A module $M$ with an indecomposable exchange decomposition has the internal exchange property. Thus we may apply 24.13 and 24.10.          $\square$

**26.6. Definition.** A module $M$ which is both $\pi$-projective and supplemented is called a *quasi-discrete module*.

For these modules we have the following characterisations.

**26.7. Quasi-discrete modules.** *For a module M the following are equivalent:*

(a) *M is quasi-discrete;*

(b) *M is amply supplemented and* $\cap$*-coclosed projective;*

(c) *M is amply supplemented and the intersection of any pair of mutual supplements in M is zero;*

(d) *M is lifting and* $\cap$*-direct projective;*

(e) *M is lifting and* $\pi$*-projective;*

(f) *M is lifting and for any direct summands* $U, V \subset M$ *with* $M = U + V$, $U \cap V$ *is a direct summand of M;*

(g) *for any submodules $U, V \subset M$ with $U + V = M$, there exists an idempotent $e \in \operatorname{End}(M)$ with $Me \subseteq U$, $M(1 - e) \subseteq V$ and $U(1 - e) \ll M(1 - e)$.*

(h) *$M$ is lifting and for every decomposition $M = K \oplus L$, $K$ and $L$ are relatively projective.*

*Proof.* (See also [363, 41.15].) We first note that in a supplemented module, if the intersection of any pair of mutual supplements is zero, then coclosed submodules (which are precisely the supplement submodules) of $M$ are direct summands.

(a)$\Rightarrow$(e). We have to prove that $M$ is lifting. By 20.9, supplements are direct summands. It is enough to prove that $M$ is amply supplemented (see 22.3).

Suppose $M = U + V$ and $X$ is a supplement of $U$ in $M$. For an $f \in \operatorname{End}(M)$ with $\operatorname{Im}(f) \subseteq V$ and $\operatorname{Im}(1 - f) \subseteq U$ we have $M = U + (X)f$ and $U \cap (X)f = (U \cap X)f \ll (X)f$. Thus $(X)f$ is a supplement of $U$ contained in $V$.

(e)$\Rightarrow$(c) follows from 20.9 and 22.3.

(c)$\Rightarrow$(d). As $M$ is supplemented, supplements are direct summands by (c) and so $M$ is lifting (see 22.3). Suppose $M = U + V$ where $U$ and $V$ are direct summands of $M$. Let $X$ be a supplement of $V$ such that $X \subseteq U$. Now (c) implies $M = X \oplus V$. As $U = X \oplus (U \cap V)$, $U \cap V$ is a direct summand of $M$. Thus $M$ is $\cap$-direct projective (see 4.29).

(d)$\Rightarrow$(b). Since coclosed submodules are direct summands (see 22.3), $M$ is $\cap$-coclosed projective (see remarks after 4.31).

(b)$\Rightarrow$(a). It is enough to prove that $M$ is $\pi$-projective. Let $K$ and $L$ be submodules of $M$ with $M = K + L$. Since $M$ is amply supplemented, there exists a submodule $K' \subseteq L$ such that $M = K + K'$ and $K \cap K' \ll K'$. There is also a submodule $L' \subseteq K$ such that $M = K' + L'$ and $K' \cap L' \ll L'$. Then both $K'$ and $L'$ are coclosed (by 20.2). Consider the map

$$f : M \to M/(K' \cap L'), \quad f(a' + b') = a' + K' \cap L', \quad \text{for } a' \in K', b' \in L'.$$

By $\cap$-coclosed projectivity, $f$ can be lifted to some $h : M \to M$ and $(M)h \subseteq K'$ and $(M)(1 - h) \subseteq L'$. Hence $M$ is $\pi$-projective.

(d)$\Leftrightarrow$(f) follows from 4.29.

(e)$\Rightarrow$(h). Since $M$ is $\pi$-projective, $M = K \oplus L$ implies the relative projectivity of $K$ and $L$ (see 4.14 (2)).

(h)$\Rightarrow$(g). Assume $U, V \subseteq M$ with $M = U + V$. Since $M$ is lifting, there is a decomposition $M = K \oplus L$ where $K \subset U$ and $U \cap L \ll L$. Clearly $K + V = M$. Since $L$ is $K$-projective, by 4.12 there is a submodule $W \subseteq V$ such that $M = K \oplus W$. Consider the canonical projection $\pi : M \to K$ and the inclusion map $\varepsilon : K \to M$ with respect to this decomposition of $M$. Then $f := \pi \diamond \varepsilon$ is an idempotent endomorphism of $M$ such that $(M)f \subseteq U$ and $(M)(1 - f) \subseteq V$.

(g)$\Rightarrow$(e). It is obvious that $M$ is $\pi$-projective. Suppose $U$ is a submodule of $M$. By (h), there exists $e^2 = e \in \mathrm{End}(M)$ with $Me \subseteq U$ and $U(1-e) \ll M(1-e)$. Hence $U = Ue \oplus U(1-e)$ where $Ue = Me$ is a direct summand of $M$ and $U(1-e) \ll M$. By 22.3, $M$ is lifting.                                                    $\square$

From 26.7 and 26.2 (d) follows immediately

**26.8. Quasi-discrete implies semi-discrete.** *Any quasi-discrete module $M$ is semi-discrete.*

Notice that the converse implication need not be true.

**26.9. A semi-discrete module which is not quasi-discrete.** Consider the $\mathbb{Z}$-modules $K = \mathbb{Z}/p\mathbb{Z}$ and $L = \mathbb{Z}/p^2\mathbb{Z}$ where $p$ is a prime. Then $K$ and $L$ are relatively generalised projective but $K$ is not $L$-projective. Then $M = K \oplus L$ is a lifting module with the finite internal exchange property but is not quasi-discrete since it is not $\pi$-projective by 4.14 (2).

The following observations are taken from [363, 41.16].

**26.10. Properties of quasi-discrete modules.** *Let $M$ be a quasi-discrete module and let $S = \mathrm{End}(M)$.*

(1) *Every direct summand of $M$ is quasi-discrete and every supplement submodule of $M$ is a direct summand.*

(2) *Let $e \in S$ be an idempotent and let $N$ be a direct summand of $M$. If $N(1-e) \ll M(1-e)$, then $N \cap M(1-e) = 0$ and $N \oplus M(1-e)$ is a direct summand in $M$.*

(3) *If $\{N_\lambda\}_\Lambda$ is a family of direct summands of $M$, upwards directed with respect to inclusion, then $\bigcup_\Lambda N_\lambda$ is also a direct summand in $M$.*

(4) *For every $0 \neq a \in M$, there is a decomposition $M = M_1 \oplus M_2$ with $M_2$ hollow and $a \notin M_1$.*

(5) *If $N$ and $H$ are direct summands of $M$ and $H$ is hollow, then either*

(i) *$N \cap H = 0$ and $N \oplus H$ is a direct summand of $M$ or*

(ii) *$N + H = N \oplus K$ where $K \ll M$ and $H$ is isomorphic to a direct summand of $N$.*

The following decomposition properties of quasi-discrete modules are shown in [363, 41.17] and [241, Theorem 4.15].

**26.11. Decomposition of quasi-discrete modules.** *Let $M$ be quasi-discrete.*

(1) *There is a decomposition $M = \bigoplus_\Lambda H_\lambda$, where the $H_\lambda$ are hollow modules; furthermore, such a decomposition complements direct summands and so (by 12.5) is unique up to isomorphism.*

(2) *If $M = \sum_\Lambda N_\lambda$ is an irredundant sum with indecomposable $N_\lambda$, then we have $M = \bigoplus_\Lambda N_\lambda$.*

(3) *If $\operatorname{Rad} M \ll M$, then $M = \bigoplus_\Lambda L_\lambda$ where each $L_\lambda$ is a local module.*

(4) *There is a decomposition $M = \bigoplus_\Lambda L_\lambda \oplus K$ where the $L_\lambda$ are local modules, $\operatorname{Rad}(\bigoplus_\Lambda L_\lambda) \ll \bigoplus_\Lambda L_\lambda$ and $K = \operatorname{Rad} K$.*

(5) *Every direct decomposition of $M/\operatorname{Rad} M$ can be lifted to $M$.*

Applying 22.17 yields the

**26.12. Corollary** *Let $f : N \to M$ be an epimorphism where $N$ is a quasi-discrete module and $M$ is lifting. Then $M$ is a direct sum of indecomposable modules.*

A $\pi$-injective module $M$ is characterised by the fact that it is invariant under idempotent endomorphisms of the self-injective hull $\widehat{M}$ of $M$ (see [85, 2.10]). A dual property for $\pi$-projective modules $M$ can be shown provided $M$ is supplemented (e.g., [241, Proposition 4.45]).

**26.13. Quasi-discrete factor modules.** *Let $K$ be a submodule of $M$.*

(1) *If $M/K$ is quasi-discrete and $K \ll M$, then $K$ is invariant under idempotents of $\operatorname{End}(M)$.*

(2) *If $M$ is quasi-discrete and $K$ is invariant under idempotents of $\operatorname{End}(M)$, then $M/K$ is quasi-discrete.*

*Proof.* (1) For $e^2 = e \in \operatorname{End}(M)$ we have $M = Me \oplus M(1-e)$ and

$$M/K = (Me + K)/K + (M(1-e) + K)/K.$$

By assumption, using 26.7 (b) there exist submodules $K \subset U \subset Me + K$ and $K \subset V \subset M(1-e)+K$ such that $M/K = U/K + V/K$ and $U \cap V = K$. Therefore $M = U + V + K = U + V$ and hence $M/K = (Me + K)/K \oplus (M(1-e) + K)/K$ implying that $Ke \subseteq K$ (and $K(1-e) \subseteq K$).

(2) Let $M/K = U/K + V/K$ where $K \subset U$, $V \subset M$ and $U + V = M$. Then, since $M$ is quasi-discrete, by 26.7 (f) there exists an idempotent $e \in \operatorname{End}(M)$ such that $Me \subset U$, $M(1-e) \subseteq V$ and $U(1-e) \ll M(1-e)$.

As $K$ is invariant under idempotents in $\operatorname{End}(M)$, $K = Ke \oplus K(1-e)$ and $e$ induces an idempotent $\bar{e} \in \operatorname{End}(M/K)$. Then $(M/K)\bar{e} = (Me + K)/K \subseteq U/K$ and $(M/K)(1 - \bar{e}) = (M(1-e) + K)/K \subseteq V/K$. Moreover $(U/K)(1-\bar{e}) = (U(1-e) + K)/K \ll (M/K)(1-\bar{e}) = (M(1-e)+K)/K$ as $U(1-e) \ll M(1-e)$. By 26.7, this shows that $M/K$ is quasi-discrete. $\qquad\square$

Recall from 11.26 that *direct (finite) decompositions of a module $M$ lift modulo a submodule $K$* if, whenever $M/K$ is expressed as a (finite) direct sum of submodules $M_i/K$, then $M = \bigoplus_I N_i$ where $(N_i)\pi = M_i/K$ for each $i \in I$ and $\pi : M \to M/K$ is the natural map.

**26.14. Lifting property.** *The following are equivalent for an R-module M:*

(a) *M is quasi-discrete;*

(b) *M is supplemented and decompositions lift modulo any submodule $K \subset M$;*

(c) *M is supplemented and for every $K \subseteq M$, any decomposition $M/K = U/K \oplus V/K$ lifts modulo K.*

*Proof.* (a)$\Rightarrow$(b). We first assume $K \ll M$. Let $\pi : M \to M/K$ be the projection map and suppose that $M/K = \bigoplus_I U_i/K$. As $M$ is lifting, we can write $U_i = V_i \oplus S_i$ where $V_i$ is a direct summand of $M$ and $S_i \ll M$, for all $i \in I$. Then $(S_i)\pi \ll M/K$ and hence is small in $U_i/K$. Thus $(V_i)\pi = U_i/K$ and $M = \sum_I V_i$. It is now enough to prove that, for any finite subset $F \subset I$, the sum $\sum_F V_i$ is direct. We can choose a maximal subset $E \subseteq F$ such that the sum $\sum_E V_i$ is direct and is a direct summand of $M$, say $M = \bigoplus_E V_i \oplus T$. If possible let $i_0 \in F \setminus E$ and set $N := \sum_{i \neq i_0} V_i$. Then $N \cap V_{i_0} \subseteq K$. Hence $N = \bigoplus_E V_i \oplus (T \cap N)$ and so $M = \bigoplus_E V_i + V_{i_0} + (T \cap N)$. As $M$ is amply supplemented, there exists a supplement $L$ of $\bigoplus_E V_i + V_{i_0}$ in $M$ contained in $N \cap T$. Now $L$ is a direct summand of $M$ and $L \subseteq T$. Thus $\bigoplus_E V_i \oplus L$ is a direct summand of $M$. Also we have $M = (\bigoplus_E V_i \oplus L) + V_{i_0}$ and $(\bigoplus_E V_i \oplus L) \cap V_{i_0} \subseteq N \cap V_{i_0} \subseteq K \ll M$. Since $M$ is quasi-discrete, this gives $M = \bigoplus_E V_i \oplus L \oplus V_{i_0}$, contradicting the choice of $E$. Thus $E = F$ and our sum $\sum_E V_i$ is direct, as required.

Now suppose that $K \not\ll M$. Since $M$ is lifting, we have $M = A \oplus B$ where $A \subseteq K$ and $S = K \cap B \ll M$. Any decomposition of $M/K$ induces a decomposition of $B/S$ which can be lifted to a decomposition of $B$ since $B$ is quasi-discrete and $S \ll B$. Now it is easy to see that any decomposition of $M/K$ can be lifted to a decomposition of $K$.

(b)$\Rightarrow$(c) is automatic.

(c)$\Rightarrow$(a). Suppose $U \subset M$ and $V$ is a supplement of $U$ in $M$. Then, by (c), the decomposition $M/(U \cap V) = U/(U \cap V) \oplus V/(U \cap V)$ can be lifted to a decomposition $M = A \oplus B$. Then $U = A \oplus (B \cap U)$ and $B \cap U \subseteq U \cap V \ll M$. Thus $M$ is a lifting module.

Now suppose that $U$ and $V$ are direct summands of $M$ and $M = U + V$. As $M$ is lifting, $M$ is amply supplemented and hence $U$ has a supplement $N \subseteq V$. Then $M = U + N$ and $U \cap N \ll N$. Since the decomposition

$$M/(U \cap N) = U/(U \cap N) \oplus N/(U \cap N)$$

can be lifted to a decomposition of $M$ and $U \cap N$ is a small submodule of both $U$ and $N$, we get $M = U \oplus N$. This gives $V = N \oplus (U \cap V)$ and so, since $V$ is a summand of $M$, it follows that $U \cap V$ is also a summand of $M$. Thus, by 26.7 (f), $M$ is quasi-discrete.                                    $\square$

Recall from 11.26 that a module $M$ is *strongly refinable* if finite decompositions lift modulo any of its submodules. Note that in general these modules need not be supplemented, in particular $M/\operatorname{Rad} M$ need not be semisimple. For more information on this class of modules the reader is referred to [364, Section 8].

It follows from 26.11 (1) that any quasi-discrete module has the internal exchange property. However we now give a simple, independent proof that a quasi-discrete module has the finite internal exchange property.

**26.15. Quasi-discrete modules and the internal exchange property.** *Any quasi-discrete module $M$ has the finite internal exchange property.*

*Proof.* By 11.38, it is enough to prove that $M$ has the 2-internal exchange property. Thus let $M = K \oplus L$. Then, by 26.7, $K$ and $L$ are relatively projective. Also, since $M$ is lifting, $K$ and $L$ are also lifting. Then, by 23.15, $M = K \oplus L$ is an exchange decomposition and hence $M$ has the 2-internal exchange property. $\square$

**26.16. Directly finite quasi-discrete modules.** *Let $M$ be a directly finite quasi-discrete module. Then*

(1) *$M$ has the internal cancellation property and*

(2) *if every hollow direct summand of $M$ is an LE-module, then $M$ has the cancellation property.*

*Proof.* (1) This follows from 26.11 (1) and 15.26.

(2) Now suppose that every hollow direct summand of $M$ is an LE-module. Then, by 26.11 (1), $M$ has an LE-decomposition that complements direct summands and so, by 14.22, $M$ has the exchange property. Thus, by (1) and 15.37, $M$ has the cancellation property. $\square$

In a $\pi$-injective module $M$, isomorphic submodules of $M$ have isomorphic closures. An analogous result is true for supplemented $\pi$-projective modules (e.g., [241, Proposition 4.24]).

**26.17. Factors of quasi-discrete modules.** *Suppose $M$ is a quasi-discrete module. Let $A$ and $B$ be direct summands of $M$ with submodules $X \ll A$ and $Y \ll B$. If $A/X \simeq B/Y$, then $A \simeq B$.*

*Proof.* Since $A$ is quasi-discrete (by 26.10 (1)), we may write $A = \bigoplus_I A_i$, where each $A_i$ is hollow (see 26.11 (1)). Let $f : A/X \to B/Y$ be the given isomorphism. As $X \ll A$, $\sum_I (A_i + X)/X$ is an irredundant sum. For each $i \in I$, let $T_i/Y = ((A_i + X)/X)f$. Then the sum $\sum_I T_i/Y$ is also irredundant. Hence $T_i \not\ll B$. As $B$ is lifting, we have $T_i = B_i \oplus S_i$ where $B_i$ is a nonzero direct summand of $B$ and $S_i \ll B$. Since $A_i$ is hollow, so is $T_i/Y$ and hence either $B_i + Y = T_i$ or $S_i + Y = T_i$. However, since $\sum_I T_i/Y$ is an irredundant sum and $(S_i + Y)/Y \ll B/Y$, we must have $B_i + Y = T_i$. Since $Y \ll B$, $\sum_I B_i$ is an irredundant sum of hollow submodules of $B$ and hence $B = \bigoplus_I B_i$ (see 26.11 (2)). As $A_i$ and $B_i$ are hollow direct summands of $M$, it follows from 26.10 (5) that either $A_i \simeq B_i$ or $A_i \oplus B_i$

is a direct summand of $M$. In the latter case, it follows from 26.7 that $A_i$ and $B_i$ are relatively projective. Since $A_i$ is $B_i$-projective, the map

$$g := f|_{(A_i+X)/X} : (A_i + X)/X \to (B_i + Y)/Y$$

can be lifted to a map $h : A_i \to B_i$. As $Y \ll B$ and $g$ is onto, $h$ is an onto map. Then, since $B_i$ is $A_i$-projective, $h$ splits. Thus, since $A_i$ is hollow, $h$ must be an isomorphism and so again $A_i \simeq B_i$. It now follows that $A \simeq B$.                    $\square$

From 26.17 and 26.16 we get the following.

**26.18. Corollary.** *Suppose $M$ is a quasi-discrete module and $M/A \simeq M/B$.*

(1) *If $C$ and $D$ are supplements in $M$ of $A$ and $B$ respectively, then $C \simeq D$.*

(2) *If further $M$ is directly finite, then $\overline{A} \simeq \overline{B}$ where $\overline{A}$ and $\overline{B}$ are coclosures in $M$ of $A$ and $B$ respectively.*

As can be seen from 26.9, the direct sum of two quasi-discrete modules may not be quasi-discrete. (As a slightly different example, consider the quasi-discrete $\mathbb{Z}$-modules $A = \mathbb{Z}/p\mathbb{Z}$ and $B = \mathbb{Z}/p^3\mathbb{Z}$. Then, as seen in 22.5, $A \oplus B$ is not even lifting.) This leads us to give the following necessary and sufficient condition for a direct sum of modules to be quasi-discrete.

**26.19. Quasi-discreteness of direct sums.** *Let $\{M_i\}_I$ be any family of $R$-modules and let $M = \bigoplus_I M_i$. Then the following statements are equivalent.*

(a) *$M$ is quasi-discrete;*

(b)  (1) *for all $i \in I$, $M_i$ is quasi-discrete,*

    (2) *$M_i$ is $M_{I \setminus \{i\}}$-projective, for all $i \in I$, and*

    (3) *the given decomposition of $M$ is exchangeable.*

*Proof.* (a)$\Rightarrow$(b). By 26.10 (1), each $M_i$ is quasi-discrete, and, by 26.11 (1), $M$ has the internal exchange property and hence the decomposition $M = \bigoplus_I M_i$ is exchangeable. Finally, (b) (2) follows from 26.7.

(b)$\Rightarrow$(a). By 26.11 (1), each $M_i$ has an indecomposable decomposition that complements direct summands and so, by (b) (3) and 13.7, $M$ has a decomposition that complements direct summands.

We next show that $M$ is lifting, and to do this it suffices to show that, for every indecomposable direct summand $H$ of $M$, $H$ is hollow and is $M/H$-projective (see 24.12). Thus suppose $H$ is an indecomposable summand of $M$. Then, since the decomposition $M = \bigoplus_I M_i$ is exchangeable and $H$ is indecomposable, it is easy to see that there exists $i_0 \in I$ such that $H \simeq M_{i_0}' \subseteq M_{i_0}$ where $M_{i_0} = M_{i_0}' \oplus M_{i_0}''$, say, and $M/H \cong M/M_{i_0}'$. Hence $H$ is hollow. Moreover, since $M_{i_0}$ is quasi-discrete, $M_{i_0}'$ and hence $H$ is $M_{i_0}''$-projective (by 26.7). Now, by (b), $H$ is also $M_{I \setminus \{i_0\}}$-projective. Thus $H$ is $M/H$-projective.

To prove that $M$ is quasi-discrete, it now suffices to show that whenever $M = A \oplus B$ is a direct sum of submodules, then $A$ is $B$-projective (see 26.7). Since $M$ has an indecomposable exchange decomposition, $A$ is a direct sum of indecomposable modules. Now if $H$ is an indecomposable direct summand of $A$, then we have shown that $H$ is $M/H$ -projective and hence $H$ is $B$-projective. Consequently $A$ is $B$-projective, as required. $\qquad\square$

We observe that the conditions can be modified in case $\operatorname{Rad} M \ll M$.

**26.20. Corollary.** *Let $\{M_i\}_I$ be any family of R-modules and let $M = \bigoplus_I M_i$. Suppose $\operatorname{Rad} M \ll M$. Then the following statements are equivalent.*

(a) *$M$ is quasi-discrete;*

(b) (1) *for all $i \in I$, $M_i$ is quasi-discrete and*

(2) *$M_i$ is $M_j$-projective for all $i \neq j$ in $I$.*

*Proof.* (a)$\Rightarrow$(b) follows from 26.19.

(b)$\Rightarrow$(a). As $M_i$ is a direct summand of $M$, $\operatorname{Rad} M_i \ll M_i$ for all $i \in I$. By 26.11, $M_i = \bigoplus_{K_i} N_{k_i}$ where each $N_{k_i}$ is a local module and $N_{j_i}$-projective for all $j_i \in K_i$, $j_i \neq k_i$. Let $i, j \in I$ such that $i \neq j$. Since $M_i$ is $M_j$-projective, $N_{k_i}$ is $N_{k_j}$-projective for all $k_j \in K_j$. Thus we can write $M = \bigoplus_\Lambda H_\lambda$ where for each $\lambda \in \Lambda$, $H_\lambda$ is local and $H_\mu$-projective for $\lambda \neq \mu$ in $\Lambda$. By 24.14, the above decomposition complements direct summands and (a) follows from 26.19. $\qquad\square$

As another corollary we have

**26.21. Quasi-discreteness of direct sums of hollow modules.** *Let $M = \bigoplus_I M_i$ where each $M_i$ is hollow. Then the following are equivalent:*

(a) *$M$ is quasi-discrete;*

(b) *for all $i \in I$, $M_i$ is $M_{I\setminus\{i\}}$-projective and the given decomposition of $M$ is exchangeable.*

Restricting 26.19 to a finite number of summands we get

**26.22. Quasi-discreteness of finite direct sums.** *Let $M = M_1 \oplus \cdots \oplus M_n$. Then the following are equivalent:*

(a) *$M$ is a quasi-discrete module;*

(b) *for each $1 \leq i \leq n$, $M_i$ is quasi-discrete and $M_j$-projective for each $j \neq i$.*

*Proof.* (a)$\Rightarrow$(b) follows from 26.19.

(b)$\Rightarrow$(a). Since $M_i$ is $M_j$-projective whenever $i \neq j$, it follows from 23.15 that the decomposition $M = \bigoplus_I M_i$ is exchangeable. Then, since $M_i$ is $\bigoplus_{j \neq i} M_j$-projective for every $1 \leq i \leq n$, 26.19 shows that $M$ is quasi-discrete. $\qquad\square$

This immediately gives the following two corollaries.

**26.23. Corollary.** *A finite direct sum of modules $M_1, \ldots, M_n$ is quasi-discrete if and only if $M_i \oplus M_j$ is quasi-discrete for all $1 \leq i < j \leq n$.*

**26.24. Corollary.** *A finite direct sum of hollow modules $M_1, \ldots, M_n$ is quasi-discrete if and only if $M_i$ is $M_j$-projective for $i \neq j$, $1 \leq i \leq n$ and $1 \leq j \leq n$.*

**26.25. Exercises.**

(1) Define a small epimorphism $f : N \to M$ to be a *quasi-discrete cover* of $M$ if $N$ is quasi-discrete, and $f$ does not split through any proper quasi-discrete factor module. Prove that if $M$ has a small cover which is supplemented and projective, then a quasi-discrete cover of $M$ exists and is unique up to isomorphism.

(2) An $R$-module $M$ is said to have the *(strong) summand sum property* if the sum of any (family of) two direct summands of $M$ is a direct summand of $M$.

   $M$ is said to have the *summand intersection property* if the intersection of any two direct summands of $M$ is again a direct summand of $M$. Prove the following:

   (i) A lifting module $M$ is UCC if and only if $M$ has the (strong) summand sum property.

   (ii) $M$ is a quasi-discrete UCC module if and only if $M$ has both the summand sum and the summand intersection property.

**26.26. Comments.** The name *semi-discrete* for lifting modules with the finite internal exchange property was suggested by Mohamed and Müller in [245] since they are dual to the concept of *semi-continuous* modules, that is, extending modules with the finite internal exchange property. (In fact Hanada, Kado and Oshiro [143] were the first to study these extending modules, calling them *finite normal extending*, and the renaming to *semi-continuous* modules was given by Mohamed and Müller in [244, Definition 17].) 26.2 is due to Mohamed and Müller [245].

As noted in the comments for section 22, Takeuchi defined *codirect modules* in [327], but these have now taken the name of our main focus, namely *lifting modules*. He also initiated the study of quasi-perfect modules, calling them *codirect modules with Condition (II)* where, in our terminology, Condition (II) is precisely $\cap$-direct projectivity.

Oshiro [269] called a module $M$ "semiperfect" if $M$ is direct projective and lifting (i.e. $M$ is discrete, see next section). He also studied "quasi-semiperfect" modules, a generalisation of semiperfect modules, these being precisely the quasi-discrete modules in our nomenclature, and established decomposition theorems for these (see 26.11).

On the other hand, Mohamed, Müller and Singh [246] referred to quasi-semiperfect modules as *quasi-dual-continuous* modules, in short qd-continuous modules, and they improved the proofs of decomposition theorems given by Oshiro [269].

**References.** Chang and Kuratomi [61]; Hanada, Kado and Oshiro [143]; Kuratomi [214]; Kutami and Oshiro [215]; Miyashita [235]; Mohamed, Müller and Singh [246]; Orhan and Keskin Tütüncü [268].

# 27   Discrete and perfect modules

The $\pi$-injective modules which are also direct injective are called *continuous* modules (see [85, 2.12]). The modules with the duals of these properties are of interest provided they are supplemented.

**27.1. Discrete modules.** A supplemented module $M$ is called *discrete* if it is $\pi$-projective and direct projective.

*The following are equivalent for M with $S = \operatorname{End}(M)$:*

(a) $M$ *is discrete;*

(b) $M$ *is lifting and direct projective;*

(c) $M$ *is quasi-discrete and surjective endomorphisms of M with small kernels are automorphisms;*

(d) $M$ *is quasi-discrete, $S/\operatorname{Jac} S$ is regular and $\operatorname{Jac} S = \{f \in S \,|\, \operatorname{Im} f \ll M\}$.*

*Proof.* (a)$\Leftrightarrow$(b) and (a)$\Rightarrow$(c) are obvious from 4.30 and the definitions.

(c)$\Rightarrow$(a). We must show that $M$ is direct projective. Consequently, let $X$ be a direct summand of $M$ and suppose that $f : M \to X$ is an epimorphism with kernel $K$. Since $M$ is quasi-discrete, there is a decomposition $M = A \oplus B$ where $A \subseteq K$ and $K \cap B \ll B$. This gives $X \simeq M/K = (K + B)/K \simeq B/(K \cap B)$ and so, by 26.17, we get $X \simeq B$. Now let $g : X \to B$ be an isomorphism and define $h \in \operatorname{End}(M)$ using $M = A \oplus B$ by setting $(a+b)h = a + (b)fg$ for all $a \in A, b \in B$. Then $h$ is an epimorphism with small kernel $K \cap B$ and so, by (c), $K \cap B = 0$. Thus $K = A \oplus (K \cap B) = A$ and so $f$ splits. Hence $M$ is direct projective.

(a)$\Rightarrow$(d). This assertion follows from [363, 41.19] or [241, Theorem 5.4].

(d)$\Rightarrow$(c). Let $f \in S$ be an epimorphism with $\operatorname{Ke} f \ll M$. Since $S/\operatorname{Jac} S$ is regular, there exists $g \in S$ with $\operatorname{Im}(f - fgf) \ll M$.

We claim that $\operatorname{Im}(1 - fg) \ll M$. To prove this assume $M(1 - gf) + V = M$ for some $V \subset M$. Then $M(1 - fg)f + Vf = Mf = M$ and hence $Vf = M$. Thus $M = V + \operatorname{Ke} f = V$ since $\operatorname{Ke} f \ll M$. Thus $\operatorname{Im}(1 - fg) \ll M$ and so $1 - fg \in \operatorname{Jac} S$. Hence $fg$ is invertible and so $f$ is a monomorphism. This establishes (c).    $\square$

As lifting and direct projective modules are closed under direct summands, discrete modules are closed under direct summands. On the other hand, the $\mathbb{Z}$-modules $A := \mathbb{Z}/8\mathbb{Z}$ and $B := \mathbb{Z}/2\mathbb{Z}$ are discrete but $A \oplus B$ is not lifting and hence not discrete. This leads us to give the following necessary and sufficient condition for a direct sum of modules to be discrete. It is a direct analogue of 26.19.

**27.2. Discreteness of direct sums.** *Let $\{M_i\}_I$ be any family of R-modules and let $M = \bigoplus_I M_i$. Then the following statements are equivalent.*

(a) $M$ *is discrete;*

(b)   (1) *for all $i \in I$, $M_i$ is discrete;*

(2) $M_i$ is $\bigoplus_{I \setminus \{i\}} M_j$-projective, for all $i \in I$;

(3) the decomposition $M = \bigoplus_I M_i$ is an exchange decomposition.

*Proof.* (a)$\Rightarrow$(b). Since direct summands of a discrete module are discrete, (b) (i) follows. As a discrete module is quasi-discrete, (b) (ii) and (iii) follow from 26.19.

(b)$\Rightarrow$(a). By 26.19, $M$ is a quasi-discrete module. Thus, by 27.1 (c), to prove (a) it is enough to show that any small epimorphism $f : M \to M$ is an isomorphism. Since each $M_i$ is discrete, we may write $M_i = \bigoplus_{k_i \in K_i} M_{k_i}$ where each $M_{k_i}$ is hollow. Let $f : M \to M$ be a small epimorphism. Then $M = \sum_I \sum_{K_i} (M_{k_i})f$ is an irredundant sum of hollow modules for the quasi-discrete module $M$ and so, by 26.11 (2), $M = \bigoplus_I \bigoplus_{K_i} (M_{k_i})f$. It now follows from 26.11 (1) that either $(M_{k_i})f \simeq M_{k_i}$ or $(M_{k_i})f \simeq M_{\ell_j}$ where $k_i \neq \ell_j$. In the first case $f|_{M_{k_i}}$ splits since $M_{k_i}$ is discrete while in the second $f|_{M_{k_i}}$ splits since $M_{k_i}$ is $M_{\ell_j}$-projective. From this it follows that $f$ is an isomorphism, proving that $M$ is discrete.   $\square$

Applying 24.14 and 27.2 we get the following whose proof is similar to that of 26.20 and hence omitted.

**27.3. Corollary.** *Let* $\{M_i\}_I$ *be any family of $R$-modules and let* $M = \bigoplus_I M_i$. *Suppose* $\operatorname{Rad} M \ll M$. *Then the following statements are equivalent.*

(a) $M$ *is discrete;*

(b)  (1) *for all* $i \in I$, $M_i$ *is discrete;*

    (2) $M_i$ *is $M_j$-projective, for all* $i \neq j$ *in* $I$.

From 23.15 and 27.2 we get

**27.4. Discreteness of finite direct sums.** *Let* $M = M_1 \oplus \cdots \oplus M_n$. *Then the following are equivalent:*

(a) $M$ *is a discrete module;*

(b) *for each* $1 \leq i \leq n$, $M_i$ *is discrete and $M_j$-projective for each* $j \neq i$.

**27.5. Corollary.** *A finite direct sum of modules* $M_1, \ldots, M_n$ *is discrete if and only if* $M_i \oplus M_j$ *is discrete for all* $1 \leq i < j \leq n$.

Equivalent conditions for a module $M$ to be quasi-discrete in terms of projectivity conditions were given in 26.7. We give below equivalent conditions for a module $M$ to be discrete in terms of projectivity conditions.

**27.6. Supplemented direct projective modules.** *Let* $M$ *be a supplemented module. Consider the following conditions:*

(i) *for* $N \subset M$ *(or* $N \ll M$*), any* $f : M \to M/N$ *with* $\operatorname{Ke} f \overset{cc}{\hookrightarrow} M$ *and* $\operatorname{Im} f = L/N$ *where* $L \overset{cc}{\hookrightarrow} M$, *can be lifted to* $M$;

(ii) *for* $N \subset M$ *(or* $N \ll M$*), any* $f : M \to M/N$ *with* $\operatorname{Ke} f \overset{cc}{\hookrightarrow} M$ *and* $\operatorname{Im} f$ *a direct summand of $M/N$, can be lifted to* $M$;

(iii) *M is direct projective and supplements in M are direct summands.*

*Then* (i) *and* (ii) $\Rightarrow$ (iii), *and* (iii) $\Rightarrow$ (i).

*Proof.* In the proof of 17.14, if $f : M \to M/N$ satisfies the conditions in (i) or (ii), then the map $fi : M \to M/\mathrm{Ke}\,g = M/N \oplus N/(N \cap T)$ also satisfies the corresponding conditions and hence the $\subset$ and $\ll$ versions of (i) (or (ii)) are equivalent.

(i),(ii)$\Rightarrow$(iii). We first prove that supplements in $M$ are direct summands. Suppose $A$ is a supplement in $M$ and $B$ is a supplement of $A$ in $M$. Then $A \cap B$ is a small submodule of both $A$ and $B$. We prove that $A \cap B = 0$. Let $f : M \to M/B$ and $\eta : M \to M/(A \cap B)$ be the natural maps. As $A + B = M$, we may define $g : M/B \to M/(A \cap B)$ by $(m + B)g = a + (A \cap B)$ where $m = a + b \in M$, $a \in A$ and $b \in B$. Now $\mathrm{Im}\,g = A/(A \cap B)$ is a direct summand of $M/(A \cap B)$ and $\mathrm{Ke}\,g = 0$. We note that $\mathrm{Ke}\,fg = B \overset{cc}{\hookrightarrow} M$ and $\mathrm{Im}\,fg = A/(A \cap B)$ where $A \overset{cc}{\hookrightarrow} M$. Hence either (i) or (ii) implies that there exists $h : M \to M$ such that $h\eta = fg$, that is, we have the commutative diagram

$$
\begin{array}{ccc}
M & \overset{f}{\longrightarrow} & M/B \\
\,\downarrow{\scriptstyle h} & & \downarrow{\scriptstyle g} \\
M & \overset{\eta}{\longrightarrow} & M/(A \cap B) \longrightarrow 0.
\end{array}
$$

Now

$$
\begin{aligned}
(A)\eta = A/(A \cap B) &= (A)fg = (A)h\eta & \text{and so} \\
A + \mathrm{Ke}\,\eta &= (A)h + \mathrm{Ke}\,\eta & \text{and so} \\
A + (A \cap B) &= (A)h + (A \cap B) & \text{and so} \\
A &= (A)h, & \text{as } A \cap B \ll A.
\end{aligned}
$$

Also, since $A/(A \cap B) = (M)h\eta = (A)\eta$ and $\mathrm{Ke}\,\eta = A \cap B \ll A$, we get $(M)h = A$. Hence $A = (A)h = (M)h$ and so $A + \mathrm{Ke}\,h = M$. Since $\mathrm{Ke}\,h \subseteq B$ and $B$ is a supplement of $A$, we must have $B = \mathrm{Ke}\,h$. Then, since $B = \mathrm{Ke}\,fg$ and $fg = h\eta$, we get $\mathrm{Ke}\,\eta = 0$, that is, $A \cap B = 0$. Hence $A$ is a direct summand of $M$, as required.

To prove that $M$ is direct projective it is enough to show that if $M/N \simeq A$, a direct summand of $M$, then $N$ is a direct summand of $M$. Let $\eta : M \to M/N$ be the natural map, $p : M \to A$ be the projection and $i : A \to M$ the inclusion map. Let $f : A \to M/N$ be the given isomorphism. For the map $pf : M \to M/N$, we have $\mathrm{Im}\,pf = M/N$ and $\mathrm{Ke}\,pf = \mathrm{Ke}\,p$ is a direct summand of $M$. By (i) or (ii), there exists a map $g : M \to M$ such that $g\eta = pf$. Then $f^{-1}ig\eta = f^{-1}ipf = 1_{M/N}$ and hence the map $\eta$ splits. Thus $N$ is a direct summand of $M$, as required.

(iii)$\Rightarrow$(i). Suppose $f : M \to M/N$ is given with $\mathrm{Ke}\,f \overset{cc}{\hookrightarrow} M$, $\mathrm{Im}\,f = L/N$ and $L \overset{cc}{\hookrightarrow} M$. Since $M$ is supplemented, coclosed submodules of $M$ are supplements and hence, by hypothesis, Ke $f$ and $L$ are direct summands of $M$. Let $M = \mathrm{Ke}\, f \oplus A = L \oplus T$. Then

$$A \simeq L/N \simeq (L \oplus T)/(N \oplus T) = M/(N \oplus T).$$

Now, since $M$ is direct projective, $N \oplus T$ is a direct summand of $M$ and hence $N$ is a direct summand of $M$. Hence the map $f$ can be lifted to $M$. $\qquad\square$

Next we give equivalent conditions for a module to be discrete in terms of some $M$-projectivity conditions.

**27.7. Discrete modules and lifting of maps.** *Suppose $M$ is an amply supplemented module. Then the following are equivalent:*

(a) *$M$ is a discrete module;*

(b) *for all $N \subset M$ (or $N \ll M$), any map $f : M \to M/N$ with $\mathrm{Ke}\, f \overset{cc}{\hookrightarrow} M$, $\mathrm{Im}\, f = L/N$ and $L \overset{cc}{\hookrightarrow} M$ can be lifted to $M$;*

(c) *for all $N \subset M$ (or $N \ll M$), any map $f : M \to M/N$ with $\mathrm{Ke}\, f \overset{cc}{\hookrightarrow} M$ and $\mathrm{Im}\, f \overset{cc}{\hookrightarrow} M/N$ can be lifted to $M$;*

(d) *for all $N \subset M$ (or $N \ll M$), any map $f : M \to M/N$ with $\mathrm{Ke}\, f \overset{cc}{\hookrightarrow} M$ and $\mathrm{Im}\, f$ a direct summand of $M/N$ can be lifted to $M$.*

*Proof.* From the proof of 17.14 we see that, in each of (b), (c) and (d), the $\subset$ and $\ll$ versions are equivalent. Note that, by 27.1 and 22.3, a module is discrete if and only if it is direct projective, amply supplemented and supplements are direct summands.

(a)$\Rightarrow$(b) follows from (iii)$\Rightarrow$(i) of 27.6.

(b)$\Rightarrow$(c). Let $\eta : M \to M/N$ be the natural map and $f : M \to M/N$ be such that $\mathrm{Ke}\, f \overset{cc}{\hookrightarrow} M$ and $\mathrm{Im}\, f = L/N \overset{cc}{\hookrightarrow} M/N$. As $M$ is amply supplemented, every submodule of $M$ has a coclosure (see 20.25). Let $L'$ be a coclosure of $L$ in $M$. Then $L' \overset{cs}{\underset{M}{\hookrightarrow}} L$ and hence $(L' + N)/N \overset{cs}{\underset{M/N}{\hookrightarrow}} L/N$ (see 3.2 (7)). Since $L/N \overset{cc}{\hookrightarrow} M/N$, this gives $L' + N = L$.

Define $f' : M \to L/N$ by $(m)f' = (m)f$ for all $m \in M$. Then $f = f'i$ where $i : L/N \to M/N$ is the inclusion map. Let $g : L/N \to M/(L' \cap N)$ be given by $(\ell + N)g = \ell' + (L' \cap N)$, where $\ell = \ell' + n \in L$, $\ell' \in L'$ and $n \in N$. Let $\eta_1 : M \to M/(L' \cap N)$, $\eta_2 : M/(L' \cap N) \to M/N$ and $\eta : M \to M/N$ be the natural maps. Then $\eta_1 \eta_2 = \eta$. We have $\mathrm{Im}\, f'g = L'/(L' \cap N)$, $L' \overset{cc}{\hookrightarrow} M$ and $\mathrm{Ke}\, f'g = \mathrm{Ke}\, f \overset{cc}{\hookrightarrow} M$. Thus by hypothesis there is a map $h : M \to M$ giving

$h\eta_1 = f'g$ and we have the commutative diagram

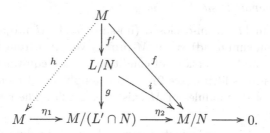

Now $h\eta = h\eta_1\eta_2 = f'g\eta_2 = f'i = f$. Hence (c) follows.

(c)$\Rightarrow$(d) is trivial, while (d)$\Rightarrow$(a) follows from (ii)$\Rightarrow$(iii) of 27.6.   $\square$

**27.8. Strongly discrete modules.** A supplemented module $M$ is called *strongly discrete* if it is self-projective.
*The following are equivalent for $M$:*

(a)  *$M$ is strongly discrete;*

(b)  *$M$ is discrete and epi-projective;*

(c)  *$M$ is quasi-discrete and epi-projective;*

(d)  *$M$ is semi-discrete and epi-projective;*

(e)  *$M$ is lifting and epi-projective.*

*Proof.* (a)$\Rightarrow$(b) is clear from the definitions and the implications (b)$\Rightarrow$(c), (c)$\Rightarrow$(d), and (d)$\rightarrow$(e) are trivial.

(e)$\Rightarrow$(a) Epi-projective implies direct projective and hence an epi-projective lifting module is discrete. Thus $M$ is a direct sum of indecomposable modules (see 26.11), more precisely, $M = \bigoplus_I M_i$ where each $M_i$ is a hollow and epi-projective module, that is, $M_i$ is self-projective (by 4.24). Moreover, each $M_i$ is $\bigoplus_{I\setminus\{i\}} M_j$-projective (see 26.7) and hence $M_i$ is $M$-projective and so $M$ is self-projective.   $\square$

As lifting modules and self-projective modules are closed under direct summands, any direct summand of a strongly discrete module is strongly discrete. However the direct sum of two strongly discrete modules need not be strongly discrete. For example, the $\mathbb{Z}$-modules $\mathbb{Z}/8\mathbb{Z}$ and $\mathbb{Z}/2\mathbb{Z}$ are strongly discrete but their direct sum is neither self-projective or lifting (see 22.5).

As in the case of discrete modules, we give some equivalent conditions for strongly discrete modules in terms of projectivity conditions.

**27.9. Strongly discrete modules and lifting of maps.** *For a supplemented module $M$, the following are equivalent:*

(a)  *$M$ is strongly discrete (i.e. self-projective);*

(b)  *coclosures of submodules of $M$ exist and, for every $N \subset M$ (or $N \ll M$), every $f : M \to M/N$ with $\operatorname{Im} f = L/N$ where $L \overset{cc}{\hookrightarrow} M$ can be lifted to $M$;*

(c) $M$ is im-coclosed (small) $M$-projective;

(d) $M$ is im-summand (small) $M$-projective.

*Proof.* As noted in 17.14, im-coclosed (im-summand) $M$-projective is equivalent to im-coclosed (im-summand) small $M$-projective. Also from the proof of 17.14 we see that the $\subset$ and $\ll$ versions of condition (b) are equivalent.

(a)$\Rightarrow$(b). As $M$ is lifting (see 27.8) $M$ is amply supplemented (see 22.3) and hence coclosures of submodules of $M$ exist (see 20.25). The rest of (b) follows as $M$ is self-projective.

(b)$\Rightarrow$(c). As $M$ is supplemented and coclosures of submodules of $M$ exist, it follows from 20.25 that $M$ is amply supplemented. Now the rest of the proof is similar to that of (b) $\Rightarrow$ (c) of 27.7 and hence is omitted.

(c)$\Rightarrow$(d) is trivial.

(d)$\Rightarrow$(a). It is obvious that (d) implies that $M$ is epi-projective. Also, by (ii) $\Rightarrow$ (iii) of 27.6, all supplements of $M$ are direct summands. Thus, by 22.3, to prove that $M$ is strongly discrete it is enough to prove that $M$ is amply supplemented.

Suppose $A + B = M$. We have to find a supplement of $A$ in $M$ contained in $B$. Since $M$ is weakly supplemented, $B$ contains a weak supplement of $A$. To see this, let $K$ be a weak supplement of $A \cap B$. Then $(A \cap B) + K = M$ and $A \cap B \cap K \ll M$. Since $B = (A \cap B) + (B \cap K)$, we have $(B \cap K) + A = M$ and it follows that $B \cap K$ is a weak supplement of $A$ contained in $B$. Hence without loss in generality we can assume $B$ is a weak supplement of $A$ in $M$.

By assumption, $A$ has a supplement $C$ in $M$, so $M = A + C$ and $A \cap C \ll C$. Consider the natural map $f : M \to M/A$ and the map $g : M/A \to M/(A \cap B)$ given by $(m + A)g = b + (A \cap B)$ where $m = a + b \in M$, $a \in A$ and $b \in B$. Let $\eta : M \to M/(A \cap B)$ be the natural map. Then $\mathrm{Ke}\,\eta \ll M$. Since $\mathrm{Im}\,fg = B/(A \cap B)$ is a direct summand of $M/(A \cap B)$, by hypothesis there exists a map $h : M \to M$ such that $h\eta = fg$. We claim that $(C)h$ is a supplement of $A$ contained in $B$. As $(\mathrm{Im}\,h)\eta = \mathrm{Im}\,fg = B/(A \cap B) = (B)\eta$ we have $\mathrm{Im}\,h \subseteq B$. Since $C + A = M$ we have $(C)h\eta = (B)\eta$ and so $(C)h + (B \cap A) = B$. Thus $(C)h + A = B + A = M$. It is easy to verify $(C \cap A)h = (C)h \cap A$. As $(C \cap A) \ll C$, we have $(C \cap A)h \ll (C)h$. Thus $(C)h$ is a supplement of $A$ contained in $B$. Hence (a) follows by 27.8.    □

**27.10. Discrete modules need not be strongly discrete.** Let $K$ be the quotient field of a discrete valuation domain $R$ which is not complete. Then the $R$-module $K$ is not self-projective but it is direct projective (see Example 23.7) and lifting and hence is discrete.

Suppose $A$ and $B$ are direct summands of a quasi-discrete module $M$ with $X \ll A$ and $Y \ll B$. We saw in 26.17 that if $A/X \simeq B/Y$, then $A \simeq B$. While in general the isomorphism between $A$ and $B$ need not be a lifting of the given isomorphism between $A/X$ and $B/Y$, we prove that there is such a lifting if $M$ is a strongly discrete modules.

**27.11. Characterisation of strongly discrete modules.** *For an R-module M, the following are equivalent:*

(a) *M is a strongly discrete module;*

(b) *M is lifting and every isomorphism $\phi : M/A \to M/B$ can be lifted to an isomorphism $\phi' : M/A' \to M/B'$ where $A'$ and $B'$ are coclosures of A and B in M, respectively.*

*Proof.* (a)$\Rightarrow$(b). Suppose $M$ is strongly discrete and hence lifting. Let an isomorphism $f : M/A \to M/B$ be given and let $A'$ and $B'$ be coclosures in $M$ of $A$ and $B$ respectively. Then $M = A' \oplus A''$, $A = A' \oplus (A'' \cap A)$ and $A'' \cap A \ll M$ and similarly $M = B' \oplus B''$, $B = B' \oplus (B'' \cap B)$ and $B'' \cap B \ll M$. Now $f$ induces an isomorphism $f' : A''/(A'' \cap A) \to B''/(B'' \cap B)$. It is enough to prove that $f'$ can be lifted to an isomorphism $f'' : A'' \to B''$.

Let $\eta_1 : A'' \to A''/(A'' \cap A)$ and $\eta_2 : B'' \to B''/(B'' \cap B)$ be the natural maps. Since $M$ is self-projective, there exists a map $f'' : A'' \to B''$ such that $f''\eta_2 = \eta_1 f'$. As $\eta_2$ and $\eta_1 f'$ are small epimorphisms, the map $f''$ is also a small epimorphism. Then, since $M$ is self-projective, $f''$ splits and hence is an isomorphism.

(b)$\Rightarrow$(a). By 27.8, it suffices to show that $M$ is epi-projective. So let $f : M \to M/N$ be an epimorphism and $\eta : M \to M/N$ the natural map. Then $f$ induces an isomorphism $\overline{f} : M/\mathrm{Ke}\, f \to M/N$. Suppose $\overline{\mathrm{Ke}\, f}$ and $\overline{N}$ are coclosures of $\mathrm{Ke}\, f$ and $N$ in $M$, respectively. Let $\eta_1 : M \to M/\mathrm{Ke}\, f$, $\eta_2 : M/\overline{\mathrm{Ke}\, f} \to M/\mathrm{Ke}\, f$, $\eta_3 : M/\overline{N} \to M/N$, and $\eta_4 : M \to M/\overline{\mathrm{Ke}\, f}$ be the natural maps. By (b), there is an isomorphism $g : M/\overline{\mathrm{Ke}\, f} \to M/\overline{N}$ such that $\eta_2 \overline{f} = g\eta_3$. As $\overline{N}$ is a direct summand of $M$, we get an inclusion map $i : M/\overline{N} \to M$. Thus we have the commutative diagram

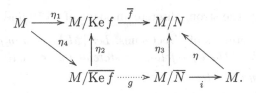

Now $\eta_4 g i \eta = \eta_4 g \eta_3 = \eta_4 \eta_2 \overline{f} = \eta_1 \overline{f} = f$, showing that $M$ is epi-projective. $\qquad \square$

**27.12. Corollary.** *For an R-module M, the following are equivalent:*

(a) *M is a strongly discrete module;*

(b) *M is lifting and any isomorphism $\phi : A/X \to B/Y$ where A and B are direct summands of M, $X \ll A$ and $Y \ll B$ can be lifted to an isomorphism $\psi : A \to B$.*

**27.13. Directly finite strongly discrete modules.** *The following are equivalent for an R-module $M$:*

(a) *$M$ is a directly finite and strongly discrete module;*

(b) *$M$ is lifting and any epimorphism from $M$ to $M/N$ can be lifted to an automorphism of $M$.*

*Proof.* (a)$\Rightarrow$(b). Suppose $M$ is strongly discrete and directly finite. It is enough to prove that any epimorphism $f : M \to M/N$ can be lifted to an automorphism of $M$. By the epi-projectivity of $M$, there exists $g : M \to M$ such that $g\eta = f$ where $\eta : M \to M/N$ is the natural map. We have an isomorphism $\overline{f} : M/\mathrm{Ke}\, f \to M/N$ induced by $f$. By 26.18, there exists an isomorphism $\phi : \overline{\mathrm{Ke}\, f} \to \overline{N}$ where $\overline{\mathrm{Ke}\, f}$ and $\overline{N}$ are coclosures of $\mathrm{Ke}\, f$ and $N$, respectively. Let $K$ be a supplement of $\mathrm{Ke}\, f$ in $M$. Then $M = \overline{\mathrm{Ke}\, f} \oplus K$ and $(K)g + N = M$ and hence $(K)g + \overline{N} = M$. Define $h : M = \overline{\mathrm{Ke}\, f} \oplus K \to M$ by $h|_{\overline{\mathrm{Ke}\, f}} = \phi$ and $h|_K = g$. Then $h$ is an epimorphism and $h\eta = f$. Since $M$ is epi-projective, there is a $k \in \mathrm{End}(M)$ such that $hk = 1_M$. Then, since $M$ is directly finite the map $h$ is an automorphism (see 15.25).

(b)$\Rightarrow$(a). Suppose $M$ is lifting and any onto map from $M \to M/N$ can be lifted to an automorphism of $M$. Then $M$ is an epi-projective lifting module and hence is self-projective (by 27.8). Now suppose that $M = A \oplus B$ and $A \simeq M$. Then we have an onto map $M = A \oplus B \to M$ whose kernel is $B$. By assumption, any epimorphism $M \to M$ is an automorphism. Hence $B = 0$ and so $M$ is directly finite.                                                                                          $\square$

Since a module with finite hollow dimension is directly finite, we get

**27.14. Corollary.** *Suppose $M$ is a strongly discrete module with finite hollow dimension and $A$ and $B$ are submodules of $M$. Then any isomorphism $\phi : M/A \to M/B$ can be lifted to an automorphism of $M$.*

We now prove the strong discrete analogue of 26.19 and 27.2.

**27.15. Strong discreteness of direct sums.** *Let $\{M_i\}_I$ be any family of R-modules and let $M = \bigoplus_I M_i$. Then the following statements are equivalent.*

(a) *$M$ is strongly discrete;*

(b)  (i)  *each $M_i$ is strongly discrete;*

(ii)  *for each $i \in I$, $M_i$ is $M_{I\setminus\{i\}}$-projective;*

(iii)  *the given decomposition of $M$ is an exchange decomposition.*

*Proof.* (a)$\Rightarrow$(b). Since direct summands of discrete modules are discrete and of epi-projective modules are epi-projective, it follows that a summand of a strongly discrete module is strongly discrete, giving (b) (i). Since a strongly discrete module is discrete, conditions (ii) and (iii) in (b) follow from 27.2.

(b)$\Rightarrow$(a). By 27.2, $M$ is a discrete module. Thus it is enough to prove that $M$ is a self-projective module and this follows from conditions (i) and (ii) in (b).   $\square$

A proof similar to that of 26.20 yields

**27.16. Corollary.** *Let $\{M_i\}_I$ be any family of $R$-modules and let $M = \bigoplus_I M_i$. Suppose* $\text{Rad}\,M \ll M$. *Then the following statements are equivalent.*

(a) *$M$ is strongly discrete;*

(b) (1) *each $M_i$ is strongly discrete;*

    (2) *for each $i \in I$, $M_i$ is $M_j$-projective for $j \neq i$.*

Since a finite direct sum of relatively projective lifting modules is an exchange decomposition (see 23.15), we get the following corollary.

**27.17. Finite direct sums of strongly discrete modules.** *Let $M = M_1 \oplus \cdots \oplus M_n$. Then the following are equivalent:*

(a) *$M$ is strongly discrete;*

(b) (i) *for $1 \leq i \leq n$, $M_i$ is strongly discrete and*

    (ii) *for $1 \leq i \neq j \leq n$, $M_i$ is $M_j$-projective.*

**27.18. Remark.** We record here that the following notation is also used in the literature for some of the properties of a module $M$ considered in this section (see, e.g., [241], p. 57).

    $(D_1)$   $M$ is lifting,

    $(D_2)$   $M$ is direct projective,

    $(D_3)$   $M$ is $\cap$-direct projective.

In this terminology quasi-discrete modules are those satisfying $(D_1)$ and $(D_3)$, while discrete modules are defined by the properties $(D_1)$ and $(D_2)$. Note that in [327] Takeuchi calls lifting modules *codirect* and denotes $D_2$ as *Condition (I)* and $D_3$ as *Condition (II)*. Thus discrete modules are *codirect modules with Condition (I)* and quasi-discrete modules are *codirect modules with Condition (II)*. In Oshiro [269] (quasi-)discrete modules figure as *(quasi-)semiperfect* modules.

To finish this section, we investigate supplemented modules with projectivity conditions that subsume all the conditions of this kind previously considered. The interest in these modules is explained by their relation to projective covers of their factor modules. In Mares [229], a projective module $M$ is called *semiperfect* provided every factor module of $M$ has a projective cover and (as we will see) this is equivalent to $M$ being supplemented. It was already observed in Kasch–Mares [193] that self-projectivity of $M$ is sufficient for parts of this relationship and this direction was further studied in Golan [121]. In contrast to projective supplemented modules, self-projective supplemented modules need not have small radicals. However, projective modules $P$ in $\sigma[M]$ are different from their radical (see [363, 22.3]) and as a consequence $\text{Rad}\,P \ll P$ provided $P$ is supplemented. We will refer to this property in the investigation to follow. First we recall a definition from [363, Section 42].

**27.19. Definition.** A module $N$ in $\sigma[M]$ is called *semiperfect in $\sigma[M]$* if every factor module of $N$ has a projective cover in $\sigma[M]$ (see 4.7).

The key to the connection between projective covers and semiperfect modules is given by the following proposition (proved in [363, 42.1]).

**27.20. Supplements and projective covers.** *Let $N$ be a projective module in $\sigma[M]$. Then for any submodule $U \subset N$ the following are equivalent:*

(a) *there exists a direct summand $V \subset N$ with $U + V = N$ and $U \cap V \ll V$;*

(b) *$N/U$ has a projective cover in $\sigma[M]$.*

Using this we now prove

**27.21. Semiperfect projective modules.** *For $M$ the following are equivalent:*

(a) *$M$ is supplemented and projective in $\sigma[M]$;*

(b) *$M$ is lifting and projective in $\sigma[M]$;*

(c) *$M$ is self-projective and semiperfect in $\sigma[M]$;*

(d) *$M$ is small-self-projective and semiperfect in $\sigma[M]$;*

(e) *$M$ is strongly discrete and $\operatorname{Rad} M \ll M$;*

(f) *$M = \bigoplus_I M_i$ where $M_i$ is a local $M$-projective module for each $i \in I$ and $\operatorname{Rad} M \ll M$;*

(g) *for any $k \in \mathbb{N}$, $M^k$ is supplemented and projective in $\sigma[M]$;*

(h) *for any $k \in \mathbb{N}$, $M^k$ is lifting and projective in $\sigma[M]$;*

(i) *$M$ is self-projective and every finitely $M$-generated module has a projective cover in $\sigma[M]$.*

*Proof.* (a)$\Rightarrow$(b) follows from 26.7. (b)$\Rightarrow$(c) is obvious by 27.20.

(c)$\Leftrightarrow$(d) follows from 17.14 (1)(i), which shows that, for every supplemented module $M$, small $M$-projective modules are $M$-projective.

(c)$\Rightarrow$(e). In view of 27.20, it suffices to show that $\operatorname{Rad} M \ll M$. This is proved in [363, 42.3 (2)].

(e)$\Rightarrow$(f). By (e), $M$ is quasi-discrete and hence the decomposition follows from 26.11.

(f)$\Rightarrow$(a). Since the $M_i$ are cyclic and $M$-projective, they are in fact projective in $\sigma[M]$ and hence their direct sum $M$ is projective in $\sigma[M]$. Moreover, since each $M_i$ is projective and local, it is the projective cover of a simple module (see [363, 19.7]). Now (a) follows by [363, 42.4 (1)].

(a)$\Leftrightarrow$(g)$\Leftrightarrow$(h)$\Leftrightarrow$(i) follow from the fact that finite direct sums of $\sigma[M]$-projective supplemented modules are again of this type.  $\square$

We note that modules with property 27.21 (e) are called *quasiperfect* in [121].

By 4.14 (2), if $M \oplus M$ is $\pi$-projective, then $M$ is self-projective. This together with 27.21 gives the following.

**27.22. Corollary.** *If* Rad $M \ll M$, *then the following are equivalent:*

(a) *$M$ is self-projective and semiperfect in $\sigma[M]$;*

(b) *$M$ is strongly discrete;*

(c) *$M \oplus M$ is quasi-discrete;*

(d) *$M \oplus M$ is discrete.*

Note that in 27.21 (i), we can not replace the finitely $M$-generated condition by factor modules of infinite sums of copies of $M$. This more general case is covered by the following notion.

**27.23. Definition.** A module $N \in \sigma[M]$ is called *perfect in $\sigma[M]$* if every $N$-generated module has a projective cover in $\sigma[M]$.

**27.24. Perfect projective modules.** *For $M$ the following are equivalent:*

(a) *$M$ is self-projective and perfect in $\sigma[M]$;*

(b) *for any index set $\Lambda$, $M^{(\Lambda)}$ is strongly discrete;*

(c) *for any index set $\Lambda$, $M^{(\Lambda)}$ is lifting and epi-projective;*

(d) *$M$ is projective in $\sigma[M]$ and every $M$-generated module is (amply) supplemented;*

(e) *$M$ is strongly discrete, and* Rad $(M^{(\Lambda)}) \ll M^{(\Lambda)}$, *$\Lambda$ any set.*

*If $M$ is a finitely generated $R$-module, then the above are equivalent to:*

(f) *$M$ is self-projective and* End$(M)$ *is a left perfect ring;*

(g) *$M$ is strongly discrete and every $M$-generated module has a small radical.*

*Proof.* The assertions follow easily from 27.21 and [363, 43.8]. □

Note that a module $N$ is (semi)perfect in $\sigma[M]$ if and only if its projective cover in $\sigma[M]$ has the corresponding property (see [363, 42.3 and 43.1]). For more details on *perfect* and *semiperfect modules in $\sigma[M]$* we refer to sections 42 and 43 in [363].

By the fact that over a left perfect ring $R$ every left $R$-module has a projective cover, 26.12 implies

**27.25. Corollary.** *Over a left perfect ring every lifting module has an indecomposable decomposition.*

Summarising some of the key concepts in this chapter, we have the following hierarchy for modules $M$:

$M$ supplemented and projective in $\sigma[M] \Rightarrow M$ strongly discrete $\Rightarrow$
$M$ discrete $\Rightarrow M$ quasi-discrete $\Rightarrow M$ semi-discrete $\Rightarrow M$ lifting.

Moreover, we have the following schema:

| The module $M$ is | if it is supplemented and |
|---:|:---|
| quasi-discrete | $\pi$-projective |
| discrete | $\pi$-projective and direct projective |
| strongly discrete | self-projective |
| semiperfect | projective in $\sigma[M]$ |

**27.26. Exercises.**

(1) Let $M$ be a self-projective module and $N \subset M$ a supplement submodule. Prove that if $\operatorname{Rad} N$ is a Hopfian module (see 4.27), then $N$ is a direct summand of $M$ (see [166]).

(2) Prove that the following are equivalent for a ring $R$:

   (a) $R$ is left perfect;

   (b) every flat left $R$-module is self-projective ([122]);

   (c) every cocyclic left $R$-module has a projective cover ([11]).

(3) Let $R$ be a left perfect ring. Prove that every self-projective module $M$ is perfect in $\sigma[M]$.

(4) Let $K_1, \ldots, K_n$ be a family of coindependent submodules of $M$ and put $K = K_1 \cap \cdots \cap K_n$. Show that if $M$ is $\cap$-coclosed projective (4.31) and every coclosed submodule is a direct summand, then every morphism $f : M \to M/K$ can be lifted to an endomorphism.

(5) Prove that for any prime number $p$, the $\mathbb{Z}$-module $\mathbb{Z}/p\mathbb{Z} \oplus \mathbb{Z}/p^2\mathbb{Z}$ is lifting but not $\cap$-coclosed projective ([201]).

(6) If $R$ is a generalised uniserial ring, prove that every quasi-discrete $R$-module is self-projective.

(7) For a ring $R$ prove that the following are equivalent ([178]):

   (a) $R$ is uniserial;

   (b) every self-injective module is quasi-discrete;

   (c) every self-injective module is discrete;

   (d) every (quasi-) continuous module is discrete;

   (e) every quasi-continuous module is quasi-discrete.

(8) Let $R$ be a semiperfect ring such that $\operatorname{Jac}(R)$ is finitely generated. Prove that the following are equivalent ([178]):

   (a) $R$ is a uniserial ring;

   (b) every (quasi-) discrete $R$-module is self-injective;

   (c) every (quasi-) discrete $R$-module is (quasi-) continuous;

   (d) every finitely generated self-projective $R$-module is (quasi-) continuous;

   (e) every self-projective $R$-module is (quasi-) continuous.

(9) For a ring $R$ prove that the following are equivalent ([178]):

   (a) $R$ is uniserial;

(b) every (quasi-) discrete $R$-module is self-injective and $R$ is left perfect;

(c) every (quasi-) discrete $R$-module is (quasi-) continuous and $R$ is left perfect;

(d) every self-projective $R$-module is (quasi-) continuous.

**27.27. Comments.** Discrete modules were first studied by Takeuchi [327] and he calls them *codirect modules with Condition (I)* (see 27.18). Oshiro [269] called a module $M$ "semiperfect" if it is direct projective and lifting (i.e. $M$ is discrete). He justified the name by the fact that a projective module $M$ is "semiperfect" if and only if it is semiperfect in the sense of Mares [229].

Mohamed and Singh [247] studied direct projective lifting modules and called them *dual-continuous* since the definition is dual to that of continuous modules. They studied the basic properties and the endomorphism ring of a discrete module. Also a decomposition theorem for discrete modules was obtained by Mohamed and Singh [247] and later improved by Mohamed and Müller [237]. A direct sum of discrete modules need not be discrete. Regarding the direct sum of discrete modules, they proved 27.4 in the case when each direct summand is indecomposable and 27.3. Keskin [199] has given different proofs for 26.22 and 27.4. Projectivity conditions for supplemented direct projective modules 27.6 and discrete modules 27.7 are given in Ganesan and Vanaja [115].

A self-projective supplemented lifting module was called strongly discrete by Ganesan and Vanaja and some of the results presented here are from [115]. A module $M$ is called *uniquely codirect* in Takeuchi [327] if it satisfies condition (b) of 27.11 and there it is proved that uniquely codirect modules are precisely epi-projective and lifting modules. It should be noted that in [327] epi-projective modules were called *pseudo-projective*. Takeuchi has also shown 27.11 and 27.14 with different proofs.

Semiperfect and perfect modules were introduced by Mares [229] and studied later by Chamard [60], Kasch and Mares [193], Cunningham and Rutter [78] and others. Semiperfect and perfect modules have been generalised in many ways. Quasi-perfect modules were defined and studied by Golan [121]. Semiregular projective modules (see 22.25) were considered by Nicholson [255]. Wisbauer [362] introduced and discussed F-semiperfect and perfect modules in the category $\sigma[M]$ where $M$ is an $R$-module.

Nakahara [250] generalises the notion of semiperfect modules in terms of preradicals and shows that many properties of semiperfect modules are preserved under this generalisation. Yousif and Zhou [374] define and study semiregular, semiperfect and perfect rings related to an ideal $I$ of $R$. When $I = \operatorname{Jac} R$ these are precisely semiregular, semiperfect and perfect rings. Zhou [382] defines a submodule $N \subseteq M$ as $\delta$-small in $M$ if, whenever $N + X = M$ with $M/X$ singular, we have $X = M$. Clearly small submodules of $M$ are $\delta$-small in $M$. An epimorphism $f : M \to N$ is called a $\delta$-cover if $\operatorname{Ke} f$ is $\delta$-small in $M$. Using this $\delta$-perfect, $\delta$-semiperfect and $\delta$-semiregular rings are defined and various characterisations of these rings are given. Some of these notions are transferred to modules in Özcan and Alkan [279].

**References.** Baba and Harada [30]; Chamard [60]; Chang and Kuratomi [61]; Ganesan and Vanaja [115]; Golan [121]; Harada [156, 161, 163]; Hausen and Johnson [166]; Jain, Lopez-Permouth and Rizvi [178]; Keskin [198, 199, 201];

Kuratomi [214]; Miyashita [235]; Mohamed and Müller [241, 238]; Mohamed, Müller and Singh [246]; Nakahara [250]; Oshiro [269, 271, 272]; Takeuchi [327]; Varadarajan [345]; Wisbauer [363, 362]; Yousif and Zhou [374]; Zhou [382]; Zöschinger [389, 391].

# 28 Injective modules lifting in $\sigma[M]$

In this section we are interested in $R$-modules $M$ for which every injective module in $\sigma[M]$ is lifting.

**28.1. Harada modules and rings.** An $R$-module $M$ is called a *Harada module* if every injective module in $\sigma[M]$ is lifting. The ring $R$ is called a *left Harada ring* (*left H-ring* for short) if every injective left $R$-module is lifting, that is, if $_R R$ is a Harada module.

The next proposition is a consequence of the properties of self-injective lifting modules (see 22.20).

**28.2. Injectives as direct sums of lifting modules.** *For a module $M$ the following are equivalent:*

(a) *every injective module in $\sigma[M]$ is a direct sum of lifting modules;*

(b) *every injective module in $\sigma[M]$ is a direct sum of hollow modules;*

(c) *$M$ is locally noetherian and every indecomposable injective module in $\sigma[M]$ is hollow.*

*Proof.* (a)$\Rightarrow$(b). If (a) holds, then any $M$-injective module $N \in \sigma[M]$ is a direct sum of $M$-injective lifting modules. By 22.20, these summands are direct sums of uniform lifting — hence hollow — modules. Thus $N$ is a direct sum of hollow modules.

(b)$\Rightarrow$(c). Since every injective module in $\sigma[M]$ is a direct sum of indecomposables, $M$ is locally noetherian (see 1.14). Now the assertion is clear.

(c)$\Rightarrow$(b) follows from 1.14 and (b)$\Rightarrow$(a) is immediate.      $\square$

**28.3. Injectives as direct sums of local modules.** *Let $M$ be a finitely generated self-projective module. Then the following assertions are equivalent:*

(a) *every injective module in $\sigma[M]$ is a direct sum of local modules;*

(b) *every injective module in $\sigma[M]$ has a small radical and is a direct sum of lifting modules;*

(c) *$M$ is locally noetherian and perfect in $\sigma[M]$ and every indecomposable injective module in $\sigma[M]$ is local.*

*Proof.* (b)$\Rightarrow$(a). By 28.2, any injective module in $\sigma[M]$ is a direct sum of hollow modules. By assumption, these have small radicals and hence are local modules.

(a)$\Rightarrow$(c). Clearly, (a) implies that $M$ is locally noetherian (see 1.14) and that the $M$-injective hull of $M$ is noetherian. By 4.11, this implies that $\mathrm{End}(M)$ is a left artinian, hence a left perfect ring and thus $M$ is perfect in $\sigma[M]$ (see 27.24).

(c)$\Rightarrow$(b). Perfectness of $M$ in $\sigma[M]$ implies that all $M$-generated modules in $\sigma[M]$ have small radicals (see 27.24). The remaining assertions are obvious consequences of the decomposition of $M$-injective modules in $\sigma[M]$.      $\square$

**28.4. Properties of Harada modules.** *Let $M$ be a Harada module. Then $M$ is locally noetherian and every injective module $N$ in $\sigma[M]$ has a decomposition $N = \bigoplus_I N_i$, where each $N_i$ is a local, uniform LE-module.*

*Proof.* By definition, every injective module in $\sigma[M]$ is lifting. Thus by 22.20, every injective module $N$ in $\sigma[M]$ is a direct sum $\bigoplus_I N_i$ of indecomposable modules and $M$ is locally noetherian. From the latter, any direct sum of copies of a fixed $N_i$ is also injective and so lifting. Then, since any indecomposable self-injective module is a uniform LE-module, it follows from 25.2 that each $N_i$ is as claimed.    $\square$

Recall that a module $N \in \sigma[M]$ is a *subgenerator in $\sigma[M]$* if every module in $\sigma[M]$ is subgenerated by $N$, that is, $\sigma[M] = \sigma[N]$. We denote by Add $M$ the class of modules isomorphic to direct summands of copies of $M$ (see 13.10).

A module $M$ is said to be *essentially countably generated* if it has an essential submodule which is countably generated.

**28.5.  Characterisations of Harada modules.** *For an $R$-module $M$, the following are equivalent:*

(a) *$M$ is a Harada module;*

(b) *$M$ is locally noetherian and every non-$M$-small module in $\sigma[M]$ contains a nonzero injective direct summand;*

(c) *every module in $\sigma[M]$ is expressible as a direct sum of an injective module in $\sigma[M]$ and an $M$-small module;*

(d) *there exists a subgenerator $N$ (a direct summand of $M$) in $\sigma[M]$ such that $N$ is $\Sigma$-lifting and $M$-injective, and the family of injective modules in $\sigma[M]$ is closed under taking small covers;*

(e) *there exists a subgenerator $N$ (a direct summand of $M$) in $\sigma[M]$ such that $N$ is locally noetherian, $\Sigma$-lifting, and $M$-injective, and every non-$M$-small factor module of any indecomposable direct summand of $N$ is injective in $\sigma[M]$;*

(f) *every essentially countably generated injective module in $\sigma[M]$ is lifting and every indecomposable injective module in $\sigma[M]$ is either countably generated or $\Sigma$-lifting.*

*Proof.* (a)$\Rightarrow$(b). By 28.4, $M$ is locally noetherian. Any non-$M$-small module $A$ in $\sigma[M]$ is a nonsmall submodule of its injective hull $\widehat{A}$. As $\widehat{A}$ is lifting, $A$ contains a nonzero direct summand of $\widehat{A}$. Hence (b) follows.

(b)$\Rightarrow$(c). Let $A \in \sigma[M]$. If $A$ is $M$-small, then we are through. Suppose $A$ is non-$M$-small. By hypothesis, $A$ contains a nonzero injective summand. Let $\{A_i\}_I$ be a maximal independent family of injective submodules of $A$. Since $M$ is locally noetherian, direct sums of injective modules in $\sigma[M]$ are injective in $\sigma[M]$ (see 1.14). Hence $A = (\bigoplus_I A_i) \oplus B$ for some $B \subset A$. From the choice of the $A_i$'s, $B$ must be an $M$-small module.

(c)⇒(a). Let $A$ be an injective module in $\sigma[M]$ and $X \subseteq A$. By condition (c), $X = B \oplus S$, where $B$ is an injective module in $\sigma[M]$ and $S$ is an $M$-small module. Hence $A = B \oplus C$ and $X = B \oplus (C \cap X)$. As $C \cap X \simeq S$, it is an $M$-small module and, since $A$ is injective in $\sigma[M]$, it is a small submodule of $A$ (see 8.2). Hence $A$ is a lifting module by 22.3.

(a)⇒(d). By (c), $M = N \oplus T$, where $N$ is injective in $\sigma[M]$ and $T$ is an $M$-small module. Since, by 28.4, $M$ is locally noetherian, direct sums of injectives are injective in $\sigma[M]$ and hence any injective module in $\sigma[M]$ is $\Sigma$-lifting. Thus $N$ is $\Sigma$-lifting. By 28.4, any injective module $A$ in $\sigma[M]$ is a direct sum of local uniform modules. We claim that $\operatorname{Rad} A \ll A$. If not, then, since $A$ is lifting, $\operatorname{Rad} A$ contains a nonzero direct summand $X$ which, being injective, is a direct sum of local modules. As $\operatorname{Rad} A$ cannot contain a finitely generated direct summand, we get a contradiction. Thus $\operatorname{Rad} A \ll A$.

By 1.12, $A$ is $M$-generated and so there is an epimorphism $f : \bigoplus_I M_i \to A$, where $M_i = N_i \oplus T_i$, $N_i = N$ and $T_i = T$, for all $i \in I$. Now, since $T_i$ is $M$-small, $(T_i)f$ is also $M$-small by 8.2 (4) and so, since $A$ is injective in $\sigma[M]$, $(T_i)f \ll A$ by 8.2 (1),·that is, $(T_i)f \subseteq \operatorname{Rad} A$. Thus $\sum_I (T_i)f \subseteq \operatorname{Rad} A \ll A$. Hence, since

$$(\bigoplus_I M_i)f = (\bigoplus_I N_i)f + (\bigoplus_I T_i)f = A,$$

we get $(\bigoplus_I N_i)f = A$. This has shown that every injective module in $\sigma[M]$ is $N$-generated, that is, $\sigma[M] = \sigma[N]$. Now let $f : B \to A$ be a small cover of an injective module $A$ in $\sigma[M]$. Then, by (c), $B = C \oplus D$, where $C$ is injective in $\sigma[M]$ and $D$ is an $M$ small module. By 8.3, $B = C$. Thus (d) follows.

(d)⇒(a). Let $L$ be any injective module in $\sigma[M]$ and $T$ be any nonsmall submodule of $L$. Since $\sigma[M] = \sigma[N]$, $L$ is injective in $\sigma[N]$. Therefore, by 1.12, $L$ is $N$-generated. Since $N$ is $\Sigma$-lifting, there exists a small epimorphism $f : A \to L$ where $A \in \operatorname{Add} N$. Let $B = (T)f^{-1}$. As $T$ is a nonsmall submodule of $L$, $B$ is a nonsmall submodule of $A$ by 8.2 (5). Since $A$ is lifting we have $A = C \oplus D$, where $0 \neq C \subseteq B$, $B = C \oplus (B \cap D)$ and $B \cap D \ll A$.

Consider the map $g : C/(\operatorname{Ke} f \cap C) \oplus D/(\operatorname{Ke} f \cap D) \to L$ induced by $f$. Then $g$ is a minimal epimorphism. Therefore, by assumption, $C/(\operatorname{Ke} f \cap C)$ is injective in $\sigma[M]$. Then, as $(C)f \simeq C/(\operatorname{Ke} f \cap C)$, $(C)f$ is a direct summand of $L$. Now $C \subseteq B = (T)f^{-1}$. Therefore $(C)f$ is a direct summand of $(B)f$. Let $(B)f = (C)f \oplus X$. Then

$$X \simeq (B)f/(C)f = [(C)f + (B \cap D)f]/(C)f \simeq (B \cap D)f/[(C)f \cap (B \cap D)f].$$

Thus $X$ is isomorphic to a factor module of $(B \cap D)f$. Now $(B \cap D)f \ll L$ and hence is an $M$-small module. Consequently $X$ is also $M$-small (see 8.2 (4)). Then, since $L$ is injective, $X$ is a small submodule of $L$. Thus we have $T = (B)f = (C)f \oplus X$, where $(C)f$ is a direct summand of $L$ and $X \ll L$. Hence, by 22.3, $L$ is a lifting module.

(a)$\Rightarrow$(e). Since (a)$\Rightarrow$(d), there exists a subgenerator $N$ in $\sigma[M]$ such that $N$ is $\Sigma$-lifting and injective in $\sigma[M]$. It is enough to prove that $N$ is locally noetherian and every non-$M$-small factor module of any indecomposable direct summand of $N$ is injective in $\sigma[M]$. By (b), $M$ is locally noetherian and hence $N$ is also locally noetherian. Let $A$ be an indecomposable direct summand of $N$. Then $A$ is $\Sigma$-injective in $\sigma[M]$ as $M$ is locally noetherian (see 1.14) and hence is $\Sigma$-lifting by (a). As $A$ is self-injective, $\text{End}(A)$ is a local ring and, by 25.2, $A$ is a local module. Suppose $A/X$ is a non-$M$-small factor module of $A$, so that $A/X$ is a nonsmall submodule of $\widehat{A/X}$. Then, since $\widehat{A/X}$ is lifting, its nonsmall submodule $A/X$ contains a nonzero direct summand of $\widehat{A/X}$. Since $A/X$ is local, this gives $A/X = \widehat{A/X}$, an injective module in $\sigma[M]$, as required.

(e)$\Rightarrow$(b). Since $N$ is locally noetherian and $\sigma[M] = \sigma[N]$, $M$ is locally noetherian. Suppose $X$ is a non-$M$-small module in $\sigma[M]$. Then $X$ is a nonsmall submodule of $\widehat{X}$. As $\sigma[M] = \sigma[N]$, $\widehat{X}$ is injective in $\sigma[N]$ and hence is $N$-generated (see 1.12). Now $N$ is $\Sigma$-lifting. There exists a module $A \in \text{Add } N$ and a small epimorphism $f : A \to \widehat{X}$. Let $Y = (X)f^{-1}$. As $X$ is nonsmall in $L$, $Y$ is nonsmall in $A$ by 8.2 (4). Since $M$ is locally noetherian and $A \in \text{Add } N$, where $N$ is an injective module in $\sigma[M]$, $A$ is injective in $\sigma[M]$. Also since $N$ is $\Sigma$-lifting, $A$ is $\Sigma$-lifting. Hence $Y$ contains a nonzero direct summand $T$ of $A$ which is injective in $\sigma[M]$ as $A$ is injective in $\sigma[M]$. Because $M$ is locally noetherian, $T$ is a direct sum of indecomposable modules (see 1.14). Hence $Y$ contains a nonzero indecomposable direct summand, say $U$, of $A$.

As $\text{End}(U)$ is a local ring, $U$ has the exchange property by 12.2 and so, since $N$ is a direct sum of indecomposable modules with local endomorphism ring, $U$ is isomorphic to an indecomposable direct summand of $N$. Thus $U$ is $\Sigma$-lifting and hence a local module by 25.2. Since the map $f$ is a small epimorphism, by 2.2 (5), the image under $f$ of any nonsmall submodule of $A$ is nonsmall in $L$. Hence $(U)f$ is nonsmall in $\widehat{X}$. Therefore $(U)f$ is a non-$M$-small factor module of an indecomposable direct summand of $N$ and hence by hypothesis is injective in $\sigma[M]$. Then, since $(U)f \subseteq X$, $X$ contains a nonzero injective direct summand, as required.

(a)$\Rightarrow$(f). By 28.4, $M$ is locally noetherian and hence any direct sum of injective modules in $\sigma[M]$ is injective in $\sigma[M]$. Thus (f) follows.

(f)$\Rightarrow$(a). Let $A = \bigoplus_{\mathbb{N}} S_n$, where each $S_n$ is a simple module in $\sigma[M]$. Then $\widehat{A}$ is essentially countably generated and hence by assumption is lifting. By 22.20, $\widehat{A}$ is a direct sum of indecomposable modules and any local direct sum of $\widehat{A}$ is a direct summand. Hence $\widehat{A} = \bigoplus_{\mathbb{N}} \widehat{S_n}$ and so, by 1.14, $M$ is locally noetherian.

Now let $D$ be an indecomposable injective module in $\sigma[M]$. Then, by assumption, either $D$ is countably generated or is $\Sigma$-lifting. Note that, by [363, 27.4], $\text{End}(D)$ is a local ring. Thus if $D$ is $\Sigma$-lifting, by 25.2, $D$ is a local module.

Now suppose, on the other hand, that $D$ is countably generated. If $B = D^{(\mathbb{N})}$, then trivially $B$ is countably generated and is injective in $\sigma[M]$ since $M$ is locally noetherian. Then by hypothesis $B$ is lifting. Thus, since $D$ is countably generated, it follows from 25.2 that $D$ is cyclic. Then, as $D$ is also lifting, it is a local module. This has shown that any indecomposable injective module is a local module.

Now let $E$ be any injective module in $\sigma[M]$. As $M$ is locally noetherian, $E = \bigoplus_I E_i$, where each $E_i$ is indecomposable and so a local LE-module. By hypothesis, $\bigoplus_K E_k$ is lifting for any countable subset $K$ of $I$. Hence the family $\{E_i\}_K$ is lsTn (see 24.7) and hence $\{E_i\}_I$ is also lsTn. Now 23.20 and 24.13 imply that $E$ is lifting, as required. $\qquad\square$

**28.6. Corollary.** *If $M$ is a Harada module, then the torsion theory $(\mathbb{S}^\circ, \mathbb{S}^{\circ\bullet})$ splits.*

*Proof.* Recall from 8.2 that $\mathbb{S}$ denotes the class of small modules in $\sigma[M]$. Suppose every injective module in $\sigma[M]$ is lifting and $N \in \sigma[M]$. Then $N = E \oplus S$ where $E$ is injective in $\sigma[M]$ and $S$ is $M$-small (see 28.5 (c)). Also $E = \bigoplus_I E_i$, where each $E_i$ is a local module (see 28.4). Let $I_1 = \{i \mid \mathrm{Re}_\mathbb{S}(E_i) = E_i\}$ and $I_2 = I \setminus I_1$ and set $A = \bigoplus_{I_1} E_i$ and $B = \bigoplus_{I_2} E_i$. Then $\mathrm{Re}_\mathbb{S}(A) = A$, $\mathrm{Re}_\mathbb{S}^2(B) = 0$ and $\mathrm{Re}_\mathbb{S}(S) = 0$. Thus $N = A \oplus (B \oplus S)$, where $A \in \mathbb{S}^\circ$ and $B \oplus S \in \mathbb{S}^{\circ\bullet}$. $\qquad\square$

Recall that a submodule $X$ is a waist of $M$ if, for every submodule $A$ of $M$, either $A \subseteq X$ or $X \subseteq A$.

**28.7. Corollary.** *Assume $M$ is a Harada module and $A$ is an indecomposable non-$M$-small summand of $M$. Then $A$ is local and has a waist $X$ such that $X$ is uniserial and $A/Y$ is injective in $\sigma[M]$ if and only if $Y \subseteq X$.*

*Proof.* By 28.5 (b), $A$ is injective in $\sigma[M]$ and so is local and noetherian by 28.4. Let $Z$ and $Y$ be submodules of $A$ such that $A/Z$ and $A/Y$ are non-$M$-small. There exists an onto map $f : A/(Z \cap Y) \to A/Z$ and so, by 8.2 (4), $A/(Z \cap Y)$ is non-$M$-small and indecomposable and so is injective in $\sigma[M]$ (see 28.5(b)). Hence $A/(Z \cap Y)$ is also uniform. Since $Z/(Z \cap Y) \oplus Y/(Z \cap Y)$ is a submodule of $A/(Z \cap Y)$, we must have either $Z \subseteq Y$ or $Y \subseteq Z$.

Consider $\mathcal{A} := \{Y \subseteq A \mid A/Y$ is non-$M$-small$\}$. Then $\mathcal{A}$ has a largest element (as $A$ is noetherian and $\mathcal{A}$ is linearly ordered), say $X$. Let $B$ be a submodule of $A$. Then $A/(B \cap X)$ is non-$M$-small by 8.2 (4) and hence is an indecomposable injective module in $\sigma[M]$ (see 28.5 (b)). Then, since

$$B/(B \cap X) \oplus X/(B \cap X) \subseteq A/(B \cap X),$$

we get that either $B \subseteq X$ or $X \subseteq B$. Thus $X$ is a waist of $A$ such that (from above) $X$ is uniserial and $A/Y$ is injective if and only if $Y \subseteq X$. $\qquad\square$

**28.8. Finitely generated Harada modules.** *Let $M$ be a finitely generated Harada module. Then there exists a direct summand $P$ of $M$ which is $\Sigma$-lifting, projective and injective in $\sigma[M]$, and such that $\sigma[P] = \sigma[M]$.*

*Proof.* By 28.5, there exists an injective direct summand $N$ of $M$ such that $\sigma[N] = \sigma[M]$. By 28.4 we have $N = \bigoplus_I N_i$, where each $N_i$ is a local LE-module and $I$ is a finite set. In this case we can choose a subset $K$ of $I$ such that $P = \bigoplus_K N_k$ generates $N$ (hence $\sigma[P] = \sigma[N]$) and, for $k \neq \ell \in K$, there exists no epimorphism from $N_k$ to $N_\ell$. Since each $N_k$ is noetherian, any epimorphism from $N_k$ to $N_k$ is an isomorphism for all $k \in K$ (see [363, 31.13 (2)]). As $P \oplus P$ is lifting, by 23.20, $N_i$ is almost $N_j$-projective for all $i, j \in K$. Thus, by 4.40, $N_i$ is $N_j$-projective for all $i, j \in K$. Now 4.3 implies that $P$ is self-projective. As $P$ is finitely generated, $P$ is then projective in $\sigma[P] = \sigma[M]$. As $N$ is injective (hence $\Sigma$-lifting) in $\sigma[M]$, $P$ is also $\Sigma$-lifting and injective in $\sigma[M]$.                                    □

A module $M$ is called *countably (finitely) embedded* if it has a countably (finitely) generated essential socle. A module M is said to be *n-embedded, $n \in \mathbb{N}$,* if it has an essential socle of length $n$.

In this terminology finitely embedded modules are the finitely cogenerated and 1-embedded modules are the cocyclic modules.

**28.9. Every countably embedded injective module in $\sigma[M]$ is lifting.** *Let $M$ be an $R$-module. The following are equivalent:*

(a) *every countably embedded injective module in $\sigma[M]$ is lifting;*

(b) *every countably embedded module in $\sigma[M]$ is a direct sum of an injective module in $\sigma[M]$ and an M-small module;*

(c) *every countably embedded non-M-small module contains a nonzero injective module in $\sigma[M]$ and $M$ is locally noetherian.*

*Proof.* (a)⇔(b) and (c)⇒(a) are easy to prove.

(a)⇒(c). Let $N = \bigoplus_{\mathbb{N}} S_n$, where each $S_n$ is a simple module in $\sigma[M]$. By hypothesis, $\widehat{N}$ is an $M$-injective lifting module and hence is a direct sum of indecomposable modules and any local direct summand of $\widehat{N}$ is a direct summand of $\widehat{N}$ (see 22.20). Hence $\widehat{N} = \bigoplus_{\mathbb{N}} \widehat{S_n}$ and so, by 1.14, $M$ is locally noetherian. The other part of (c) clearly follows from (a).                                    □

We now apply 28.5 to the case $M = R$. Recall from 4.11 (2) that a left noetherian ring is left artinian if the injective hull $E(R)$ of $R$ is finitely generated.

**28.10. Left Harada rings.** *The following are equivalent for a ring $R$:*

(a) *$R$ is a left Harada ring;*

(b) *$R$ is left artinian and every nonsmall module contains a nonzero injective direct summand;*

(c) *every $R$-module is a direct sum of an injective module and a small module;*

(d) *$R$ is left perfect and the family of all injective modules is closed under taking small covers;*

(e) $R$ is left artinian and every nonsmall factor module of an indecomposable direct summand of $R$ is injective;

(f) every essentially countably generated injective $R$-module is lifting and every indecomposable injective module is either countably generated or $\Sigma$-lifting;

(g) $R$ is left and right perfect and $E((R/\mathrm{Rad}\,R)^{(\mathbb{N})})$ is lifting;

(h) $R$ is left artinian and any 2-embedded injective $R$-module is lifting;

(i) $R$ has an essential left socle and any injective $R$-module with essential socle is lifting;

(j) $R$ has an essential left socle, $\mathrm{Rad}\,E(S) \neq E(S)$, for every simple left ideal $S$ of $R$, and every countably embedded injective $R$-module is lifting.

*Proof.* (a)$\Rightarrow$(b). By (a)$\Rightarrow$(b) of 28.5, $R$ is left noetherian and every nonsmall module contains a nonzero injective summand. By 28.4, every injective $R$-module is a direct sum of local modules. Hence $E(R)$, the injective hull of $R$, is finitely generated and so, since $R$ is noetherian, $R$ is left artinian by 4.11 (2).

(b)$\Rightarrow$(c)$\Rightarrow$(a) follows from (b)$\Rightarrow$(c)$\Rightarrow$(a) of 28.5. Thus the conditions (a), (b) and (c) are equivalent.

(c)$\Rightarrow$(d) follows from condition (b) and (c)$\Rightarrow$(d) of 28.5.

(d)$\Rightarrow$(a). As $R$ is left perfect we can write

$$R = \bigoplus_1^n A_i \oplus \bigoplus_1^m B_k,$$

where each $A_i$ is an indecomposable nonsmall projective module, for $1 \leq i \leq n$, and each $B_k$ is an indecomposable small projective module, for $1 \leq k \leq m$. Consider $A_i$, where $1 \leq i \leq n$. Suppose $f : P \to E(A_i)$ is the projective cover of $E(A_i)$. By assumption, $P$ is injective. As $A_i$ is projective, we get an injective map from $A_i$ to $P$. As $P$ is lifting, any indecomposable nonsmall submodule of $P$ is a direct summand of $P$. Hence $A_i$ is injective. Thus $A := \bigoplus_{i=1}^n A_i$ is injective, projective and hence a $\Sigma$-lifting module.

Consider $B_k$, where $1 \leq k \leq m$. Suppose $Q \to E(B_k)$ is the projective cover of $E(B_k)$. By 8.3, any nonzero direct summand of $Q$ is a nonsmall module. Thus $Q \subseteq A^{(I)}$ for some indexing set $I$. As $B_k$ is projective, there exists an injective map from $B_k$ to $Q$. Hence $A^{(I)}$ contains a copy of $B_k$. Therefore, for some index set $K$, $A^{(K)}$ contains a copy of $E(R)$ and so $\sigma[R] = \sigma[A]$. Now (a) follows from (d)$\Rightarrow$(a) of 28.5.

(b)$\Rightarrow$(e). Any indecomposable direct summand of $R$ is a local module.

(e)$\Rightarrow$(b). Suppose $N$ is a nonsmall $R$-module. Let $f : P \to E(N)$ be the projective cover of $E(N)$. As $N$ is a nonsmall module, $N$ is a nonsmall submodule of $E(N)$. As $f$ is a small cover, $(N)f^{-1}$ is a nonsmall submodule $P$. Since $P$ is lifting, $(N)f^{-1}$ contains an indecomposable direct summand, say $A$, of $P$. Because $A$ is not small in $P$ and $f$ is a small map, $(A)f$ is not small in $E(N)$ and hence is a nonsmall module. As $A$ is isomorphic to a direct summand of $R$, by hypothesis, $(A)f$ is an injective direct summand of $N$.

(a)$\Leftrightarrow$(f) follows from the equivalence of (a) and (f) in 28.5.

(b)$\Rightarrow$(g) follows easily from the equivalence of (b) and (a), since any left artinian ring is both left and right perfect (see, e.g., [13, 28.8]).

(g)$\Rightarrow$(h). By 28.9, $R$ is left noetherian and so, since $R$ is right perfect, $R$ is left artinian.

(h)$\Rightarrow$(i). It is enough to prove that any injective module with essential socle is lifting and since $R$ is left noetherian, it is enough to show that any nonsmall submodule of an injective module with essential socle contains a nonzero injective direct summand. Suppose $N$ is an injective module with essential socle and $A$ is a nonsmall submodule of $N$. As $R$ is left artinian, Rad $N \ll N$ and hence $A$ contains a cyclic nonsmall submodule of $N$, say $B$. Then $B$ is finitely embedded and any 2-embedded injective module is lifting implies $E(B)$ is lifting (see 23.20). As $B$ is a nonsmall submodule of $E(B)$ and $E(B)$ is lifting, $B$ contains a nonzero injective direct summand.

(i)$\Rightarrow$(j). By 28.9, $R$ is left noetherian. Therefore, for any simple $R$-module $S$, $E(S)$ is $\Sigma$-lifting and hence is a cyclic module (see 25.2) and so Rad $E(S) \neq E(S)$.

(j)$\Rightarrow$(b). By 28.9, $R$ is noetherian and thus it is left artinian by 4.11. Let $N$ be a nonsmall $R$-module. As Rad $E(N) \ll E(N)$ and $N$ is a nonsmall submodule of $E(N)$, $N$ contains a cyclic nonsmall submodule, say $A$, of $E(N)$. Then $A$ is finitely embedded and by assumption, $E(A)$ is lifting. Since $A$ is nonsmall in $E(A)$, $A$ contains a nonzero direct summand of $E(A)$. Thus $N$ contains a nonzero injective direct summand, as required.                                                       $\square$

**28.11. Co-Harada modules and rings.** We call an $R$-module $M$ a *co-Harada module* if every projective module in $\sigma[M]$ is $\Sigma$-extending. $R$ is called a *left co-Harada ring* (*left co-H-ring* for short) if $_RR$ is a co-Harada module, that is, $_RR$ is a $\Sigma$-extending module.

Notice that there may be no projective modules in $\sigma[M]$ and in this case the condition on co-Harada modules is empty. To avoid this we will assume the existence of projective subgenerators in $\sigma[M]$; indeed, we will assume that $M$ is finitely generated, self-projective and a generator in $\sigma[M]$, that is, $M$ is a *progenerator in* $\sigma[M]$.

From 25.6 we get that if $M$ is a co-Harada module which is projective in $\sigma[M]$, then $M$ is $\Sigma$-lifting and hence a direct sum of local modules. Also $\widehat{M}$ is $\Sigma$-$M$-injective and belongs to Add $M$ and so is projective in $\sigma[M]$.

Recall that a ring $R$ is a *quasi-Frobenius ring*, in short *QF-ring*, if $R$ is noetherian and injective as a left (or right) $R$-module. Such rings are left and right artinian. A well-known result of Faith and Walker [103] states that $R$ is a

QF-ring if and only if every projective left $R$-module is injective, or — equivalently — every injective left $R$-module is projective.

**28.12. Quasi-Frobenius modules.** A module $M$ is called *quasi-Frobenius*, or *QF module* for short, if $M$ is weakly $M$-injective (see 1.12) and a weak cogenerator (in $\sigma[M]$), that is, for finitely generated submodules $K \subseteq M^{(\mathbb{N})}$, the factor modules $M^{(\mathbb{N})}/K$ are $M$-cogenerated.

The following recalls characterisations of a self-projective noetherian QF module (see [363, 48.14]). In particular, we see that $R$ is a quasi-Frobenius ring if and only if $_RR$ is a noetherian QF module.

**28.13. Self-projective, noetherian QF modules.** *For a finitely generated, self-projective $R$-module $M$, the following are equivalent:*

(a) *$M$ is noetherian and a QF module;*

(b) *$M$ is noetherian, self-injective and a self-generator;*

(c) *every injective module in $\sigma[M]$ is projective in $\sigma[M]$;*

(d) *$M$ is a self-generator and every projective module in $\sigma[M]$ is injective.*

We show that QF modules are closely related to Harada modules.

**28.14. Self-injective progenerators as Harada modules.** *Let $M$ be an $R$-module that is a progenerator in $\sigma[M]$. Then the following are equivalent:*

(a) *$M$ is a noetherian QF module;*

(b) *$M$ is a self-injective Harada module;*

(c) *$M$ is a self-injective co-Harada module.*

*Proof.* (a)$\Rightarrow$(b). $M$ is injective in $\sigma[M]$ (see 28.13) and hence is $\Sigma$-injective in $\sigma[M]$. As $M$ is projective in $\sigma[M]$, by 25.6, $M$ is $\Sigma$-lifting. Since $M$ is a generator in $\sigma[M]$, every projective module in $\sigma[M]$ is lifting. By 28.13, every injective module in $\sigma[M]$ is projective and hence $M$ is a Harada module.

(b)$\Rightarrow$(a). A Harada module is locally noetherian (see 28.5) and hence every injective module in $\sigma[M]$ is $\Sigma$-injective and $\Sigma$-lifting. As $M$ is self-injective and finitely generated, $M$ is noetherian $\Sigma$-injective and $\Sigma$-lifting. Since $M$ is a generator in $\sigma[M]$, projectives are injectives in $\sigma[M]$ and hence (a) follows from 28.13.

(a)$\Rightarrow$(c). By 28.13, $M$ is injective in $\sigma[M]$ and so, as $M$ is noetherian, $M$ is a $\Sigma$-injective module in $\sigma[M]$ and hence a co-Harada module.

(c)$\Rightarrow$(a). Since $M$ is injective in $\sigma[M]$, $M = \widehat{M}$. By 25.6, $M$ is $\Sigma$-lifting and $\Sigma$-injective in $\sigma[M]$. As any projective module in $\sigma[M]$ is $M$-generated, every projective module in $\sigma[M]$ is injective in $\sigma[M]$. Thus (a) follows from 28.13. $\square$

Next we characterise Harada modules for which the radical and the singular submodule coincide.

**28.15. Harada modules with Rad $M = Z_M(M)$.** *Let $M$ be a progenerator in $\sigma[M]$. Then the following are equivalent:*

(a) *$M$ is a noetherian QF module;*

(b) *$M$ is a Harada module with $\operatorname{Rad} M = Z_M(M)$;*

(c) *$M$ is a co-Harada module with $\operatorname{Rad} M = Z_M(M)$.*

*Proof.* (a)$\Rightarrow$(b), (c). By 28.14, $M$ is a self-injective Harada as well as co-Harada module. Thus $M$ is $\Sigma$-injective and $\Sigma$-lifting and belongs to Add $M$ (see 25.6). Therefore $M$ is a direct sum of local modules which are injective and projective in $\sigma[M]$. Now Rad $M$ is generated by $M$ and hence $\operatorname{Rad} M \simeq P/K$, where $P$ is a direct sum of copies of $M$. Then $P$ is injective and projective in $\sigma[M]$. If $K$ is not essential in $P$, then Rad $M$ will contain a direct summand injective in $\sigma[M]$ and hence a cyclic direct summand of $M$, which is a contradiction. Hence $K \trianglelefteq P$ and so $\operatorname{Rad} M \subseteq Z_M(M)$. Now $Z_M(M)$ is an $M$-singular module and hence cannot contain a projective direct summand. Then, since $M$ is a lifting module, $Z_M(M)$ must be small in $M$ and hence $Z_M(M) \subseteq \operatorname{Rad} M$. Therefore $\operatorname{Rad} M = Z_M(M)$, as required.

(b)$\Rightarrow$(a). By 28.5, we have a decomposition $M = N \oplus T$, where $N$ is injective in $\sigma[M]$, $T$ is $M$-small and $\sigma[M] = \sigma[N]$. Now $\widehat{T}$ is $N$-generated (see 1.12) and so there is an epimorphism $f : P \to \widehat{T}$, where $P$ is a direct sum of copies of $N$. Since Rad $M = Z_M(M)$, we have Rad $P = Z_M(P)$. Since $T$ is projective, the inclusion map $T \to \widehat{T}$ can be lifted to $g : T \to P$ and $g$ must be injective. As $T \ll \widehat{T}$, $T$ is $M$-small and so $(T)g$ is also $M$-small. Since $N$ is $\Sigma$-injective in $\sigma[M]$, $P$ is injective in $\sigma[M]$ and so $(T)g \ll P$. Therefore $(T)g \subseteq \operatorname{Rad} P = Z_M(P)$. Since $(T)g$ is projective and projective modules in $\sigma[M]$ cannot be $M$-singular, this gives $(T)g = 0$ and hence $T = 0$. Thus $M = N$. Now (a) follows from 28.14.

(c)$\Rightarrow$(a). By 25.6, $M$ is $\Sigma$-lifting and $\widehat{M} \in$ Add $M$. Hence $M$ is a direct sum of indecomposable modules. Suppose $A$ is an indecomposable direct summand of $M$. Then $\widehat{A}$ is a direct summand of $\widehat{M}$ and hence belongs to Add $M$. By 25.6, $A$ is also $\Sigma$-lifting. Then, since $\operatorname{End}(\widehat{A})$ is a local ring, $\widehat{A}$ is a local module (see 25.2). Now if $A$ is a non-$M$-small module, then $A = \widehat{A}$ and so $A$ is injective in $\sigma[M]$. On the other hand, if $A$ is an $M$-small module, then, since $\widehat{A} \in$ Add $M$ and Rad $M = Z_M(M)$, we have Rad $\widehat{A} = Z_M(\widehat{A})$ and so $A \subseteq \operatorname{Rad} \widehat{A} = Z_M(\widehat{A})$. As $A$ is projective in $\sigma[M]$, it cannot be $M$-singular and consequently $A = 0$. Thus $M$ is an injective module. By 28.14, (a) follows.                                              $\square$

Putting $M = R$ in 28.15 we get the following.

**28.16. Characterisations of QF rings.** *The following are equivalent for any ring $R$.*

(a) *$R$ is a QF-ring;*

(b) *$R$ is a left H-ring with $\operatorname{Rad} R = Z(R)$;*

(c) *$R$ is a left co-H-ring with $\operatorname{Rad} R = Z(R)$.*

Next we consider the case when $R$ is a commutative ring. Suppose $M$ is a finitely generated self-generator in $\sigma[M]$. We prove that $M$ is a noetherian QF module if and only if $M$ is a Harada module, if and only if $M$ is a co-Harada module.

**28.17. Lemma.** *Let $R$ be commutative and $M$ be a $\Sigma$-lifting $R$-module that is projective in $\sigma[M]$. Then any nonzero projective module in $\sigma[M]$ contains a nonzero indecomposable summand isomorphic to a direct summand of $M$.*

*Proof.* Suppose $N$ is a nonzero projective module in $\sigma[M]$. Then $N$ is isomorphic to a submodule of a direct sum of copies of $M$. Note that as $M$ is $\Sigma$-lifting and projective, $M$ is a direct sum of local LE-modules (see 25.2). Thus $N \subseteq V = \bigoplus_I M_i$, where each $M_i$ is isomorphic to an indecomposable summand of $M$ and hence a local module. For each $i \in I$, let $p_i : V \to M_i$ be the projection map. There exists some $i \in I$ such that $L := (N)p_i \neq 0$. Suppose $M_i = Ra$. Then $L = \sum_K Rar_k$, where each $r_k \in R$. As $R$ is commutative, $L$ is $M_i$-generated and hence has a small cover $f : T \to L$, where $T \in \mathrm{Add}\, M_i$. From the map $p_i|_N : N \to L$ and the projectivity of $N$ in $\sigma[M]$, we get a map $g : N \to T$ such that $fg = p_i|_N$. As $f$ is a small map and $p_i|_N$ is onto, $g$ is an epimorphism. Since $T$ is projective in $\sigma[M]$, $g$ splits. Therefore $N$ contains a direct summand isomorphic to $T$ and hence an indecomposable summand of $M$ as a direct summand. $\square$

**28.18. Projective (co-)Harada modules over commutative rings.** *Let $R$ be a commutative ring and assume that the $R$-module $M$ is projective in $\sigma[M]$.*

(1) *If $M$ is a Harada module, then $M$ is locally artinian, $\Sigma$-lifting (hence perfect) and $\Sigma$-injective in $\sigma[M]$.*

(2) *If $M$ is a co-Harada module, then $M$ is $\Sigma$-lifting (hence perfect) and $\Sigma$-injective in $\sigma[M]$.*

*Proof.* (1) Let $M$ be a Harada module. From the proof of the implication (a)$\Rightarrow$(d) of 28.5, $M = N \oplus L$, where $N$ is $\Sigma$-injective $\Sigma$-lifting, $\sigma[M] = \sigma[N]$ and $L$ is an $M$-small module. Applying 28.17 to $\sigma[N]$, we get that if $L \neq 0$, then $L$ contains a nonzero indecomposable direct summand of $N$. However, since $N$ is injective in $\sigma[N]$ and $L$ is $N$-small we must have $L = 0$. Thus $M = N$ is $\Sigma$-injective $\Sigma$-lifting and projective in $\sigma[M]$. Therefore $M$ is a direct sum of local modules (see 28.5). By 4.11, over a commutative ring any noetherian self-injective module is artinian. Hence $M$ is locally artinian.

(2) Let $M$ be a co-Harada module. By 25.6, $M$ is $\Sigma$-lifting and hence a direct sum of local modules. We can write

$$M = N \oplus K \text{ with } N = \bigoplus_I N_i \text{ and } K = \bigoplus_K L_k,$$

where $N_i$ is an indecomposable non-$M$-small local module, for every $i \in I$, and $L_k$ is an indecomposable $M$-small local module, for every $k \in K$. By 25.6, $\widehat{M}$ is

$\Sigma$-injective and belongs to Add $M$ and hence is $\Sigma$-lifting. Therefore, for all $i \in I$, $N_i$ is injective in $\sigma[M]$ and hence $N$ is injective in $\sigma[M]$. Given $k \in K$, since $\widehat{L}_k$ is $M$-generated and $M$ is $\Sigma$-lifting, there exists a small epimorphism $f : T \to \widehat{L}_k$ where $T \in$ Add $M$. By 8.3, $T$ cannot contain any $M$-small direct summand. As $M$ is a direct sum of LE-modules, it is easy to see that $T \in$ Add $N$. Thus $\widehat{M} = N \oplus \widehat{L}$ is $N$-generated and hence $\sigma[N] = \sigma[M]$. Hence $L$ is an $N$-small module. Applying 28.17 to $\sigma[N]$ we get $L = 0$. Thus $M = \widehat{M}$ and so $M$ is $\Sigma$-injective in $\sigma[M]$.    $\square$

Using 28.18 and 28.14 we get the following.

**28.19. Characterisation of noetherian QF modules.** *Let $R$ be a commutative ring and assume that the $R$-module $M$ is a progenerator in $\sigma[M]$. Then the following are equivalent:*

(a) *$M$ is a noetherian QF module;*

(b) *$M$ is a Harada module;*

(c) *$M$ is a co-Harada module.*

Next we ask for conditions on $M$ which make every essentially finitely generated injective module in $\sigma[M]$ lifting.

**28.20. Essentially finitely generated injectives lifting I.** *For a module $M$ the following are equivalent:*

(a) *every essentially finitely generated injective module in $\sigma[M]$ is lifting;*

(b) *every essentially finitely generated module in $\sigma[M]$ is a direct sum of an injective module in $\sigma[M]$ and an $M$-small module, and every finitely generated module in $\sigma[M]$ has finite uniform dimension;*

(c) *every non-$M$-small essentially finitely generated module has a nonzero summand which is injective in $\sigma[M]$, and every finitely generated module in $\sigma[M]$ has finite uniform dimension.*

*Proof.* (a)$\Rightarrow$(b). Let $E$ be an essentially finitely generated injective module in $\sigma[M]$. Since it is lifting, by 22.20, $E = \bigoplus_I E_i$, where each $E_i$ is an indecomposable uniform module. As $E$ is essentially finitely generated, $I$ is a finite set. Thus any essentially finitely generated injective module in $\sigma[M]$ has finite uniform dimension. If $A$ is a finitely generated module in $\sigma[M]$, then $\widehat{A}$ is an essentially finitely generated injective module and hence has finite uniform dimension. As $A \trianglelefteq \widehat{A}$, it follows that $A$ has finite uniform dimension, as required.

Now let $A$ be an essentially finitely generated module in $\sigma[M]$. Then $\widehat{A}$ is also essentially finitely generated and hence lifting. By 22.3, $A = B \oplus S$, where $B$ is a direct summand of $\widehat{A}$ (and hence is $M$-injective) and $S$ is a small submodule of $\widehat{A}$ (and hence is $M$-small).

(b)$\Rightarrow$(c) is trivial.

(c)$\Rightarrow$(a). As every finitely generated module in $\sigma[M]$ has finite uniform dimension, every essentially finitely generated module has likewise. Let $E$ be an essentially finitely generated injective module in $\sigma[M]$. To prove that $E$ is lifting, it is enough to show that every submodule of $E$ is a direct sum of a direct summand of $E$ and a small submodule of $E$ (see 22.3). Suppose $A$ is a submodule of $E$. If $A$ is small in $E$, then we are through. Thus suppose that $A$ is nonsmall in $E$. As $E$ is injective in $\sigma[M]$, $A$ is a non-$M$-small module. Moreover, $A$ is also essentially finitely generated and so, by (c), $A = A_1 \oplus B_1$, where $A_1$ is a nonzero injective direct summand of $A$. Since $A_1$ is injective in $\sigma[M]$, $A_1$ is a direct summand of $E$. If $B_1 \ll E$, then we are through. Otherwise $B_1 = A_2 \oplus B_2$ where $A_2$ is a nonzero direct summand of $E$. This procedure must stop after a finite number of steps as $E$ has finite uniform dimension. Thus $E$ is a lifting module. $\qquad\square$

Now we consider the case when $M$ is $\Sigma$-lifting and projective in $\sigma[M]$. Note that, by 27.24, a projective module in $\sigma[M]$ is $\Sigma$-lifting if and only if it is a perfect module in $\sigma[M]$. We need the following lemma.

**28.21. Lemma.** *Suppose that the $R$-module $M$ is projective and perfect in $\sigma[M]$, and non-$M$-small factor modules of any indecomposable direct summand of $M$ are $M$-injective. Let $A$ be an $M$-generated module with finite uniform dimension such that any submodule which is not small in $A$ is a non-$M$-small module. Then $A$ is finitely generated and $M$-injective.*

*Proof.* Since $M$ is $\Sigma$-lifting, there exists a small epimorphism $f : N \to A$, where $N \in \text{Add } M$. Note that $N = \bigoplus_I N_i$, where each $N_i$ is a local module isomorphic to a direct summand of $M$. We show that for any finite subset $K$ of $I$, $(\bigoplus_K N_k)f$ is injective in $\sigma[M]$ with uniform dimension $|K|$. We use induction on $|K|$. For each $i \in I$, $(N_i)f$ is not small in $A$ (since $f$ is a small cover) and hence by assumption on $A$, is non-$M$-small. By hypothesis, each $(N_i)f$ is injective in $\sigma[M]$ and has uniform dimension 1. Let $K \subseteq I$ be a finite subset such that $(N_K)f$ is injective in $\sigma[M]$ with uniform dimension $= |K|$, where $N_K = \bigoplus_K N_k$. Suppose $i \in I \setminus K$. We have

$$(N_K \oplus N_i)f = (N_K)f + (N_i)f = (N_K)f \oplus T,$$

where $T$ is a homomorphic image of $(N_i)f$. $T$ is a nonsmall submodule of $A$ since $f$ is a small cover. Then, by our assumption on $A$, $T$ is non-$M$-small. Thus $T$ is a non-$M$-small factor module of $N_i$ and hence, by hypothesis, is injective in $\sigma[M]$. Therefore $(N_K \oplus N_i)f$ is injective in $\sigma[M]$ with uniform dimension $|K| + 1$. $\qquad\square$

**28.22. Essentially finitely generated injective lifting II.** *Let $M$ be an $R$-module that is projective and perfect in $\sigma[M]$. Then the following are equivalent:*

(a) *every essentially finitely generated injective module in $\sigma[M]$ is lifting;*

(b) *every non-M-small factor module of any indecomposable summand of M is injective in σ[M], and every finitely generated module in σ[M] has finite uniform dimension;*

(c) *every small cover of an essentially finitely generated injective module in σ[M] is injective in σ[M].*

*Proof.* (a)⇒(b). By 28.20, every finitely generated module in $σ[M]$ has finite uniform dimension. Since $M$ is a projective and perfect module in $σ[M]$, $M$ is a direct sum of local modules (see 27.21). Suppose $A$ is an indecomposable direct summand of $M$ and $A/X$ is a non-$M$-small module. Then, since $A/X$ is local, 28.20 (b) implies that $A/X$ is injective in $σ[M]$. Thus (b) follows.

(b)⇒(c). Suppose $g : B → C$ is a small cover, where $C$ is an essentially finitely generated injective module in $σ[M]$. As every finitely generated module has finite uniform dimension, $C$ has finite uniform dimension. Moreover, since $C$ is injective in $σ[M]$, $C$ is $M$-generated and any submodule of $C$ which is not small in $C$ is a non-$M$-small module. By 28.21, $C$ is finitely generated. As $g$ is a small epimorphism, $B$ is also finitely generated and so, by (b), $B$ has finite uniform dimension. Since $M$ is projective in $σ[M]$, $C$ is $M$-generated and $g$ is a small map, it follows that $B$ is also $M$-generated. As $B$ is a small cover of an injective module, any submodule of $B$ which is not small in $B$ is a non-$M$-small module (see 8.3). Thus, by 28.21, $B$ is injective in $σ[M]$.

(c)⇒(a). The proof is similar to the proof of (d)⇒(a) of 28.5.                □

### 28.23. Exercises.

(1) Let $R$ be a left artinian ring whose indecomposable injective $R$-modules are finitely generated. Prove that the following are equivalent:

(a) $R$ is a left $H$-ring;

(b) every finitely generated injective module is a lifting module;

(c) every finitely generated nonsmall $R$-module contains a nonzero injective submodule;

(d) the family of all finitely generated injective $R$-modules is closed under taking small covers;

(e) every finitely generated $R$-module can be expressed as a direct sum of an injective module and a small module.

(2) Let $M$ be an $R$-module such that $M$ is projective in $σ[M]$. Show that for any simple module $S ∈ σ[M]$, $\widehat{S}$ is not $M$-singular if and only if $S$ is isomorphic to a submodule of $M$.

(3) Recall that a ring $R$ is a *left QF-3 ring* if it has a minimal faithful module $N$, minimal in the sense that any faithful module contains a direct summand isomorphic to $N$.

Suppose $R$ is a semiperfect ring with essential socle and $\mathrm{Rad}\,E(S) ≠ E(S)$, for every simple submodule $S$ of $R$. Prove that the following are equivalent:

(a) $R$ is a left QF-3 ring;

(b) any finite direct sum of indecomposable not singular injective $R$-modules is lifting;

(c) the direct sum of any two indecomposable not singular injective $R$-modules is lifting;

(d) any indecomposable not singular injective $R$-module is projective.

(4) Suppose $R$ is a semiperfect ring with essential socle and $\operatorname{Rad} E(S) \neq E(S)$, for every simple submodule $S$ of $R$. Prove that the following are equivalent:

(a) $R$ has a faithful $\Sigma$-injective left ideal;

(b) any injective module in $\sigma[M]$ with essential socle whose indecomposable direct summands are not singular is lifting;

(c) any countably embedded injective module in $\sigma[M]$ whose indecomposable direct summands are not singular is lifting.

(5) Let $R$ be a perfect ring. Recall that $R$ is *semiprimary* if $R/J(R)$ is semisimple and $J(R)$ is nilpotent. Prove that the following are equivalent:

(a) $R$ is a semiprimary QF-3 ring;

(b) any projective $R$-module whose indecomposable direct summands are non-small is extending;

(c) any countably generated projective $R$-module whose indecomposable direct summands are nonsmall is extending;

(d) any injective $R$-module whose indecomposable direct summands are not singular is lifting;

(e) any countably embedded injective $R$-module whose indecomposable direct summands are not singular is lifting.

(6) Let $M$ be an $R$-module. Prove that the following are equivalent:

(a) every finitely embedded injective module in $\sigma[M]$ is lifting;

(b) any 2-embedded injective module in $\sigma[M]$ is lifting;

(c) every finitely embedded module in $\sigma[M]$ contains a nonzero injective module in $\sigma[M]$;

(d) every finitely embedded module in $\sigma[M]$ is a direct sum of an injective module in $\sigma[M]$ and an $M$-small module.

**28.24. Comments.** Various characterisations of rings have been given when a class of modules defined by some projectivity conditions coincides with a class of modules defined by some injectivity conditions. For example, a ring $R$ is a QF ring if the class of projective modules coincide with the class of injective modules. Faith and Walker [103] proved that a ring $R$ is a QF if and only if every projective right $R$-module is injective, if and only if every injective right $R$-module is projective. It is easy to see that $R$ is a QF ring if and only if every projective module is quasi-injective (continuous, quasi-continuous), if and only if every injective module is quasi-projective (discrete, quasi-discrete) (see [49]).

Recall that a left and right artinian serial principal ideal ring is called a *uniserial ring*. It follows from Fuller's work (e.g. [110], [13, Exercise 9, page 359]) that a ring $R$ is a uniserial ring if and only if every factor ring of $R$ is QF. Using this Byrd [49] proved that a ring $R$ is uniserial if and only if every quasi-projective $R$-module is quasi-injective, if and only if every quasi-injective $R$-module is quasi-projective.

Colby and Rutter [74] characterised semiprimary QF-3 rings as those rings for which the injective envelope of any projective module is projective. They also showed that these are precisely the left perfect rings with the property that the projective cover of any injective module is injective.

It is obvious that any module containing a nonzero direct summand is nonsmall. Harada considered the rings satisfying the converse of this in [152] as well as the rings satisfying the dual property. He characterised them in terms of ideals in [152], [153] and [154]. Oshiro [271] defined a left H-ring (in honour of Harada) as a left artinian ring for which every nonsmall module contains a nonzero injective module and defined a co-H-ring as a ring with ACC on annihilator left ideals and such that every module which is not singular has a nonzero projective direct summand. He studied the module theoretic properties of left H-rings and left co-H-rings. A ring is a left H-ring if and only if $R$ is left perfect and any small cover of an injective module is injective, if and only if every injective module is lifting. A ring is a left co-H-ring if the class of projective modules is closed under essential extensions; this holds if and only if all projective modules are extending.

The class of all H-rings (co-H-rings) lies strictly in between the class of all semiprimary QF-3 rings and the class of all QF rings. Unlike QF rings and semiprimary QF-3 rings, a left H-ring (co-H-ring) need not be a right H-ring (co-H-ring). In [271], Oshiro showed that if $R$ is a left nonsingular ring, then $R$ is a left H-ring if and only if it is a left co-H-ring. For commutative rings the classes of QF rings, H-rings and co-H-rings coincide. Somewhat unexpectedly it turns out that a ring is a left H-ring if and only if it is a right co-H-ring (see Oshiro [273]). Jayaraman and Vanaja [182] defined and studied H-modules and co-H-modules.

Knowing that uniserial rings are those for which every factor ring is QF, the question arises for which rings the factor rings are (semiprimary) QF-3 rings, H-rings, or co-H-rings. Kawada [197] has proved that an artinian ring is a generalised uniserial ring if and only if every factor ring is QF-3. In [342], Vanaja and Purav have improved this result. They showed that for a ring $R$ the following are equivalent: (a) $R$ is a generalised uniserial ring; (b) every factor ring of $R$ is a left H-ring; (c) every factor ring of $R$ is a left co-H-ring; (d) every factor ring of $R$ is a semiprimary QF-3 ring. They also characterise generalised uniserial rings in terms of the quasi-projective cover of quasi-injective modules and the quasi-injective envelope of quasi-projective modules.

It should be noted that "Harada modules" are defined by Kasch [188] as modules with an LE-decomposition which complement direct summands and in [208] Gupta and Khurana study the endomorphism rings of Harada modules as defined by Kasch.

**References.** Byrd [49]; Colby and Rutter [74]; Fuller [110]; Gupta and Khurana [208]; Harada [155, 153, 154, 156, 158]; Harada and Mabuchi [161]; Harada and Tozaki [163]; Jayaraman and Vanaja [182]; Kasch [188]; Kawada [197]; Oshiro [269, 271, 273, 274, 275, 272]; Oshiro and Wisbauer [277]; Vanaja [340].

# 29  Extending modules lifting in $\sigma[M]$

In this section we will consider modules $M$ for which extending — or even all — modules in $\sigma[M]$ are lifting and we give various characterisations of modules with this property. We begin with some conditions under which the extending property of a noetherian serial module $N$ implies the lifting property for $N$ and vice versa.

**29.1. Extending and lifting property of noetherian uniserial modules.** *Let $M$ be an R-module such that every indecomposable injective module in $\sigma[M]$ is noetherian and uniserial. Let $A \in \sigma[M]$ be a finitely generated serial module.*

(1)  *If $A$ is a lifting module, then $A$ is an extending module.*

(2)  *If $A$ is an extending module and $\widehat{A}$ is a lifting module, then $A$ is a lifting module.*

*Proof.* Let $U$ be a uniform module in $\sigma[M]$. As $\widehat{U}$ is uniserial noetherian, $U$ is self-injective (see 2.18), noetherian, uniserial and $\mathrm{End}(U)$ is a local ring.

(1) Suppose $A = \bigoplus_{i=1}^{n} A_i$, where each $A_i$ is cyclic uniserial, is a lifting module. Then, for $1 \leq i \leq n$, $A_i$ is a noetherian uniserial LE-module. To prove that $A$ is extending it is enough to prove that $A_i$ is almost $A_j$-injective for $i \neq j$ (see 1.19). Since both $A_i$ and $A_j$ are uniform, to show that $A_j$ is almost $A_i$-injective it is enough to show that given $f : C \to A_j$ with $C \subseteq A_i$, either $f$ has an extension to $A_i$ or $f$ is injective and $f^{-1} : (C)f \to A_i$ can be extended to $A_j$.

Case (i). Suppose $f$ is an injective map. Then $\widehat{A_i} \simeq \widehat{A_j}$. Also $\widehat{A_i}$ is uniform and hence by assumption a noetherian uniserial module. Either $A_j$ is isomorphic to a submodule of $A_i$ or $A_i$ is isomorphic to a submodule of $A_j$ and both $A_j$ and $A_i$ are self-injective. Thus either $A_i$ is $A_j$-injective or $A_j$ is $A_i$-injective. Hence either $f^{-1} : (C)f \to A_i$ can be extended to $A_j$ or $f$ can be extended to $A_i$.

Case (ii). Suppose $f$ is not injective. There exists a map $g : \widehat{A_i} \to \widehat{A_j}$ such that $(x)g = (x)f$, for all $x \in C$. We claim that $(A_i)g \subseteq A_j$. Suppose not. Then, as $\widehat{A_j}$ is uniserial, we must have $A_j \subset (A_i)g$. Thus if $X = (A_j)g^{-1}$, then by assumption $A_i \not\subseteq X$ and so $X \subset A_i$. Now define $Y = \{(x, (x)g) \in \widehat{A_i} \oplus \widehat{A_j} \,|\, x \in X\}$. Then $Y$ is uniserial, as it is isomorphic to $X$, and it is a submodule of $A_i \oplus A_j$. Since $(X)g = A_j$, we have $Y + A_i = A_i \oplus A_j$. Hence $Y$ is a nonsmall submodule of $A_i \oplus A_j$. Since $A_i \oplus A_j$ is a lifting module, it follows that $Y$ is a direct summand of $A_i \oplus A_j$. From our introductory remark, $\mathrm{End}(Y)$ is a local ring and so $Y$ has the exchange property (see 12.2). Thus either $A_i \oplus A_j = Y \oplus A_j$ or $A_i \oplus A_j = Y \oplus A_i$. As $\mathrm{Ke}\, g \neq 0$, $Y \cap A_i \neq 0$ and hence $A_i \oplus A_j = Y \oplus A_j = X \oplus A_j$. This implies that $X = A_i$, a contradiction to our assumption. Thus $(A_i)g \subseteq A_j$ and $g\,|_{A_i}$ is an extension of $f$.

(2) Now suppose that $A$ is an extending module for which $\widehat{A}$ is a lifting module and write $A = \bigoplus_{i=1}^{n} A_i$, where each $A_i$ is cyclic uniserial. By 23.20, to prove that $A$ is a lifting module it is enough to prove that for $i \neq j$ and $1 \leq i$,

$j \leq n$, $A_j$ is almost $A_i$-projective. As $A_i$ and $A_j$ are hollow modules it is enough to show that given an epimorphism $f : A_i \to A_i/L$ and $g : A_j \to A_i/L$ either $g$ can be lifted to a map $h : A_j \to A_i$ or $g$ is an onto map and $f$ can be lifted to a map $h : A_i \to A_j$.

Consider the diagram

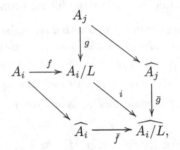

where $i : A_i/L \to \widehat{A_i/L}$ is the inclusion map, $\overline{f} : \widehat{A_i} \to \widehat{A_i/L}$ is an extension of $f \diamond i$, and $\overline{g} : \widehat{A_j} \to \widehat{A_i/L}$ is an extension of $g \diamond i$. As $\widehat{A_i/L}$ is uniserial, either $(\widehat{A_i})\overline{f} \subseteq (\widehat{A_j})\overline{g}$ or $(\widehat{A_i})\overline{f} \supseteq (\widehat{A_j})\overline{g}$. Also $\widehat{A_i} \oplus \widehat{A_j}$ is a lifting module since it is a direct summand of $\widehat{A}$.

Case (i). Let $(\widehat{A_j})\overline{g}$ be a proper submodule of $(\widehat{A_i})\overline{f}$. As $\widehat{A_i} \oplus \widehat{A_j}$ is a lifting module, $\widehat{A_j}$ is almost $\widehat{A_i}$-projective by 23.20 and so there exists $\overline{h} : \widehat{A_j} \to \widehat{A_i}$ such that $\overline{h} \diamond \overline{f} = \overline{g}$ (for, as $\widehat{A_i}$ and $\widehat{A_j}$ are hollow modules, there cannot exist a map $\overline{h} : \widehat{A_i} \to \widehat{A_j}$ such that $\overline{h} \diamond \overline{g} = \overline{f}$) yielding the commutative diagram

$$
\begin{array}{ccc}
 & \widehat{A_j} & \\
{\scriptstyle \overline{h}} \swarrow & \downarrow {\scriptstyle \overline{g}} & \\
\widehat{A_i} \xrightarrow{\ \overline{f}\ } (\widehat{A_i})\overline{f} & \longrightarrow & 0.
\end{array}
$$

Now $((A_j)\overline{h})\overline{f} = (A_j)\overline{g} = (A_j)g \subseteq A_i/L$ and $(A_i)\overline{f} = (A_i)f = A_i/L$. Thus, since $\widehat{A_j}$ is uniserial, we get $(A_j)\overline{h} \subseteq A_i$. Hence $g$ can be lifted to $A_i$.

Case (ii). Let $(\widehat{A_i})\overline{f}$ be a proper submodule of $(\widehat{A_j})\overline{g}$. In this case there exists a map $\overline{h} : \widehat{A_i} \to \widehat{A_j}$ such that $\overline{h}\,\overline{g} = \overline{f}$, that is, we have the commutative diagram

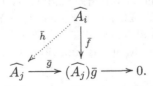

We claim that $(A_i)\overline{h} \subseteq A_j$ and hence $g$ is an onto map and $\overline{h}\,|_{A_i}$ is a lifting of the map $f$. Suppose instead that $A_j$ is a proper submodule of $(A_i)\overline{h}$ and $X := (A_j)\overline{h}^{-1}$.

Then $X$ is a proper submodule of $A_i$. Let $h = \overline{h}\,|_X \colon X \to A_j$. As $X$ is a small submodule of $A_i$, the map $h$ cannot be extended to $A_i$. Since $A_i \oplus A_j$ is an extending module, $A_j$ is almost $A_i$-injective and hence $h$ must be injective. This implies that $\overline{h}$ is also injective. Now $(\widehat{A_i})\overline{h} \simeq \widehat{A_i}$ and hence is a direct summand of $\widehat{A_j}$. However, since $\widehat{A_j}$ is indecomposable, this gives $(\widehat{A_i})\overline{h} = \widehat{A_j}$. Thus $\overline{h}$ is an onto map and so $(\widehat{A_i})\overline{f} = (\widehat{A_j})\overline{g}$, a contradiction.

Case (iii). Finally suppose that $(\widehat{A_i})\overline{f} = (\widehat{A_j})\overline{g}$. In this case either $\overline{g}$ has a lifting or $\overline{f}$ has a lifting. As in the proofs of cases (i) and (ii), we can show that either the map $g$ has a lifting or the map $f$ has a lifting.

Using 23.20 we now conclude that $N$ is a lifting module.                    $\square$

**29.2. Descending Loewy series.** The *descending Loewy series* of $M$ is the chain of submodules

$$\mathrm{Rad}_0 M = M \supseteq \mathrm{Rad}_1 M = \mathrm{Rad}\, M \supseteq \cdots \supseteq \mathrm{Rad}_\alpha M \supseteq \mathrm{Rad}_{\alpha+1} M \supseteq \cdots ,$$

where for each ordinal $\alpha \geq 0$, $\mathrm{Rad}_{\alpha+1} M = \mathrm{Rad}\,(\mathrm{Rad}_\alpha M)$ and, for a limit ordinal $\alpha$, $\mathrm{Rad}_\alpha M = \bigcap_{\beta < \alpha} \mathrm{Rad}_\beta M$.

**29.3. Lemma.** *Suppose* $M = \bigoplus_I M_i$, *where each* $M_i$ *is noetherian uniserial and, for every ordinal* $\alpha$, $M/\mathrm{Rad}_\alpha M$ *is extending. Then any uniform homomorphic image of* $M$ *is uniserial.*

*Proof.* Let $g : M \to U$ be an epimorphism, where $U$ is a uniform module. By 2.17, it is enough to prove that every finitely generated submodule of $U$ is local. As each $M_i$ is noetherian, this will follow if for any finite subset $K \subseteq I$, $(\bigoplus_K M_k)f$ is uniserial. Thus it is enough to prove the case when $I$ is a finite set and we now assume this.

Suppose $\mathrm{Ke}\,g$ is not essential in $M$. As $M$ is extending, $M = A \oplus B$, where $\mathrm{Ke}\,g \trianglelefteq A$. Since $U$ is uniform, $A = \mathrm{Ke}\,g$ and $U \simeq B$. For all $i \in I$, $M_i$ is an LE-module and so, by 12.3 and 11.35, $B \simeq M_i$, for some $i \in I$. Thus $U$ is a noetherian uniserial module.

Suppose $\mathrm{Ke}\,g \trianglelefteq M$ and $X_i = \mathrm{Ke}\,g \cap M_i$, for $i \in I$. For each $i \in I$, since $M_i$ is noetherian uniserial and $X_i \neq 0$, there exists an ordinal $\alpha_i$ such that $X_i = \mathrm{Rad}_{\alpha_i} M_i$. As $I$ is a finite set, we can choose $i_0 \in I$ such that $\alpha_{i_0} \geq \alpha_i$, for all $i \in I$. Then $g$ induces an epimorphism $f : M/\mathrm{Rad}_{\alpha_{i_0}} M \to U$. Now $\mathrm{Ke}\,f$ is not essential in $M/\mathrm{Rad}_{\alpha_{i_0}} M$ and by hypothesis, $M/\mathrm{Rad}_{\alpha_{i_0}} M$ is extending. Then proceeding as in the previous paragraph, we can prove that $U$ is a uniserial noetherian module.   $\square$

**29.4. Extending modules are lifting I.** *For $M$, the following are equivalent:*

(a) *every $M$-generated extending module is lifting;*

(b) *every $M$-generated self-injective module is lifting;*

(c) *every M-generated module is a Harada module;*

(d) *every extending module in σ[M] is a lifting module;*

(e) *every self-injective module in σ[M] is lifting;*

(f) *every module in σ[M] is a Harada module;*

(g) *every finitely generated module is a Harada module and* $\text{Rad}\,N \ll N$ *for every module* $N \in \sigma[M]$;

(h) *every finitely generated module in σ[M] is a direct sum of noetherian uniserial modules and* $\text{Rad}\,N \ll N$ *for every module* $N \in \sigma[M]$;

(i) *M is locally noetherian and every module in σ[M] has a cyclic uniserial direct summand;*

(j) *every lifting module is extending in* $\sigma[M]$, $\widehat{M}$ *is a Σ-lifting module, and* $\text{Rad}\,U \ll U$ *for all uniform modules* $U \in \sigma[M]$;

(k) *every M-generated lifting module is extending,* $\widehat{M}$ *is a Σ-lifting module, and* $\text{Rad}\,U \ll U$ *for all uniform modules* $U \in \sigma[M]$.

*Proof.* We first recall that if $M$ is a nonzero Harada module, then, by 28.5 (d), $M$ has a direct summand $N$ which is injective in $\sigma[M]$ and such that $\sigma[M] = \sigma[N]$. Also $M$ is locally noetherian and every injective module in $\sigma[M]$ is a direct sum of local modules (see 28.4). Hence any nonzero Harada module $M$ contains a nonzero direct summand which is local, uniform, noetherian and injective in $\sigma[M]$.

(a)⇒(b) is trivial.

(b)⇒(c). Suppose $N$ is an $M$-generated module. Then any injective module in $\sigma[N]$ is $M$-generated (see 1.12). Thus any injective module in $\sigma[N]$ is an $M$-generated self-injective module. Therefore, by (b), $N$ is a Harada module.

(c)⇒(d). By (c), $M$ is a Harada module and hence is locally noetherian. If $N$ is an indecomposable injective module in $\sigma[M]$, then $N$ is a local module (see 28.4) and is $M$-generated. By (c), every factor module of $N$ is an indecomposable Harada module and hence is a uniform module. Then, by 2.17, $N$ is uniserial. Thus any uniform module $U$ in $\sigma[M]$ is a uniserial noetherian module.

Now let $A$ be an extending module in $\sigma[M]$. As $M$ is a locally noetherian module, by [363, 27.3] $A$ is also locally noetherian. From [85, 8.3 and 8.6] we get that every local direct summand of $A$ is a direct summand and that $A$ is a direct sum of uniform modules. Thus $A = \bigoplus_I A_i$ where, for all $i \in I$, $A_i$ is a noetherian uniserial module and hence an LE-module. Since $M$ is a Harada module, for $i \neq j \in I$, $\widehat{A_i} \oplus \widehat{A_j}$ is lifting and hence, by 29.1, $A_i \oplus A_j$ is a lifting module. It now follows from 23.20 and 24.13 that $A$ is a lifting module.

(d)⇒(e) is obvious.

(e)⇒(f). Suppose $N \in \sigma[M]$. Every injective module in $\sigma[N]$ is self-injective and hence lifting. Thus $N$ is a Harada module.

(f)$\Rightarrow$(g). If possible, let $\operatorname{Rad} N$ be a nonsmall submodule of $N \in \sigma[M]$. Then there exists a proper submodule $L$ of $N$ such that $\operatorname{Rad} N + L = N$. Then $N/L = (\operatorname{Rad} N + L)/L \subseteq \operatorname{Rad}(N/L)$ and hence $N/L = \operatorname{Rad}(N/L)$. As $N/L$ is a nonzero Harada module, it contains a nonzero direct summand which is a local module. This contradicts $N/L = \operatorname{Rad}(N/L)$. Thus for any module $N \in \sigma[M]$, $\operatorname{Rad} N \ll N$. Hence (g) follows.

(g)$\Rightarrow$(h). Since Harada modules are locally noetherian, every finitely generated module in $\sigma[M]$ is noetherian. Also any nonzero Harada module contains a nonzero direct summand which is local, uniform and noetherian. Hence if $A$ is a finitely generated module in $\sigma[M]$, then $A = \bigoplus_{i=1}^{n} A_i$, where each $A_i$ is local and uniform. For $1 \leq i \leq n$, any factor module of $A_i$ is uniform (as it is an indecomposable Harada module) and hence by 2.17, $A_i$ is a uniserial module. Thus $A$ is a noetherian serial module and (h) follows.

(h)$\Rightarrow$(i). By assumption, every finitely generated module in $\sigma[M]$ is noetherian and so $M$ is locally noetherian. We now show that any uniform module $U$ in $\sigma[M]$ is a noetherian uniserial module. Any finitely generated submodule of $U$ is uniform and hence, by (h), is cyclic uniserial and so has a unique maximal submodule. Thus $U$ is uniserial by 2.17. As $\operatorname{Rad} U \ll U$ by hypothesis, $U$ is in fact a cyclic noetherian uniserial module.

Let $N \in \sigma[M]$ and let $\overline{N}$ be the injective hull of $N$ in $\sigma[N]$. Now $\operatorname{Rad} \overline{N} \ll \overline{N}$ and $\overline{N}$ is an epimorphic image of a direct sum of copies of $N$, say $\phi : \bigoplus_I N_i \to \overline{N}$, where $N_i = N$ for each $i \in I$. If $N$ is an $N$-small module, then for all $i \in I$, $(N_i)\phi \ll \overline{N}$ (see 8.2 (4)). In this case $\overline{N} = \operatorname{Im} \phi \subseteq \operatorname{Rad} \overline{N}$, contradicting $\operatorname{Rad} \overline{N} \ll \overline{N}$. Hence $N$ is not a small submodule of $\overline{N}$ (see 8.2). Then, since $\operatorname{Rad} \overline{N} \ll \overline{N}$ and $N \not\ll \overline{N}$, there must exist an $x \in N$ such that $Rx$ is not small in $\overline{N}$. As $Rx$ is a finite direct sum of cyclic uniserial modules (by hypothesis), there exists a cyclic uniserial summand $X$ of $Rx$ which is not small in $\overline{N}$. The injective hull $\overline{X}$ of $X$ in $\sigma[N]$ is a direct summand of $\overline{N}$ and hence $X$ is not a small submodule of $\overline{X}$. Since $X$ is uniform, $\overline{X}$ is also uniform and so, from above, is a cyclic uniserial module. As $X$ is not small in $\overline{X}$, $X = \overline{X}$ and hence a direct summand of $N$.

(i)$\Rightarrow$(j). We first show that $\operatorname{Rad} N \ll N$ for all modules $N \in \sigma[M]$. Let $N \in \sigma[M]$. Suppose $\operatorname{Rad} N$ is not small in $N$. Then there exists a proper submodule $L$ of $N$ such that $\operatorname{Rad} N + L = N$. Then $N/L = \operatorname{Rad}(N/L)$, a contradiction since by hypothesis $N/L$ contains a nonzero cyclic uniserial direct summand. Thus $\operatorname{Rad} N \ll N$, for every $N \in \sigma[M]$.

We claim that any injective module $A$ in $\sigma[M]$ is lifting. Since every module in $\sigma[M]$ has a cyclic uniserial direct summand, every module in $\sigma[M]$ is a sum of cyclic uniserial submodules. Suppose $X$ is not a small submodule in $A$. Then, since $X$ is a sum of cyclic uniserial submodules and $\operatorname{Rad} A \ll A$, it follows that $X$ contains a uniserial cyclic submodule $Y$ which is not small in $A$ and hence a non-$M$-small module. As any uniform module in $\sigma[M]$ is cyclic uniserial, $Y$ is injective in $\sigma[M]$ and hence a direct summand of $A$. Now consider a maximal independent family $\{A_i\}_I$ of direct summands of $A$ such that $A_i \subseteq X$ for each $i \in I$. Then each $A_i$ is $M$-injective and, as $M$ is locally noetherian, $V = \bigoplus_I A_i$ is $M$-injective (see 1.14) and hence a direct summand of $A$, say $A = V \oplus W$. Then $X = V \oplus (W \cap X)$. If $W \cap X$ is not small in $A$, then, by what we have proved, it will contain a cyclic uniserial injective (in $\sigma[M]$) direct summand, contradicting the choice of the $A_i$'s. Thus $W \cap X$ is small in $A$ and so $A$ is a lifting module (see 22.3). It now follows that $M$ is a Harada module and in particular $\widehat{M}$ is a $\Sigma$-lifting module.

Next we show that any lifting module $N$ in $\sigma[M]$ is serial. By hypothesis and Zorn's Lemma, we can find a maximal family $\{N_i\}_I$ of independent submodules of $N$ which are cyclic uniserial and $T = \bigoplus_I N_i$ is a local direct summand of $N$. Then $T$ is not small in $N$ (as it contains cyclic direct summands of $N$) and we have $T + L = N$, where $L$ is a proper submodule of $N$. As $N$ is lifting we can write $T = A \oplus B$, where $A$ is a summand of $N$ and $B \ll N$. Let $x \in B$. Then there exists a finite subset $K$ of $I$ such that $Rx \subseteq \bigoplus_{i \in K} N_i$, a direct summand of $N$. Thus $Rx$ is a small submodule of a direct summand of $T$ and hence of $T$. As $\operatorname{Rad} T \ll T$, this shows that $B \ll T$ and so $B = 0$. Thus $T$ is a direct summand of $N$. By the choice of $T$, we have $T = N$. Thus $N$ is a direct sum of cyclic uniserial modules.

Continuing with our lifting module $N$, we now have $N = \bigoplus_I N_i$, where each $N_i$ is a cyclic uniserial noetherian module and hence an LE-module. To prove that $N$ is extending, it is enough to prove that $N$ is uniform-extending (see [85, 8.13]). As any uniform module in $\sigma[M]$ is cyclic, without loss in generality we can assume $I$ is finite. Then using 29.1 it follows that $N$ is an extending module.

(j)$\Rightarrow$(k) is trivial.

(k)$\Rightarrow$(a). **Step 1.** We first show that an $M$-generated indecomposable $\Sigma$-lifting module is noetherian and uniserial and that $M$ is locally noetherian.

Let $A$ be an $M$-generated indecomposable $\Sigma$-lifting module. By hypothesis, $A$ is also $\Sigma$-extending. By 25.4, $A$ is self-injective and so an LE-module. Hence, by 25.2, $A$ is a cyclic module and so a local uniform module. If $X \subseteq A$, then $A/X$ is a lifting (as it is local) $M$-generated module and hence, by (k), an extending module. Thus the socle of $A/X$ is either zero or simple. Then, by 2.17, $A$ is a uniserial module. Now let $X$ be a submodule of $A$. Then $X$ is uniform uniserial and hence, by hypothesis, $\operatorname{Rad} X \ll X$. Thus $X$ is cyclic. As every submodule of $A$ is cyclic, $A$ is noetherian. Thus $A$ is a noetherian uniserial module.

Now $\widehat{M}$ is an injective lifting module and hence is a direct sum of indecomposable modules by 22.20. Each indecomposable direct summand of $\widehat{M}$ is $M$-generated and $\Sigma$-lifting (by hypothesis) and so, from above, a noetherian uniserial module. Thus $\widehat{M}$ and hence $M$ is a locally noetherian module.

**Step 2.** Next we show that every uniform module $U$ in $\sigma[M]$ is a uniserial noetherian module. Using the injective hull, it is enough to prove that any indecomposable injective module $A$ in $\sigma[M]$ is noetherian uniserial. By Step 1, $\widehat{M}$ is a direct sum of noetherian uniserial modules. As $A$ is injective in $\sigma[M]$, by 1.12, $A$ is $\widehat{M}$-generated and hence there exists an epimorphism $f : N \to A$, where $N$ is a direct sum of copies of $\widehat{M}$. By hypothesis $N$ is lifting. For any ordinal $\alpha$, $\mathrm{Rad}_\alpha N$ is fully invariant in $N$ and hence $N/\mathrm{Rad}_\alpha N$ is a lifting module (see 22.2) and therefore extending by (k). It now follows from 29.3 that $A$ is uniserial. By hypothesis, $\mathrm{Rad}\, U \ll U$ for all uniform modules $U$ in $\sigma[M]$. Hence $A$ is a noetherian uniserial module.

**Step 3.** We now show that the direct sum $V = S \oplus T$ of any two indecomposable injective modules $S$ and $T$ in $\sigma[M]$ is lifting. As $V$ is $\widehat{M}$-generated (see 1.12) and $\widehat{M}$ is $\Sigma$-lifting, there exists a small cover $f : N \to V$, where $N \in \mathrm{Add}\, \widehat{M}$. Note that $N$ is $\Sigma$-lifting and is a direct sum of uniserial noetherian modules. Let $X$ be a nonsmall proper submodule of $V$. Then $Y = (X)f^{-1}$ is a nonsmall submodule of $N$. As $N$ is lifting, $Y$ contains a nonzero direct summand of $N$ and hence a nonzero uniserial noetherian direct summand $Z$. As $f$ is a small epimorphism, $(Z)f$ is a nonsmall submodule of $V$. Thus, since $V$ is an injective module in $\sigma[M]$, $(Z)f$ is a non-$M$-small uniserial module. Since every indecomposable injective in $\sigma[M]$ is uniserial, $(Z)f$ is injective in $\sigma[M]$ and hence is a direct summand of $V$, say $V = (Z)f \oplus T$. Then $X = (Z)f \oplus (T \cap X)$. By considering the uniform dimension of $V$, we see that $T$ is uniform and hence, by Step 2, a uniserial module. As $X$ is a proper submodule of $V$, $T \cap X \neq T$ and so $T \cap X \ll T$. Thus $V$ is a lifting module (see 22.3).

**Step 4.** We finally prove that every $M$-generated extending module $A$ is lifting. Since $M$ is locally noetherian, by [85, 8.3], $A$ is a direct sum of indecomposable modules, say $A = \bigoplus_I A_i$, where each $A_i$ is a uniform module. By Step 3, $\widehat{A_i} \oplus \widehat{A_j}$ is a lifting module for all $i \neq j$ in $I$. Then Step 2 and 29.1 imply that $A_i \oplus A_j$ is lifting. Moreover, by Step 2, each $A_i$ is noetherian uniserial and so a cyclic LE-module. Then, by [85, 8.13], every local direct summand of $A$ is a direct summand and so, by 23.20 and 24.13, $A$ is a lifting module. $\qquad\square$

Let $M$ be a finitely generated module over a commutative ring $R$ such that every extending module in $\sigma[M]$ is lifting. Then, by 29.4, $M$ is noetherian and every finitely generated indecomposable module in $\sigma[M]$ is uniserial. Since $R$ is commutative, any noetherian self-injective $R$-module is artinian (see 4.11) and hence has finite length. Thus it follows from [363, 55.14] that every module in $\sigma[M]$ is serial. Thus we have the following.

**29.5. Corollary.** *Let $R$ be a commutative ring and $M$ be a finitely generated $R$-module. Then the following are equivalent:*

(a) *every extending module in $\sigma[M]$ is lifting;*

(b) *every module in $\sigma[M]$ is a direct sum of cyclic uniserial modules;*

(c) *every lifting module is extending and $\widehat{M}$ is a $\Sigma$-lifting module.*

Next we consider the case when $M$ is a projective generator in $\sigma[M]$.

**29.6. Extending modules are lifting II.** *Let $M$ be a projective generator in $\sigma[M]$. Then the following assertions are equivalent:*

(a) *every self-injective module in $\sigma[M]$ is lifting and $M$ is a direct sum of finitely generated modules;*

(b) *every extending module in $\sigma[M]$ is lifting and $M$ is a direct sum of finitely generated modules;*

(c) *every lifting module in $\sigma[M]$ is extending and $M$ is $\Sigma$-lifting;*

(d) *every self-projective module in $\sigma[M]$ is extending.*

*Proof.* (a)$\Leftrightarrow$(b) follows from the equivalence of (d) and (e) of 29.4.

(b)$\Rightarrow$(c). $M$ satisfies the condition (d) of 29.4. By 29.4 (j), every lifting module in $\sigma[M]$ is extending. By 29.4 (i), $M$ is a locally noetherian module and by 29.4 (h), every finitely generated module is a direct sum of noetherian uniserial modules. By hypothesis, $M$ is a direct sum of finitely generated modules and hence $M = \bigoplus_I M_i$, where each $M_i$ is a noetherian uniserial module. Hence the family $\{M_i\}_I$ is lsTn. It is now straightforward to see from 24.13 that $M$ is a $\Sigma$-lifting module.

(c)$\Rightarrow$(d). Suppose $A$ is a self-projective module in $\sigma[M]$. Since $M$ generates $A$ and $M$ is $\Sigma$-lifting, $A$ has a small cover $f : P \to A$, where $P \in \mathrm{Add}\, M$. As $\mathrm{Ke}\, f \ll P$ and $P/\mathrm{Ke}\, f$ is self-projective, $\mathrm{Ke}\, f$ is fully invariant in $P$ (see 4.9). Thus, by 22.2 (2) $A$ is also lifting. Hence $A$ is an extending module.

(d)$\Rightarrow$(a). Let $P \in \mathrm{Add}\, M$. As $P$ is projective in $\sigma[M]$, $P$ is $\Sigma$-extending. Hence $P$ is $\Sigma$-lifting (see 25.6) and so is a direct sum of local, uniform LE-modules (see 27.24). By 27.21, $\mathrm{Rad}\, P \ll P$.

Suppose $N \in \sigma[M]$. We show that $\mathrm{Rad}\, N \ll N$ and $N/\mathrm{Rad}\, N$ is semisimple. By hypothesis $N$ is $M$-generated and hence $N$ has a projective cover $f : P \to N$, where $P \in \mathrm{Add}\, M$. Since $f$ is a small epimorphism, $(\mathrm{Rad}\, P)f = \mathrm{Rad}\, N$. As $\mathrm{Rad}\, P \ll P$, $\mathrm{Rad}\, N \ll N$. Also, since $P/\mathrm{Rad}\, P$ is semisimple, $N/\mathrm{Rad}\, N$ is also semisimple.

Next we prove that any indecomposable direct summand $A$ of $M$ is a noetherian uniserial module. We see from above that $A$ is a uniform local module. Also, from the previous paragraph, $\mathrm{Rad}_\alpha A/\mathrm{Rad}_{\alpha+1} A$ is semisimple for any ordinal $\alpha$. since $\mathrm{Rad}_{\alpha+1} A$ is fully invariant in $A$, it follows from 4.9 that $A/\mathrm{Rad}_{\alpha+1} A$ is self-projective and hence is extending and therefore a uniform module. Thus $\mathrm{Rad}_\alpha A/\mathrm{Rad}_{\alpha+1} A$ is a uniform semisimple module and hence is simple. Therefore $A$ is a uniserial module. Since $\mathrm{Rad}\, B \ll B$ for every submodule $B$ of $A$, it follows that $A$ is a noetherian uniserial module.

Now we claim that any uniform module $U$ in $\sigma[M]$ is a noetherian uniserial module. By hypothesis, $U$ is $M$-generated. If $N$ is a direct sum of copies of $M$, then for any ordinal $\alpha$, $\mathrm{Rad}_\alpha N$ is fully invariant in $N$ and hence $N/\mathrm{Rad}_\alpha N$ is self-projective (see 4.9) and so is extending by (d). As $M$ is a direct sum of noetherian uniserial modules, it now follows from 29.3 that $U$ is a uniserial module. Since $\mathrm{Rad}\, V \ll V$ for every submodule $V$ of $U$, it follows that $U$ is a noetherian uniserial module, as claimed. Consequently, any non-$M$-small uniform module is injective in $\sigma[M]$.

Let $N$ be a self-injective module in $\sigma[M]$. Since $M$ is locally noetherian so too is $N$ (by [363, 27.3]). Thus any local direct summand of $N$ is a direct summand (see 1.14). Therefore to prove that $N$ is lifting it is enough to show that any nonsmall submodule $X$ of $N$ contains a nonzero direct summand of $N$. There exists a small cover $f : P \to N$, where $P \in \mathrm{Add}\, M$. As $(X)f^{-1}$ is nonsmall in $P$, it contains a nonzero direct summand $L$ of $P$ and hence a noetherian uniserial direct summand $A$ of $P$. Then, since $f$ is a small map, $(A)f \not\ll N$ and so, since $N$ is injective in $\sigma[M]$, $(A)f$ is a non-$M$-small module and is also uniserial. Thus $(A)f \subseteq X$ is injective in $\sigma[M]$ and hence a direct summand of $N$. $\square$

A ring $R$ is called an *artinian serial ring* if both $_RR$ and $R_R$ are artinian and serial modules. Classically such rings are also known as *generalised uniserial rings*. A ring $R$ is an artinian serial ring if and only if every left $R$-module is a direct sum of finitely generated uniserial modules, or — equivalently — $R$ is left artinian and every finitely generated left $R$-module is serial (see, e.g. [101, 25.4.2], [363, 55.16]).

For the case $M = R$, the foregoing results yield the following.

**29.7. All extending $R$-modules are lifting.** *For $R$ the following are equivalent:*

(a) *every self-injective $R$-module is a lifting module;*

(b) *every extending $R$-module is a lifting module;*

(c) *every lifting $R$-module is extending and $R$ is a left perfect ring;*

(d) *every self-projective $R$-module is an extending module;*

(e) *every lifting $R$-module is extending, $E(R)$ is a $\Sigma$-lifting module, and $\mathrm{Rad}\, U \ll U$ for every uniform $R$-module $U$;*

(f) *every finitely generated $R$-module is a noetherian serial module, and $\mathrm{Rad}\, M \ll M$ for every $R$-module $M$;*

(g) *$R$ is an artinian serial ring.*

*Proof.* The equivalence of (a) through (d) follows from 29.6.

(a)$\Leftrightarrow$(e) follows from the equivalence of (a) and (j) of 29.4.

(a)$\Leftrightarrow$(f) follows from the equivalence of (a) and (h) of 29.4.

(f)$\Rightarrow$(g). By (f), $R$ is left noetherian serial. Since $\mathrm{Rad}\, N \ll N$ for all $R$-modules $N$, it follows that $R$ is left perfect and hence $R$ is left artinian (see [363, 43.9]). By [363, 55.16], $R$ is an artinian serial ring.

(g)$\Rightarrow$(f) is obvious. $\square$

We end this section by considering the case when every module in $\sigma[M]$ is lifting. For this the following observations — generalising properties of semisimple modules — will be crucial.

**29.8. Modules of length at most 2.** *Let $M$ be an $R$-module such that every module in $\sigma[M]$ is a direct sum of uniserial modules of length at most 2. Then:*

(1) *Every module of length 2 in $\sigma[M]$ is $M$-injective and $M$-projective.*

(2) *Every module in $\sigma[M]$ is a direct sum of*

    (i) *a semisimple module and an $M$-injective, $M$-projective module;*

    (ii) *an $M$-injective module and an $M$-small module;*

    (iii) *an $M$-projective module and an $M$-singular module.*

(3) *$\sigma[M]$ has a projective generator.*

(4) *For any indecomposable modules $U, V \in \sigma[M]$, $U$ is almost $V$-projective.*

*Proof.* Notice that our assumption implies that $M$ is locally noetherian.

(1) Let $N \in \sigma[M]$ be an indecomposable (hence uniserial) module of length 2 and $f : P \to N$ be an epimorphism in $\sigma[M]$. By assumption, $P = \bigoplus_I P_i$ with uniserial modules $P_i$ where $\ell(P_i) \leq 2$. Clearly there must be some $i \in I$ for which $(P_i)f = N$ and $\ell(P_i) = 2$. Then $f|_{P_j} : P_j \to N$ is an isomorphism and hence $f$ splits. This shows that $N$ is projective in $\sigma[M]$.

Our assumption also implies that $N$ has no proper essential extension in $\sigma[M]$ and hence $N$ is injective in $\sigma[M]$.

(2) Let $N = \bigoplus_I N_i$ where the $N_i$ are indecomposables of length at most 2.

(i) Grouping the modules of length 1 together yields a semisimple summand of $N$ while those with length 2 form a direct summand which is projective and injective in $\sigma[M]$ by (1).

(ii) Take the sum $\overline{N} = \bigoplus_K N_k$, $K \subset I$, of all $M$-injective summands $N_k$ of $N$. The remaining summands $N_j$, $j \in I \setminus K$ are not injective, hence have an $M$-injective hull $\widehat{N}_j \neq N_j$ which is uniserial and thus $N_j \ll \widehat{N}_j$. By the Harada–Sai Lemma 12.17, the direct sum of the $\widehat{N}_i$'s has a small radical and hence the direct sum of the $N_j$'s is $M$-small.

(iii) Take the sum $\widetilde{N} = \bigoplus_L N_l$, $L \subset I$, of all $M$-projective summands $N_l$ in $N$. The remaining summands $N_j$, $j \in I \setminus L$, are not $M$-projective and hence there is a uniserial module $P \in \sigma[M]$ of length 2 with a non-splitting epimorphism $P \to N_j$. Thus $N_j$ is $M$-singular and the direct sum of these summands yields an $M$-singular summand of $N$.

(3) Let $P_1$ be the direct sum of all $M$-projective simple modules in $\sigma[M]$ (up to isomorphisms). For any non-projective simple module $S$ in $\sigma[M]$ there is a uniserial module $P$ of length 2 with an epimorphism $P \to S$. Such a $P$ is $M$-projective by (1) and if $P_2$ denotes the direct sum of these modules, then $P_1 \oplus P_2$ is a projective generator in $\sigma[M]$ (see [363, 18.5]).

(4) If $\ell(U) = 2$, then $U$ is projective by (1). So assume $U$ to be simple and consider the diagram of module morphisms with exact bottom line

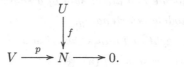

If $p$ is not an isomorphism and $f \neq 0$, then $\ell(V) = 2$ and $f$ is surjective (in fact bijective). Then $V$ is $M$-projective and there exists an epimorphism $g : V \to U$ with $p = g \diamond f$. This shows that $U$ is almost $V$-projective. $\qquad\square$

From these results we easily derive the following.

**29.9. Every module in $\sigma[M]$ is lifting.** *For $M$ the following are equivalent:*

(a) *every module in $\sigma[M]$ is lifting;*

(b) *every module in $\sigma[M]$ is a direct sum of uniserial modules of length $\leq 2$;*

(c) *every module in $\sigma[M]$ is a direct sum of a semisimple and an $M$-injective module;*

(d) *every module in $\sigma[M]$ is a direct sum of a semisimple and a projective module in $\sigma[M]$;*

(e) *every module in $\sigma[M]$ is extending.*

*Proof.* (a)$\Rightarrow$(b). By 29.4, every indecomposable module in $\sigma[M]$ is cyclic and uniserial, and it follows from 22.5 that all these uniserial modules have length at most 2. Clearly if $U$ is a uniserial modules of length 2, then $\mathrm{Rad}\,U = \mathrm{Soc}\,U$.

Since $M$ is a Harada module, every module in $\sigma[M]$ is a direct sum of an injective module and an $M$-small module (see 28.5). From the preceding observations we know that the injective part is a direct sum of uniform modules which are uniserial of length at most 2, and every $M$-small module $N$ is contained in the radical of its $M$-injective hull $\widehat{N}$ which obviously is semisimple.

(b)$\Rightarrow$(a). Any $N \in \sigma[M]$ is of the form $N = \bigoplus_I N_i$ where the $N_i$ are uniserial of length at most 2 and $\{N_i\}_I$ is lsTn by the Harada–Sai Lemma 12.17. By 29.8 (4), for any $i, j \in I$, $N_i$ is almost $N_j$-projective and hence $N$ is lifting by 24.13.

(b)$\Rightarrow$(c), (d) is shown in 29.8 (2).

For the remaining implications we refer to [85, 13.3]. $\qquad\square$

Applied to $M = R$, 29.9 yields the characterisations of the corresponding rings given by (a)–(e) following, while the equivalence of these to (f) and (g) is documented in [85, 13.5].

**29.10. Every $R$-module lifting.** *For a ring $R$ the following are equivalent:*

(a) *every left $R$-module is lifting;*

(b) *every left R-module is a direct sum of uniserial modules of length at most 2;*

(c) *every left R-module is a direct sum of a semisimple and an injective module;*

(d) *every left R-module is a direct sum of a semisimple and a projective module;*

(e) *every left R-module is extending;*

(f) *R is a left and right artinian serial ring with $J(R)^2 = 0$;*

(g) *the right-handed versions of* (a)–(e).

**29.11. Comments.** Continuing the work in [271], Oshiro considered in [272] the rings for which every quasi-injective (quasi-projective) module is lifting (extending). He proved that these are indeed generalised uniserial rings. Also he showed that a ring $R$ is a generalised uniserial ring if and only if every extending module is lifting, if and only if every lifting module is extending and $R$ is left perfect. Jayaraman and Vanaja [182] extended the above results to the category $\sigma[M]$ where $M$ is an $R$-module.

A ring $R$ is left semisimple if and only if all $R$-modules are quasi-continuous (continuous, injective), if and only if all $R$-modules are quasi-discrete (discrete, projective). However, rings for which every left module is lifting (extending) need not be semisimple. Those rings were characterised by Vanaja and Purav in [342]. This has been extended to modules $M$ and the category $\sigma[M]$ by Oshiro and Wisbauer in [277] (see 29.8).

Abyzov [1] defines a module to be *weakly regular* if, whenever $N$ is a submodule of $M$ which is not contained in Rad $M$, then $N$ contains a nonzero direct summand of $M$. He shows that if $R$ is a semiperfect ring for which $J(R)^2 \neq J(R)^3$, then there is an $R$-module which is not weakly regular. Furthermore, he proves that if $R$ is a left perfect ring, then every left $R$-module is weakly regular if and only if $R$ is artinian serial with $J(R)^2 = 0$.

**References.** Abyzov [1]; Baba and Harada [30]; Harada [155, 156]; Harada and Mabuchi [161]; Jayaraman and Vanaja [182]; Oshiro [271, 272]; Oshiro and Wisbauer [277]; Vanaja [340, 341]; Vanaja and Purav [342, 343].

# Appendix

In this appendix we provide details of two graph-theoretical techniques, namely Hall's Marriage Theorem and König's Graph Theorem.

## 30  Hall's Marriage Theorem

**30.1. Bipartite graphs.** A *bipartite graph* is a triple $G = (X, Y, E)$ where

(a) $X$ and $Y$ are disjoint nonempty finite sets and

(b) $E$ is a (possibly empty) subset of the set of unordered pairs of the form $(x, y)$ where $x \in X$ and $y \in Y$.

In this case,

(i) the disjoint union $V := X \cup Y$ is called the *vertex set* of $G$ and its elements are called *vertices*;

(ii) $E$ is called the *edge set* of $G$ and, if $e = (x, y) \in E$ where $x \in X$ and $y \in Y$, then $e$ is said to be an *edge joining $x$ and $y$* and $x$ and $y$ are called the *ends* of the edge $e$;

(iii) $V$ is said to *bipartition* into $X$ and $Y$.

**30.2. Matchings.** Let $G = (X, Y, E)$ be a bipartite graph, as above, and let $M$ be a collection of its edges in which no two edges have an end in common. Then

(a) $M$ is called a *matching* in $G$,

(b) if a vertex $v$ of $G$ is an end vertex of some edge $e \in M$, then $v$ is said to be *saturated* or *matched* by $M$, and

(c) if no matching in $G$ has more edges than $M$, then $M$ is called *maximum*.

Of particular interest are matchings $M$ in the bipartite graph $G$ which saturate all vertices in the bipartition subset $X$ (or $Y$). Such matchings may not exist, but to see when they do we introduce more terminology and notation.

**30.3. Neighbours.** Let $v$ and $w$ be vertices of the bipartite graph $G$. If there is an edge $e$ joining $v$ and $w$, then $w$ is said to be a *neighbour* of $v$ (and so $v$ is a neighbour of $w$).

If $S$ is any subset of the vertex set $V$ of $G$, its *neighbour set* $N(S)$ is defined by

$$N(S) = \{w \in V : w \text{ is a neighbour of some vertex } v \in S\}.$$

It is easily seen that if $M$ is a matching in $G$ which saturates every vertex $x$ in the bipartition subset $X$, then, for every $S \subseteq X$, we must have

$$|S| \leq |N(S)|,$$

where $|\ |$ denotes cardinality. In fact, this condition is also sufficient, as the following result, well-known to graph-theorists, shows. For its proof we refer to [71, Theorem 4.3].

**30.4. Hall's Marriage Theorem.** *A bipartite graph $G = (X, Y, E)$ has a matching $M$ which saturates all vertices of $X$ if and only if*

$$|S| \leq |N(S)|$$

*for all subsets $S$ of $X$.*

We now modify the definition of a bipartite graph so that edges are assigned a direction.

**30.5. Bipartite digraphs.** A *bipartite directed graph*, or *bipartite digraph* for short, is a triple $D = (X, Y, E)$ where

 (a) $X$ and $Y$ are disjoint nonempty finite sets and

 (b) $E$ is a subset of the cartesian product union $(X \times Y) \cup (Y \times X)$.

In this case,

 (i) the disjoint union $V := X \cup Y$ is called the *vertex set* of $D$ and its elements are called *vertices*;

 (ii) $E$ is called the *directed edge set* of $D$ and, if $e = (u, v) \in E$ (so that either $u \in X$ and $v \in Y$ or $u \in Y$ and $v \in X$), then $e$ is said to be a *directed edge from $u$ to $v$*, $u$ is called the *tail* of $e$, and $v$ is called the *head* of $e$;

 (iii) $V$ is said to *bipartition* into $X$ and $Y$.

If $e = (u, v) \in E$, we say that $v$ is an *out-neighbour* of $u$ and $u$ an *in-neighbour* of $v$. If $v$ is any vertex in $D$, its *indegree* (*outdegree*) is defined to be the number of its in-neighbours (out-neighbours, respectively).

If $S$ is any subset of the vertex set $V$ of $G$, its *out-neighbourhood* $N^+(S)$ is defined by

$$N^+(S) = \{v \in V : v \text{ is an out-neighbour of some vertex } u \in S\}.$$

**30.6. Directed paths.** A (*directed*) *path* in the bipartite digraph $D$ is a finite sequence of directed edges $p = e_1, e_2, \ldots, e_n$ where

 (a) for each $i = 1, \ldots, n - 1$, the head of $e_i$ is the tail of $e_{i+1}$ and

 (b) the heads of $e_1, e_2, \ldots, e_n$ are distinct.

In this case, if $u$ and $v$ are the tail of $e_1$ and the head of $e_n$ respectively, then $p$ is said to be a path *from u to v*.

In the particular case where $u = v$, $p$ is called a *cycle*. In this case, there is an even number of edges in $p$, with half having heads in $X$, the other half in $Y$.

We may now define an equivalence relation $\sim$ on the vertex set $V$ of $D$ by setting, for any $u, v \in V$,

$u \sim v$ if $u = v$ or there is a path from $u$ to $v$ and a path from $v$ to $u$.

It is easy to see that $u \sim v$ if $u$ and $v$ are the heads of edges in a cycle.

We're now ready to state and prove the key result of [81].

**30.7. The Krull–Schmidt Theorem for bipartite digraphs.** *Let $D = (X, Y, E)$ be a bipartite digraph, where $X = \{x_1, \ldots, x_m\}$ and $Y = \{y_1, \ldots, y_n\}$. If*

$$|U| \le |N^+(U)|$$

*for all subsets $U$ of $V = X \cup Y$, then $m = n$ and, after renumbering the vertices of $X$ if necessary,*

$$x_i \sim y_i \text{ for each } i.$$

*Proof.* We first note that $|U| \le |N^+(U)|$ for all subsets $U$ of $V = X \cup Y$ if and only if $|S| \le |N^+(S)|$ for all subsets $S$ of $X$ and $|T| \le |N^+(T)|$ for all subsets $T$ of $Y$. In particular, under our hypothesis we have $|X| \le |N^+(X)| \le |Y|$ and $|Y| \le |N^+(Y)| \le |X|$, giving $|X| = |Y|$, that is, $m = n$ as required.

Now let $G = (X, Y, E)$ be the bipartite graph obtained from $D$ by taking $E$ to be the set of unordered pairs $(x, y)$ where $x \in X$, $y \in Y$, and the ordered pair $(x, y)$ is a directed edge in $D$. Then clearly $G$ satisfies Hall's Marriage Theorem criterion and so has a matching $M_1$ which saturates all vertices in $X$ and so, since $m = n$, all vertices in $Y$. Let $E_1$ be the set of directed edges in $D$ which, under our construction of $G$, correspond to the edges in $M_1$. Then each $x \in X$ is the tail of exactly one directed edge in $E_1$ and each $y \in Y$ is the head of exactly one directed edge in $E_1$. Reversing the rôles of $X$ and $Y$ similarly produces a subset $E_2$ of $E$ such that each $x \in X$ is the head of exactly one directed edge in $E_2$ and each $y \in Y$ is the tail of exactly one directed edge in $E_2$.

Next let $D_0$ be the bipartite digraph $(X, Y, E_1 \cup E_2)$. Then, from above, every vertex in $D_0$ has indegree 1 and outdegree 1. From this it follows that the directed edge set of $D_0$ can be expressed as the union of cycles no two of which have a vertex in common. A reordering of the vertices in $X \cup Y$ going through each cycle in turn then gives the desired $x_i \sim y_i$ for each $i = 1, \ldots, n$. $\qquad\square$

# 31 König's Graph Theorem

Let $I$ be a nonempty set. For each $n \in \mathbb{N}$, let $I^n$ denote the cartesian product of $n$ copies of $I$ (with $I^1 = I$).

We say that a countably infinite sequence $G = (P_1, P_2, \ldots, P_n, \ldots)$ is a *König graph on* $I$ if

(i) each $P_n$ is a finite subset of $I^n$ and

(ii) for each $n \in \mathbb{N}$, $(x_1, x_2, \ldots, x_n, x_{n+1}) \in P_{n+1}$ only if $(x_1, x_2, \ldots, x_n) \in P_n$ (and so only if $(x_1, x_2, \ldots, x_k) \in P_k$ for each $1 \le k \le n$).

Note that if $P_n = \emptyset$ for some $n \in \mathbb{N}$, then $P_t = \emptyset$ for all $t \ge n$. We say that the König graph $G$ is *infinite* if $P_n \ne \emptyset$ for each $n \in \mathbb{N}$.

An element $p_n = (x_1, x_2, \ldots, x_n) \in P_n$ is called a *path* in $G$ of *length* $n$ *(from $x_1$ to $x_n$)* and we say that each of the elements $x_1, \ldots, x_n$ *belongs to* $p_n$.

An infinite sequence $(x_1, x_2, \ldots, x_n, \ldots)$ is called a *path of infinite length* in $G$ if $(x_1, x_2, \ldots, x_k) \in P_k$ for each $k \in \mathbb{N}$.

A path $p_n = (a_1, \ldots, a_n) \in P_n$ is said to have *infinite order in $G$* if, for every $t \in \mathbb{N}$ with $t \ge n$, there is a path $p_t \in P_t$ of the form $p_t = (a_1, \ldots, a_n, b_{n+1}, \ldots, b_t)$.

**31.1. König's Graph Theorem.** *If $G = (P_1, P_2, \ldots, P_n, \ldots)$ is an infinite König graph, then $G$ has a path of infinite length.*

*Proof.* Suppose first that there is no path of length 1 in $G$ which is also of infinite order in $G$. Then, since $P_1$ is finite, this implies that there is a path of maximal length, say $n$, in $G$. However, since $P_{n+1} \ne \emptyset$, we have a contradiction. This shows that there is a path $p_1 = (a_1)$ in $P_1$ which is of infinite order in $G$.

Proceeding inductively, let $p_n = (a_1, \ldots, a_n)$ be a path of length $n$ which is of infinite order in $G$. Then, since $P_{n+1}$ is finite, the same reasoning as before shows that this extends to an element $p_{n+1} = (a_1, \ldots, a_n, a_{n+1})$ of infinite order in $P_{n+1}$. Continuing in this way produces an infinite path $(a_1, \ldots, a_n, \ldots)$ in $G$.  $\square$

# Bibliography

[1] Abyzov, A. N., *Weakly regular modules*, Russ. Math. 48(3), 1–3 (2004); transl. from Izv. Vyssh. Uchebn. Zaved. Mat. 2004, No. 3, 3–6.

[2] Al-attas, A. O. and Vanaja, N., *On Σ-extending modules*, Comm. Algebra 25, 2365–2393 (1997).

[3] Albu, T. and Năstăsescu, C., *Relative Finiteness in Module Theory*, Marcel Dekker, New York, Basel (1984).

[4] Ali, M. H. and Zelmanowitz, J. M., *Discrete implies continuous*, J. Algebra 183, 186–192 (1996).

[5] Alizade, R., Bilhan, G. and Smith, P. F., *Modules whose maximal submodules have supplements*, Comm. Algebra 29, 2389–2405 (2001).

[6] Alizade, R. and Büyükasik, E., *Cofinitely weak supplemented modules*, Comm. Algebra 31, 5377–5390 (2003).

[7] Alkan, M. and Özcan, A. C., *Semiregular modules with respect to a fully invariant module*, Comm. Algebra 32, 4285–4301, (2005).

[8] Al-Khazzi, I. and Smith, P. F., *Modules with chain conditions on superfluous submodules*, Comm. Algebra 19, 2331–2351 (1991).

[9] Al-Takhman, K., Lomp, Ch. and Wisbauer, R., *τ-complemented and τ-supplemented modules*, to appear.

[10] Amin, I. A. and Soueif, L. M., *Exchange property in Abelian categories and exchange rings*, manuscript (1982)

[11] Amini, A., Amini, B. and Ershad, M., *A new characterization of perfect rings*, preprint (2004).

[12] Anderson, F. W. and Fuller, K. R., *Modules with decompositions that complement direct summands*, J. Algebra 22, 241–253 (1972).

[13] Anderson, F. W. and Fuller, K. R., *Rings and Categories of Modules, 2. ed.*, Springer Verlag, Berlin (1992).

[14] Angeleri-Hügel, L., *Covers and envelopes via endoproperties of modules*, Proc. London Math. Soc. 86, 649–665 (2003).

[15] Angeleri Hügel, L. and Saorín, M., *Modules with perfect decompositions*, Math. Scand. 98, 1–25 (2006).

[16] Ánh, P. N., Herbera, D., and Menini, C., *AB-5\* and linear compactness*, J. Algebra 200, 99–117 (1998).

[17] Ara, P., *Strongly π-regular rings have stable range one*, Proc. Amer. Math. Soc. 124, 3293–3298 (1996).

[18] Ara, P., *Extensions of exchange rings*, J. Algebra 197, 409–423 (1997).

[19] Ara, P., *Stability properties of exchange rings*, International Symposium on Ring Theory (Kyongju, 1999), 23–42, Trends Math., Birkhäuser (2001).

[20] Ara, P., Goodearl, K. R., O'Meara, K. C., and Pardo, E., *Separative cancellation for projective modules over exchange rings*, Israel J. Math. 105, 105–137 (1998).

[21] Ara, P., O'Meara, K. C., and Perera, F., *Gromov translation algebras over discrete trees are exchange rings*, Trans. Amer. Math. Soc. 356, 2067–2079 (2004).

[22] Auslander, M., Reiten, I. and Smalø, S. O., *Representation Theory of Artin Algebras*, Cambridge University Press, Cambridge (1995).

[23] Averdunk, J., *Moduln mit Ergänzungseigenschaft*, Dissertation, Ludwig-Maximilians-Universität München (1996).

[24] Azumaya, G., *Corrections and supplementaries to my paper concerning Krull-Remak-Schmidt's theorem*, Nagoya Math. J. 1, 117–124 (1950).

[25] Azumaya, G., *On generalized semi-primary rings and Krull-Remak-Schmidt's theorem*, Japan J. Math. 19, 525–547 (1960).

[26] Azumaya, G., *Locally split submodules and modules with perfect endomorphism rings*, Noncommutative ring theory (Athens, OH, 1989), 1–6, LNM 1448, Springer, Berlin (1990).

[27] Azumaya, G., *F-semi-perfect modules* J. Algebra 136, 73–85 (1991).

[28] Azumaya, G., *A characterization of semi-perfect rings and modules*, Jain, S. K. (ed.) et al., Ring theory. Proc. Ohio State-Denison conf., World Scientific, Singapore, 28-40 (1993).

[29] Azumaya, G. and Facchini, A., *Rings of pure global dimension zero and Mittag-Leffler modules*, J. Pure Appl. Algebra 62, 109–122 (1989).

[30] Baba, Y. and Harada, M., *On almost M-projectives and almost M-injectives*, Tsukuba J. Math. 14, 53–69 (1990).

[31] Baccella, G., *Exchange property and the natural preorder between simple modules over semi-Artinian rings*, J. Algebra 253, 133–166 (2002).

[32] Baccella, G. and Ciampella, A., *$K_0$ of semi-Artinian unit-regular rings*, Abelian groups, module theory, and toplogy (Padua, 1997), 69–78, Dekker, New York (1998).

[33] Barioli, F., Facchini, A., Raggi, F., and Ríos, J., *Krull-Schmidt theorem and homogeneous semilocal rings*, Comm. Algebra 29, 1649–1658 (2001).

[34] Bass, H., *Finitistic dimension and a homological generalization of semiprimary rings*, Trans. Amer. Math. Soc. 95, 466–488 (1960).

[35] Bass, H., *K-theory and stable algebra*, Inst. Hautes Études Sci. Publ. Math. 22, 5–60 (1964).

[36] Beck, I., *An independence structure on indecomposable modules*, Ark. Mat. 16, 171–178 (1978).

[37] Beck, I., *On modules whose endomorphism ring is local*, Israel J. Math. 29, 393–407 (1978).

[38] Beidar, K. I. and Kasch, F., *Good conditions for the total*, International Symposium on Ring Theory (Kyongju, 1999), 43–65, Birkhäuser Boston, Boston (2001).

[39] Bessenrodt, C., Brungs, H.-H., and Törner, G., *Right chain rings. Part 1*, Schriftenreihe des Fachbereichs Math. 181, Universität Duisburg (1990).

[40] Bican, L., *Corational extensions and pseudo-projective modules*, Acta Math. Acad. Sci. Hung. 28, 5–11 (1976).

[41] Bican, L., Kepka, T. and Nemec, P., *Rings, modules, and preradicals*, Lect. Notes Pure and Appl. Math., 75, Marcel Dekker, New York-Basel (1982).

[42] Birkenmeier, G. F., *Quasi-projective modules and the finite exchange property*, Internat. J. Math. Math. Sci. 12, 821–822 (1989).

[43] Brodskiĭ, G. M., *Annihilator conditions in endomorphism rings of modules*, Math. Notes 16 (1974), 1153–1158 (1975).

[44] Brodskiĭ, G. M., *On modules that are lattice-isomorphic to linearly compact modules*, Math. Notes 59, 123–127 (1996).

[45] Brodskiĭ, G. M. and Wisbauer, R., *General distributivity and thickness of modules*, Arabian J. Sci. Eng. 25 (2C), 95–128 (2000).

[46] Brzeziński, T. and Wisbauer, R., *Corings and Comodules*, London Math. Soc. Lect. Note Ser. 309, Cambridge Univ. Press, Cambridge (2003).

[47] Buchsbaum, D. A., *A note on homology in categories*, Ann. Math. 69, 66–74 (1959).

[48] Bumby, R. T., *Modules which are isomorphic to submodules of each other*, Arch. Math. 16, 184–185 (1965).

[49] Byrd, K. A., *Some characterization of uniserial ring*, Math. Ann. 186, 163–170 (1970).

[50] Camillo, V. P., *Modules whose quotients have finite Goldie dimension*, Pacific J. Math. 69, 337–338 (1977).

[51] Camillo, V. P., *On Zimmermann-Huisgen's splitting theorem*, Proc. Amer. Math. Soc. 94, 206–208 (1985).

[52] Camillo, V. P. and Yu, H.-P., *Exchange rings, units and idempotents*, Comm. Algebra 22, 4737–4749 (1994).

[53] Camillo, V. P. and Yu, H.-P., *Stable range 1 for rings with many idempotents*, Trans. Amer. Math. Soc. 347, 3141–3147 (1995).

[54] Camillo, V. P. and Zelmanowitz, J. M., *On the dimension of a sum of modules*, Comm. Algebra 6, 345–352 (1978).

[55] Camillo, V. P. and Zelmanowitz, J. M., *Dimension modules*, Pacific J. Math. 91, 249–261 (1980).

[56] Camps, R. and Dicks, W., *On semilocal rings*, Israel J. Math. 81, 203–221 (1993).

[57] Canfell, M. J., *A note on right equivalence of module presentations*, J. Austral. Math. Soc. (Ser. A) 52, 141–142 (1992).

[58] Canfell, M. J., *Completions of diagrams by automorphisms and Bass's first stable range condition*, J. Algebra 176, 480–503 (1995).

[59] Çelik, C., *Modules satisfying a lifting condition*, Turk. J. Math. 18, 293–301 (1994).

[60] Chamard, J.-Y., *Modules quasi-projectifs, projectifs et parfaits*, Séminaire Dubreil-Pisot, Algèbre et Théorie des Nombres, 21e année, Nr. 8 (1967/68).

[61] Chang, Ch. and Kuratomi, Y., *Lifting modules over right perfect rings*, Proc. 36th Symp. Ring Theory and Representation Theory, Yamanashi, 38–41 (2004).

[62] Chen, Huanyin, *On stable range conditions*, Comm. Algebra 28, 3913–3924 (2000).

[63] Chen, Huanyin, *Exchange rings having stable range one*, Int. J. Math. Math. Sci. 25, 763–770 (2001).

[64] Chen, Huanyin, *Idempotents in exchange rings*, New Zealand J. Math. 30, 103–110 (2001).

[65] Chen, Huanyin, *On partially one-sided unit-regularity*, Comm. Algebra 30, 5781–5793 (2002).

[66] Chen, Huanyin, *On partially unit-regularity*, Kyungpook Math. J. 42, 13–19 (2002).

[67] Chen Jianlong, Li Yuanlin, and Zhou Yiqiang, *Constructing morphic rings*, Advances in Ring Theory, Proceedings of the 4th China-Japan-Korea International Conference, 2004, 26–32, World Scientific, Singapore (2005).

[68] Chu Trong Thanh, *A note on quasi-continuous and quasi-discrete modules*, Kyungpook Math. J. 39, 185–189 (1999).

[69] Clark, J., *Locally semi-T-nilpotent families of modules*, Advances in Ring Theory, Proceedings of the 4th China-Japan-Korea International Conference, 2004, 41–54, World Scientific, Singapore (2005).

[70] Clark, J. and Nguyen Viet Dung, *On the decomposition of nonsingular CS-modules*, Canad. Math. Bull. 39, 257–265 (1996).

[71] Clark, J. and Holton, D. A., *A First Look at Graph Theory*, World Scientific, Singapore (1991).

[72] Cohn, P. M., *The complement of a finitely generated direct summand of an abelian group*, Proc. Amer. Math. Soc. 7, 520–521 (1956).

[73] Cohn, P. M., *Algebra, Vol. 1*, Wiley, Chichester (1974).

[74] Colby, R. R. and Rutter, E. A., Jr., *Generalization of QF-3 algebras*, Trans. Amer. Math. Soc. 153, 371-381 (1971).

[75] Corisello, R. and Facchini, A., *Homogeneous semilocal rings*, Comm. Algebra 29, 1807–1819 (2001).

[76] Courter, R. C., *The maximal co-rational extension by a module*, Canad. J. Math. 18, 953–962 (1966).

[77] Crawley, P. and Jónnson, B., *Refinements for infinite direct decompositions of algebraic systems*, Pacific J. Math. 91, 249–261 (1980).

[78] Cunningham, R. S. and Rutter, E. A., Jr., *Perfect modules*, Math. Z. 140, 105–110 (1974).

[79] Del Valle, A., *Goldie dimension of a sum of modules*, Comm. Algebra 22, 1257–1269 (1994).

[80] Dinh, H. Q., *A note on pseudo-injective modules*, Comm. Algebra 33, 361–369 (2005).

[81] Diracca, L. and Facchini, A., *Uniqueness of monogeny classes for uniform objects in abelian categories*, J. Pure Appl. Algebra 172, 183–191 (2002).

[82] Diracca, L. and Facchini, A., *Descending chains of modules and Jordan-Hölder theorem*, Semigroup Forum 68, 373–399 (2004).

[83] Dung, N. V., *Modules with indecomposable decompositions that complement maximal direct summands*, J. Algebra 197, 449–467 (1997).

[84] Dung, N. V. and Facchini, A., *Weak Krull-Schmidt for infinite direct sums of uniserial modules*, J. Algebra 193, 102–121 (1997).

[85] Dung, N. V., Huynh, D. V., Smith, P. F. and Wisbauer, R., *Extending modules*, Pitman Research Notes in Mathematics Series 313, Longman Scientific & Technical, Harlow (1994).

[86] Ehrlich, G., *Units and one-sided units in regular rings*, Trans. Amer. Math. Soc. 216, 81–90 (1976).

[87] Eilenberg, S., *Homological dimension and syzygies*, Ann. of Math. 64, 328–336 (1956).

[88] Eilenberg, S. and Nakayama, T., *On the dimension of modules and algebras. V. Dimension of residue rings*, Nagoya Math. J. 11, 9–12 (1957).

[89] Eisenbud, D. and de la Peña, J. A., *Chains of maps between indecomposable modules*, J. Reine Angew. Math. 504, 29–35 (1998).

[90] Evans, E. G., Jr., *Krull-Schmidt and cancellation over local rings*, Pacific J. Math. 46, 151–121 (1973).

[91] Facchini, A., *Krull-Schmidt fails for serial modules*, Trans. Amer. Math. Soc. 348, 4561–4575 (1996).

[92] Facchini, A., *Module Theory. Endomorphism rings and direct sum decompositions in some classes of modules*, Birkhäuser, Basel (1998).

[93] Facchini, A., *Krull-Schmidt theorem and semilocal endomorphism rings*, Ring theory and algebraic geometry (León, 1999), 187–201, Lect. Notes Pure Appl. Math., 221, Dekker, New York (2001).

[94] Facchini, A., *Direct sum decompositions of modules, semilocal endomorphism rings, and Krull monoids*, J. Algebra 256, 280–307 (2002).

[95] Facchini, A., *The Krull-Schmidt theorem*, Handbook of algebra, Vol. 3, 357–397, North-Holland, Amsterdam (2003).

[96] Facchini, A., Herbera, D., Levy, L. S. and Vámos, P., *Krull-Schmidt fails for artinian modules*, Proc. Amer. Math. Soc. 123, 3587–3592 (1995).

[97] Facchini, A., Herbera, D. and Sakhaev, I., *Finitely generated flat modules and a characterization of semiperfect rings*, Comm. Algebra 31, 4195–4214 (2003).

[98] Facchini, A. and Levy, L. S., *Infinite progenerators sums*, Proc. International Conference on Algebras, Modules and Rings (Lisbon, 2003), 61–64, World Scientific, Singapore (2005).

[99] Facchini, A. and Salce, L., *Uniserial modules: Sums and isomorphisms of subquotients*, Comm. Algebra 18(2), 499–517 (1990).

[100] Faith, C., *Algebra: Rings, Modules and Categories I*, Springer, Berlin (1973).

[101] Faith, C., *Algebra II*, Springer, Berlin (1976).

[102] Faith, C., *Rings whose modules have maximal submodules*, Publ. Mat. 39, 201–214 (1995).
Addendum, Publ. Mat. 42, 265–266 (1998).

[103] Faith, C. and Walker, E. A., *Direct-sum representations of injective modules*, J. Algebra 5, 203–221 (1967).

[104] Faticoni, T., *On quasiprojective covers*, Trans. Amer. Math. Soc. 278, 101–113 (1983).

[105] Fieldhouse, D. J., *Regular modules over semi-local rings*, Rings, modules and radicals (Proc. Colloq., Keszthely, 1971), 193–196. Colloq. Math. Soc. Janos Bolyai, 6, North-Holland, Amsterdam (1973).

[106] Fleury, P., *A note on dualizing Goldie dimension*, Canad. Math. Bull. 17, 511–517 (1974).

[107] Fuchs, L., *On quasi-injective modules*, Ann. Scuola Norm. Sup. Pisa 23, 541–546 (1969).

[108] Fuchs, L., *On a substitution property of modules*, Monatsh. Math. 75, 198–204 (1971).

[109] Fuchs, L. and Salce, L., *Modules over Non-Noetherian Domains*, American Mathematical Society, Providence (2001).

[110] Fuller, K. R., *Generalized uniserial rings and their Kupisch series*, Math. Z. 106, 248–260 (1968).

[111] Fuller, K. R., *On rings whose left modules are direct sums of finitely generated modules*, Proc. Amer. Math. Soc. 54, 39–44 (1976).

[112] Fuller, K. R. and Reiten, I., *Note on rings of finite representation and decompositions of modules*, Proc. Amer. Math. Soc. 50, 92–94 (1975).

[113] Fuller, K. R. and Shutters, W. A., *Projective modules over non-commutative semilocal rings*, Tôhoku Math. J. 27, 303–311 (1975).

[114] Ganesan, L. and Vanaja, N., *Modules for which every submodule has a unique coclosure*, Comm. Algebra 30, 2355–2377 (2002).

[115] Ganesan, L. and Vanaja, N., *Strongly discrete modules*, Comm. Algebra, to appear.

[116] Garcia Hernández, J. L. and Gómez Pardo, J. L., *On endomorphism rings of quasiprojective modules*, Math. Z. 196, 87–108 (1987).

[117] Generalov, A. I., *On weak and ω-high purity in the category of modules*, Math. USSR, Sb. 34, 345-356 (1978).

[118] Generalov, A. I., *ω-cohigh purity in the category of modules*, Math. Notes 33, 402-408 (1983); translation from Mat. Zametki 33, 785-796 (1983).

[119] Generalov, A. I., *A radical that is dual to Goldie's torsion*, Abelian groups and modules, 11, 12, 70–77, 254, Tomsk. Gos. Univ., Tomsk (1994).

[120] Gerasimov, V. N. and Sakhaev, I. I., *A counterexample to two hypotheses on projective and flat modules*, Sib. Math. J. 25, 855–859 (1984).

[121] Golan, J. S., *Quasiperfect modules*, Quart. J. Math. Oxford (2), 22, 173–182 (1971).

[122] Golan, J. S., *Characterization of rings using quasiprojective modules. II*, Proc. Amer. Math. Soc. 28, 337–343 (1971).

[123] Golan, J. S., *Colocalization at idempotent ideals*, Ring Theory, Proc. of 1978 Antwerpen Conference, LNPAP 51, Marcel Dekker, 631–668 (1979).

[124] Golan, J. S. and Wu Quanshui, *On the endomorphism ring of a module with relative chain conditions*, Comm. Algebra 18, 2595–2609 (1990).

[125] Gómez Pardo, J. L. and Guil Asensio, P. A., *Big direct sums of copies of a module have well behaved indecomposable decompositions*, J. Algebra 232, 86–93 (2000).

[126] Gómez Pardo, J. L. and Guil Asensio, P. A., *Indecomposable decompositions of $\aleph$-$\sum$-CS-modules*, Algebra and its applications (Athens, OH, 1999), 467–473, Contemp. Math., 259, Amer. Math. Soc., Providence (2000).

[127] Goodearl, K. R., *Singular torsion and the splitting properties*, Mem. Amer. Math. Soc. 124 (1972).

[128] Goodearl, K. R., *Ring Theory. Nonsingular Rings and Modules*, Monographs and Textbooks in Pure and Appl. Math., 33, Marcel Dekker, New York (1976).

[129] Goodearl, K. R., *Cancellation of low-rank vector bundles*, Pacific J. Math. 113, 289–302 (1984).

[130] Goodearl, K. R., *Surjective endomorphisms of finitely generated modules*, Comm. Algebra 15, 589–609 (1987).

[131] Goodearl, K. R., *Von Neumann Regular Rings*, Second edition, Krieger Publ. Co., Malabar, Florida (1991).

[132] Goodearl, K. R., *Torsion in $K_0$ of unit-regular rings*, Proc. Edinburgh Math. Soc. 38, 331–341 (1995).

[133] Goodearl, K. R. and Warfield, R. B., Jr., *An Introduction to Noncommutative Noetherian Rings*, London Math. Soc. 16, Cambridge Univ. Press, Cambridge (1989).

[134] Goodearl, K. R. and Warfield, R. B., Jr., *Algebras over zero-dimensional rings*, Math. Ann. 223, 157–168 (1989).

[135] Grezeszcuk, P. and Puczyłowski, E. R., *On Goldie and dual Goldie dimension*, J. Pure Appl. Algebra 31, 47–54 (1984).

[136] Guil Asensio, P. A., *Every $\aleph_1$-$\sum$-CS module is $\sum$-CS*, J. Algebra 279, 720–725 (2004).

[137] Guil Asensio, P. A. and Herzog, I., *Left cotorsion rings*, Bull. London Math. Soc. 36, 303–309 (2004).

[138] Güngöroğlu, G., *Copolyform modules*, Commun. Fac. Sci. Univ. Ank., Sér. A1, Math. Stat. 49(1-2), 101–110 (2000).

[139] Güngöroğlu, G. and Keskin Tütüncü, D., *Copolyform and lifting modules*, Far East J. Math. Sci. (FJMS) 9(2), 159–165 (2003).

[140] Gupta, A. K. and Varadarajan, K., *Modules over endomorphism rings*, Comm. Algebra 8, 1291–1333 (1980).

[141] Haack, J. K., *The duals of the Camillo-Zelmanowitz formulas for Goldie dimension*, Canad. Math. Bull. 25, 325–334 (1982).

[142] Hall, F. J., Hartwig, R., Katz, I. J. and Neuman, M., *Pseudo-similarity and partial unit regularity*, Czech. Math. J. 33(108), 361–372 (1983).

[143] Hanada, K., Kuratomi, Y. and Oshiro, K., *On direct sums of extending modules and internal exchange property*, Proc. 32nd Symposium on Ring Theory and Representation Theory (Yamaguchi, 1999), 1–17, Tokyo Univ. Agric. Technol., Tokyo (2000).

[144] Hanada, K., Kuratomi, Y. and Oshiro, K., *On direct sums of extending modules and internal exchange property*, J. Algebra 250, 115–133 (2002).

[145] Handelman, D., *Perspectivity and cancellation in regular rings*, J. Algebra 48, 1–16 (1977).

[146] Hanna, A. and Shamsuddin, A., *Duality in the category of modules. Applications*, Algebra Berichte 49, Verlag Reinhard Fischer, München (1984).

[147] Hanna, A. and Shamsuddin, A., *Dual Goldie dimension*, Rend. Istit. Mat. Univ. Trieste 24, 25–38 (1994).

[148] Harada, M., *On categories of indecomposable modules. II*, Osaka J. Math. 8, 301–321 (1971).

[149] Harada, M., *Supplementary remarks on categories of indecomposable modules*, Osaka J. Math. 9, 49–55 (1972).

[150] Harada, M., *Small submodules in a projective module and semi-T-nilpotent sets*, Osaka J. Math. 14, 355–364 (1977).

[151] Harada, M., *A note on hollow modules*, Rev. Un. Mat. Argentina 28, 186–194 (1977/78).

[152] Harada, M., *Non-small modules and non-cosmall modules*, Ring Theory, Proceedings of 1978 Antwerp conference, 669–689, Marcel Dekker (1979).

[153] Harada, M., *On one sided QF-3 rings I*, Osaka J. Math. 17, 421–431 (1980).

[154] Harada, M., *On one sided QF-3 rings II*, Osaka J. Math. 17, 433–438 (1980).

[155] Harada, M., *On lifting property on direct sums of hollow modules*, Osaka J. Math. 17, 783–791 (1980).

[156] Harada, M., *On modules with lifting properties*, Osaka J. Math. 19, 189–201 (1982).

[157] Harada, M., *On modules with extending properties*, Osaka J. Math. 19, 203–215 (1982).

[158] Harada, M., *Factor categories with applications to direct decomposition of modules*, Lect. Notes Pure Appl. Math. 88, Marcel Dekker, New York (1983).

[159] Harada, M. and Ishii, T., *On perfect rings and the exchange property*, Osaka J. Math. 12, 483–491 (1975).

[160] Harada, M. and Kanbara, H., *On categories of projective modules*, Osaka J. Math. 8, 471–483 (1971).

[161] Harada, M. and Mabuchi, T., *On almost M-projectives*, Osaka J. Math. 26, 837–848 (1989).

[162] Harada, M. and Sai, Y., *On categories of indecomposable modules. I*, Osaka J. Math. 7, 323–344 (1970).

[163] Harada, M. and Tozaki, A., *Almost M-projectives and Nakayama rings*, J. Algebra 122, 447–474 (1989).

[164] Harmanci, A., Keskin, D., and Smith, P. F., *On ⊕-supplemented modules*, Acta Math. Hung. 83, 161-169 (1999).

[165] Hausen, J., *On characterizations of semi-simple and semi-local rings by properties of their modules*, Ann. Soc. Math. Pol., Ser. I, Commentat. Math. 24, 277–280 (1984).

[166] Hausen, J. and Johnson, J. A., *On supplements in modules*, Comm. Math. Univ. Sancti Pauli 37, 29–31 (1982).

[167] Hausen, J. and Johnson, J. A., *A new characterisation of perfect and semi-perfect rings*, Bull. Calcutta Math. Soc. 75, 57–58 (1983).

[168] Herbera, D. and Shamsuddin, A., *Modules with semi-local endomorphism ring*, Proc. Amer. Math. Soc. 123, 3593–3600 (1995).

[169] Herzog, I., *A test for finite representation type*, J. Pure Appl. Algebra 95, 151–182 (1994).

[170] Hirano, Y., *On rings over which each module has a maximal submodule*, Comm. Algebra 26, 3435–3445 (1998).

[171] Hirano, Y., Huynh, D. V. and Park, J. K., *Rings characterised by semiprimitive modules*, Bull. Austral. Math. Soc. 52, 107–116 (1995).

[172] Hiremath, V. A., *Cofinitely projective modules*, Houston J. Math. 11, 183–190 (1985).

[173] Huisgen-Zimmermann, B. and Saorín, M., *Direct sums of representations as modules over their endomorphism rings*, J. Algebra 250, 67–89 (2002).

[174] Idelhadj, A. and Tribak, R., *Modules for which every submodule has a supplement which is a direct summand*, Arabian J. Sci. Eng. 25 (2C), 179–189 (2000).

[175] Inoue, T., *Sum of hollow modules*, Osaka J. Math. 20, 331–336 (1983).

[176] Ishii, T., *On locally direct summands of modules*, Osaka J. Math. 12, 473–482 (1975).

[177] Jain, S. K., *Generalizations of QF-rings, and continuous and discrete modules*, Math. Stud. 57, 219–224 (1989).

[178] Jain, S. K., López-Permouth, S. R. and Tariq Rizvi, S., *A characterization of uniserial rings via continuous and discrete modules*, J. Aust. Math. Soc., Ser. A 50, 197–203 (1991).

[179] Jain, S. K. and Singh, S., *Quasi-injective and pseudo-injective modules*, Canad. Math. Bull. 18, 359–366 (1975).

[180] Jambor, P., *Commutative rings with no super-decomposable modules*, Arch. Math. 42, 517–519 (1984).

[181] Jambor, P., Malcolmson, P. and Shapiro, J., *Localizations and colocalizations via limits*, Comm. Algebra 10, 817–846 (1982).

[182] Jayaraman, M. and Vanaja, N., *Harada modules*, Comm. Algebra 28, 3703–3726 (2000).

[183] Kamal, M. A. and Müller, B. J., *The structure of extending modules over Noetherian rings*, Osaka J. Math. 25, 539–551 (1988).

[184] Kanbara, H., *Note on Krull-Remak-Schmidt-Azumaya's theorem*, Osaka J. Math. 9, 409–413 (1972).

[185] Kaplansky, I., *Elementary divisors and modules*, Trans. Amer. Math. Soc. 66, 464–491 (1949).

[186] Kaplansky, I., *Infinite abelian groups*, Univ. Michigan Press, Ann Arbor (1954).

[187] Kaplansky, I., *Bass's first stable range condition*, mimeographed notes (1971).

[188] Kasch, F., *Moduln mit LE-Zerlegung und Harada Moduln*, Lecture notes, Univ. of Munich (1982).

[189] Kasch, F., *Modules and Rings*, Academic Press, London, New York (1982).

[190] Kasch, F., *Partiell invertierbare Homomorphismen und das Total*, Algebra Berichte 60, Verlag Reinhard Fischer, Munich (1988).

[191] Kasch, F., *Locally injective modules and locally projective modules*, Proceedings of the Second Honolulu Conference on Abelian Groups and Modules (2001), Rocky Mountain J. Math. 32, 1493–1504 (2002).

[192] Kasch, F. and Mader, A., *Rings, modules, and the total*, Birkhäuser Verlag, Basel (2004).

[193] Kasch, F. and Mares, E. A., *Eine Kennzeichnung semi-perfekter Moduln*, Nagoya Math. J. 27, 525–529 (1966).

[194] Kasch, F. and Schneider, W., *The total of modules and rings*, Algebra Berichte 69, Verlag Reinhard Fischer, Munich (1992).

[195] Kasch, F. and Schneider, W., *Exchange properties and the total*, Advances in ring theory (Granville, OH, 1996), 161–174, Trends Math., Birkhäuser Boston, Boston (1997).

[196] Kashu, A. I., *Radicals and torsions in modules* (Russian), Shtiinza, Kishinev (1983).

[197] Kawada, Y., *A generalisation of Morita's theorem concerning generalized uniserial algebras*, Proc. Japan Acad. 34, 404–406 (1958).

[198] Keskin, D., *Supplemented modules and endomorphism rings*, PhD Thesis, Hacettepe University, Ankara (1999).

[199] Keskin, D., *On lifting modules*, Comm. Algebra 28, 3427–3440 (2000).

[200] Keskin, D., *An approach to extending and lifting modules by modular lattices*, Indian J. Pure Appl. Math. 33, 81–86 (2002).

[201] Keskin, D., *Discrete and quasi-discrete modules*, Comm. Algebra 30, 5273–5282 (2002).

[202] Keskin, D. and Lomp, Ch., *On lifting LE-modules*, Vietnam J. Math. 30, 167–176 (2002).

[203] Keskin, D., Smith, P. F. and Xue, W., *Rings whose modules are $\oplus$-supplemented*, J. Algebra 218, 470–487 (1999).

[204] Keskin Tütüncü, D. and Tribak, R., *When $M$-cosingular modules are projective*, Vietnam J. Math. 33, 214–221 (2005).

[205] Keskin Tütüncü, D., and Tribak, R., *On lifting modules and weak lifting modules*, Kyungpook Math. J. 45(3), 445–453 (2005).

[206] Keskin, D. and Xue, W., *Generalizations of lifting modules*, Acta. Math. Hung. 91, 253–261 (2001).

[207] Ketkar, R. D. and Vanaja, N., *A note on FR-perfect modules*, Pacific J. Math. 96, 141–152 (1981).

[208] Khurana, D. and Gupta, R. N., *Endomorphism rings of Harada modules*, Vietnam J. Math. 28, 173–175 (2000).

[209] Khurana, D. and Gupta, R. N., *Lifting idempotents and projective covers*, Kyungpook Math. J. 41, 217–227 (2001).

[210] Khurana, D. and Lam, T. Y., *Rings with internal cancellation*, J. Algebra 284, 203–235 (2005).

[211] Kosan, M. T. and Harmanci, A., *Decompositions of modules supplemented relative to a torsion theory*, Int. J. Math. 16, 43–52 (2005).

[212] Krause, H., *Uniqueness of uniform decompositions in abelian categories*, J. Pure Appl. Algebra 183, 125–128 (2003).

[213] Krempa, J. and Okniński, J., *Semilocal, semiperfect and perfect tensor products*, Bull. Acad. Pol. Sci., Sér. Sci. Math. 28, 249–256 (1980).

[214] Kuratomi, Y., *On direct sum of lifting modules and internal exchange property*, Comm. Algebra 33, 1795 1804 (2005).

[215] Kutami, M. and Oshiro, K., *An example of a ring whose projective modules have the exchange property*, Osaka J. Math. 17, 415–420 (1980).

[216] Lam, T. Y., *A First Course in Noncommutative Rings*, Springer, New York (1991).

[217] Lam, T. Y., *A crash course on stable range, cancellation, substitution, and exchange*, J. Algebra Appl. 3, 301–343 (2004).

[218] Lemonnier, B., *Dimension de Krull et codéviation. Application au théorème d'Eakin.*, Comm. Algebra 6, 1647–1665 (1978).

[219] Lenzing, H., *Direct sums of projective modules as direct summands of their direct product*, Comm. Algebra 4, 681–691 (1976).

[220] Leonard, W. W., *Small modules*, Proc. Amer. Math. Soc. 17, 527–531 (1966).

[221] Leron, U., *Matrix methods in decompositions of modules*, Linear Algebra Appl. 31, 159–171 (1980).

[222] Leszczyński, Z. and Simson, D., *On pure semisimple Grothendieck categories and the exchange property*, Bull. Acad. Polon. Sci. Sér. Sci. Math. 27, 41–46 (1979).

[223] Li, M.S. and Zelmanowitz, J.M., *On the generalizations of injectivity*, Comm. Algebra 16, 483–491 (1988).

[224] Lin, You, *Cotorsion Theories*, Northeastern Math. J. 8(1), 110-120 (1992).

[225] Lie, T., *On modules with LE-decompositions*, Math. Scand. 49, 156–160 (1981).

[226] Lomp, Ch., *On dual Goldie dimension*, Diploma Thesis, University of Düsseldorf (1996).

[227] Lomp, Ch., *On semilocal modules and rings*, Comm. Algebra 27, 1921–1935 (1999).

[228] Lomp, Ch., *On the splitting of the dual Goldie torsion theory*, Contemp. Math. 259, 377–386 (2000).

[229] Mares, E.A., *Semi-perfect modules*, Math. Z. 82, 347–360 (1963).

[230] Mbuntum, F.F. and Varadarajan, K., *Half-exact pre-radicals*, Comm. Algebra 5, 555-590 (1977).

[231] McConnell, J.C. and Robson, J.C., *Noncommutative Noetherian rings*, Revised edition, Amer. Math. Soc., Providence, RI (2001).

[232] McMaster, R.J., *Cotorsion theories and colocalization*, Canad. J. Math. 27, 618–628 (1975).

[233] Mermut, E., *Homological approach to complements and supplements*, PhD thesis, Dokuz Eylül University, Izmir (2004).

[234] Mishina, A.P. and Skornjakov, L.A., *Abelian groups and modules*, Amer. Math. Soc., Providence (1976).

[235] Miyashita, Y., *Quasi-projective modules, perfect modules, and a theorem for modular lattices*, J. Fac. Sci. Hokkaido 19, 86–110 (1966).

[236] Mohamed, S.H. and Abdul-Karim, F.H., *Semidual continuous abelian groups*, J. Univ. Kuwait Sci. 11, 23–28 (1984).

[237] Mohamed, S.H. and Müller, B.J., *Decomposition of dual-continuous modules*, Module theory (Proc. Special Session, Amer. Math. Soc., Seattle, 1977), 87–94, Lect. Notes Math. 700, Springer, Berlin (1979).

[238] Mohamed, S.H. and Müller, B.J., *Direct sums of dual continuous modules*, Math. Z. 178, 225–232 (1981).

[239] Mohamed, S.H. and Müller, B.J., *Dual continuous modules over commutative Noetherian rings*, Comm. Algebra 16, 1191–1207 (1988).

[240] Mohamed, S.H. and Müller, B.J., *Continuous modules have the exchange property*, Abelian group theory (Perth, 1987), 285–289, Contemp. Math. 87, Amer. Math. Soc., Providence (1989).

[241] Mohamed, S. H. and Müller, B. J., *Continuous and Discrete Modules*, London Math. Soc. Lect. Notes Ser. 147, Cambridge Univ. Press, Cambridge (1990).

[242] Mohamed, S. H. and Müller, B. J., *On the exchange property for quasi-continuous modules*, Abelian groups and modules (Padova, 1994), 367–372, Kluwer Acad. Publ., Dordrecht (1995).

[243] Mohamed, S. H. and Müller, B. J., ℵ-*exchange rings*, Abelian groups, module theory, and topology (Padua, 1997), 311–317, Dekker, New York (1998).

[244] Mohamed, S. H. and Müller, B. J., *Ojective modules*, Comm. Algebra 30, 1817–1827 (2002).

[245] Mohamed, S. H. and Müller, B. J., *Cojective modules*, J. Egyptian Math. Sci. 12, 83–96 (2004).

[246] Mohamed, S. H., Müller, B. J., and Singh, S., *Quasi-dual-continuous modules*, J. Aust. Math. Soc., Ser. A 39, 287–299 (1985).

[247] Mohamed, S. and Singh, S., *Generalizations of decomposition theorems known over perfect rings*, J. Australian Math. Soc. 24, 496–510 (1977).

[248] Mohammed, A. and Sandomierski, F. L., *Complements in projective modules*, J. Algebra 127(1), 206–217 (1989).

[249] Monk, G. S., *A characterization of exchange rings*, Proc. Amer. Math. Soc. 35, 349–353 (1972).

[250] Nakahara, S., *On a generalization of semiperfect modules*, Osaka J. Math. 20, 43-50 (1983).

[251] Nebiyev, C. and Pancar, A., *Strongly ⊕-supplemented modules*, Intern. Journal of Computational Cognition (http:/www.YangSky.com/yangijcc.htm) 2(3), 57–61 (2004).

[252] von Neumann, J., *Continuous geometry*, Proc. Nat. Acad. Sci. 22, 92–100 (1936).

[253] Nicholson, W. K., *I-rings*, Trans. Amer. Math. Soc. 207, 361–373 (1975).

[254] Nicholson, W. K., *On semiperfect modules*, Canad. J. Math. 18, 77–80, (1975).

[255] Nicholson, W. K., *Semiregular modules and rings*, Canad. J. Math. 28(5), 1105–1120 (1976).

[256] Nicholson, W. K., *Lifting idempotents and exchange rings*, Trans. Amer. Math. Soc. 229, 269–278 (1977).

[257] Nicholson, W. K., *An elementary proof of a characterization of semiperfect rings*, New Zealand J. Math. 25, 195–197 (1996).

[258] Nicholson, W. K., *On exchange rings*, Comm. Algebra 25, 1917–1918 (1997).

[259] Nicholson, W. K., *A survey of morphic modules and rings*, Advances in Ring Theory, Proceedings of the 4th China-Japan-Korea International Conference, 2004, 167–180, World Scientific, Singapore (2005).

[260] Nicholson, W. K. and Sánchez Campos, E., *Rings with the dual of the isomorphism theorem*, J. Algebra 271, 391–406 (2004).

[261] Nicholson, W. K. and Sánchez Campos, E., *Morphic modules*, Comm. Algebra 33, 2629–2647 (2005).

[262] Nicholson, W. K. and Zhou, Y., *Strong lifting*, J. Algebra 285, 795–818 (2005).

[263] Nielsen, P. P., *Abelian exchange modules*, Comm. Algebra 33, 1107–1118 (2005).

[264] Oberst, U. and Schneider, H.-J., *Die Struktur von projektiven Moduln*, Invent. Math. 13, 295–304 (1971).

[265] Ohtake, K., *Colocalization and localization*, J. Pure Appl. Algebra 11, 217–241 (1978).

[266] Okado, M. and Oshiro, K., *Remarks on the lifting property of simple modules*, Osaka J. Math. 21, 375–385 (1984).

[267] O'Meara, K. C., *The exchange property for row and column-finite matrix rings*, J. Algebra 268, 744–749 (2003).

[268] Orhan, N. and Keskin Tütüncü, D., *Characterizations of lifting modules in terms of cojective modules and the class of* $\mathcal{B}(\mathcal{M}, \mathcal{X})$, Int. J. Math. 16(6), 647–660 (2005).

[269] Oshiro, K., *Semiperfect modules and quasi-semiperfect modules*, Osaka J. Math. 20, 337–372 (1983).

[270] Oshiro, K., *Projective modules over von Neumann regular rings have the finite exchange property*, Osaka J. Math. 20, 695–699 (1983).

[271] Oshiro, K., *Lifting modules, extending modules and their applications to QF-rings*, Hokkaido Math. J. 13, 310–338 (1984).

[272] Oshiro, K., *Lifting modules, extending modules and their applications to generalized uniserial rings*, Hokkaido Math. J. 13, 339–346 (1984).

[273] Oshiro, K., *On Harada rings I*, Math. J. Okayama Univ. 31, 161–178 (1989).

[274] Oshiro, K., *On Harada rings II*, Math. J. Okayama Univ. 31, 179–188 (1989).

[275] Oshiro, K., *On Harada rings III*, Math. J. Okayama Univ. 32, 111–118 (1990).

[276] Oshiro, K. and Rizvi, S. T., *The exchange property of quasi-continuous modules with the finite exchange property*, Osaka J. Math. 33, 217–234 (1996).

[277] Oshiro, K. and Wisbauer, R., *Modules with every subgenerated module lifting*, Osaka J. Math. 32, 513–519 (1995).

[278] Özcan, A. C., *Characterizations of some rings by small modules and decomposition of modules*, PhD Thesis, Hacettepe University, Ankara (1999).

[279] Özcan, A. C. and Alkan, M., *Semiperfect modules with respect to a preradical*, Comm. Algebra, to appear.

[280] Özcan, A. C. and Smith, P. F., *The $Z^*$ functor for rings whose primitive images are Artinian*, Comm. Algebra 30, 4915–4930 (2002).

[281] Page, S. S., *Relatively semiperfect rings and corank of modules*, Comm. Algebra 21, 975–990 (1993).

[282] Pareigis, B., *Radikale und kleine Moduln*, Bayer. Akad. Wiss. Math.-Natur. Kl. S.-B. 1965, 1966 Abt. II, 185–199 (1966).

[283] Park, Y. S. and Rim, S. H., *Hollow modules and corank relative to a torsion theory*, J. Korean Math. Soc. 31, 439–456 (1994).

[284] Prest, M. and Wisbauer, R., *Finite presentation and purity in categories $\sigma[M]$*, Colloq. Math. 99(2), 189–202 (2004).

[285] Příhoda, P., *On uniserial modules that are not quasi-small*, J. Algebra, to appear.

[286] Puczyłowski, E. R., *On the uniform dimension of the radical of a module*, Comm. Algebra 23, 771–776 (1995).

[287] Puninski, G., *Serial Rings*, Kluwer Academic Publishers, Dordrecht (2001).

[288] Puninski, G., *Some model theory over a nearly simple uniserial domain and decompositions of serial modules*, J. Pure Appl. Algebra 163, 319–337 (2001).

[289] Puninski, G., *Projective modules over the endomorphism ring of a biuniform module*, J. Pure Appl. Algebra 188(1-3), 227–246 (2004).

[290] Raggi, F., Ríos Montes, J. and Wisbauer, R., *Coprime preradicals and modules*, J. Pure Appl. Algebra 200, 51–69 (2005).

[291] Ramamurthi, V. S., *The smallest left exact radical containing the Jacobson radical*, Ann. Soc. Sci. Bruxelles Sér. I 96, 201–206 (1982).

[292] Ramamurthi, V. S. and Rangaswamy, K. M., *Finitely injective modules*, J. Austral. Math. Soc. 16, 239–248 (1973).

[293] Rangaswamy, K. M., *Modules with finite spanning dimension*, Canad. Math. Bull. 20, 255–262 (1977).

[294] Rayar, M., *On small and cosmall modules*, Acta Math. Acad. Sci. Hungar. 39, 389–392 (1982).

[295] Reiter, E., *A dual to the Goldie ascending chain condition on direct sums of submodules*, Bull. Calcutta Math. Soc. 73, 55–63 (1981).

[296] Renault, G., *Étude de certains anneaux A lies aux sous-modules complements d'un A-module*, C. R. Acad. Sc. Paris, Sér. A 259, 4203–4205 (1964).

[297] Renault, G., *Étude des sous-modules complements dans un module*, Bull. Soc. Math. France Mém. 9 (1967).

[298] Renault, G., *Sur les anneaux A tels que tout A-module à gauche non nul contient un sous-module maximal*, C. R. Acad. Sci., Paris, Sér. A 267, 792–794 (1968).

[299] Rim, S. H. and Takemori, K., *On dual Goldie dimension*, Comm. Algebra 21, 665–674 (1993).

[300] Robson, J. C., *Idealizers and hereditary noetherian prime rings*, J. Algebra 22, 45–81 (1972).

[301] Rososek, S. K., *The correctness of modules* (Russian), Izv. Vyssh. Uchebn. Zaved. Mat. 22, 77–82 (1978).

[302] Rososek, S. K., *Pure semisimple rings and modules* (Russian), in *Abelian groups and modules. No. 5* (Russian) Edited by L. A. Skornyakov, Tomsk. Gos. Univ., Tomsk, 80–87 (1985).

[303] Rotman, J. J., *An Introduction to Homological Algebra*, Academic Press, New York (1979).

[304] Rudlof, P., *On the structure of couniform and complemented modules.*, J. Pure Appl. Algebra 74(3), 281–305 (1991).

[305] Rudlof, P., *On minimax and related modules*, Canad. J. Math. 44, 154–166 (1992).

[306] Sandomierski, F. L., *On semiperfect and perfect rings*, Proc. Amer. Math. Soc. 21, 205–207 (1969).

[307] Sarath, B. and Varadarajan, K., *Dual Goldie dimension II*, Comm. Algebra 7, 1885–1899 (1979).

[308] Satyanarayana, B., *On modules with finite spanning dimension*, Proc. Japan Acad., 61, Ser. A, 23-25 (1985).

[309] Satyanarayana, B., *A note on E-direct and S-inverse systems*, Proc. Japan Acad. 64, Ser. A, 292–295 (1988).

[310] Satyanarayana, B., *Modules with finite spanning dimension*, J. Austral. Math. Soc. (Ser. A) 57, 170–178 (1994).

[311] Schneider, W., *Das Total von Moduln und Ringen*, Algebra Berichte 55, Verlag Reinhard Fischer, Munich (1987).

[312] Shock, R. C., *Dual generalizations of the Artinian and Noetherian conditions*, Pacific J. Math. 54, 227–235 (1974).

[313] Simson, D., *On right pure semisimple hereditary rings and an Artin problem*, J. Pure Appl. Algebra 104, 313–332 (1995).

[314] Singh, S., *Semidual continuous modules over Dedekind domains*, J. Univ. Kuwait Sci. 11, 33–40 (1984).

[315] Singh, S., *Dual continuous modules over Dedekind domains*, J. Univ. Kuwait Sci. 7, 1–10 (1980).

[316] Sklyarenko, E. G., *Relative homological algebra in the category of modules*, Russian Math. Surv. 33, 97-137 (1978); translation from Uspekhi Mat. Nauk 33, (201), 85-120 (1978).

[317] Smith, P. F., *Finitely generated supplemented modules are amply supplemented*, Arabian J. Sci. Eng. 25 (2C), 69–79 (2000).

[318] Smith, P. F., de Viola-Prioli, A. M. and Viola-Prioli, J. E., *Modules complemented with respect to a torsion theory*, Comm. Algebra 25, 1307–1326 (1997).

[319] Song, Guang-tian, Chu, Cheng-hao and Zhu, Min-xian, *Regularly stable rings and stable isomorphism of modules*, J. Univ. Sci. Technol. China 33, 1–8 (2003).

[320] Steinitz, E., *Rechteckige Systeme und Moduln in algebraischen Zahlkörpern*, Math. Ann. 71, 328–354 (1912).

[321] Stenström, B., *High submodules and purity*, Ark. Mat. 7, 173–176 (1967).

[322] Stenström, B., *Direct sum decompositions in Grothendieck categories*, Ark. Mat. 7, 427–432 (1968).

[323] Stenström, B., *Rings of Quotients*, Springer-Verlag, Berlin (1975).

[324] Stock, J., *On rings whose projective modules have the exchange property*, J. Algebra 103, 437–453 (1986).

[325] Tachikawa, H., *QF-3 rings and categories of projective modules*, J. Algebra 28, 408–413 (1974).

[326] Takeuchi, T., *The endomorphism ring of an indecomposable module with an Artinian projective cover*, Hokkaido Math. J. 4, 265–267 (1975).

[327] Takeuchi, T., *On cofinite-dimensional modules*, Hokkaido Math. J. 5, 1–43 (1976).

[328] Takeuchi, T., *Coranks of a quasi-projective module and its endomorphism ring*, Glasgow Math. J. 36, 381–383 (1994).

[329] Talebi, Y. and Vanaja, N., *The torsion theory cogenerated by $M$-small modules*, Comm. Algebra 30, 1449–1460 (2002).

[330] Talebi, Y. and Vanaja, N., *Copolyform modules*, Comm. Algebra 30, 1461–1473 (2002).

[331] Talebi, Y. and Vanaja, N., *Copolyform Σ-lifting modules*, Vietnam J. Math. 32, 49-64 (2004).

[332] Tiwary, A. K. and Pandeya, B. M., *Pseudo projective and pseudo injective modules*, Indian J. Pure Appl. Math. 9, 941–949 (1978).

[333] Tuganbaev, A. A., *Endomorphism rings of strictly indecomposable modules* (Russian), Uspekhi Mat. Nauk 53, 207–208 (1998); translation in Russian Math. Surveys 53, 642–643 (1998).

[334] Tuganbaev, A. A., *Quasiprojective modules with the finite exchange property* (Russian), Uspekhi Mat. Nauk 54, 191–192 (1999); translation in Russian Math. Surveys 54, 459–460 (1999).

[335] Tuganbaev, A. A., *Modules with the exchange property and exchange rings*, Handbook of algebra, Vol. 2, 439–459, North-Holland, Amsterdam, 2000.

[336] Tuganbaev, A. A., *Rings whose nonzero modules have maximal submodules*, J. Math. Sci. (New York) 109(3), 1589–1640 (2002).

[337] Tuganbaev, A. A., *Rings and modules with exchange properties*, J. Math. Sci. (New York) 110(1), 2348–2421 (2002).

[338] Tuganbaev, A. A., *Rings close to regular*, Kluwer Academic Publishers, Dordrecht (2002).

[339] Utumi, Y., *On continuous rings and self-injective rings*, Trans. Amer. Math. Soc. 118, 158–173 (1965).

[340] Vanaja, N., *Characterization of rings using extending and lifting modules*, Ring theory (Granville, OH, 1992), 329–342, World Sci. Publ., River Edge (1993).

[341] Vanaja, N., *All finitely generated M-subgenerated modules are extending*, Comm. Algebra 24, 543–572 (1996).

[342] Vanaja, N. and Purav, V. M., *Characterisations of generalised uniserial rings in terms of factor rings*, Comm. Algebra 20, 2253–2270 (1992).

[343] Vanaja, N. and Purav, V. N., *A note on generalised uniserial ring*, Comm. Algebra 21, 1153–1159 (1993).

[344] Varadarajan, K., *Dual Goldie dimension*, Comm. Algebra 7, 565–610 (1979).

[345] Varadarajan, K., *Modules with supplements*, Pacific J. Math. 82, 559–564 (1979).

[346] Varadarajan, K., *Dual Goldie dimension of certain extension rings*, Comm. Algebra 10, 2223–2231 (1982).

[347] Varadarajan, K., *Properties of endomorphism rings*, Acta Math. Hungar. 74, 83–92 (1997).

[348] Varadarajan, K., *Study of Hopficity in certain classes of rings*, Comm. Algebra 28, 771–783 (2000).

[349] Vasconcelos, W., *On finitely generated flat modules*, Trans. Amer. Math. Soc. 138, 505–512 (1969).

[350] Vaserstein, L.N., *Stable rank of rings and dimensionality of topological spaces*, (in Russian) Funktsional. Anal. i Prilozhen. 5 (1971), 17–27 (1971); translated in Functional Anal. Appl. 5, 102–110 (1971).

[351] Vaserstein, L.N., *Bass's first stable range condition*, J. Pure Appl. Algebra 34, 319–330 (1984).

[352] Vinsonhaler, C.I., *Orders in QF-3 rings*, J. Algebra 14, 83–90 (1970). Supplement to this paper, J. Algebra 17, 149-151 (1971).

[353] Walker, E.A., *Cancellation in direct sums of groups*, Proc. Amer. Math. Soc. 7, 898–902 (1956).

[354] Wang, Dingguo and Yao, Zhongping, *Rings characterized by discrete modules*, Bull. Inst. Math. Acad. Sinica 28, 207–214 (2000).

[355] Wang, Yongduo and Ding, Nanqing, *Generalized supplemented modules*, preprint.

[356] Ware, R. *Endomorphism rings of projective modules*, Trans. Amer. Math. Soc. 155, 233–256 (1971).

[357] Warfield, R.B., Jr., *Decompositions of injective modules*, Pacific J. Math. 31, 263–276 (1969).

[358] Warfield, R.B., Jr., *A Krull-Schmidt theorem for infinite sums of modules*, Proc. Amer. Math. Soc. 22, 460-465 (1969).

[359] Warfield, R.B., Jr., *Exchange rings and decompositions of modules*, Math. Ann. 199, 31–36 (1972).

[360] Warfield, R.B., Jr., *Serial rings and finitely presented modules*, J. Algebra 37, 187–222 (1975).

[361] Warfield, R.B., Jr., *Cancellation of modules and groups and stable range of endomorphism rings*, Pacific J. Math. 91, 457–485 (1980).

[362] Wisbauer, R., *F-semiperfekte und perfekte Moduln in σ[M]*, Math. Z. 173, 229–234 (1980).

[363] Wisbauer, R., *Foundations of Module and Ring Theory*, Gordon and Breach, Reading (1991).
*Grundlagen der Modul- und Ringtheorie*, Verlag Reinhard Fischer, München (1988).

[364] Wisbauer, R., *Modules and Algebras: Bimodule Structure and Group Actions on Algebras*, Pitman Monogr. 81, Addison Wesley, Longman, Essex (1996).

[365] Wisbauer, R., *On module classes closed under extensions*, Rings and radicals, Gardner, B., Liu Shaoxue, Wiegandt, R. (eds.), Pitman Research Notes 346, 73-97 (1996).

[366] Wisbauer, R., *Correct classes of modules*, Algebra Discrete Math. (Kiev) 4, 106–118 (2004).

[367] Wu, Tongsuo, *Exchange rings with general aleph-nought-comparability*, Comm. Algebra 29(2), 815–828 (2001).

[368] Xin, L., *A note on dual Goldie dimension over perfect rings*, Kobe J. Math. 11, 21–24 (1994).

[369] Xue, Weimin, *Characterization of rings using direct-projective modules and direct- injective modules*, J. Pure Appl. Algebra 87(1), 99–104 (1993).

[370] Xue, Weimin, *Characterizations of semiperfect and perfect rings*, Publ. Mat., Barc. 40(1), 115-125 (1996).

[371] Yamagata, K., *On projective modules with the exchange property*, Sci. Rep. Tokyo Kyoiku Daigaku Sect. A 12, 149–158 (1974).

[372] Yamagata, K., *The exchange property and direct sums of indecomposable injective modules*, Pacific J. Math. 55, 301–317 (1974).

[373] Yamagata, K., *On rings of finite representation type and modules with the finite exchange property*, Sci. Rep. Tokyo Kyoiku Daigaku Sect. A 13, 1–6 (1975).

[374] Yousif, M. F. and Zhou, Yiqiang, *Semiregular, semiperfect and perfect rings relative to an ideal*, Rocky Mt. J. Math. 32, 1651–1671 (2002).

[375] Yu, Hua-Ping, *On modules for which the finite exchange property implies the countable exchange property*, Comm. Algebra 22, 3887–3901 (1994).

[376] Yu, Hua-Ping, *Stable range one for exchange rings*, J. Pure Appl. Algebra 98, 105–109 (1995).

[377] Zelinksy, D., *Linearly compact modules and rings*, Amer. J. Math. 75, 75–90 (1953).

[378] Zelmanowitz, J. M., *Representation of rings with faithful polyform modules*, Comm. Algebra 14, 1141–1169 (1986).

[379] Zelmanowitz, J. M., *On the endomorphism ring of a discrete module: a theorem of F. Kasch*, Advances in ring theory (Granville, 1996), 317–322, Birkhäuser Boston (1997).

[380] Zelmanowitz, J. M., *Radical endomorphisms of decomposable modules*, J. Algebra 279, 135–146 (2004).

[381] Zeng, Qing-yi and Shi, Mei-hua, *On closed weak supplemented modules*, J. Zheijang Univ. Science A 7(2), 210–215 (2006).

[382] Zhou, Yiqiang, *Generalizations of perfect, semiperfect and semiregular rings*, Algebra Colloq. 7, 305–308 (2000).

[383] Zimmermann-Huisgen, B., *Rings whose right modules are direct sums of indecomposable modules*, Proc. Amer. Math. Soc. 77, 191–197 (1979).

[384] Zimmermann-Huisgen, B., *The sum-product splitting property and injective direct sums of modules over von Neumann regular rings*, Proc. Amer. Math. Soc. 83, 251–254 (1981).

[385] Zimmermann-Huisgen, B. and Zimmermann, W., *Classes of modules with the exchange property*, J. Algebra 88, 416–434 (1984).

[386] Zöllner, A., *Lokal-direkte Summanden*, Algebra Berichte 51, Verlag Reinhard Fischer, Munich, (1984).

[387] Zöllner, A., *On modules that complement direct summands*, Osaka J. Math. 23, 457–459 (1986).

[388] Zöschinger, H., *Komplementierte Moduln über Dedekindringen*, J. Algebra 29, 42–56 (1974).

[389] Zöschinger, H., *Komplemente als direkte Summanden*, Arch. Math. 25, 241–253 (1974).

[390] Zöschinger, H., *Über Torsions- und $\kappa$-Elemente von* Ext$(C, A)$, J. Algebra 50, 295–336 (1978).

[391] Zöschinger, H., *Projektive Moduln mit endlich erzeugtem Radikalfaktormodul*, Math. Ann. 255, 199–206 (1981).

[392] Zöschinger, H., *Gelfandringe und koabgeschlossene Untermoduln*, Sitzungsber. Bayer. Akad. Wiss. Math.-Naturwiss. Kl. 1982, 43–70 (1983).

[393] Zöschinger, H., *Linear-kompakte Moduln über noetherschen Ringen*, Arch. Math. 41, 121–130 (1983).

[394] Zöschinger, H., *Minimax-Moduln*, J. Algebra 102, 1–32 (1986).

[395] Zöschinger, H., *Kosinguläre und kleine Moduln*, Comm. Algebra 33(10), 3389–3404 (2005).

# Index